EDWIN D. MORGAN, JR.

rare & curious

with col^d plates &c

5/

fine copy

Frontispiece

Dodd delin.

Key to Physic, and the Occult Sciences

Prattent sculp

A

KEY TO PHYSIC

AND THE

OCCULT SCIENCES.

OPENING TO MENTAL VIEW,

The SYSTEM and ORDER of the Interior and Exterior HEAVENS;

The ANALOGY betwixt ANGELS and the SPIRITS of MEN;

AND THE

SYMPATHY between CELESTIAL and TERRESTRIAL BODIES.

FROM WHENCE IS DEDUCED,

An obvious Discrimination of Future Events, in the Motions and Positions of the Luminaries, Planets, and Stars; the universal Spirit and Economy of Nature in the Production of all Things; the principles of ethereal and atmospherical Influx, in constituting the proper Recipient of Life; the active and passive Tinctures requisite in the Generation of Men and Brutes; the Properties of Vegetable, Mineral, and ANIMAL MAGNETISM: the fundamental Causes and Qualities, visible or occult, of all DISEASES, both of Mind and Body; and the simple Modes prescribed by NATURE for their Prevention and Cure.

TO WHICH ARE ADDED,

LUNAR TABLES, calculated from Sidereal Motion; exhibiting, upon the most simple yet unerring Construction, the actual Moment of the CRISIS of every DISEASE, and the consequent Termination thereof, whether for LIFE or DEATH.

THE WHOLE FORMING

An interesting Supplement to CULPEPER's FAMILY PHYSICIAN, and Display of the OCCULT SCIENCES. Published for the good of all who search after Truth and Wisdom; to preserve to all the Blessings of Health and Life; and to give to all the Knowledge of Primitive Physic, and the Art of Healing.

By E. SIBLY, M.D. F.R.H.S.

The FIFTH EDITION. Illustrated with elegant COPPER PLATES.

LONDON:

Printed by W. Lewis, St. John's Square;

FOR G. JONES, (LATE WILKES,) N° 17, AVE-MARIA LANE, St. PAUL's.

1814.

DEDICATION,

TO THE NUMEROUS

SUBSCRIBERS TO MY FORMER WORKS.

THE Liberality, Attention, and Confidence, I have received from YOU, demand the earliest fulfilment of my Promise, in publishing the following Sheets. Not conscious of having deviated from the line of REASON or of TRUTH, not challenged by Critics, nor accused by the Faculty, of leading you astray, I feel more than common Gratification in submitting myself once more to your Patronage; and no longer than I can render myself useful to SOCIETY, and worthy of YOUR Protection, do I wish to retain the Power of subscribing myself

Your much-honoured Brother,

Friend, and Servant,

No. 1, Upper Titchfield-street,
Cavendish-Square.

E. SIBLY.

A KEY TO PHYSIC AND THE OCCULT SCIENCES.

WISDOM is the light of Reafon, and the bond of Peace. It affimilates Man to God, and elevates his mind above unworthy purfuits. It is the principal excellence which diftinguifhes him from brutes, and the chief ornament that dignifies his character. Whatever is founded in Wifdom's laws, defies the mouldering hand of time, and ranks with immortality. Hence it is that a thirft after knowledge is natural to man; and, if the cares and follies of this world could be eftranged from his concerns, his defire of information would be infeparable from his exiftence. Ignorance and fuperftition may be confidered as the curfe of God, which chains its votaries to unworthy objects; whilft, on the contrary, wifdom and underftanding provide us with wings, whereby to foar above the earth; to contemplate the works of creation—to difcern the myfteries of divinity, and converfe with angels.

The beautiful defcription given by Solomon of his acquirements in wifdom, is highly deferving the attention of all men; but particularly of thofe who profefs the fcience of phyfic, and the cure of fouls. *I prayed,* fays Solomon, *and underftanding was given me; I called upon God, and the fpirit of wifdom came to me. All good things came with her, and innumerable riches in her hand,* Wifd. vii. 7, 11. What greater reward could any one defire? And though the intellectual faculties of all men are not alike ftrong and apt for occult fpeculations; yet it is manifeft that all perſons are capable of deriving great improvements from reading; and, that it is not ſo much the want of natural ability, as of induſtry and application, that ſo many men diſgrace the image of the Deity, and degrade the venerable profeſſions of Divinity, Phyſic, and Law.

It rarely happens that the want of intellect, or natural endowments of the mind, are the rocks on which men ſplit in their profeſſional character. Indocile and unapt indeed muſt that man be, whom education, experience, obſervation, reading, or enquiry, will not ſet generally right in his progreſs through life. Yet, without induſtry, and an anxious deſire of knowledge and improvement, neither education, nor all the advantages of natural ability, can ſave us from the wreck of error, or the diſgrace of ignorance. Obſtinate men, though of the firſt capacity in the world, are a forlorn hope, and often irrecoverably loſt, by unadviſedly purſuing the phantom of their own brain; whilſt others, enriched by dignity of ſenſe, and qualified by depth of underſtanding to form the brighteſt characters amongſt ſociety, ſurrender up their talents for diſcernment and enquiry, and content themſelves with taking upon truſt whatever they ſee or hear; particularly in the practice of phyſic, in the law, and in the church. The miſchiefs attendant on this general conduct of mankind, are great and many; for by thus implicity ſubſcribing to the vague notions, and falſe doctrines, of others; by ſhutting their eyes againſt the light of reaſon and enquiry, and refuſing to receive the conviction of their own ſenſes, they transfer error from one generation to another, until the unlettered multitude, dazzled by the ſplendid ignorance of the learned few, become diſciples to their miſtakes, and make error and enthuſiaſm an hereditary diſeaſe.

Hence, then, we ſee the neceſſity of conſulting our own reaſon, and employing our own underſtanding, in the diſcrimination of all our temporal and eternal affairs; and of acting and judging for ourſelves on all occaſions which immediately regard our health, our happineſs, or our life; and under all thoſe afflictions and misfortunes wherewith we ſtruggle in this world, in our paſſage to a better; to one more glorious and permanent; the ultimate end and reward of all our labours! Our ſenſes, on theſe occaſions, are ever ready to ſupport our endeavours, and perform their office; and it is unqueſtionably the duty of all men to exerciſe, to improve, and employ them. Yet it is aſtoniſhing in general to ſee how diſtruſtful we are of thoſe very faculties Nature has given us for our guide, and how fondly we ſubmit to the opinions of others, whoſe nerves cannot feel for us, and whoſe judgment is often founded upon erroneous principles, and ſometimes on no principles at all. This, however, is a conduct by no means fitted to the dignity and office aſſigned to man; who being placed at the head of all God's works upon earth,

walking

walking in his image, and exerciſing dominion over his creatures, is bound to improve that intellect of reaſon and underſtanding, whereby he is to govern and direct them, according to the dictates of truth, of juſtice, and of mercy. For this purpoſe he ought, like Solomon, to ſtudy the occult properties and qualities of all things: "from the cedar-tree that is in Lebanon, even unto the hyſſop that ſpringeth out of the wall;" with whatever relates to a proper knowledge of himſelf, "and of beaſts, and of fowls, and of creeping things, and of fiſhes"—not to worſhip the ſun, nor the moon, nor the ſtars, nor any of the hoſt of heaven; but to conſider, to admire, and to inveſtigate their characters, fixed by the hand of God for ſigns and for ſeaſons, and for days and years. They, in fact, contain no more than what every man ought to be acquainted with, to the beſt of his abilities; becauſe they lead to a comprehenſive idea of thoſe occult cauſes and effects, which act the moſt, though they are the leaſt ſeen; and whereby the human underſtanding is enlightened and improved, and the mind enriched with thoſe divine precepts, which lead to a manifeſtation of that FIRST and omnipotent CAUSE to whoſe power all ſecond cauſes are ſubſervient, and operate but as the agents of his Will; and under whoſe provident care and ſufferance we ſee, feel, move, ſpeak, and have our being! The ten thouſand bleſſings which reſult from this ſtudy, are found in our enquiries after truth, and the myſteries which ſurround us; of the aſtoniſhing ſympathy and antipathy betwixt heavenly and earthly ſubſtances; of the wonderful harmony and conſtruction of the celeſtial bodies; of the nature and qualities of our own exiſtence, and the propagation of our ſpecies; of the occult properties implanted in all created beings; and the end for which they are and were created!

To ſuch enquiries all men are alike competent, and may boaſt the ſame pretenſions, unleſs obſtinacy or indolence is ſubſtituted to prevent them. There is certainly implanted in the human mind, a power which perceives truth, and commands belief, in all the occult properties of nature, not by the force of argument, nor learning, nor ſcience; but by an inſtantaneous, inſtinctive, and irreſiſtible, impulſe, derived neither from education nor from habit, but from the peculiar gift of Providence, acting independently of our will, whenever theſe objects are preſented, bearing evidence of their reality, even when the pride of our external deportment, and our very words, affect to deny them. This is an intellectual ſenſation, which I will venture to affirm, is felt more or leſs by all mankind; and I know the hearts of all my readers, if not their tongues, will admit the fact. It is therefore evident that the humble cottager, the claſſical curate, the regular phyſician, and the village doctor, ſtand on the ſame level in this reſpect. The ſtudy of Nature's laws, of the occult properties in medicine, and in the frame and temperature of our bodies, is no

leſs

less simple, than important to our welfare; and without knowing these, we know nothing that can place us beyond the sagacity of the brute creation. We can neither foresee danger, nor shun it when it is near—we are subject to misguided treatment, and mistake, in our medical applications, and advice—we receive intuitive signs and tokens of misfortune or advantage, without knowing how to benefit by the admonition.—In short, without this study, our enquiries are vain—our perceptions are clouded—our views limited, and all our pursuits are vanity, vexation, and disappointment. The weakness of our reason, and the avocations arising from the infirmities and necessities of our situations, require the most powerfu linstructions, and the clearest perceptions of heavenly and earthly things, for the preservation of our souls and bodies, and for the illumination of our minds; advantages that can in no wise be more completely obtained than by an intimate acquaintance with the Occult Sciences, or, in other words, by a contemplation

OF GOD.

THOUGH God has given us no innate ideas of himself, yet having furnished us with those faculties our minds are endowed with, he hath not left himself without a witness; since we have sense, perception, and reason, and cannot want a clear proof of him, as long as we carry any thought of ourselves about us. To show, therefore, that we are capable of knowing, that is, being certain, that there is a God; and how we may come by this certainty, I think we need go no farther than ourselves, and that undoubted knowledge we have of our own existence. I think it is beyond question, that man has a clear perception of his own being: he knows certainly that he exists, and that he is something. In the next place, man knows, by an intuitive certainty, that bare nothing can no more produce any real being, than it can be equal to two right angles. If, therefore, we know there is some real being, it is an evident demonstration, that from eternity there has been something; since what was not from eternity had a beginning; and what had a beginning must be produced by something else. Next it is evident, that what has its being from another must also have all that which is in and belongs to its being from another too; all the powers it has must be owing to, and received from, the same source. This eternal source of all being, must be also the source and original of all power; and so this eternal being must be also the most powerful.

Again, man finds in himself perception and knowledge: we are certain then that there is not only some being, but some knowing intelligent being, in the world. There was a time when there was no knowing being, or else there has been a know-

ing

ing being from eternity. If it be ſaid, there was a time when that eternal being had no knowledge; I reply, that then it is impoſſible there ſhould have ever been any knowledge; it being as impoſſible that things wholly void of knowledge, and operating blindly, and without any perception, ſhould produce a knowing being, as it is impoſſible that a triangle ſhould make itſelf three angles bigger than two right ones. Thus, from the conſideration of ourſelves, and what we infallibly find in our own conſtitutions, our reaſon leads us to the knowledge of this certain and evident truth, that there is an eternal, moſt powerful, and knowing, being, which whether any one will call God, it matters not. The thing is evident; and from this idea, duly conſidered, will eaſily be deduced all thoſe other attributes we ought to aſcribe to this eternal Being.

From what has been ſaid, it is plain that we have a more certain knowledge of the exiſtence of a God than of any thing our ſenſes have not immediately diſcovered to us. Nay, I preſume I may ſay, that we more certainly know that there is a God than that there is any thing elſe without us. When I ſay, we *know*, I mean, there is ſuch a knowledge within our reach, which we cannot miſs, if we will but apply our minds to that, as we do to other inquiries.

It being then unavoidable for all rational creatures to conclude, that ſomething has exiſted from eternity, let us next ſee what kind of thing that muſt be. There are but two ſorts of beings in the world, that man knows or conceives: ſuch as are purely material without ſenſe or perception; and ſenſible perceiving beings, ſuch as we find ourſelves to be. Theſe two ſorts we ſhall call cogitative and incogitative beings; which, to our preſent purpoſe, are better than material and immaterial.

If then there muſt be ſomething eternal, it is very obvious to reaſon, that it muſt neceſſarily be a cogitative being; becauſe it is as impoſſible to conceive that bare incogitative matter ſhould ever produce a thinking intelligent being, as that nothing of itſelf ſhould produce matter. Let us ſuppoſe any parcel of matter eternal, we ſhall find it in itſelf unable to produce any thing. Let us ſuppoſe its parts firmly at reſt together; if there were no other being in the world, muſt it not eternally remain ſo, a dead unactive lump? is it poſſible to conceive that it can add motion to itſelf, or produce any thing? Matter then, by its own ſtrength, cannot produce in itſelf ſo much as motion. The motion it has, muſt alſo be from eternity, or elſe added to matter by ſome other being, more powerful than matter. But let us ſuppoſe motion eternal too; yet matter, incogitative matter, and motion, could never produce thought. Knowledge will ſtill be as far beyond the power of nothing to produce. Divide matter into as minute parts as you will, vary its figure and motion as much as you pleaſe, it will operate no otherwiſe upon other bodies of proportionable bulk than it did before this diviſion. The minuteſt particles of matter, knock, repel, and reſiſt, one another, juſt as the greater do; and that is all they can do: ſo that, if we ſuppoſe nothing eternal, matter can never begin to be: if we ſuppoſe bare matter without motion eternal, motion can never begin to be: if we ſuppoſe only matter and motion eternal, thought can never begin to be: for it is impoſſible to conceive, that matter, either with or without motion, could have originally, in and from itſelf, ſenſe, perception, and knowledge; as is evident from hence, that then ſenſe, perception, and knowledge, muſt be a property eternally inſeparable from matter, and every particle of it. Since, therefore, whatſoever is the firſt eternal being, muſt neceſſarily be cogitative; and whatſoever is firſt of all things, muſt neceſſarily contain in it, and actually have at leaſt all the perfections that can ever after exiſt; it neceſſarily follows, that the firſt eternal being cannot be matter. If, therefore, it be evident, that ſomething muſt neceſſarily exiſt from eternity, it is alſo as evident, that that ſomething muſt be a cogitative being. For it is as impoſſible that incogitative matter ſhould produce a cogitative being, as that nothing, or the negation of all being, ſhould produce a poſitive being, or matter.

This diſcovery of the neceſſary exiſtence of an eternal mind ſufficiently leads us to the knowledge of God; for it will hence follow, that all other knowing beings that have a beginning, muſt depend on him, and have no other ways of knowledge or extent of power, than what he gives them; and therefore, if he made thoſe, he made alſo the leſs-excellent pieces of this univerſe, all inanimate bodies, whereby his omniſcience, power, and providence, will be eſtabliſhed; and from thence all his other attributes neceſſarily follow.

Thus, a manifeſtation of the Deity is viſible in all his works. There is not the ſmalleſt part of that immenſe ſpace our eyes behold, or our imaginations conceive, that is not filled with His preſence. The worlds which revolve with ſo much order, beauty, and harmony, through the immenſity of ſpace, the ſun, moon, ſtars, and planets, are upheld by the light of his countenance; but for which they would drop from their orbs, and, plunged into the vaſt abyſs, would return to their primitive chaos. To the mercy of God we owe all the bleſſings of this life, as the reward of good and virtuous actions. To his anger, we juſtly attribute all violent concuſſions of the elements, famine, plague, peſtilence, &c. brought on a wicked and abandoned people, like the ſtorm of fire and brimſtone on Sodom and Gomorrah. The vengeance of the Deity cannot be more awfully deſcribed, than by David in his Pſalms; which ſhould act as a timely warning to thoſe atheiſts and unbelievers, and to thoſe wicked, idolatrous, and polluted, countries, againſt whoſe deteſtable crimes theſe terrible ſcourges have been ſo often ſent. The ſhaking of the earth; the

trembling

trembling of the hills and mountains; the flames of devouring fire darting through the firmament; the heavens bending down with forked thunderbolts; their riding on the clouds, and flying on the wings of a whirlwind, the burfting of the lightnings from the horrid darknefs; the tremendous peals of thunder; the ftorms of fiery hail; the melting of the heavens, and diffolving into floods of tempeftuous rains; the earth opening and fwallowing up her inhabitants; the rocks and mountains cleaving afunder, and difclofing their fubterraneous channels, their torrents of water, and bituminous fire, at the very breath of the noftrils of the Almighty; are all of them circumftances which fill the guilty mind with horror and difmay, and admirably exprefs the power, the prefence, and the omnifcience, of God!

To what has been ftated above, I would earneftly recommend an attentive perufal of what I have written in the firft volume of my Complete Illuftration of the Occult Sciences, from page 61 to 71; whence it will be manifeft to the full conviction of the moft obftinate atheift, (if fuch a thing can really exift,) that there is a God, all-powerful and intelligent; fupremely perfect; eternal and infinite; omnipotent and omnifcient; who endures from eternity to eternity, and is prefent from infinity to infinity!

But though, from the nature and perfections of the Deity, he is invifibly prefent in all places, and nothing happens without his knowledge and permiffion; yet it is exprefsly revealed in Scripture, and admitted by all wife and intelligent authors, that he is vifibly prefent with the angels and fpirits, and bleffed fouls of the departed, in thofe manfions of blifs called Heaven. There he is pleafed to afford a nearer and more immediate view of himfelf, and a more fenfible manifeftation of his glory, and a more adequate perception of his attributes, than can be feen or felt in any other parts of the univerfe; which place, for the fake of pre-eminent diftinction, and as being the feat, and centre, from whence all things flow, and have their beginning, life, light, power, and motion, is called the *interior*, or *empyrean*, heaven.

The pofition and order of this interior heaven, or centre of the Divinity, has been varioufly defcribed, and its locality fomewhat difputed among the learned; but all agree as to the certainty of its exiftence. Hermes Trifmegiftus defines heaven to be an intellectual fphere, whofe centre is every where, and circumference no where; but by this he meant no more than to affirm, what we have done above, that God is prefent every where, and at all times, from infinity to infinity, that is to fay, without limitation, bounds, or circumference. Plato fpeaks of this internal heaven in terms which bear fo ftrict a refemblance with the books of Revelation, and in fo elevated and magnificent a ftyle, that it is apparent the heathen philofophers, notwithftanding their worfhipping demi or falfe gods, poffeffed an unfhaken confidence

fidence in one omnipotent, ſupreme, over-ruling, Power, whoſe throne was the centre of all things, and the abode of angels and bleſſed ſpirits.

To deſcribe this interior heaven, in terms adequate to its magnificence and glory, is utterly impoſſible. The utmoſt we can do, is to collect from the inſpired writers, and from the words of Revelatión, aſſiſted by occult philoſophy, and a due knowledge of the celeſtial ſpheres, that order and poſition of it, which reaſon, and the divine lights we have, bring neareſt to the truth. That God muſt be ſtrictly and literally the centre, from whence all ideas of the Divine Mind flow, as rays in every direction, through all ſpheres, and through all bodies, cannot admit of a doubt. That the inner circumference of this centre is ſurrounded, filled, or formed, by arrangements of the three hierarchies of angels, is alſo conſonant to reaſon and Scripture, and forms what may be termed the entrance or inner gate of the empyrean heaven, through which no ſpirit can paſs without their knowledge and permiſſion; and within which we muſt ſuppoſe the vaſt expanſe or manſions of the Godhead, and glory of the Trinity, to be. This is ſtrictly conformable to the idea of all the prophets and evangelical writers. From this primary circle, or gate of heaven, Lucifer, the grand Apoſtate, as Milton finely deſcribes it, was hurled into the bottomleſs abyſs; whoſe office, as one of the higheſt order of angels, having placed him near the eternal throne, he became competitor for dominion and power with God himſelf! But,

> Him the Almighty Pow'r
> Hurl'd headlong flaming from th' ethereal ſky,
> With hideous ruin and combuſtion, down
> To bottomleſs perdition, there to dwell
> In adamantine chains and penal fire!
>
> *Milton, Parad. Loſt*, b. i. l. 44, &c.

The circles next ſurrounding the hierarchies, are compoſed of the miniſtering angels and ſpirits, and meſſengers of the Deity. In poſitions anſwering to the ideas of the holy Trinity, and interſecting all orders of angels, are ſeated, in fulneſs of glory and ſplendor, thoſe ſuperior angels, or intelligent ſpirits, who anſwer to the divine attributes of God, and are the pure eſſences or ſtream through which the will or fiat of the Godhead is communicated to the angels and ſpirits, and inſtantaneouſly conducted to the *Anima Mundi*. Round the whole, as an atmoſphere round a planet, the *Anima Mundi*, or univerſal Spirit of Nature, is placed; which, receiving the impreſſions or ideas of the Divine Mind, conducts them onward, to the remoteſt parts of the univerſe; to infinity itſelf; to, and upon, and through, all bodies, and to all God's works. This *Anima Mundi* is therefore what we underſtand of Nature, of Providence,

Dodd delin.

System of the Interior, or Empyrean Heaven,
Shewing the fall of Lucifer.

Prattent sculp.

Providence, of the preſence of God, and the fountain or feat of all ſecond cauſes, being, as it were, the Eye of God, or medium between God and all created things. Next to the *Anima Mundi*, is that vaſt region or expanſe, called the ethereal heaven, or firmament, wherein the fixed ſtars, planets, and comets, are diſpoſed; and wherein the celeſtial bodies, and the comets, move freely in all directions, and towards all parts of the heavens.

To illuſtrate what has been ſtated above, I have ſubjoined a plate of the Interior Heaven, with the different orders of the Spirits and Eſſences of the Divine Mind, diſtinguiſhed by their proper names and characters, in the original Hebrew text, as pointed out in holy writ, and in the manuſcripts of ancient and learned philoſophers. As theſe names and characters are printed at length, and fully explained in the firſt volume of my Illuſtration of the Occult Sciences, p. 69, 70, and 71, it is unneceſſary to repeat the ſame here; but for a more perfect explanation of what is there written, the annexed plate is abſolutely neceſſary to aſſiſt the inquiſitive reader in forming a competent underſtanding of the ſubject. It will alſo appear from this plate, in what manner the rays or beams of Divine Providence paſs from the centre or ſeat of the Godhead, through all the different orders of angels and ſpirits, to the *Anima Mundi*, and from thence to all the celeſtial bodies, planets, and ſtars; to our earth, and to the remoteſt parts of infinite ſpace, conſtituting what is termed *celeſtial influx*, or that faculty in nature by which the quality and temperature of one body is communicated to another.

OF NATURE.

No one expreſſion, uſed by authors, or ſpoken amongſt men, is in general more variouſly applied, or ſo little underſtood, as the word Nature. When ſpeaking of the *nature* of a thing, we moſt commonly mean its eſſence; that is, the attributes or cauſes which make it what it is, whether the thing be corporeal or not; as when we attempt to define the nature of a fluid, of a triangle, &c. oftentimes we confound that which a man has by nature with what accrues to him by birth; as when we ſay, that ſuch a man is noble by nature. Sometimes we take nature for an internal principle of motion; as when we ſay, that a ſtone by nature falls to the ground. Sometimes we underſtand by nature, the eſtabliſhed courſe and order of things. Sometimes we take nature for an aggregate of powers belonging to the ſame body, eſpecially a living one; in which ſenſe phyſicians ſay, that nature is ſtrong, weak, or ſpent; or that, in ſuch and ſuch diſeaſes, nature left to herſelf will perform the cure. Sometimes we uſe the term nature for the univerſe, or whole ſyſtem of the corporeal works of God, as when it is ſaid of a

 phœnix,

phœnix, or any imaginary being, that there is no ſuch thing in nature. Sometimes too, and that not unfrequently, we expreſs by the word nature, a kind of ſemi-deity, or ſupernatural ſpirit preſiding over all things.

This general abuſe of the word nature, is by no means peculiar to the Engliſh people or language; it prevails more or leſs in all countries, and amongſt all ſects; and ſeems to have been copied from the fabulous ideas of the ancients. Ariſtotle has written a whole chapter, expreſsly to enumerate the various acceptations of the Greek word φυσις rendered in Engliſh *nature;* and, among Latin writers, there are no leſs than fifteen or ſixteen different acceptations of the ſame word, with advocates out of number for their interpretation. The bulk of them inſiſt, that the word nature radically means the ſyſtem of the world; the machine of the univerſe; or the aſſemblage of all created beings; in which ſenſe they ſpeak of the Author of Nature; and call the ſun the *eye* of nature, becauſe he illuminates the univerſe; and the *father* of nature, becauſe he warms the earth, and makes it fruitful. Others, underſtanding the word in a more confined ſenſe, apply it to each of the ſeveral kinds of beings, created and uncreated; ſpiritual and corporeal; thus they ſay *divine* nature, *angelical* nature, and *human* nature, meaning all men together who poſſeſs the ſame ſpiritual reaſonable ſoul. In this ſenſe the ſchoolmen and divines ſay, *natura naturans,* and *natura naturata;* ſpeaking of God, who is the *natura naturans,* as giving *being* and *nature* to all others; in oppoſition or diſtinction to the creatures, who are the *natura naturata,* as receiving their nature from the hands of another.

Nature, in a ſtill more limited ſenſe, is uſed for the eſſence of a thing; according to which the Carteſians ſay, it is the nature of the ſoul to think; and that the nature of matter conſiſts in extenſion. Others more properly uſe the word Nature for the eſtabliſhed order and courſe of material things; the ſeries of ſecond cauſes; or the laws which God has impoſed on every part of the creation; in which ſenſe it is they ſay, *nature* makes the night ſucceed the day; *nature* has rendered reſpiration neceſſary to life, &c. According to which, St. Thomas ſpeaks of nature as a kind of divine art, communicated to beings, which direct and carry them to the ends they were intended for; in which ſenſe nature can be neither more or leſs than a concatenation of cauſes and effects, or that order and œconomy which God has eſtabliſhed in all parts of his creation. Others ſtill more ſtrictly conſider nature as the action of Providence, and the principle of all things; or that ſpiritual power or being, which is diffuſed throughout the creation, and moves and acts in all bodies, and gives them peculiar properties, and produces peculiar effects. In this ſenſe, our modern philoſopher Mr. Boyle conſiders nature as nothing elſe but God,

acting

acting himſelf according to certain laws he himſelf has fixed. This correſponds very much with the opinion of a ſect of ancient philoſophers, who made Nature the god of the univerſe, *Το Παν*, whom they conceived to preſide over and govern all things; but this they acknowledged to be only an imaginary being; and that *nature* meant no more than the qualities, or virtues, which God implanted in his creatures, but which their poets and orators had figuratively perſonified as a god. Hence F. Malebranche was aggravated to ſay, "that the *nature* ſo much ſpoken of in the ſchools, is only fit to lead us back to pagan idolatry; ſince it taught us to underſtand ſomething, which, without being God, acts continually throughout the univerſe;" according to which, he ſuppoſes nature would be adored as an idol, conceived to poſſeſs an actual principle, which, in concurrence with God, was the next and immediate cauſe of all the changes which befal matter.

Ariſtotle, with a view of concentrating theſe ideas of nature into one point, as beſt adapted to the works of an infinitely-perfect and all-powerful Being, defines nature, *principium et cauſa motus et ejus in quo eſt primo per ſe, et non per accidens*; which definition being miſtaken by the Peripatetics and Stoics, they from hence conceived the principle of nature to be a certain ſpirit or virtue diffuſed throughout the univerſe, which gave every thing its motion by the invariable order of an inevitable neceſſity, without liberty or knowledge. This induced the idea of a *plaſtic nature*, which ſeveral learned modern writers have deſcribed to be an incorporeal created ſubſtance, indued with a vegetative life, but not with ſenſation or thought, penetrating the whole created univerſe, being co-extended with it, and, under God, moving matter ſo as to produce the phenomena which cannot be ſolved by mechanical laws; active for ends unknown to itſelf, not being conſcious of its own actions, and yet having an obſcure idea of the action to be entered upon. In ſupport of this *plaſtic nature*, Dr. Cudworth argues thus: "Since neither all things are produced fortuitouſly, or by the unguided mechaniſm of matter, nor God himſelf may reaſonably be thought to do all things immediately and miraculouſly, it may well be concluded, that there is a *plaſtic* or *formative* nature under him, which, as an inferior and ſubordinate inſtrument, executes that part of his providence which conſiſts in the regular motion of matter; yet ſo as that there is alſo, beſides this, a higher providence to be acknowledged, which, preſiding over it, doth often ſupply the defects of it, and ſometimes over-rule it, foraſmuch as this plaſtic nature cannot act electively, nor with diſcretion." This doctrine, he conceives, had the ſuffrage of the beſt philoſophers of all ages, Ariſtotle, Plato, Empedocles, Heraclitus, Hippocrates, Zeno, and the Stoics, and the latter Platoniſts and Peripatetics, as well as the chemiſts and Paracelſians, and ſeveral modern writers.

Now

Now I am clearly of opinion, that, notwithſtanding theſe great authors have ſo obſtinately contended for the definition of the word, and for the principles and conſtruction of Nature, yet they all in reality meant one and the ſame thing, only giving different explanations of the ſame ideas; and if their arguments are cloſely purſued, and compared with each other, they will all tend to ſhow, that the *anima mundi*, or ſoul of the univerſe, was what they meant by Nature.

This *anima mundi*, as we have before ſeen, is a medium inveſting the whole interior heavens, and conſiſts of a pure ethereal ſubſtance, or ſpirit: which, as it more immediately reſides in the celeſtial regions, is the ſecond or next cauſe under God, that moves and governs the heavens, and heavenly bodies, ſtars, and planets; which bodies, having received their firſt exiſtence from the fecundity of the ſame ſpirit in the act of creation, are by an influx of ſympathetic rays, and by light, heat, gravity, and motion, nouriſhed and ſuſtained, upheld and continued, in the ſame regular courſe, and in the beautiful order we ſee them. From the celeſtial regions, the ſame influx of pure ethereal ſpirit deſcends into every part of the immeaſurable ſpace, and is diffuſed through the maſs of this world, informing, actuating, and uniting the different parts thereof into various ſubſtances; and being the primary ſource of life, every-where breathing a ſpirit like itſelf, it pervades all elementary bodies, and, intimately mixing with all the minute atoms thereof, conſtitutes the power or inſtrument we call Nature, forming, faſhioning, and propagating, all things, conformably to the ideas or will of the Divine Mind in the firſt act of creation. And ſo the poet:

> *Spiritus intus alit, totoſque infuſa per artus*
> *Mens agitat molem, et magno ſe corpore miſcet.*

The only thing that has been objected to the notion of an *anima mundi*, is, that it mingles the Deity too much with his creatures; confounds, in ſome meaſure, the workman with his work, making this, as it were, a part of that, and the ſeveral portions of the univerſe ſo many parts of the Godhead.—Yet is the ſame principle aſſerted by Seneca, Epiſt. 92. *Totum hoc quo continemur, et unum eſt, et Deus. Et ſocii ejus ſumus, et membra.* M. du Hamel thinks, that thoſe who deny it, object without a reaſon; of which every one will be ſenſible, who reads the deſcription above given, ſince it in no reſpect confounds our comprehenſion of an infinitely-wiſe and Supreme Being with that of the *anima mundi;* but, on the contrary, proves it to be as diſtinct from the Deity as are the angels and ſpirits in heaven. And we may further obſerve, from what is above premiſed, that thoſe who deny the *anima mundi* on one hand generally admit it on the other. Thus the Peripatetics have

have recourſe to celeſtial influxes, which is partly the ſame thing, in order to account for the origin of forms, and the occult power of bodies. The Carteſians have their ſubtle matter, which anſwers to the active ſpirit of the *anima mundi.* Others ſuppoſe an actuating ſpirit to flow from the ſun, and the other heavenly bodies, which is diffuſed over all parts of the world, and is the ſource or principle of life, motion, &c. which is ſtill the ſame thing. Some philoſophers, in the place of theſe, have ſubſtituted the idea of fire, or an etherial elaſtic ſpirit, diffuſed through all parts of ſpace, as the medium by which elementary bodies are nouriſhed and ſuſtained; which nearly comes to the ſame thing. Even thoſe who have contended for a *plaſtic* nature, fall in with every principle of the *anima mundi;* only they inſiſt that the *formative* power is lodged in the earth; whereas the truth is, that it dwells in the heavens, and is conveyed to the earth, to the elements, and into all matter, by the medium I have deſcribed.

I ſhall conclude this definition of Nature, with remarking, agreeable to the opinion of the ingenious Mr. Boyle, that, in order to regulate our conceptions of the word in common, and to render the application of it leſs ambiguous, we ſhould diſtinguiſh between the univerſal and the particular nature of things. *Univerſal nature* we ſhould conſider to be the aggregate of all the bodies that make up the world, under the *anima mundi,* conſidered as a principle by virtue whereof they act and ſuffer, according to the laws originally preſcribed by the Author of all things. And this makes way for the other ſubordinate notion; ſince the *particular nature* of an individual conſiſts in the general nature, but only applied to that diſtinct part or portion of the univerſe; or, which is the ſame thing, it is a particular aſſemblage of the mechanical properties of matter, motion, &c. of that ſubject which immediately engages our attention.

Of the VISIBLE and OCCULT PROPERTIES of NATURE.

HAVING thus far explained the foregoing ſubject, we come next to an inveſtigation of *Cauſes,* and their *Effects,* or the means whereby Nature acts in the fructification of the univerſe. We derive the idea of cauſes and effects from our obſervations of the viciſſitudes of things, while we perceive ſome qualities or ſubſtances begin to exiſt, and that they receive their exiſtence from the due application and operation of other beings; in all which circumſtances, that which produces is the Cauſe, and that which is produced is the Effect. There is ſuch a relation and connection between the cauſe and the effect, that we cannot have a true notion of the one, unleſs at the ſame time we have a conception of the other. So in general we ſay that a cauſe is nothing elſe but that which gives being to another thing, which is the effect of it, which way ſoever it happens, according to the various cauſes.

The *Firſt Cauſe*, which acts of itſelf, and of its own ſupreme power and will, is God. This is a truth ſo evident, and ſo conſpicuous, that it cannot be denied. The exiſtence of a Firſt Cauſe may be deduced from the certainty of our own exiſtence; for that we exiſt in the world, is a ſelf-evident truth; but that we came into it of ourſelves, or by caſualty, neceſſity, or chance, is abſolutely impoſſible. The ſource of our exiſtence muſt therefore be derived from ſome being, who, as the author, muſt alſo be the free principle, of that eſſence, or life, we poſſeſs. To ſay that we ourſelves were the cauſe of our being, would be ridiculous; becauſe from thence it would follow that we exiſted before we had a being; that we gave ourſelves that which we were not in the poſſeſſion of; and that the cauſe and the effect were one and the ſame thing; which is likewiſe impoſſible. It is no leſs an error to affirm that we are in the world by neceſſity; for, if ſuch were the caſe, our exiſtence would have never had a beginning, and we ſhould have been immutable and independent, and infinite in every kind of perfection; but, as theſe qualities are only applicable to a firſt cauſe, it follows that ſuch cauſe muſt be Almighty.

Thoſe who are convinced of the exiſtence of a firſt cauſe, muſt neceſſarily attribute to it all the perfections which are or can be in the world; that it is not only moſt perfect, and moſt noble, but alſo, that all the effects which it hath produced, or is capable of producing, are contained within itſelf, in the utmoſt perfection; and that every one of them is infinite, in the unity of its being; for it is neceſſary it ſhould poſſeſs the perfections of thoſe beings it hath or can produce, otherwiſe it would be ſaid to communicate that which it neither hath nor can have. The firſt cauſe would not be abſolutely perfect, if it was not Eternal; for ſo it would have had a beginning, and might have an end; and then it could not have been the firſt cauſe, inſomuch as it derives its exiſtence from that which was pre-exiſtent to it; and by conſequence this cauſe, which we ſuppoſe to be the firſt, would be a ſecond cauſe, limited in its being and perfections, as in its duration; and it would ſeem to have a dependence upon another; whereas, when we ſuppoſe it to be the firſt, all others muſt depend upon, and be ſubordinate to, it. Whence it follows, that theſe qualities muſt be inſeparable from it, independence, eternity, infinity, and ſupreme authority; andthat we cannot conceive any firſt cauſe, but at the ſame time we muſt acknowledge the exiſtence of God.

This *firſt cauſe*, which is God, muſt neceſſarily have that *perfect unity*, which admits no multiplication either of nature or perfections. Certainly if God was not one in his being, but had ſeveral natures, the number of them ought to be infinite, and yet none of these beings in particular would be infinite, because, when the perfection of one cannot be the perfection of another, there will not be one to be found but will stand in need of the other, that is, in whom there would not be requisite that perfection which the other beings possess. Therefore we may add, that all these supposed beings would be opposite, independent, and all supreme, which is impossible; or that all would be subject to one or other of them, which is ridiculous; whence it follows, that there is but one only God, who is one in his existence, incapable of any multiplication, and who is the primary and universal cause of all things. The great number, or rather the infinity, of perfections which we apprehend to be in the first cause, is not repugnant to the Trinity, because that does not divide the being: and these perfections are but one and the same thing, though we give them several names, and consider them under several ideas, which we are forced to correct; since without that unity there would be necessarily a composition of parts, which would be the materials of the whole compound, and would precede its existence; therefore could not be the ingredients of that composition, without something else intervening. They may also be divided and separated; so that, by the dissolution of the parts, the compound would cease; which is plainly inconsistent with that idea we have of God, who is not simple in his nature, independent in his will, and every way incorruptible. The first cause is only one, and without its like, in its essence, but also one sole, and without a second, in that action by which this world was produced; and for this reason the action is called *creation*, supposing nothing but mere nothing, out of which all things were made, by the only power of God, without the help of any other, having either the quality of an agent or a subject. The world, being produced by this first cause, remains subject to the will and pleasure of it; and, in the same manner as it was produced by the sole act of this first cause, so it is preserved in the same state, by the sole influence of the same cause; which, as it did not want any help in the creation of the universe, so neither doth it stand in need of any assistance in the conservation of it. If the first cause was free in the creation of the world, thence it follows that all things were made by direction of reason and understanding, and by consequence according to a certain idea and rule: and, since the first cause operates after an independent manner, it could not have the type of its production any where else but from itself; neither could it act by a rule distinct from its own being; so that in truth God is not only the first, but the exemplary, cause of all things. For the same reason it may be said, that the first cause, which is God, is the final cause of all things; for when he, as an intelligent and free being, produced this world, he proposed to himself an end answerable to his dignity, that is, to himself and his own proper glory. So that

the

the First Cause is, necessarily, the ultimate end of all its effects. This is a nice subject, wherein all preachers and writers seem at a loss; but the case is clear and beautiful to those who pursue the lessons of wisdom and science.

Second Causes are those which derive the power or faculty of acting from the influence of the First Cause. Hence the *anima mundi* is the seat of all second causes, which are also termed natural causes, because they have implanted in them, by the First Cause, the occult power of diffusing through all space, and of communicating to all bodies, that universal soul, or etherial spirit, whereby every particle of matter is moved, and made to act upon one another, so as to produce the various phenomena in nature, in the animal, vegetable, and mineral, worlds; in the elements, and in the firmament of heaven. This universal spirit or cause acts universally with particular causes; but after a manner agreeing with the nature of every particular thing, and according to the power which was given it when it was created; which neither alters the quality of the causes, nor the necessity or liberty of their actions. This power of acting, which is communicated by second causes, is not a quality different from those things impowered to act; whence the power which the atoms have of moving in all directions doth not differ from the atoms themselves; the power of burning or heating doth not differ from the fire to which it is inherent, unless it be in the manner of our conceiving things, or of speaking of them according to our conceptions. So it is of an action which terminates from the cause to the effect, and which is nothing else than a certain relation, or an actual subordination, which is always found betwixt the cause and the effect. Hence we perceive that second causes are what all philosophers, ancient and modern, have contended for under different forms; and are neither more nor less than that universal spirit, or inherent law, implanted in nature at the creation, whereby all God's works are regulated and preserved, and the ends and purposes of that creation conducted to God's glory and manifestation, and to the good of all his creatures, the study of which opens our eyes to the bright beams of true wisdom; to the mutual harmony and dependence one thing has upon another; to the sympathy and antipathy of material bodies; to the perceptions of sense, reason, and intellectual vision; to the nice faculties and exquisite connection of soul and body; and ultimately to the knowledge of ourselves, of our progress through this world; of our sublunary fate and fortune; and of the things calculated to preserve life, or to destroy it.

Efficient Causes are all those actions of bodies or things, which are the agents or direct means whereby any effects are produced. Thus a painter, painting a picture, is an efficient cause; and the picture itself, when finished, is the effect thereof. Efficient causes comprehend a number of compound or subordinate causes, which alſo contribute towards the production of their effect. If the efficient cauſe acts by a power proper to itſelf, then it is called the principal cauſe; but, if only by the force and ſtrength of another, then it is termed the inſtrumental cauſe. So we diſtinguiſh between the painter and the pencil, though both contribute to the production of the picture. Again, the ſubject whereon the agent works, or whereof the thing is formed, is called the *material* cauſe; thus the marble out of which a ſtatue is carved, is called the material cauſe; as is likewiſe the paint and canvas of a picture, as being the matter, or materials, of which they are made; the ſculptor and painter being the efficient cauſes. There is alſo an efficient cauſe in the ſun, moon, ſtars, and elements, whereby they act upon ſublunary matter, and produce a variety of effects in the fructification and phenomena of this world. From theſe ariſe neceſſary and free cauſes: the firſt of which act neceſſarily and without choice; as fire, the ſun, and all created beings, except angels and men; for they act by a free will, wherein conſiſts the eſſence of liberty. The efficient cauſe is likewiſe either phyſical, or moral; the *phyſical* is that which produces a ſenſible corporeal effect, and acts obviouſly and immediately; thus fire is the phyſical cauſe of burning, and the ſun the phyſical cauſe of heat. The *moral* cauſe alſo produces a real effect, but in things immaterial; thus, repentance is the moral cauſe of forgiveneſs. Again, we define a *phyſical* cauſe to be that which produces its effect by a phyſical virtue; and a *moral* cauſe, that which *determines* the phyſical cauſe, though not *neceſſarily*. to produce the effect: thus, the ſun is the phyſical cauſe of light; a ſtone that falls from an eminence, and fractures the ſkull, is a phyſical cauſe of death; and thus the advice, intreaty, commands, or menaces, which determine us, though not *neceſſarily*, to do, or not to do, any thing, are moral cauſes. In this ſenſe it is obvious a moral cauſe is only applicable to a free intelligent agent; and it is this notion of a moral and phyſical cauſe that is the moſt juſt, clear, and diſtinct.

Whenever the efficient cauſe applies to a free intelligent being, and acts from knowledge, all the ſubordinate cauſes concur to the production of one and the ſame effect; as for example, the painter drawing his picture is the principal cauſe; the pencil the inſtrumental; the end propoſed by the painter is the final cauſe; the idea directing the performance is the exemplary cauſe; the form and diſpoſition of its parts is the formal cauſe; the colours and the canvas are the material cauſe, as being the conſtituent matter of the work; and if the artiſt, by any accidental touch, improves his picture, like Agillaus, who labouring many days to draw a foaming horſe, and could not, in a paſſion flung his bruſh at the painting, and the thing he wanted was thus accompliſhed when it was not intended—it is termed an accidental cauſe, So that we ſee all ſubordinate cauſes are in effect under the efficient cauſe; and are ſubſervient to it, when applied by reaſon, and directed by ſkill.

Form and *formal caufe* is one and the fame thing; and, when we fay there are two forts of forms, that is only according to our manner of conceiving things; fo we fay there are two forts of formal caufes, the fubftantial and accidental. But all forms are imaginary; neither do true philofophers acknowledge any other fubftance to be in natural compounds than matter, except only in man; nor any other form than the difpofition of the parts, becaufe all thefe forms are altogether ufelefs. Moreover, the exemplary caufe may be referred to the formal, becaufe it is the idea and inward form of that which we frame in our fpirit; fo the formal caufe of a picture is the difpofition of its parts, according to the difpofition and ordination which it then had in the mind of the painter. The fame may be faid of all rational agents which are endued with underftanding.

There is no difference betwixt matter and a material caufe; and there are two forts of material caufes, as well as of matter. That is the *firft matter* out of which all bodies are compofed, and into which, by an univerfal divifion, they may be reduced; the fecond is nothing elfe but bodies made of the firft, and upon which the efficient caufes exercife their action. Therefore it is apparent, that there is nothing in the world but what is a compound, and that there is no compound without matter. It is alfo certain that there is nothing made without an efficient caufe, which acts upon compounds, and deftroys them, that of them others may be made; becaufe the matter of the firft ferves for the compofition of the fecond. The matter which goes to the compofition of the firft and fecond is the *firft matter*, or material caufe, of the compound; and that matter which ferves the efficient caufe for a fubject and patient is called the *fecond matter*. Both of them may be an efficient caufe; for compounds act upon one another like the elements, which drive one another backwards and forwards. That which drives another is called the agent, as that which is driven is called the patient; and if there be any thing which refifts it, and drives back another, this regrefs of the motion is called *re-action*. So that one and the fame thing may be the fubject and caufe of motion; and that, to give and receive being the principle of agent and patient, both may happen at the fame time, but in divers refpects.

Efficient caufes, in folid and fluid bodies, we often fee, act in a moft wonderful manner; and, if they were not vifible to our eyes, we fhould fcarcely believe any of thefe occult properties exifted in them. Thus, the action of oil, in ftopping the violent ebullition of various fubftances, is truly furprifing. It is well known, that if a mixture of fugar, honey, or the like, be boiling on the fire, and in danger of rifing over the fides of the veffel, the pouring in a little oil immediately makes it fubfide. In many cafes the marking a circle round the infide of a veffel, in which a liquor of this kind is to be boiled, with a piece of hard ſoap, ſhall, like a magic ring, confine the ebullition to that height, and not ſuffer it to ſtir any farther. This is wholly owing to the oil, or fat, contained in the ſoap: but there is, beſides theſe, another very important uſe of oil, on a like occaſion, which is the pouring a little of it on any metallic ſolution while making; this reſtrains the aſcent of the noxious vapours; preſerves the operator from danger; and, at the ſame time, by keeping down the evaporating matter, gives redoubled ſtrength to the menſtruum. Pliny has mentioned an extraordinary effect of oil, in ſtilling the ſurface of water when it is agitated with waves, and the uſe made of it, by the divers, for this purpoſe: *Omne*, ſays he, *oleo tranquillari, &c.* lib. ii. cap. 103. And Plutarch, in Quæſt. Natur. aſks, *Cur mare oleo conſperſum per lucidum ſit et tranquillum?* Pliny's account ſeems to have been either diſcredited or diſregarded by our writers on experimental philoſophy, till it was confirmed by ſeveral curious experiments of Dr. Franklin, which were publiſhed in the year 1774. The property of oil above mentioned has, however, been well known to modern divers and dredgers for oyſters at Gibraltar, and elſewhere. The divers in the Mediterranean, in particular, deſcend, as in Pliny's time, with a little oil in their mouths, which they now and then let out; and which, on riſing to the ſurface of the ſea, immediately renders it ſmooth, ſo as to permit the light to paſs through the water undiſturbed by various and irregular refractions. The Bermudans are enabled to ſee and ſtrike fiſh, which would be concealed from their view through the roughneſs of the ſea, by pouring a little oil upon it. And the Liſbon fiſhermen effect a ſafe paſſage over the bar of the Tagus, by emptying a bottle or two of oil into the ſea, when the ſurf is ſo great as to endanger its filling their boats. Our ſailors have alſo obſerved, that the water is always much ſmoother in the wake of a ſhip that hath been newly tallowed than it is in one that is foul. Dr. Franklin was led, by an accidental obſervation made at ſea in 1757, to attend particularly to Pliny's account; and the various informations which he afterwards received relating to it induced him to try ſome experiments on the ſubject. Standing on the windward ſide of a large pond, the ſurface of which was rendered very rough with the wind, he poured a tea-ſpoonful of oil on the water. This ſmall quantity produced an inſtant calm over a ſpace of ſeveral yards ſquare, which ſpread amazingly, and extended itſelf gradually, till it reached the lee ſide, making all that quarter of the pond, perhaps half an acre, as ſmooth as a looking-glaſs. On repeating this experiment, which conſtantly ſucceeded, one circumſtance ſtruck him with particular ſurpriſe; this was the ſudden, wide, and forcible, ſpreading of a drop of oil on the face of the water, which, he adds, "I do not know that any body has conſidered." When a drop of oil is put on a looking-glaſs, or poliſhed marble, it ſpreads very little; but

on

on water it inftantly expands into a circle extending feveral feet in diameter, becoming fo thin as to produce the prifmatic colours for a confiderable fpace, and beyond them fo much thinner as to be invifible, except in its effects of fmoothing the waves at a much greater diftance. It feems, fays Dr. Franklin, as if a mutual repulfion between its particles took place as foon as it touched the water, and a repulfion fo ftrong as to act on other bodies fwimming on the furface, as ftraws, leaves, &c. forcing them to recede every way from the drop, as from a centre, leaving a large clear fpace. In endeavouring to account for the fingular effects of oil, Dr. Franklin obferves, that there feems to be no natural repulfion between water and air, fuch as to keep them from coming into contact with each other.—Therefore, air in motion, which is wind, in paffing over the fmooth furface of water, may rub, as it were, on that furface, and raife it into wrinkles, which, if the wind continues, are the elements of future waves. The fmalleft wave does not immediately fubfide, but in fubfiding raifes nearly as much of the water next to it. A fmall power, continually operating, will produce a great action: fo that the firft-raifed waves, being continually acted upon by the wind, are, though the wind does not increafe in ftrength, continually increafed in magnitude, rifing higher and extending their bafes, fo as to include a vaft mafs of water in each wave, which, in its motion, acts with great violence. But, if there be a mutual repulfion between the particles of oil, and no attraction between oil and water, oil dropt on water will not be held together by adhefion to the fpot on which it falls: it will not be imbibed by the water, but be at liberty to expand itfelf and fpread on a furface, that prevents, perhaps, by repelling the oil, all immediate contact; the expanfion will continue till the mutual repulfion between the particles of oil is weakened, and reduced to nothing by their diftance. Dr. Franklin imagines, that the wind, blowing over water thus covered with a film of oil, cannot eafily catch upon it, fo as to raife the firft wrinkles, but flides over it, and leaves it fmooth as it finds it. It moves a little the oil, indeed, which, being between it and the water, ferves it to flide with, and prevents friction: hence the oil, dropt on the windward fide of the pond, proceeds gradually to leeward, as may be feen by the fmoothnefs it carries with it quite to the oppofite fide; for the wind, being thus prevented from raifing the firft wrinkles, which he calls the elements of waves, cannot produce waves, which are to be made by continually acting upon and enlarging thofe elements; and thus the whole pond is calmed. Upon the whole, there is great room to fuppofe (notwithftanding the partial failure of an experiment made at Portfmouth by Dr. Franklin and others), that feafaring people may derive advantages from ufing oil on particular occafions, in order to moderate the violence of the waves, or to leffen the furf which fometimes renders the landing

on

on a lee-ſhore dangerous or impracticable. To this purpoſe we are imformed, that the captain of a Dutch Eaſt-India ſhip, being overtaken by a ſtorm, found himſelf obliged, for greater ſafety in wearing the ſhip, to pour oil into the ſea, to prevent the waves breaking over her; which had an excellent effect, and ſucceeded in preſerving her. Phil. Tranſ. vol. lxiv. part 2. p. 445, &c. It is alſo obſervable, on the coaſt of Sutherland, when the lump-fiſh abounds in ſpring, and are devoured by the ſeals, that it may be known by the ſmoothneſs of the water above the ſpot; the oil ſerving to ſtill the agitation of the waves.

Occaſional cauſe is applied to the ſoul and body of man, and are only the *occaſions*, not the direct *cauſes*, of their effects. The ſoul is not able to act on the body, nor the body reciprocally on the ſoul: to keep up an intercourſe between them, God, on occaſion of a motion of the body, impreſſes a ſenſation on the ſoul; and, on occaſion of a ſentiment of the ſoul, impreſſes a motion on the body: the motions, therefore, of the ſoul and body, are only *occaſional cauſes* of what paſſes in the one or the other. Thus, we ſay, the ſtroke, or percuſſion, is only the occaſional cauſe of the motion produced in the body ſtruck; it is God is the direct efficient cauſe. And thus the action of objects on our organs is not the efficient cauſe of our ideas and perceptions, but merely the occaſional cauſe, which determines God to act on the mind, according to the laws of the union of ſoul and body.

In a *medical ſenſe*, whatever produces a diſeaſe is called the *cauſe* thereof. This operates either by inducing a new ſtate of the ſolids and fluids, or by taking away ſomething which is abſolutely requiſite to the exerciſe of ſome function. If a cauſe pre-exiſted in ſome meaſure in the body before the effect produced, it is called an internal cauſe; but if it exiſted out of the body, and by its application to it produced the diſeaſe, it is called external.

Internal cauſes generally injure firſt the humours, and then the ſolid parts; whereas, the external cauſes affect the ſolids, and, in conſequence of that, the humours; and this holds univerſally, unleſs perhaps in ſome few diſeaſes produced by poiſon or contagion. The immediate cauſe is that which taken altogether immediately conſtitutes the preſent diſeaſe; this is always adequate, and ſufficient to the formation of the diſeaſe, whether ſimple or complicated. The preſence of this conſtitutes and continues the diſeaſe: and the abſence of it removes the diſorder, being very little different from the diſeaſe itſelf. The inveſtigation, therefore, of this is extremely uſeful and very neceſſary. The remote cauſe is that which changes the body in ſuch a manner, as to diſpoſe it for the reception of a diſeaſe upon the acceſſion of another cauſe; but it is never adequate or ſufficient to produce a diſeaſe alone; nor would that other cauſe, the accceſſion of which is neceſſary, be of itſelf ſufficient for

 the

the production of the diſeaſe; but both muſt concur. The buſineſs of phyſic, therefore, is to eradicate both together, which, in conjunction, conſtitute the proximate or immediate cauſe. The remote cauſe inherent in the body, is called prediſponent, antecedent, and by the Greeks προηγουμενη, and conſiſts principally in temperaments, plethora, and cacochymy. The cauſe whoſe acceſſion to the remote cauſe excites, and in conjunction with it forms, the diſeaſe, is called the procatarctic cauſe, or the προφασις, or occaſion of the diſeaſe. It is ſometimes internal, ſometimes external. Theſe Boerhaave reduces to four claſſes. Firſt, the *ingeſta*, or things entering the body; ſuch as the air, aliments, drink, medicines, poiſons, ſuch things as enter by the pores of the ſkin and noſtrils; by the ſeveral paſſages of the mouth, lungs, œſophagus, ſtomach, inteſtines, &c. either in a viſible or inviſible manner; whether by ſteam, draught, deglutition, clyſter, or injection. Secondly, the *geſta*, or things acted, as motion of the whole or any part of the body, affections of the mind, reſt, both of the body and mind, ſleeping and watching. Thirdly, things retained, or excreted, whether ſalubrious, ſecrementious, or morbid. Fourthly, things applied to the body; as air, vapours, fomentations, cloths, liniments, ointments, and plaſters; together with whatever wounds, contuſes, or corrodes: all which circumſtances ſhould be well conſidered by medical men.

Of the FIRST MATTER.

ALL philoſophers agree that there is a *firſt matter*, which was produced from the beginning; and, though it can never undergo any change, yet it is to be ſeen in all the generations and corruptions which are in nature. Hence it appears, that the firſt matter exiſted before the generation of the compounds wherein it is found, and that it ſtill remains and ſurvives the corruption of it; as, for inſtance, in fire which is made of chips; the matter of the fire was in the chips, and is found partly in the fire, partly in the ſmoke, and partly alſo in the aſhes. It is agreed by all, that nothing produces nothing, and that there is not any thing in nature that can be reduced into nothing, but that the firſt principles remain in all revolutions which can happen; therefore, in reſpect of matter, we may juſtly ſay, that there is nothing new in the world ſince the creation, and that this matter in its nature is incorruptible; ſo that to explain the eſſence of this firſt matter is the only difficulty.—Ariſtotle makes it the ſubject of all forms, and nothing but a paſſive power or a mere capacity of producing them. He ſays alſo, that matter in itſelf hath neither quality nor quantity, nor any eſſence beſides that which it received from that form which perfected it: but this explication gives us no clear idea of matter, neither doth it teach us any thing of the nature of it. Thoſe were nearer the truth who ſaid that the firſt matter was

was nothing elſe but the firſt elements into which compounds by a total diſſolution are reduced ; alſo that theſe elements ought to be ſimple and indiviſible, for otherwiſe the firſt elements are not ſuch as we ſuppoſe them to be. It follows from this doctrine, that neither water, air, earth, nor fire, are firſt elements, becauſe they are compounds, as by the new chemiſtry is evidently proved ; therefore, we muſt look for other elements which are ſimple and indiviſible. Now, it is obvious, that ſimple and indiviſible atoms are the only firſt matter, and the principle and elements whereof bodies are compoſed: out of theſe atoms are corpuſcles made, out of theſe corpuſcles ſmall maſſes, out of maſſes greater parts, and of theſe parts greater bodies, whereof the univerſe itſelf conſiſts. And *vice verſa*, going backward analytically, the world is divided into great bodies, thoſe bodies are reduced by mortality and time into parts, parts into ſmall maſſes, maſſes into corpuſcles, and, laſtly, corpuſcles into atoms.

OF ATOMS, AND THEIR NATURE.

TO demonſtrate the exiſtence of *atoms*, we may ſuppoſe that every compound may be divided into ſo many parts as they are which make the compound; and diviſion muſt neceſſarily ceaſe when there is a failure of parts to be divided. On the other hand, there is no end of diviſion as long as there are particles to be divided; for we muſt allow, either that a body cannot be ſo exactly divided but that there always remain diviſible parts *in infinitum*, or that there are parts after a certain number of diviſions which will not admit any further diviſions. Ariſtotle holds the former, but Gaſſendus and the ancient philoſophers defend the latter; and, according to this laſt doctrine, after all the diviſions are made, nothing can remain beſides atoms, that is, indiviſible beings, which are the firſt elements of natural bodies.

I confeſs it is hard to imagine a corporeal thing to be indiviſible, becauſe we ſee nothing in this world which is not diviſible; but this makes nothing againſt atoms which are corporeal, becauſe they compoſe bodies, and are indiviſible, becauſe they are the firſt and moſt ſimple elements of bodies. Hence ariſes another difficulty, becauſe it cannot be eaſily explained after what manner a thing that is diviſible is compoſed of parts which are indiviſible. Impartial minds do not find ſo much difficulty in conceiving this matter, as thoſe do who follow the prejudices they have received; ſome people do not conſider, firſt, that there are many things which eſcape our ſenſes, and yet are moſt real; ſecondly, that which compoſes a body is not a compound, as we ſee that unity makes number, though itſelf be not a number; letters, whereof nouns and words are framed, yet are neither one or the other; the drops of water, whereof rivers conſiſt, are not rivers: ſo atoms, though they are inviſible and indiviſible

viſible, yet they compoſe bodies which are viſible and diviſible. They are alſo immutable, in order to the world's continuing in the ſame ſtate, and bodies being of the ſame nature now as formerly.

OF THE PROPERTIES, MAGNITUDE, FIGURE, WEIGHT, AND MOTION, OF ATOMS.

AN *atom* is a corporeal being, ſimple, inviſible, and indiviſible; ſolidity conſtitutes its eſſence or eſſential property, which diſtinguiſhes it from ſpirit and vacuity, which hath no power of reſiſting. Atoms neceſſarily avoid all our ſenſes, becauſe they are compoſed of many diſtinct and groſs parts, whoſe object ought to be compoſed ere it can be perceived by the external organ; this, however, does not deſtroy the truth and reality of atoms, becauſe ſmall corpuſcles eſcape our ſenſes; as we obſerve in duſt, which ſticks to our clothes; in the corpuſcles of a ſtone, which is made hollow by drops of water; in divers occult parts in a mite, which cannot be ſeen without the help of a microſcope; and, laſtly, in ſmall corpuſcles which are ſeen to move in a chamber by the ſun-beams: we may omit many others that are ſmaller, which, without doubt, we could ſee if our ſight were more acute.

Though atoms are moſt ſubtle and imperceptible, yet they have their particular extenſion, magnitude, and figure, from whence their differences ariſe: for the figure of ſome of them is round, as the atoms of water, oil, and quickſilver; others have cubicular figures, ſuch as the atoms of ſea-water; and others are pyramidal, as thoſe whereof nitre conſiſts; there are ſome which have ſharp points like needles, as thoſe of fire; whence we have to ſuppoſe there are others variouſly figured. This difference is neceſſary to diſtinguiſh compounds: and as theſe atoms, as to their ſolidity or inviſibility and indiviſibility, (which are their inſeparable properties,) are alike, ſo alſo, if they did not differ in their figure and thickneſs, all bodies would be alike.

Weight is the principle of the ſaid natural motion, inaſmuch as it doth reſiſt a violent motion. I mention this, that we may know whether motion in atoms hath an internal or external principle, or whether weight be determined only to one motion, or that it be indifferently inclined to many; and whether the motion of atoms doth tend to ſome centre; and whether it be continuant or interrupted, and, laſtly, whether it be perpendicular or horizontal, parallel or declined, right, or parabolical, or circular.

In order to ſolve this difficulty, I ſuppoſe that atoms may be conſidered in a double ſtate: the firſt ſtate is before the compoſition of the bodies which are made of them, and may be called the ſtate of liberty; the other is that which they have in the bodies which conſiſt of them, which may be termed the ſtate of obligation or ſervitude.

4 If

If atoms be considered in their first state, their motion is perpetual; so that an atom that is loose, and freed from any composition, is essentially in motion, which ought not in the least to be wondered at; for motion in respect of a free atom is the same that understanding is in respect of an angel, which is never without knowing, unless his intellect be bound and clouded.

From this principle it is evident, that atoms are in continual motion, unless they are hindered, or that they meet with some obstruction, or that there are other atoms resisting or repelling them, or that they find such as will stick to them, or that they insinuate themselves into the atoms of certain bodies, or that they enter into some composition whereby their motion is stopped. Nevertheless, atoms in compounds are not altogether void of motion, because they are not so straitly embodied together but that they have some motion, like vibrations and palpitations, according to the liberty which is granted them by the disseminated vacuities; nay, some of them do sometimes attempt their escape, especially in porous bodies, which therefore sooner corrupt and perish than bodies which are more solid and close. It is yet more evident in living bodies, out of which the animal spirits, which are but the bodies of atoms, and most subtle corpuscles, are dissipated by transpiration, whence aliment is necessarily requisite to supply to the whole body those spirits which are dissipated by motion and agitation. This motion of atoms, or the least corpuscles, in living bodies, may be deservedly accounted the image of their first liberty; and, though they do but seldom enjoy their full liberty, yet they are apt to raise the greatest commotions in order to be freed, and to gain their liberty. This is the origin of many distempers; as, in acute fevers, the atoms or corpuscles of the boiling blood, or obstructed choler, are carried and driven into the brain, where they produce watchfulness, deliriums, and phrensies. According to this principle, that which we said before may be concluded, *That many distempers arise from minute corpuscles and emancipated atoms;* for these, being driven forwards by other atoms, and forced back, run into the membranes, periostium, meninges, or intestines, and cause the cholic, head-ach, gout, and rheumatism; so that this solution of corpuscles and emancipation of atoms in our bodies is much to be dreaded; and, to prevent their danger, all motions of the body which are too violent must be avoided; for these are the external cause of the confusion of the spirits and the emancipation of the atoms.

The emancipation of the atoms, and also of the small corpuscles which are composed of those atoms, are to be seen no less in the great world than in the little: for the winds are nothing less than emancipated atoms, which, by their impetuosity being driven backwards and forwards, force all bodies that stand in their way:

it is thefe atoms which agitate the air, and overturn all things which refift their motion; therefore the motion of atoms is not equal, nor every where alike, but varies according to the diverfity of bodies whereby they are driven, or as the figures of them are more or lefs fitted for motion, or otherwife, according to the proportion of vacuities which are difperfed in bodies; fo that fome atoms are moved quicker, and others flower, not becaufe fome are heavier than others, but becaufe they are driven backwards or forwards, or are ftopped, by others which happen to fix them with greater or lefs violence.

An atom is not a body according to the notion we have, that it is a compound being; but a fimple being, and alfo corporeal; that is, fimple, becaufe it is indivifible; and corporeal, becaufe it hath a certain extenfion, and makes up the compofition of bodies, which in the total divifion of them are reduced again into atoms. Two other difficulties arife from the former opinion: for, if an atom be indivifible, after what manner can we propofe to ourfelves that it hath extenfion, or how can it be an ingredient in the compofition of divifible bodies? To this we anfwer in few words, that extenfion is according to the nature of the thing extended; for, if the thing extended be divifible, in the fame manner is the extenfion; and fo, on the other fide, it is of the rational foul, which is poffeffed of the whole body, and exercifes its operations in all the parts of it; neverthelefs it is, like an atom, indivifible; and, though it be divifible in refpect of the fpace it occupies, yet it hath an internal extenfion which is indivifible; it is the fame thing which divines are forced to fay of angels, and fome philofophers about their phyfical tumid points. But fome will fay, that atoms are neither like fouls, angels, nor phyfical points, becaufe they have parts, and thefe have none; and fince that which confifts of parts is divifible, it follows alfo that an atom is divifible. To this difficulty I anfwer, with the divines, that angels and our fouls, which are fpirits—and alfo, with philofophers, that phyfical points, which are material—have no real but only potential parts; that is, an angel and rational foul in refpect of the operations which they exercife and the fpace which they occupy, and the tumid points in refpect of the fpace which they fill up. Indeed, an angel and the foul have two powers, whereof the one is the intellect, the other the will, which, being only an indivifible fubftance, capable of underftanding and willing, yet no man will deny but they, notwithftanding their indivifibility, (which at leaft is equal to the indivifibility of an atom,) do fill up a divifible fpace; as no man can doubt but that an angel can be at the fame time in the four corners of the room and in the middle of it, and that it hath a four-fquare figure by communication of the four angles or corners, and that it can quit this and affume another figure at its pleafure; which cannot be faid of tumid points and atoms, which are destitute of underſtanding and will. The rational ſoul, being equally indiviſible with an atom, angel, or point, doth wholly poſſeſs a great body, no leſs than it did when the body was little; therefore it dilates itſelf without being divided, becauſe in its nature it is ſimple and indiviſible, and is without diſtinct parts. This is the opinion of Ariſtotle, and indeed it is the moſt common opinion. But, if the ſoul were not by its own ſubſtance extended through the whole body, and had its ſeat only in the heart, as Empedocles would have it; or in the ſpleen and the ſtomach, as Van Helmont places it; or in the *glandula pinealis* of the brain, according to Carteſius; or in the ſtriate bodies of the brain where the common ſenſe is, or the ſenſe itſelf, as it is called by way of excellency; and in the callous parts, becauſe there it forms the ideas of things and judges of them; and in the cineritious part of the brain, becauſe there it performs the functions of the memory, according to the opinion of Duncan—it is certain that all theſe parts, which are taken to be the ſeat of the ſoul, are diviſible, and that thay have diſtinct parts and figures: ſo the ſoul, as it is indiviſible, occupies a ſpace or place which is diviſible; whence I conclude, that indiviſibility does not hinder but that a ſubſtance may have a certain indiviſible extenſion, but diviſible as to the place which it poſſeſſes, or that it may have angles and figure in reſpect of place, though its ſubſtance eſſentially remain one, ſimple, and indiviſible.

Hence it follows that there is one catholic or univerſal matter, called corpuſcles or atoms, filling all ſpace, which is an extended, impenetrable, and diviſible, ſubſtance, common to all bodies, and capable of all forms; infinitely harder than any of the ſenſible porous bodies compounded of them; even ſo hard, as never to wear, or break in pieces; no other power being able to divide what God made one in the firſt creation. While theſe corpuſcles remain entire, they may compoſe bodies of one and the ſame nature and texture in all ages; but, ſhould they wear away or break in pieces, the nature of things depending on them would be changed. Water and earth, compoſed of old worn particles and fragments of particles, would not be of the ſame nature and texture now with water and earth compoſed of entire particles at the beginning; and therefore, that nature may be laſting, the changes of corporeal things are only to be placed in the various ſeparations and new aſſociations of theſe permanent corpuſcles; that, in order to form the vaſt variety of natural bodies, this matter muſt have motion in all its aſſignable parts, and act in all manner of directions and tendencies. Theſe corpuſcles have therefore not only a *vis inertiæ*, accompanied with ſuch paſſive laws of motion as naturally reſult from that force; but alſo are moved by certain active principles, ſuch as that of gravity, and that which cauſes fermentation, and the coheſion and ſympathy of bodies. That this

this matter muſt alſo be actually divided into parts, and each of theſe primitive particles, fragments, or atoms of matter, muſt have their proper magnitude, figure, and ſhape; and muſt have different orders, poſitions, ſituations, and poſtures, whence all the varieties of compound bodies ariſe. This view of the firſt principles of matter accounts for an infinity of phenomena, otherwiſe inexplicable; and points out all the occult operations in nature, by ſympathy, antipathy, faſcination, coheſion, coagulation, diſſolution, &c. for, ſince theſe corpuſcles are every-where and at all times in motion, iſſuing from and cohering to all bodies that fall in their way; and ſince they are operated upon and diverſly altered by the four elements proper to this world; and theſe elements again by rays of light, heat, and influx of the anima mundi, and celeſtial bodies—all the viciſſitudes of nature are deduced from them; and, according to the qualities and temperature of the matter ſo formed, and of thoſe they come in contact with, are the affections of the mind, the functions of the body, the paſſion of love, and a thouſand inexplicable circumſtances attendant on human affairs, regulated and governed; as we ſhall now proceed to ſhow.

Of SYMPATHY and ANTIPATHY in NATURAL BODIES.

THE wonderful effects we ſee in nature, whoſe true and efficient cauſes are not eaſily found out, obliged philoſophers heretofore to have recourſe to occult cauſes, and to attribute all thoſe effects to natural ſympathy and antipathy, which happen amongſt the ſeveral bodies whereof the world is compounded.

That we may the better underſtand what may be ſaid upon a ſubject ſo nice and delicate, and give a reaſon for thoſe wonderful effects which are attributed to ſympathy and antipathy, in the firſt place we muſt ſuppoſe, that the difficulty which occurs in explaining an effect of this nature, is becauſe the mind is not able to know the truth of things but by the ſenſes, which are the gates through which the objects enter, and form their ideas in our underſtanding; but, becauſe there are an abundance of things that eſcape our ſenſes, it is no wonder that it is ſo hard to give a reaſon for thoſe cauſes which are ſo remote from our view: as for example, iron moves itſelf, and that by way of local motion, and joins itſelf to the loadſtone; we do not ſee that which draws the iron to it, though we ſee it attracted; and therefore, that we may give a ſolid reaſon for this and other phenomena of the like nature, we declare, according to our philoſophy, that there are no bodies but what continually emit certain ſubtle particles and imperceptible corpuſcles which are diſperſed through the air, and are ſometimes carried to a great diſtance, unleſs they are ſurrounded by other bodies in their way. By this principle we find the reaſon why a dog follows the

Dodd delin.

Pratten sculp.

Symbol of the Universal Spirit of Nature.

the footstep of the hare, or from a heap of a thousand stones, he readily knows the stone his master threw, and picks it out, and by his command brings it to him; from this dispersion of corpuscles, we find the reason how the contagion of the plague, either from the person infected, or from the wind blowing from that region, is carried a great way off; by this appears the reason why wounds may be cured at a hundred miles distance by means of the sympathetic powder, the astonishing properties of which are fully described in my Illustration of the Occult Sciences; so likewise of the fermentation of Canary wine brought into England, which ferments here at the same time of the vintage there.

We must suppose, farther, that these small corpuscles differ in figure and magnitude, and are not equally received by this or that body; so one man is infected by the plague in the same place where many others escape; for the same reason the beams of the sun melt wax, but not lead unless they are collected and united by the help of a burning lens, or the like; and the heat of fire melts metals in a very different manner.

Again, the palm-tree of the male kind is barren unless the female be planted near it; and, if separated by a river, they lean to each other as if they would embrace; if you strike the string of a lute in one corner of a room, it shall cause the string of another lute, tuned to the same height, and placed in an opposite corner, to give a sound; the cock always crows and claps his wings in the same moment the sun ascends the horizon. All effects which we see from sympathy afford us matter of admiration; but the loadstone demonstrates the affinity of corpuscles more palpably to our senses than most other experiments. The loadstone is found in iron-mines, and is not much different from the nature of iron; wherefore the particles which proceed from the loadstone have a kind of agreement with the pores of iron; and these small corpuscles, going out of the loadstone, and meeting with the iron in the way, rush into the pores of it; but, because all cannot enter at once, a great many remain without, and these are as strongly beaten back by the particles of the iron which they meet with, as if they were of the number of those corpuscles, which, being at liberty, return of their own accord, and at length send these by a reflective motion to the loadstone, from whence they first came. Hence it is that the iron is drawn towards the loadstone, principally by the agitation of those minute magnetic corpuscles moved in the concavities of the iron; and, being shaken together by the sundry motion of those corpuscles which are twisted one within another, those corpuscles which return by reflection are complicated and annexed to those which are in the pores of the iron, and cannot be returned or moved towards the loadstone, unless they draw along with them those corpuscles to which they are annexed, and which cannot follow unleſs by their motion the iron be carried with them; ſo the iron follows and is moved toward the loadſtone, except the iron be bigger than the loadſtone; for then the corpuſcles which proceed from the loadſtone are not ſo many, nor by conſequence ſo powerful, as to draw the iron. The reaſon the loadſtone draws no other body but iron, is becauſe other bodies do not return the atoms, neither are their pores well fitted for thoſe magnetic corpuſcles. By the ſame reaſon it appears, that the loadſtone does not approach to the iron, but the iron to the loadſtone. It may be ſaid that hard and ſolid bodies, ſuch as iron, cannot emit ſuch a great number of corpuſcles as other bodies, which, like the loadſtone, are leſs ſolid and more porous. There may be a reaſon given alſo why the loadſtone, being rubbed with garlic or oil, does not ſo eaſily draw iron to it; which is, that theſe ſtrange corpuſcles, by their oilineſs, hinder the emiſſion of the corpuſcles out of the loadſtone, and alſo their entrance into the pores of the iron, and thus break their elaſtic force.

We may obſerve many other effects of the loadſtone: as for example, that iron put upon a table is moved by the virtue of this ſtone placed under the table; for it is certain that the corpuſcles of the loadſtone which moves the iron penetrate through the vacuity or pores of the table, as if by ſmall and inviſible threads it had been tied to the loadſtone. It is the ſame thing if the table be of marble or glaſs, provided it be not greaſy nor too thick; which proves the poroſity of bodies in general. Another effect of the loadſtone is ſeen in a needle, which, being touched by it, always turns towards the north pole; the reaſon is, becauſe there are mountains of loadſtones under the poles, diſperſing their attractive ſpirits through the univerſe, ſpirits which are entangled with thoſe which adhere to the magnetical needle, whoſe force is leſſened as the ſpirits of it are diſſipated; eſpecially if the compaſs be ſet in a place where there are pieces of iron to which the ſpirits ſtick, and leave the needle, which had taken no greater quantity of them than what was requiſite according to its capacity. But the moſt wonderful property in this ſtone is, that it draws iron on one ſide, and rejects it on the other; ſo that it appears in every loadſtone that there are two poles of the world, the north pole attracts iron, the ſouth pole repels it; becauſe the ſpirit of the north pole enters in at the pores of the iron, but the ſouthern cannot, for it ſtrikes againſt the iron, and drives back too much its elaſtic particles. This explication preſuppoſes the being of ſpirits and atoms, and their figures and motions, as alſo ſmall occult vacuities, which are diſperſed through all bodies.

There are obſerved to be many effects for which no reaſon can be given, without the help of the word *antipathy*. We will inſtance ſome few: and, firſt, of the baſiliſk, who kills all whom he ſees, which is by the antipathy ſubſiſting betwixt it and other animals; but this is rather done by the emiſſion of certain venomous ſpirits, which

which penetrate the eyes of thoſe ſeen by the baſiliſk; the nature of this poiſon cannot be explained, unleſs we know the occult property of poiſon, becauſe poiſon kills only by a contrariety betwixt us and it; whence we diſcover the principle of this contrariety of the baſiliſk, that the ſpirits iſſuing out of the pores of its eyes kill thoſe animals which they meet with, becauſe the ſpirits penetrate them by their ſubtilty, or ſharp figure, like needles piercing the heart. The poiſon of vipers, and ſuch like, is not ſo acute nor ſo deadly, nor ſo ready in its effects, as that of the baſiliſk. In reference to this matter, there are many things that are worth conſidering. In the firſt place, it is certain that the baſiliſk is not engendered but in moiſt deep places, as in the bottom of pits where there is nothing but muddy, thick, ſtinking, water. In the ſecond place, it is to be obſerved, that, if you take a glaſs, and hold it againſt the baſiliſk's eyes, thoſe ſpirits which iſſue from his eyes reflecting upon the glaſs are ſent back from whence they came, and will kill the baſiliſk. Now it cannot be ſaid that the baſiliſk hates itſelf; but that the venomous ſpirits, reflecting from the glaſs, receive a more violent motion, and forcibly drive back the other ſpirits which are iſſuing from his eyes, ſo that they penetrate his brain and heart, and thence occaſion his death; in the ſame manner as vapours often ariſe with ſo great violence from the hypochondria, the meſentery, and the ſtomach, into the head, that they cauſe an apoplexy, epileptic dizzineſs, or lethargy; and ſometimes they are carried with ſuch ſubtilty and violence into the heart, that the perſons ſo afflicted die ſuddenly.

A ſtrange antipathy ſubſiſts in ſome vegetables, as between the colewort and the vine, which, if planted near together, will inſenſibly give back and lean ſideways, as if they really hated one another. This effect cannot be aſcribed to any thing but the emiſſion of the corpuſcles and material ſpirits of both of them, which ruſh upon one another, and mutually repel, by the irregularity of their figures. This is apparent in the juice of coleworts, which if taken by a man when he is drunk, he preſently comes to himſelf and is ſober; becauſe the corpuſcles of the juice of coleworts blunt the corpuſcles of the juice of the vine. In the ſame manner, we find by experience, that ſpirit of opium or laudanum cures the cholic, head-ach, tooth-ach, and other kinds of pains; it alſo cures the phrenſy, and procures ſleep. But there is need of the greateſt care in uſing theſe narcotic medicines, becauſe it often happens that the vital ſpirits are ſo ſtupified by them, that they are deprived of their motion, which cauſes a deadly ſleep. The colewort and the vine have not ſo powerful an effect on each other as narcotic medicines have on the animal ſpirits; for neither the vine nor the colewort will lean ſideways if there be cloth or paper ſet betwixt them, becauſe the corpuſcles flowing from each are then ſtopped in their way.

A third effect, which is ascribed to antipathy, is observed in the use of medicines, as well internal as external; the external, of which we now speak, are those we carry about us, which by their quality take away the malign air, and preserve us from the plague and other contagions; as prepared quicksilver, or a toad dried and shut up in a box. This phenomenon is ascribed to the pestiferous spirits or corpuscles, which, approaching towards us, find subjects more apt for their reception, and fix in them, but not in us, at least in such a quantity as to hurt us; which most evidently appears in this, that the prepared quicksilver, or the toad, being once filled with these contagious atoms, becomes useless, and ought to be changed and renewed; and I know by experience that quicksilver, prepared white and shining like an adamant, or polished silver, being carried about a person who is frequently with sick people, in time becomes black, so that afterwards it is useless to him who carries it, because there are no small vacuities left to retain the airy poisons; but it may be renewed by another preparation, whereby it may be made as white, transparent, and useful, as before. Quicksilver turns black more or less, sooner or later, according to the proportion of the greater or less malignity in the air where the person goes who carries it about him. And these antidotes can never hurt; nay, if rightly prepared, they not only withstand the contagious air, when they hinder its nearer approach towards us, but also suppress inward vapours, which, ascending into the head, occasion many distempers. These consequences, properly speaking, are the joint effects of sympathy and antipathy acting together; for the animal effluvia or corpuscles issuing from our bodies repel as much as possible the malignity of the ambient matter, by antipathy; whilst bodies composed of poisonous or noxious particles, draw to themselves, by sympathy, the foul or poisonous atoms which surround them, just the same as the loadstone draws iron. In this we see, and shall hereafter prove, that amulets or charms, worn about the body; that electricity, animal magnetism, and other occult properties, acting upon our bodies; though attributed to witchcraft, or some inexplicable cause, are nothing more than the natural effects of sympathy and antipathy, pre-ordained at the beginning of all things. Aromatic herbs and sweet-smelling flowers, used against infectious air, act by repulsion, or antipathy; whereas nightshade, hemlock, and all poisonous herbs, act by sympathy, drawing into their pores the infectious atoms, just the same as horse-radish draws in vinegar; and hence it follows that both these classes of plants are useful in preserving the animal juices from infection.

I now proceed to effects internal. Rhubarb, and the leaves of senna, purge melancholy; jalop and diagridium, phlegm and waterish humours. It is a constant and certain truth, that every purgative medicine comprehends in it certain spirits

or

or corpuſcles which are venomous, that is, acute, pungent, and biting; ſo that, nature being ſtirred up by them, and thereby the internal parts and membranes being touched and agitated, the animal ſpirits ruſh together in order to aſſiſt the part affected, and draw with them the foreign humours which are leſs fixed; and then nature, by the help of theſe ſpirits, expels them by the proper paſſages: ſo that, after a purgation by rhubarb, the urine is yellow, but, after ſenna or caſſia, it is dark and high-coloured.

Thus ſympathy and antipathy exiſt in all ſubſtances, whether animal, vegetable, or mineral; and things of one claſs have affinity or correſpondence with things of another claſs, or contrariwiſe, according to the nature and quality of the atoms or corpuſcles whereof they are formed. Hence it is that ſo ſtrong a ſympathy exiſts between rue and the fig-tree; and that the elm rejoiceth to cohabit with the vine: and hence it is likewiſe that ſerpents preſerve their ſight by fennel; and that the hind draws out the piercing dart with dittany, or garden-ginger. Hence alſo, by antipathy, water and oil, and wine and the juice of hemlock, repel each other; as do the vine and braſſic plants; for the vine, which uſually entwines round every thing it is near, ſhuns and inclines another way from theſe. Rue, and the aſh-tree, are ſo inimical to ſerpents, that they cannot exiſt under their branches; and the herb polypody is ſo obnoxious to crabs, that, if they are covered over with its leaves, they will in a ſhort time caſt off their ſhell and claws.

From an inveſtigation of theſe properties in nature, medicine and the art of healing were firſt diſcovered. All things temperate in quality concord ſympathetically with our bodies; as ſweet marjoram and nutmeg to the head, and wormwood to the belly. Thoſe which exceed this medium in their temperaments are noxious and hurtful; and are the more dangerous or deadly, in proportion as they recede from the temperate quality, which we perceive in opium, arſenic, and the like. From this cauſe, we likewiſe find that ſympathy and ſimilitude are ſynonymous; and that all ſubſtances which have reſemblance by ſignature, have ſympathy and agreement by nature, and ſerve for the conſervation of each other. Thus ſulphur is found to preſerve wine, which hath great analogy with our blood; and, if wood, or cables, or any thing whoſe uſe is in the water, be done over with a preparation of the oil of ſulphur, they will be preſerved, in a moſt ſingular and remarkable manner, from injury or decay. From hence Paracelſus concludes, that the particles of ſulphur are of themſelves a balſam, ſufficient to prevent wine or any inanimate ſubſtance from putrefaction; and ſo conſerves the body, that no adverſe qualities can prejudice or affect it. Querintius in his Pharmacy, aſſures us, that ſulphur rightly prepared is the true balſam of the lungs, and the principal ingredient uſed by the ancient Egyptians to embalm their mummies, or bodies of their deceaſed nobles, whereby they are preſerved even to this day from putrefaction, as may be ſeen in the Britiſh Muſeum, and in moſt other magazines of curioſities. Sulphur is certainly the efficient cauſe of all mineral ſprings; of all cryſtallizations, ſtones, pebbles, &c. by which they concrete, and are held together, as is evident from ſtriking them againſt ſteel, the ſparks of fire produced being the ſulphurous or inflammable part. All volcanos, burning mountains, and ſubterraneous fires, are occaſioned by ſulphur; as are likewiſe earthquakes, thunder and lightning, meteors, &c. The active properties of ſulphur are indeed wonderful; and, were I to inſtance many that I have diſcovered in the various chemical preparations I have made of it, few of my readers would be diſpoſed to give me credit. For ten ſucceſſive years I applied myſelf to the daily toil of making chemical experiments; and there is ſcarcely an herb or a mineral ſubſtance to be found, that I have not paſſed through the retort, or the crucible, in order to aſcertain their native qualities, and power of action, previous to the invention of my Solar and Lunar Tinctures; and I muſt confeſs that the occult properties of ſulphur coſt me more labour to fix, to inveſtigate, and to aſcertain, than all things elſe together, except mercury. And I do in conſequence affirm, that there are no mineral ſubſtances in the bowels of the earth, whoſe virtues are not communicated to plants and herbs growing on the earth's ſurface; and that the correſpondent virtues found in theſe herbs are infinitely more pure, innocent, balſamic, nutritive, and better adapted to medicine, than any groſs or earthy particles whatever. Even from the common herb borage, we can extract nitre, ſea-ſalt, tartarum vitriolatum, and the common fixed alkali; and it is no trivial information to the medical world, to know, that the three mineral acids are all to be found in one humble plant. Indeed vegetables appear to be the medium contrived by an all-wiſe and omnipotent Creator, for ſelecting, concocting, and combining, the medical and occult virtues of the different ſubſtances found in the bowels of the earth, and for adapting their virtues by an eaſy and natural concoction to the alleviation of human infirmities; according to that paſſage in ſcripture which ſays, that *the Lord hath cauſed medicine to grow out of the earth, and he that is wiſe will not abhor them, for with ſuch doth he heal men, and taketh away their pains.* Eccl. xxxviii. 4, 7. Whence I conclude, that all diſorders incident to mankind are to be cured, preferably, and more elegantly, more ſafely, and certainly, by preparations from medical plants and herbs, than from any mineral ſubſtances whatever; and that mercury, for the *lues venerea,* for the ſcrophula, and impurities of the blood, ought to be wholly expunged from our practice. Its baneful effects are every day more or leſs experienced, in the rotten bones and ruined conſtitutions of thoſe who have habitually

habitually taken it in advertiſed noſtrums for a certain complaint, until it has fixed itſelf, and the diſeaſe likewiſe, ſo ſtrongly in the habit, as to be almoſt beyond the reach of a proper remedy, which in reality and truth can only be found in the vegetable world.

Vegetables bear relation to the ſeven planets, and have form and affinity with the microcoſm, or parts of man; and conſtitute the original aliment intended by the Creator for the ſuſtenance of our bodies. And whatever ſignature or ſimilitude a plant has with any member or part of our body, it will promote a cure in that part, and tend ſympathetically to its comfort and preſervation. For example, thoſe herbs which in any reſpect reſemble the form of the eyes, are ſalubrious and healing to the eyes; as eyebright, ſcabious, marigold, chamelion, ſempervivum, nardum, and ſtar-wort. So plants which reſemble the head are cephalic, and help the diſorders and infirmities thereof; the walnut reſembles the brain, ſo that, if the oil or ſpirit of the nut be applied to the head, it ſtrengthens the fibres and comforts the brain. Maidenhair and the moſs of quinces have the figure of the hair of our head; and a decoction of theſe herbs, in reſtoring hair loſt by the *lues venerea*, is wonderfully efficacious. So plants, which in root, leaves, or fruit, reſemble the figure of the heart, have a power of comforting and ſuſtaining the heart: as the citron-apple, fuller's thiſtle, ſpikenard, mint, balm, white-beet, trefoil, and mother-wort. Herbs which reſemble the lungs promote reſpiration, and ſtrengthen the lungs; as hounds-tongue, lung-wort, ſage, camphor, wall-wort, &c. Plants which reſemble the ears conduce much to the relief of all diſorders of the ears; as fools-foot, or wild ſpikenard, which are a ſpecific for deafneſs; and ſo an oil extracted from the ſhell of ſea-ſnails, which reſembles the ear, has been found of wonderful efficacy in reſtoring the faculty of hearing, even after ſeveral years deafneſs. The ſenſe of ſmelling is greatly promoted by the application of thoſe herbs which reſemble the noſe, as water-mint, &c. So plants that bear reſemblance with the womb, conduce much to ſtrengthen and comfort the ſame, to purge the uterus, and promote fecundity, as the round birth-wort, briony, ladies-ſeal, heart-wort, ſatyrium, and mandrakes, which have round and hollow roots. Plants which bear ſimilitude with the gall and bladder contribute to the benefit of thoſe parts, by breaking the ſtone, ſtrengthening the urinary paſſages, and bringing away the gravel; as particularly pointed out in my edition of CULPEPER'S HERBAL. So likewiſe herbs and roots which bear affinity with the generative parts contribute much to their virility, ſtrength, and vigour; and truffles, potatoes, and the capſula of the caſhew-tree, which have ſimilitude with the teſticles, wonderfully ſtir up and promote the ſemen; as do the parſnip, the root of rag-wort, and the *mangel wurzel*, or root of

ſcarcity,

ſcarcity, contribute much to ſtimulate the virile member. Herbs having formation like the milt, nouriſh and preſerve the ſame, ſuch as ſpleen-wort, milt-wort, lupines, and ivy. Plants which in leaves or roots bear ſignature with the liver, do wonderfully concur to promote a good digeſtion and concoction of the blood, to prevent the liver from decay, and to heal and cure all infirmities thereof: ſuch virtue has the herb trinity, agaric, liver-wort, fumitory, lent-figs, &c. Herbs and ſeeds reſembling the teeth confer much to the good and preſervation of them; as tooth-wort, the pine-kernel, and the ſeeds of hemlock. Thoſe plants which have reſemblance with the knuckles and joints of the body, are wonderfully efficacious againſt the gout, white ſwellings, and all pains whatever in the joints; ſuch as galingal and knotty odoriferous ruſhes, &c. Plants and herbs expreſſing a natural fatneſs or oilineſs increaſe corpulency, or fatneſs of the body; as all pulſe, almonds, and kernels of every kind; and, by the ſame rule, thoſe vegetables which have a lean and ſpare deſignation, macerate and reduce the body, ſuch as ſarſaparilla, long-leafed roſa-ſolis, &c. Nervous or ſtringy plants ſupple and fortify the nerves and ſinews; as fennel, flax, hemp, the nettle, the herb neuras, and the root of mallows. Vegetables poſſeſſing a milky juice propagate milk in all female animals; and thoſe poſſeſſing a ſerous quality purge the noxious humours between the fleſh and ſkin, as ſperage, ſcammony, and the like. Plants that are hollow, as the ſtalks of corn, reeds, leeks, mallows, hollyocks, garlic, and bugloſs, are ſingularly good to purge, open, and comfort, the porous and hollow organs of the body. St. John's wort, having its leaves perforated, is ſanative to wounds; and palma Chriſti, having in its root a ſtrong reſemblance of the hands and fingers, is remarkably healing to all cuts, burns, ſcalds, and injuries, thereof.

There is another ſimilitude found between ſome vegetables and the brute ſpecies, which directs us to a very curious occult virtue, in curing hurts or injuries received from thoſe creatures they bear affinity with. Thus, the herb dragon, which in form reſembles a ſnake, and the bramble called Chriſt's-thorn, having its thorns ſet like the teeth of ſerpents, are an abſolute cure for the bite of thoſe animals. Ragwort, which is like a bee, is the beſt cure for the ſting of bees. Fleabane, which grows as if covered with vermin, cauſeth all fleas to avoid the room. Scorpion-graſs, dart-wort, and the flowers of turnſol, having ſimilitude with the tail of a ſcorpion, have ſurpriſing efficacy in curing hurts by all venomous creatures.

The properties and virtues of plants are alſo known by the analogy of their form; thoſe of the ſame or like figure having the ſame or like virtues and uſes. Thus the umbelliferous tribe have all a carminative taſte, or ſmell, and are conſequently powerful expellers of wind, and good in all flatulent diſorders. The galeate or verticillate kind are all of them a degree warmer, and more potent, and therefore may be reputed aromatic, and proper for nervous diſorders. The tetrapetalous kind are hot and biting, and exert their power by means of a diuretic volatile ſalt with which they abound, and are therefore good in chronical diſeaſes, obſtructions, cacochymias, &c.

The colours of plants and herbs likewiſe bear ſimilitude or ſympathy, and direct us to a knowledge of their temperature and uſe; thoſe of a light colour, ſuch as briony and water-lily, are profitable for the cure of phlegmatic diſeaſes. Thoſe of a yellow aſpect purge choler, and remove obſtructions occaſioned thereby, as is the effect of rhubarb, celandine, &c. Thoſe of a ſanguine hue purify the blood and juices, and contribute greatly to a good complexion, as the root of fernbrake, agrimony, germander, and ſorrel. And this rule is to be obſerved with reſpect to plants in general; that ſo many diſtinct colours as it hath commixed, ſo many virtues will it poſſeſs; and whatever diſeaſe it bears analogy or ſympathy with, that diſeaſe it will cure. The flower of the water-lily, bearing the ſignature of a drop of water, is a preſervative againſt the apoplexy. The root of ſaſſafras, and the ſtones of cherries, are good againſt the ſtone and gravel in the bladder and kidneys. The ſeeds of marigolds have reſemblance with the canker, and are a certain cure for that complaint. All plants of a glutinous nature, having their ſtalks ſignated with cuts and ſtabs, are ſanative to cuts, ſcars, and wounds. The root of galangale growing in marſhy grounds, and taken up in May or June, and worn as an amulet againſt the belly, will perform moſt aſtoniſhing cures in the dyſentery and flux; it has a ſtrong reſemblance to the natural excrements, both in figure and colour. All the excreſcences of trees ariſing above the branches, are good againſt excreſcences of the arteries. The ſtrawberry very much reſembles puſtules of the leproſy; and the diſtilled water of ſtrawberries is a moſt admirable cure for that complaint, as well as for red and pimpled faces.

We may further remark, that the more ſignatures or ſimilitudes are found cohering in a plant to one and the ſame ſignification, ſo much the more powerful and efficacious will its operation be, in any of the purpoſes for which it is applied; for the ſpirit is in quality the ſame in all bodies, but different in quantity, which conſtitutes that variety or difference perceivable by our ſenſes. In ſome bodies this ſpirit is more copious and active; in others, more ſparing and debilitated; ſo that, by how much the more the ſame ſpirit produces a convenient form and figure, in divers things or ſubjects under the ſame climate, by ſo much the more the ſame ſubjects are enabled to ſympathize with, and aſſiſt, each other. For ſympathy is by the ſpirit; and ſimilitude points out the things that act by ſympathy. Hence it is that ſimilitude of affection increaſes ſtrength, and the contrary, hatred. So it is that plants whoſe parts reſemble the ſcorpion, as libards-bane, hellebore, and aconitum, will cure the bite of that reptile; and that the flowers of plants, having the reſemblance of butterflies, conduce to fruitfulneſs and virility, as gandergooſe, the flower of beans, woodbine, and ragwort. Plants ſpotted like ſerpents, as cow-garlic, wake-robin, dragon-wort, ſea-dragon, &c. are ſanative againſt the bite of ſerpents; and plants which reſemble the head of ſuch animals are alſo good againſt their poiſon; as the flowers of wild bugloſs, which reſemble the head of a viper, Dioſcorides affirms to be a certain cure.

The virtues of plants and herbs are, however, variable, and liable to be injured by change of climate, which will alter or deſtroy them, as we ſee in many of the medical plants of other countries brought into England; which, though they ſeem to flouriſh with us, never poſſeſs their virtues in the ſame excellent degree as in their own climate, which is the reaſon that CULPEPER *recommends Engliſh herbs for an Engliſh conſtitution.* The bodies of different animals alſo render the effect of the ſame plant different; the tythmals, or ſpurges, being all very violent cathartics when taken by us; but yet they are eaten by goats and ſeveral other animals, without any purgative effect, and ſeem to give them a particular ſhare of vigour and ſpirits. Fiſhes, on the contrary, are more ſtrongly affected by them than we are; for the juice of ſpurge, made into a paſte with flour and honey, will ſo much intoxicate them, that they may be taken out of the water with one's hand. Again, bitter almonds are of no ill conſequence to us, yet they kill all ſorts of birds that touch them. The Buceros, or hornbill, feeds upon the *nux vomica,* which is known to be a moſt deadly poiſon to man and all quadrupeds and fiſhes. This is a late diſcovery. See my Syſtem of Nat. Hiſt. vol. viii. p. 153. and ENCYCLOPÆDIA LONDINENSIS, vol. iii. p. 478.

The foregoing notions may be ridiculed, from their extreme ſimplicity; yet where is the man, at all converſant with natural philoſophy and phyſic, and poſſeſſed of an impartial mind, that will dare to controvert theſe facts? Every wayfaring man knows ſomething of the herbs I have mentioned, and is capable of judging of their ſimilitude and ſympathy. Let him apply them for the purpoſes I have pointed out, and their occult properties will ſoon be viſible to his ſenſes. The track of nature is a plain and obvious road, abounding with moſt pleaſing proſpects, and the ſureſt guides. God, in the plenitude of his omniſcience and mercy, ſeems to have ſet a mark on the minuteſt particles of his creation, for man's information and benefit; in the contemplation of which, our happineſs, as well as our health, will invariably be found.

OF THE OCCULT PROPERTIES OF GENERATION IN PLANTS AND HERBS.

ALL plants are produced from ſeeds, as all animals are produced from eggs, and the proceſs of nature is very ſimilar in both kinds of generation. The ſmalleſt vegetables have ſeeds, though often not diſcoverable by the naked eye. Miſtletoe is known to be produced from ſeed; and the ſeſſile and flat funguſes, which ſome conſider as morbid excreſcences, are true ſpecies of thoſe agarics which are furniſhed with caps and ſtems, and grow on the ground, whoſe ſeeds, falling on a moiſt tree, produce, as it were, half-caps without ſtems. Beſides, that ſeeds are the eggs of plants appears from hence, that, as every egg produces an offspring ſimilar to the parent, ſo do alſo the ſeeds of vegetables; and therefore they alſo are eggs. A ſeed reſembles the egg of an hen; as this, as well as the egg, has a ſhell, external membrane or film, a membrane including the yolk, the yolk itſelf, and the ſcar, or point of life. In ſeeds, the white is wanting, becauſe the moiſture of the earth ſupplies its place, and nouriſhes the embryo of the plant. When the flower is going off, the ſeed begins to ſwell, and on the outſide there appears a veſicle, which is the amnion, furniſhed with an umbilical chord, or navel-ſtring, which is introduced through the chorion to the oppoſite ſide of the egg. While with the egg the amnion increaſes, on its top is obſerved another ſmall body, which likewiſe increaſes continually, till it has filled the whole chorion and egg; and the amnion and chorion are turned into the external ſhell or coat of the ſeed. Thus, as the ſame changes are brought about on the ſeed as in the egg, the ſeeds muſt be the eggs of plants. Farther, that plants ſpring from eggs, is plain from the lobes, which, when we ſpeak of cows and ſimilar quadrupeds, are nothing elſe than ſeveral ſecundines, always adhering to the fœtus, drawing their ſupply or fluids from the matrix, which fluids they prepare for the nouriſhment of the tender fœtus. That moſt plants have ſeminal leaves, or lobes, is very well known. Theſe ſeminal leaves once conſtituted the whole ſeed, except the *hilum*, or little heart, in which is the point of life; and theſe lobes prepare the nouriſhment for the very tender plant, until it be able to ſtrike root in the earth; in the ſame manner as the yolk in an egg, becoming the placenta, prepares the nouriſhment, and ſends it by the navel-ſtring to the chick; after which they drop off. Hence it appears, that the ſeminal leaves are the lobes; but, ſince all lobes proceed from the egg or ſeed, we may fairly conclude that plants are produced from eggs. But, as no egg can produce an animal till it be impregnated or fecundated by the male, it will be neceſſary to inveſtigate the ſituation of the genital organs deſtined by nature for this purpoſe in plants.

It is plain that the genital organs of plants muſt be ſituated where the ſeeds are produced; now the ſeeds are produced where the flower and fruit are; therefore in the flower and fruit are the genital organs of plants. And, as there was never a clear and evident example produced of any plant which wanted flowers and fruit, though they might not be diſtinctly known on account of their exceeding minuteneſs, we may juſtly ſay, that the eſſence of plants conſiſts in their fructification. Moreover, as generation precedes the birth in animals, the flower in plants always precedes the fruit; and, therefore, we are neceſſarily led to aſcribe fecundation to the flower, and the birth or excluſion of the ſeed to the ripe fruit. The flower may, conſequently, be defined to be the genital organs of a plant, ſerving for fecundation, and the fruit to be the genital organs ſerving for the birth or maturation of the ſeed. And, ſince we know that there are many plants, ſome of which want the calyx, others the corolla, others the filaments of the ſtamina, and others the ſtyle; but that all flowers, the moſſes only excepted, are furniſhed with the antheræ, or ſtigmata, or both together; theſe parts muſt conſtitute the eſſence of the flower. If we find a flower with antheræ, but no ſtigmata, we may alſo aſſuredly find another flower, either in the ſame or a different plant of the ſame ſpecies, which has ſtigmata with the antheræ, or without them. The act of fecundation is performed in the flower; and therefore the genital organs of both ſexes muſt be preſent in the flower; not indeed always in one and the ſame flower; but it is ſufficient that thoſe of the male be in one flower, and thoſe of the female in another; and theſe genital organs are the antheræ and ſtigmata. The antheræ, or male organs of generation in flowers, are nothing elſe but the bodies which prepare and contain the male ſperm; therefore theſe antheræ are the teſticles together with the ſeminal veſicles, and their duſt the genuine male ſperm of plants, anſwering to thoſe particles which are called animalcules in the male ſperm of animals. This propoſition may be evinced by the following arguments: the antheræ and the duſt always come before the fruit; and, when they ſhed their duſt, which they do before the flower has attained its full vigour, they have performed their office, and then drop and become uſeleſs. Beſides, the antheræ are ſo ſituated in the flower, that their duſt, which is the male ſperm, may reach the piſtil or female organs; for the ſtamina either ſurround the piſtil, as in moſt flowers, or, if the piſtil incline to the upper ſide of the flower, the ſtamina do the ſame; or, if the piſtil nods, the ſtamina aſcend.

Farther, the antheræ and ſtigmata are in full vigour at the ſame time, both when they are in the ſame flower, and when they are in ſeparate flowers. Moreover, if we cut aſunder the antheræ before they have ſhed their duſt, their ſtructure will be found altogether as wonderful and curious as that of the ſeed-

veſſels

veſſels themſelves; for, within, they conſiſt of one, two, three, or four, cells; and they open either longitudinally or at the baſe, ſeparating into pieces or valves, or from the top, or at the two points or horns. And, if we cut off the antheræ of any plant which bears but one flower, taking care at the ſame time that no other plant of the ſame ſpecies is near it, the fruit proves abortive, or at leaſt produces ſeeds which will not vegetate. Finally, the figure of the fertilizing duſt will clearly convince any one, that this fine powder is not accumulated by chance, or from the dryneſs of the antheræ.

The powder of the antheræ, in point of fecundation, anſwers to the animalcules in the male ſperm; and the ſtigma which receives this duſt is always moiſtiſh, that the duſt may inſtantly adhere or ſtick to it. That the ſtigmata, which are the other eſſential parts of the flower, are the female organs of generation, may be proved by the following conſiderations: the parts of the piſtillum are the germen, the ſtyle, and the ſtigma; the germen, or ſeed-bud, while the plant is in flower, is always imperfect and immature, being only the rudiment of the future fœtus; the ſtyle is no eſſential part, for it is wanting in many ſpecies of plants; but the germen can never bring the fruit to maturity, except it be within the flower along with the ſtigma. Hence it follows, that the ſtigma is that part of the flower which receives the impregnating duſt. This will farther appear, if we conſider that the ſtigma is always ſo ſituated, that the antheræ, or their impregnating duſt, can reach it: moreover, it has always a figure peculiar to itſelf, ſo that in moſt (though not all) plants it is double when the fruit conſiſts of two cells; triple when the ſeed-veſſel has three cells; quadruple when it has four cells, &c. Again, the ſtigmata are always in full vigour at the ſame time with the antheræ: beſides, the ſtigmata in moſt plants, when they have diſcharged their office, drop off in the ſame manner as the antheræ do; which proves, that the ſtigmata contribute nothing to the ripening of the fruit, but ſerve only for the purpoſe of generation. If the ſtigmata be cut off before they have received the impregnating duſt of the antheræ, the plant is caſtrated as to the female organs, and the fruit periſhes. The ſtigma of the flower has, beſides, two other ſingular properties: viz. that it is always diveſted of the cuticle or film, nor has it any bark as the other parts have, and it is always bedewed with a moiſture.

Upon the whole it appears, that the generation of plants is accompliſhed by the antheræ ſhedding their duſt on the ſtigmata. In the generation of animals, we are certain, that the male ſperm muſt come in contact with the female organ, if there be any impregnation. In the vegetable kingdom, the genital duſt is carried by the air to the moiſt ſtigmata, where the particles burſt and diſcharge their exceeding fine or ſoluble contents, which impregnate the ovary. This will appear if it be conſidered, that, when a plant is in flower, and the duſt of the antheræ flying about, part of this duſt viſibly lights upon and clings to the ſtigma; the ſtamina and piſtillum are generally of the ſame height, that the male duſt may more eaſily come at the ſtigma; and, in thoſe plants where this is not the caſe, a ſingular proceſs of fecundation may be obſerved; thus in the African tree crane's-bill, or Geranium inquinans, where the piſtillum is ſhorter than the ſtamina, the flowers before they blow are pendulous, but upon their opening they ſtand upright, that the powder may fall upon the ſtigma; after which they again nod till the fruit is ripe, and then ſtand upright a ſecond time, that their ſeeds may be more eaſily ſcattered about. In ſome of the pinks, the piſtilla, which are longer than the ſtamina, are bent back like rams-horns towards the antheræ.

Again, the ſtamina for the moſt part ſurround the piſtillum, ſo that ſome of the duſt is always blown by the wind on the ſtigma. Moreover, the ſtamina and piſtillum come at the ſame time, not only in one and the ſame flower, but alſo where ſome are male and others female on the ſame plant, very few excepted.

Farther, in almoſt all ſorts of flowers we ſee how they expand or open by the heat of the ſun; but in the evening, and in a moiſt ſtate of the air, they cloſe or contract their flowers, leſt the moiſture, getting to the duſt of the antheræ, ſhould coagulate the ſame, and render it incapable of being blown on the ſtigma; but when once the fecundation is over, the flowers neither contract in the evening, nor yet againſt rain. The wind on many occaſions ſerves as a vehicle for bringing the farina of the males to the females. M. Geoffroy cites a ſtory from Jovius Pontanus, who relates that in his time there were two palm-trees, the one male, cultivated at Brindiſi, the other female, in the wood of Otranto, fifteen leagues apart; that this latter was ſeveral years without bearing any fruit; till at length, riſing above the other trees of the foreſt, ſo as it might ſee (ſays the poet) the male palm-tree at Brindiſi, it then began to bear fruit in abundance. M. Geoffroy makes no doubt but that the tree then only began to bear fruit, becauſe, it was in a condition to catch on its branches the farina of the male, brought thither by the wind. In the male and female of the piſtachia-nut-tree they obſerve the ſame method as in thoſe of the date-tree. We may obſerve farther, that, ſince the male duſt is generally of greater ſpecific gravity than the air, in moſt plants that have the piſtillum longer than the ſtamina, the all-wiſe Creator has made the flowers nodding, that the powder may more eaſily reach the ſtigma. With reſpect to thoſe plants whoſe ſtems grow under water, the flowers, a little before they blow, emerge or riſe above the ſurface of the water; and thoſe, all whoſe parts

grow

grow under water, about the time of flowering raiſe their genital ſtems above the water, which ſtems ſink again as ſoon as the time of generating is over. A ſimilar concluſion may be farther eſtabliſhed from the conſideration of all ſorts of flowers; but enough has been ſaid to prove that the generation of plants is performed by the genital duſt of the antheræ falling on the moiſt ſtigma, or female organ; which duſt, by the help of the moiſture, adheres and burſts, diſcharging its contents, the ſubtile particles of which are abſorbed by the ſtyle into the ovarium, germen, or ſeed-bud. However, the duſt of the antheræ does not penetrate through the ſtyle to the germen and rudiments of the ſeed, as ſome writers have ſuppoſed: the contrary appears to be the caſe from opening a flower of the oriental rough poppy, with a large flower, cutting its piſtillum perpendicularly downwards; and the lamellæ or folds, the placentæ, and the ſmall ſeeds ſticking to them, will be found of a pure white colour, though at the ſame time the ſtyle and all the ſtigmata are wholly tinged with a purple hue from the duſt of the antheræ. Hence we conclude, that not one grain or particle of the farina enters the folds of the receptacle or ſeeds themſelves.

We may cloſe this account with obſerving upon the whole, that the calyx is the marriage-bed, in which the ſtamina and piſtilla, the male and female organs, celebrate the nuptials of plants, and where they are cheriſhed and defended from external injuries: the corolla or petals are the curtains cloſely ſurrounding the genital organs, in order to keep off ſtorm, rain, or cold; but, when the ſun ſhines bright, they freely expand, to give acceſs both to the ſun's rays and to the fecundating duſt: the filaments are the ſpermatic veſſels, by which the juice, ſecreted from the plant, is carried to the antheræ; the antheræ are the teſticles, and may not improperly be compared to the ſoft roe or milt of fiſhes: the duſt of the antheræ anſwers to the ſperm and ſeminal animalcules; for, though it is dry, that it may be the more eaſily conveyed by the wind, yet it gets moiſture upon touching the ſtigma: the ſtigma is that external part of the female organ which receives the male duſt, and on which the male duſt acts: the ſtyle is the vagina, or tube, through which the effluvia of the male duſt paſs to the germen, or ſeed-bud: the germen is the ovary, for it contains the unimpregnated or unfertilized ſeeds: the pericarpium, or ſeed-veſſel, anſwers to the impregnated ovary, and in fact is the ſame with the germen, or ſeed-bud, only increaſed in bulk, and loaded with fertile ſeeds; the ſeeds are the eggs. Moreover, the calyx is a production of the external bark of the plant; the corolla of the inner bark; the ſtamina of the alburnum, or white ſap; the pericarpium, or ſeed-veſſel, of the woody ſubſtance; and the ſeeds of the pith of the tree; for in this manner they are placed, and in this

this they are alfo unfolded; fo that in the flower we find all the internal parts of a plant unfolded.

The ftomach of plants is the earth, from which they receive their nourifhment; and the fineft and moft fubtile parts of its foil is their chyle: the root, which carries the chyle from the ftomach to the body of the plant, is analogous to the lacteals or chyliferous veffels of animals: the trunk, which fupports and gives ftrength to the whole plant, is analogous to the bones; the leaves, by which plants tranfpire, are inftead of lungs, and they may be alfo compared to the mufcles of animals, for by their agitation with the wind the plant is put in motion; on which account, herbs furnifhed with leaves cannot thrive except they have air; but fucculent plants, which have no leaves, though fhut up in green-houfes and quite deprived of the external air, thrive very well. Heat is to plants analogous to the heart in animals; for they have no heart, nor have they occafion for any; becaufe they live like polypes in the animal kingdom; their juices mixed with air being propelled through their veffels, but not circulated back again by returning veffels. Plants have generally their genital organs placed at their ramifications, as animals have theirs at the ramification of the iliac veffels, with this difference, that the ramifications of plants afcend, whereas thofe of animals go downwards or backwards; whence the ancients called a plant *an inverted animal.* Pliny obferves, that there is in plants a natural inftinct to generation; and that the males, by a certain blaft and a fubtle powder, do confummate the nuptials on the females.

For the manner wherein the farina fecundifies, M. Geoffroy advances two opinions:—1. That the farina being always found of a fulphureous compofition, and full of fubtile penetrating parts (as appears from its fprightly odour), falling on the piftils of the flowers, there refolves, and the fubtileft of its parts, penetrating the fubftance of the piftil and the young fruit, excite a fermentation to open and unfold the young plant, inclofed in the embryo of the feed. In this hypothefis the feed is fuppofed to contain the plant in miniature, and only to want a proper juice to unfold its parts, and make them grow. 2. The fecond opinion is, that the farina of the flower is the firft germ or bud of the new plant, and needs nothing to unfold it, and enable it to grow, but the juice it finds prepared in the embryos of the feed. Thefe two theories of vegetable generation, the reader will obferve, bear a ftrict analogy to thofe two of animal generation: viz. either that the young animal is in the femen mafculinum, and only needs the juice of the matrix to cherifh and bring it forth; or that the animal is contained in the female ovum, and needs only the male feed to excite a fermentation, &c. M. Geoffroy rather takes the proper feed to be in the farina; inafmuch as the beft microfcopes do not difcover the leaft appearance of any bud in the little embryos of the grains, when examined before the apices have ſhed their duſt. In leguminous plants, if the leaves and ſtamina be removed, and the piſtil, or that part which becomes the pod, be viewed with the microſcope, before the flower be opened; the little green tranſparent veſiculæ, which are to become the grains, will appear in their natural order; but ſtill ſhowing nothing elſe but the mere coat or ſkin of the grain. If the obſervation be continued for ſeveral days ſucceſſively, in other flowers, as they advance, the veſiculæ will be found to ſwell, and by degrees to become replete with a limpid liquor; wherein, when the farina comes to be ſhed, and the leaves of the flower to fall, we obſerve a little greeniſh ſpeck, or globule, floating about at large. At firſt there is not any appearance of organization in this little body; but in time, as it grows, we begin to diſtinguiſh two little leaves like two horns. The liquor diminiſhes inſenſibly, as the little body grows, till at length the grain becomes quite opaque; when, upon opening it, we find its cavity filled with a young plant in miniature; conſiſting of a little germ or *plumula*, a little root, and the lobes of the bean, or pea, &c.

The manner wherein this germ of the apex enters the veſicula of the ſeed, is not very difficult to determine. For, beſides that the cavity of the piſtil reaches from the top to the embryos of the grains, thoſe grains or veſiculæ have a little aperture correſponding to the extremity of the cavity of the piſtil, ſo that the ſmall duſt, or farina, may eaſily fall through the aperture into the mouth of the veſicula, which is the embryo of the grain. This cavity, or cicatricula, is much the ſame in moſt grains, and it is eaſily obſerved in peaſe, beans, &c. without the microſcope. The root of the little germ is juſt againſt this aperture, and through this it paſſes out when the little grain comes to germinate.

From what has been ſaid, it becomes evident, that, unleſs the female plant is impregnated by the male, it can bring forth no fruit, nor ſeed, that will grow. This holds good throughout the whole ſyſtem of vegetation. But, as trees and plants are immoveably fixed, and cannot like animals rove about in ſearch of a mate, the all-wiſe Creator has compenſated this, by means of little inſects, the bee, and the winds, which doubtleſs carry the pollen, or fecundating matter of the male, to the piſtilla of the female, whereby impregnation and generation follow. But as this, in the production of fruits, is rather a fortuitous event, which ſometimes happens in profuſion, and at others but ſparingly, thoſe who cultivate fruits have been led by art to aſſiſt nature in this neceſſary contact of the ſexes. While in Arabia, I was taken to ſee this curious operation performed on the date-tree, by which the Arabs always ſecure to themſelves a plentiful harveſt of that fruit, which is of ſo much importance to their traffic, and amongſt whom this art appears to have been known long before any botanift dreamed of the difference of fexes in vegetables. Of this the gardener informed me, but was furprifed to find I knew the circumftance before; "for (fays he) all who come from Europe to this country have regarded this operation as a fable. When they obferve a tree where the fpadix has female flowers, they fearch on a tree that has male flowers (which they know by cuftom and experience) for a male fpadix which has not yet burft out of its fpatha or hufk; this they open, take out the fpadix, and cut it lengthwife in feveral pieces, taking care not to hurt the flowers. Thefe pieces of fpadix with male flowers they put lengthwife between the fmall branches of the fpadix which hath female flowers, and then cover them over with a palm-leaf; in this fituation the piftilla of the female flower becomes impregnated by the male, which foon after withers and dies; and, unlefs the natives thus wed and fecundate the female date-tree, it bears no fruit. Or even if they permit the fpadix of the male flower to burft, or come out, before it is taken, it is ufelefs for fecundation; it muft for this purpofe have its *maidenhead*, as the Arabs term it, or it will not do; and this is loft the fame moment the bloffoms burft out of their cafe." From this curious procefs of nature in the generation of vegetables, and from a contemplation of the apparatus fhe has contrived for that purpofe, many ufeful hints may be derived how to alter, improve, enrich, and vary, the tafte, form, and quality, of fruits, &c. by impregnating the flower of one with the farina of another of the fame clafs; and to this artificial coupling and mixing it is, that the numberlefs varieties of new fruits, flowers, &c. produced every year by our nurfery-men and gardeners, with many other phenomena in the vegetable kingdom, are to be afcribed.

In the cultivation of many of our home-plants, we fometimes meet with circumftances not unfimilar to thofe of the date-tree, which become barren when deprived of the males. Thus, if the flowers of the male hemp are pulled off before thofe of the female are fully expanded, the females do not produce fertile feeds. But, as a male flower is fometimes found upon a female plant, this may be the reafon why fertile feeds are fometimes produced even after this precaution has been obferved. The tulip affords another experiment to the fame purpofe.—Cut off all the antheræ of a red tulip before the pollen is emitted; then take the ripe antheræ of a white tulip, and throw the pollen of the white one upon the ftigma of the red; the feeds of the red tulip, being thus impregnated by one of a different complexion, will next feafon produce fome red, fome white, but moftly variegated, flowers.

In the year 1744, Linnæus publifhed a defcription of a new genus, which he called *peloria*, on the fuppofition of its being a hybrid or mule plant, i. e. a plant produced by an unnatural commixture of two different genera. The root, leaves, caulis, &c. of of this plant, are exceedingly ſimilar to thoſe of the Antirrhinum linaria, or common yellow toad-flax; but the flower and other parts of the fructification are totally different. On account of its ſimilarity to the linaria in every part but the flower, Linnæus imagined it to have been produced by a fortuitous commixture of the linaria with ſome other plant; and from this doctrine he ſuppoſes that only two ſpecies of each genus of plants exiſted *ab origine*, and that all the variety of ſpecies which now appear have been produced by unnatural embraces betwixt ſpecies of different genera. Under this head he defends the caſe of Richard Baal, gardener at Brentford. This Baal ſold a large quantity of the ſeeds of the Braſſica florida to ſeveral gardeners in the ſuburbs of London. Theſe gardeners, after ſowing their ſeeds in the uſual manner, were ſurpriſed to find them turn out to be plants of a different ſpecies from that which Baal made them believe they had purchaſed; for, inſtead of the Braſſica florida, the plants turned out to be the Braſſica longifolia. The gardeners, upon making this diſcovery, commenced a proſecution of fraud againſt Baal in Weſtminſter-hall. The court found Baal guilty of fraud, and decerned him not only to reſtore the price of the ſeeds, but likewiſe to pay the gardeners for their loſt time, and the uſe of their ground. "Had theſe judges (ſays Linnæus) been acquainted with the ſexual generation of plants, they would not have found Baal guilty of any crime, but would have aſcribed the accident to the fortuitous impregnation of the Braſſica florida by the pollen of the Braſſica longifolia."

With reſpect to the nouriſhment of plants, we need only recur to the analogy that is known to ſubſiſt between plants and animals. It is highly probable that the radical fibres of plants take up their nouriſhment from the earth, in the ſame manner that the lacteal veſſels abſorb the nutriment from the inteſtines; and, as the oily and watery parts of our food are perfectly united into a milky liquor, by means of the ſpittle, pancreatic juice, and bile, before they enter the lacteals, we have all the reaſon imaginable to keep up the analogy, and ſuppoſe that the oleaginous and watery parts of the ſoil are alſo incorporated, previous to their being taken up by the abſorbing veſſels of the plant. To form a perfect judgment of this, we muſt reflect that every ſoil, in a ſtate of nature, has in itſelf a quantity of abſorbent earth, ſufficient to incorporate its inherent oil and water; but, when we load it with fat manures, it becomes eſſentially neceſſary to beſtow upon it, at the ſame time, ſomething to aſſimilate the parts. Lime, ſoap-aſhes, kelp, marl, and all the alkaline ſubſtances, perform that office. In order to render this operation viſible to the ſenſes, diſſolve one dram of Ruſſia pot-aſh in four ounces of water; then add one ſpoonful of oil; ſhake the mixture, and it will inſtantly become an uniform maſs of a whitiſh colour, adapted to all the purpoſes of vegetation. This eaſy and familiar experiment is a juſt repreſentation of what happens after the operation of burn-baking, and conſequently may be conſidered as a confirmation of the hypotheſis advanced. In this proceſs, the ſward being reduced to aſhes, a fixed alkaline ſalt is produced; the moiſture of the atmoſphere ſoon reduces that ſalt into a fluid ſtate, which, mixing with the ſoil, brings about an union of the oily and watery parts, in the manner demonſtrated in the experiment. When the under ſtratum conſiſts of a rich vegetable mould, the effects of burn-baking will be laſting; but, when the ſoil happens to be thin and poor, the firſt crop frequently ſuffers before it arrives at maturity. The farmer, therefore, who is at the expenſe of paring and burning a thin ſoil, ſhould beſtow upon it a portion of rotten dung, or ſhambles manure, before the aſhes are ſpread, in order to ſupply the deficiency of oily particles: in this way the crop will be ſupported during its growth, and the land will be preſerved in health and vigour. But plants not only receive nouriſhment by their roots, but alſo by their leaves. Vegetables that have a ſucculent leaf, ſuch as vetches, peaſe, beans, and buck-wheat, draw a great part of their nouriſhment from the air, and on that account impoveriſh the ſoil leſs than wheat, oats, barley, or rye, the leaves of which are of a firmer texture. Rape and hemp are oil-bearing plants, and, conſequently, impoveriſhers of the ſoil; but the former leſs ſo than the latter, on account of the greater ſucculency of its leaf. The leaves of all kinds of grain are ſucculent for a time, during which period the plants take little from the earth; but, as ſoon as the ear begins to be formed, they loſe their ſoftneſs, and diminiſh in their attractive power. The radical fibres are then more vigorouſly employed in extracting the oily particles of the earth for the nouriſhment of the ſeed. The leaves of plants ſerve, not only as excretory ducts to ſeparate and carry off the redundant watery fluid, which, by being long detained in the plants, would turn rancid and prejudicial to them, but likewiſe to imbibe the dew and rain, which contain ſalt, ſulphur, &c. and to be of the ſame uſe to plants that the lungs are to animals. But, as plants have not a dilating and contracting thorax, their inſpirations and expirations will not be ſo frequent as thoſe of animals, but depend wholly on the alternate changes from hot to cold for inſpiration, and *vice verſa* for expiration. But the greater part of their nouriſhment is derived from the roots. Theſe, therefore, are found to bear a conſiderable proportion to the body of the plant above ground; the ſuperficies of the former being above four-tenths of that of the latter. Hence appears the neceſſity of cutting off many branches from a tranſplanted tree; becauſe, in digging it up, a great part of the root is cut off.

It is a curious occult fact, with reſpect to vegetables, that they thrive beſt from putrefaction, and flouriſh moſt in putrid air. Manure, though it has a ſtench almoſt ſufficient

ſufficient to infect the blood, yet, if placed round plants and herbs, will make them grow ſurpriſingly; and we every day ſee how luxuriantly they will thrive upon a dunghill. Yet it is as true, that though theſe vegetables eagerly ſuck in and imbibe ſo foul a moiſture, and thrive in air ſo ſtrongly tainted with putrefaction, even in ſuch as would prove fatal to human life, yet thoſe very plants exhale a direct oppoſite effluvium, tending to refreſh and ſweeten the atmoſphere, and to render it wholeſome, when it is become noxious in conſequence of epidemical complaints, or of animals dying and putrifying in it; whence it follows that vegetables draw in the foul or infected corpuſcles, as favourable to their ſuſtenance, which being concocted, altered, and changed, in the body of the plant, it again emits them purified and ſweet. This I have proved by the following experiment: A quantity of air was made thoroughly noxious, by ſome mice breathing and dying in it. This I divided into two parts, in glaſs receivers. Into one I put a mouſe with a ſprig of mint, which lived very well, and the mint alſo flouriſhed; but in the other, where there was no mint, the mouſe died almoſt immediately. This experiment I have many times repeated with different kinds and portions of infected air, and have always found the reſult nearly the ſame; wherefore this plain reaſoning follows; that, as vegetables draw in by their leaves and roots the putrid effluvium of the air, ſo their emiſſion of purified corpuſcles contributes to make the remaining air more fit and wholeſome for reſpiration; and from this circumſtance I recommend all perſons who viſit the ſick, or have putrid diſorders in their families, to uſe as many freſh vegetables as poſſible, and never to be without ſome ſprigs of mint about them.

Of SYMPATHY, ANTIPATHY, SAGACITY, and OCCULT INSTINCT, in BRUTES.

BRUTE is a general name given to all animals except man; and an animal muſt be an organized living body, endowed with ſenſe; for minerals are ſaid to grow and increaſe, plants to grow and live; but animals alone are endowed with ſenſation. It is this property of ſenſation alone, that conſtitutes the eſſential characteriſtic of an animal; and by which the animal and vegetable kingdoms ſeem to be ſo materially ſeparated. Thoſe naturaliſts, who have ſuppoſed the dictinction between animals and vegetables to conſiſt in any thing elſe than the gift of ſenſation, have found themſelves greatly embarraſſed; and have generally agreed, that it was extremely difficult, if not impoſſible, to ſettle the boundaries between the animal and vegetable kingdoms. But this difficulty will be eaſily ſeen to ariſe from their taking the characteriſtic marks of the animal kingdom from ſomething that was evidently common to both. Thus Boerhaave attempted to diſtinguiſh an animal from a vegetable,

table, by the former having a mouth, which the latter has not: but here, as the mouth of an animal is only the inſtrument by which nouriſhment is conveyed to its body, it is evident that this can be no eſſential diſtinction, becauſe vegetables, as we have above demonſtrated, require nouriſhment, and have inſtruments proper for conveying it into their bodies; and where the end is the ſame, a difference in the means can never be eſſential. The fixing the difference in an animal's having a gula, ſtomach, and inteſtines, as is done by Dr. Tyſon, is as little to the purpoſe.

The power of moving from one place to another, hath by many been thought to conſtitute their eſſential difference; and indeed, in moſt caſes, it is the obvious mark by which we diſtinguiſh an animal from a vegetable; but Lord Kames hath given us ſeveral very curious inſtances of the locomotive power of plants; ſome of which would doubtleſs do honour to an animal.—Upon the ſlighteſt touch, the ſenſitive plant ſhrinks back, and folds up its leaves,* ſimilar to a ſnail, which on the ſlighteſt touch retires within its ſhell. If a fly perch upon one of its flower-leaves, it cloſes inſtantly, and cruſhes the inſect to death. There is not an article in botany more admirable than a contrivance, viſible in many plants, to take advantage of good weather, and to protect themſelves againſt bad. They open and cloſe their flowers and leaves in different circumſtances; ſome cloſe before ſunſet, ſome after: ſome open to receive rain, ſome cloſe to avoid it. The petals of many flowers expand in the ſun; but contract at night, or on the approach of rain. After the ſeeds are fecundated, the petals no longer contract. All the trefoils may ſerve as a barometer to the huſbandman; they always contract their leaves on an impending ſtorm. Some plants follow the ſun, others turn from it. Many plants, on the ſun's receſs, vary the poſition of their leaves, which is ſtyled the *ſleep of plants.* A ſingular plant was lately diſcovered in Bengal: its leaves are in continual motion all day long: but, when night approaches, they fall down from an erect poſture to reſt.

A plant has a power of directing its roots for procuring food. The red whortleberry, a low ever-green plant, grows naturally on the tops of our higheſt hills, among ſtones and gravel. This ſhrub was planted in an edging to a rich border, under a fruit-wall. In two or three years, it over-ran the adjoining deep-laid gravel walk; and ſeemed to fly from the border, which was not congenial to its nature, and in which not a ſingle runner appeared. An effort to come at food in a bad ſituation, is extremely remarkable in the following inſtance: Among the ruins of New Abbey, formerly a monaſtery in Galloway, there grows on the top of a wall a plane-tree about twenty feet high. Straitened for nouriſhment in that barren ſituation, it ſeveral years ago directed roots down the ſide of the wall, till they reached

* See a particular deſcription of the ſenſitive plant, in the Appendix to Culpeper, p. 27.

the

the ground ten feet below; and now the nouriſhment it afforded to thoſe roots during the time of their deſcending was amply repaid, having every year ſince that time made vigorous ſhoots. From the top of the wall to the ſurface of the earth, theſe roots have not thrown out a ſingle fibre; but are now united in a ſingle root.

Plants, when forced from their natural poſition, are endowed with a power to reſtore themſelves. A hop-plant, twiſting round a ſtick, directs its courſe from ſouth to weſt, as the ſun does. Untwiſt it, and tie it in the oppoſite direction; it dies. Leave it looſe in the wrong direction, it recovers its natural direction in a ſingle night. Twiſt a branch of a tree ſo as to invert its leaves, and fix it in that poſition; if left in any degree looſe, it untwiſts itſelf gradually, till the leaves be reſtored to their natural poſition. What better can an animal do for its welfare? A root of a tree, meeting with a ditch in its progreſs, is laid open to the air. What follows? It alters its courſe like a rational being; dips into the ground, undermines the ditch, riſes on the oppoſite ſide to its wonted diſtance from the ſurface, and then proceeds in its original direction. Lay a wet ſponge near a root laid open to the air; the root will direct its courſe to the ſponge. Change the place of the ſponge; the root varies its direction. Put a pole into the ground at a moderate diſtance from a ſcandent plant: the plant directs its courſe to the pole, lays hold of it, and riſes on to its natural height. A honeyſuckle proceeds in its courſe, till it be too long for ſupporting its weight; and then ſtrengthens itſelf by ſhooting into a ſpiral. If it meet with another plant of the ſame kind, they coaleſce for mutual ſupport; the one ſcrewing to the right, the other to the left. If a honeyſuckle-twig meets with a dead branch, it ſcrews from the right to the left. The claſpers of briony ſhoot into a ſpiral, and lay hold of whatever comes in their way for ſupport. If, after completing a ſpiral of three rounds, they meet with nothing, they then try again for further ſupport, by altering their courſe.

By comparing theſe and other inſtances of ſeeming voluntary motion in plants, with that ſhare of life wherewith ſome of the inferior kinds of animals are endowed, we can ſcarcely heſitate in aſcribing the ſuperiority to the former; that is, putting ſenſation out of the queſtion. Muſcles, for inſtance, are fixed to one place, as much as plants are; nor have they any power of motion, beſides that of opening and ſhutting their ſhells; and in this reſpect they have no ſuperiority over the motion of the ſenſitive plant; nor doth their action diſcover more ſagacity, or even ſo much, as the roots of the plane-tree, or the action of other vegetables.

M. Buffon, who ſeems to be deſirous of confounding the animal and vegetable kingdoms, denies ſenſation to be any eſſential diſtinction: "Senſation (ſays he) more eſſentially diſtinguiſhes animals from vegetables: but ſenſation is a complex idea,

idea, and requires ſome explication. For if ſenſation implied no more than motion conſequent upon a ſtroke or an impulſe, the ſenſitive plant enjoys this power. But if, by ſenſation, we mean the faculty of perceiving and comparing ideas, it is uncertain whether brute animals are endowed with it. If it ſhould be allowed to dogs, elephants, &c. whoſe actions ſeem to proceed from motives ſimilar to thoſe by which men are actuated, it muſt be denied to many ſpecies of animals, particularly to thoſe which appear not to poſſeſs the faculty of progreſſive motion. If the ſenſation of an oyſter, for example, differed only in degree from that of a dog; why do we not aſcribe the ſame ſenſation to vegetables, though in a degree ſtill inferior? This diſtinction, therefore, between the animal and vegetable, is neither ſufficiently general nor determined. Hence we are led to conclude that there is no abſolute and eſſential diſtinction between the animal and vegetable kingdoms; but that nature proceeds, by imperceptible degrees, from the moſt perfect to the moſt imperfect animal, and from that to the vegetables; and the freſh-water polypus may be regarded as the laſt of animals and the firſt of plants."

It were to be wiſhed, that philoſophers would on ſome occaſions conſider, that a ſubject may be dark as well on account of their inability to ſee, as when it really affords no light. This great author boldly concludes, that there is no eſſential difference between a plant and an animal, becauſe we aſcribe ſenſation to an oyſter, and none to the ſenſitive plant; but we ought to remember, that, though we cannot perceive a diſtinction, it may neverthelеſs exiſt. Before M. Buffon, therefore, had concluded in this manner, he ought to have proved that ſome vegetables were endowed with ſenſation.

It is no doubt, however, as much incumbent on thoſe who take the contrary ſide of the queſtion, to prove that vegetables are not endowed with ſenſation, as it was incumbent on M. Buffon to prove that they are. But a little attention will ſhow us, that the difficulty here proceeds entirely from our inability to ſee the principle of ſenſation. We perceive this principle in ourſelves, but no man can perceive it in another. Why then does every individual of mankind conclude, that his neighbour has the ſame ſenſations with himſelf? It can only be from analogy. Every man perceives his neighbour formed in a manner ſimilar to himſelf; he acts in a ſimilar manner on ſimilar occaſions, &c. Juſt ſo it is with brute animals. It is no more doubtful that they have ſenſations, than that we have them ourſelves. If a man is wounded with a knife, for inſtance, he expreſſes a ſenſe of pain and endeavours to avoid a repetition of the injury. Wound a dog in the ſame manner, he will alſo expreſs a ſenſe of pain; and, if you offer to ſtrike him again, will endeavour to eſcape,

escape, before he feels the stroke. To conclude here, that the action of the dog proceeded from a principle different from that of the man, would be absurd and unphilosophical to the last degree.

We must farther take notice, that there are sensations essentially distinct from one another; and in proportion as an animal is endowed with more or fewer of these different species, it is more or less perfect as an animal: but, as long as only one of them remains, it makes not the least approach to the vegetable kingdom; and, when they are all taken away, is so far from becoming a vegetable, that it is only a mass of dead matter. The senses of a perfect animal, for instance, are five in number. Take away one of them, suppose sight, he becomes then a less perfect animal; but is as unlike a vegetable as before. Suppose him next deprived of hearing, his resemblance to a vegetable would be as little as before; because a vegetable can neither feel, taste, nor smell; and we suppose him still to enjoy these three senses. Let us, lastly, suppose him endowed only with the sense of feeling, which however seems to include that of taste, and he is no more a vegetable than formerly, but only an imperfect animal. If this sense is then taken away, we connect him not with the vegetable kingdom, but with what M. Buffon calls brute-matter. It is to this kingdom, and not to the vegetable, that animals plainly approximate as they descend. Indeed, to suppose an approximation between the vegetable and animal kingdoms, is very absurd; for, at that rate, the most imperfect animal ought to be the most perfect plant: but we observe no such thing. All animals, from the highest to the lowest, are possessed of vegetable life; and that, as far as we can perceive, in an equal degree, whether the animal life is perfect or imperfect: nor doth there seem to be the smallest connection between the highest degree of vegetation and the lowest degree of sensation. Though all animals are possessed of vegetable life, these two seem to be as perfectly distinct and incommensurate to one another, as any two things we can possibly imagine.

The power of vegetation, for instance, is as perfect in an onion or leek, as in a dog, an elephant, or a man: and yet, though you threaten a leek or an onion ever so much, it pays no regard to your words, as a dog would do: nor, though you wound it, does it avoid a second stroke. It is this principle of self-preservation in animals, which, being the most powerful one in their nature, is generally taken, and with very good reason, as the true characteristic of animal life. This principle is undoubtedly a consequence of sensation; and, as it is never observed to take place in vegetables, we have a right to say that the foundation of it, namely, sensation, belongs not to them. There is no animal, which makes any motion in consequence of external impulse where danger is threatened, but what puts itself in a poſture of defence; but no vegetable whatever does ſo. A muſcle, when it is touched, immediately ſhuts its ſhell; and, as this action puts it in a ſtate of defence, we conclude that it proceeded from the principle of ſelf-preſervation. When the ſenſitive plant contracts from a touch, it is no more in a ſtate of defence than before; for whatever would have deſtroyed it in its expanded ſtate, will alſo do ſo in its contracted ſtate. The motion of the ſenſitive plant proceeds only from a certain property called *irritability;* and which, though our bodies poſſeſs it in an eminent degree, is a characteriſtic neither of animal nor vegetable life, but belongs to us in common with brute-matter. It is certain, that an electrified ſilk thread ſhows a much greater variety of motions than any ſenſitive plant. If a bit of ſilk thread is dropt on an electrified metal plate, it immediately erects itſelf; ſpreads out the ſmall fibres like arms; and, if not detained, will fly off. If a finger is brought near it, the thread ſeems greedily to catch at it. If a candle approaches, it claſps cloſe to the plate as if afraid of it. Why do we not conclude that the thread in this caſe is really afraid of the candle? For this plain reaſon, that its ſeeming flight is not to get away from the candle, but to get towards the electrified metal; and, if allowed to remain there, will ſuffer itſelf to be burnt without offering to ſtir. The ſenſitive plant, in like manner, after it has contracted, will ſuffer itſelf to be cut to pieces, without making the leaſt effort to eſcape. The caſe is not ſo with the meaneſt animal. An hedge-hog, when alarmed, draws its body together, and expands its prickles, thereby putting itſelf in a poſture of defence. Throw it into water, and the ſame principle of ſelf-preſervation prompts it to expand its body and ſwim. A ſnail, when touched, withdraws itſelf into its ſhell, but if a little quicklime is ſprinkled upon it, ſo that its ſhell is no longer a place of ſafety, it is thrown into agonies, and endeavours to avail itſelf of its locomotive power in order to eſcape the danger. In muſcles and oyſters, indeed, we cannot obſerve this principle of ſelf-preſervation ſo ſtrongly, as nature has deprived them of the power of progreſſive motion: but, as we obſerve them conſtantly to uſe the means which nature has given them for ſelf-preſervation, we can have no reaſon to think that they are deſtitute of that principle upon which it is founded.

But there is no need of arguments drawn from the inferior creation. We ourſelves are poſſeſſed both of the animal and vegetable life, and certainly muſt know whether there is any connection between vegetation and ſenſation or not. We are conſcious that we exiſt; that we hear, ſee, &c. but of our vegetation we are abſolutely unconſcious. We feel a pleaſure, for inſtance, in gratifying the calls of hunger and thirſt; but of the proceſs by which our aliment improves our growth and vigour, we are altogether ignorant. If we, then, who are more perfect than other vegetables,

vegetables, are utterly infenfible of our own vegetable life, why fhould we imagine that the lefs perfect vegetables are fenfible of it?

To illuftrate our reafoning here by an example. The direction of the roots of the plane-tree, mentioned above, fhows as much fagacity, if we are to look only to the outward action, as can be obferved in any motion of the moft perfect animal whatever; neverthelefs, we have not the leaft fufpicion, either that the tree faw the ground at a diftance, or that it was informed of its being there by the reft of its roots. If a wound is made in the body of a man, and a lofs of fubftance is to be repaired, the fame fagacity will be obferved in the arrangement of the fibres, not only as if they were animated, but they will difpofe of themfelves feemingly with a degree of wifdom far fuperior to what we have any idea of; yet this is done without our having the leaft knowledge either how it is done, or of its being done at all. We have therefore in ourfelves a demonftration, that vegetable life acts without our knowing what it does: and, if vegetables are ignorant of their moft fagacious actions, why fhould we fufpect that we have a fenfation, let it be ever fo obfcure, of any of their inferior ones, fuch as contracting from a touch, turning towards the fun, or advancing to meet a pole? Thus we may eafily give M. Buffon a reafon why we afcribe fenfation to an oyfter, and none to a vegetable; namely, becaufe we perceive the vegetable to do nothing but what is alfo performed in our own bodies without our having the leaft fenfation of it; whereas an oyfter puts itfelf in a defenfive pofture on the approach of danger; and, this being an action fimilar to our own upon a like occafion, we conclude that it proceeds from the fame principle of fenfation. Here it may alfo be obferved, that, though the inferior animals are deficient in the number, they are by no means fo in the acutenefs of their fenfation; on the contrary, though a mufcle or an oyfter is probably endowed with no other fenfe than that of feeling, yet this fenfe is fo exquifite, that it will contract upon the flighteft touch, fuch as we fhould altogether be infenfible of.

As to that power of contractility, or irritability, which is obferved in fome plants; our folids have it, when deprived both of vegetable and animal life; for the human heart, or a mufcle, cut out of an animal body, will continue to contract, if it is irritated by pricking it, after it has neither fenfation nor vegetation.

A very good moral reafon may alfo be adduced, why vegetables are not endowed with fenfation. Had they been fo, we may fuppofe them to fuffer pain when they are cut or deftroyed; and, if fo, what an unhappy ftate muft they be in, who have not the leaft power to avoid the injuries daily offered them? In fact, the goodnefs of the Deity is very confpicuous in not giving to vegetables the fame fenfations as to animals; and, as he hath given them no means of defence, though we had not

been

been told it by himfelf, we might have known that he gave them for food to animals; and, in this cafe, to have endowed them with fenfation would have been cruelty. Though animals without number prey upon one another, yet all of them have fome means of defence; from whence we juftly conclude, that their mutual deftruction was not an original appointment of the Creator, but what followed from the fall of Adam, and what he forefaw would happen in a courfe of time, and which he therefore gave every one of them fome means of guarding againft. It may no doubt be here objected, that the giving fome means of felf-defence to every animal cannot be reckoned a fufficient proof that it was not the original defign of the Creator that they fhould be deftroyed, feeing thefe means are not always effectual for their prefervation. This objection, however, cannot be completely obviated without a folution of the queftion concerning the origin of evil among the works of a perfectly-good Being. But, whatever difficulty there may be in folving this queftion, it is certain, that, as fome means of felf-defence is given to every animal, it has been the original defign of the Creator, that in all cafes one fpecies of animals fhould not be deftroyed at the pleafure or will of any other fpecies: and, as no means of felf-defence is given to any vegetable, it is plain that they have been deftined for a prey to every fpecies of animals that had accefs to them. Philofophers have infifted much on the neceffity of one animal's devouring another, that there might be room fufficient for all; but this, fo far from being a fyftem worthy of the divine wifdom, feems to be a reflection upon it, as if the Author of Nature could not have found means to preferve the life of one part of his creatures, without the deftruction and mifery of the reft. The facred writings leave us at no lofs to fee how this carnivorous difpofition came in; and in the next world, this piece of perfection (as the fanguinary philofophers above-mentioned would have it to be) feems to be left out; for there, it is faid, "They fhall not hurt nor deftroy; the lion fhall eat ftraw like the ox; and there fhall be no more pain." Ifa. xi. 7, 9. Rev. xxi. 4.

Of ANIMAL FLOWERS.

The grand argument for animal life in vegetables, was inferred from the curious conftruction of the frefh-water polypus, and the *actinia* genus, called animal flowers, fea-anemone, fea-funflower, &c. which, having indeed the external form and figure of vegetables, with fcarcely any progreffive motion, might eafily deceive fuperficial obfervers; but, when more minutely examined, the polypus, and all the actinia clafs, turn out to be abfolute animals, of the viviparous kind, and feed on fifh; the heads or mouths of which, when open, refemble a full-blown flower, whence they

they are called flower-fifh. There is one fpecies of them, in which the pureft white, carmine, and ultramarine, are fcarcely fufficient to exprefs their brilliancy. The bodies of fome of them are hemifpherical, of others cylindrical, and of others fhaped like a fig. Their fubftance likewife differs; fome are ftiff and gelatinous, others flefhy and mufcular; but all of them are capable of altering their figure when they extend their bodies and claws in fearch of food. They are found on many of the rocky coafts of the Weft-India iflands, and on fome parts of the coaft of England. They have only one opening, which is in the centre of the uppermoft part of the animal; round this are placed rows of flefhy claws; this opening is the mouth of the animal, and is capable of great extenfion. The animals themfelves, though exceedingly voracious, will bear long fafting. They may be preferved alive a whole year, or perhaps longer, in a veffel of fea-water, without any vifible food; but, when food is prefented, one of them will fucceffively devour two mufcles in their fhells, or even fwallow a whole crab as large as a hen's egg. In a day or two the crab-fhell is voided at the mouth, perfectly cleared of all the meat. The mufcle-fhells are likewife difcharged whole, with the fhells joined together, but entirely empty, fo that not the leaft particle of fifh is to be perceived on opening them. An anemone of one fpecies will even fwallow an individual of another fpecies; but, after retaining it ten or twelve hours, will throw it up alive and uninjured. Through this opening alfo it produces its young ones alive, already furnifhed with little claws, which, as foon as they fix themfelves, they begin to extend in fearch of food. One of the extremities of the fea-anemone exactly refembles the outward leaves of that flower; while its limbs are not unlike the fhag or inner part of it. By the other extremity it fixes itfelf, as by a fucker, to the rocks or ftones lying in the fand; but it is not totally deprived of the power of progreffive motion, as it can fhift its fituation, though very flowly.

A peculiar fpecies of animal-flowers, called the clustered animal-flower, has been found in fome of the Weft-India iflands, an account of which was publifhed in the Philofophical Tranfactions, vol. lvii. by Mr. Ellis, in a letter to Lord Hillfborough. This compound animal, which is of a tender flefhy fubftance, confifts of many tubular bodies, fwelling gently towards the upper part, and ending like a bulb or very fmall onion; on the top of each is its mouth, furrounded by one or two rows of tentacles, or claws, which when contracted look like circles of beads. The lower part of all thefe bodies has a communication with a firm flefhy wrinkled tube, which fticks faft to the rocks, and fends forth other flefhy tubes, which creep along them in various directions. Thefe are full of different fizes of thefe remarkable animals, which riſe up irregularly in groups near to one another. This adhering tube, that ſecures them faſt to the rock, or ſhelly bottom, is worthy of our notice. The knobs are formed into ſeveral parts of it by its inſinuating itſelf into the inequalities of the coral rock, or by graſping pieces of ſhells, part of which ſtill remain in it, with the fleſhy ſubſtance grown over them. This ſhows us the inſtinct of nature, that directs theſe animals to preſerve themſelves from the violence of the waves, not unlike the anchoring of muſcles by their fine ſilken filaments that end in ſuckers; or rather like the ſhelly baſis of the ſerpula, or worm-ſhell, the tree-oyſter, and the ſlipper barnacle, &c. whoſe baſes conform to the ſhape of whatever ſubſtance they fix themſelves to, graſping it faſt with their teſtaceous claws, to withſtand the fury of a ſtorm. When we view the inſide of this animal diſſected lengthwiſe, we find a little tube leading from the mouth to the ſtomach, from whence there riſe eight wrinkled ſmall guts, in a circular order, with a yellowiſh ſoft ſubſtance in them; theſe bend over in the form of arches towards the lower part of the bulb, whence they may be traced downwards, to the narrow part of the upright tube, till they come to the fleſhy adhering tube, where ſome of them may be perceived entering into the papilla, or the beginning of an animal of the like kind, moſt probably to convey it nouriſhment till it is provided with claws; the remaining part of theſe ſlender guts are continued on in the fleſhy tube, without doubt for the purpoſe of producing and ſupporting more young from the ſame common parent.

The Abbé Dicquemarre, by many curious though cruel experiments, related in the Phil. Tranſ. for 1773, has ſhown that theſe animals poſſeſs, in a moſt extraordinary degree, the power of reproduction; ſo that ſcarcely any thing more is neceſſary to produce as many ſea-anemones as we pleaſe, than to cut a ſingle one into as many pieces. A ſea-anemone being cut in two by a ſection through the body, that part, where the limbs and mouth are placed, ate a piece of muſcle offered to it ſoon after the operation, and continued to feed and grow daily for three months after. The food ſometimes paſſed through the animal; but was generally thrown up again, conſiderably changed, as in the perfect ſea-anemone. In about two months, two rows of limbs and a mouth were perceived growing out of the part where the inciſion was made. On offering food to this new mouth, it was laid hold of and eaten; and, the limbs continually increaſing, the animal gradually became as perfect as thoſe which had never been cut. In ſome inſtances, however, when one of theſe creatures was cut through, new limbs would be produced from the cut place, thoſe at the mouth remaining as before; ſo that a monſtrous animal was the conſequence, having two mouths, and feeding at both ends.

Under

Under a large hollow cliff, in the iſland of Barbadoes, where the ſea flows up, and forms a baſon, there is a fixed ſtone, or piece of rock, in the middle, which is always under water. Round its ſides, at different depths, ſeldom exceeding eighteen inches, are ſeen at all times of the year, iſſuing out of little holes, certain ſubſtances that have the appearance of fine radiated flowers, of a pale yellow, or a bright ſtraw colour, ſlightly tinged with green, having a circular border of thick-ſet petals, about the ſize of, and much reſembling, thoſe of a ſingle garden-marigold, except that the whole of this ſeeming flower is narrower at the diſcus, or ſetting-on of the leaves, than any flower of that kind. Mr. Hughes, in his Hiſtory of Barbadoes, gives the following curious account of them: "I have attempted to pluck one of theſe animal-flowers from the rock to which they are fixed; but never could effect it; for, as ſoon as my fingers came within two or three inches of it, it would immediately contract together its yellow border, and ſhrink back into the hole of the rock; but, if left undiſturbed for about four minutes, it would come gradually in ſight, expanding, though at firſt very cautiouſly, its ſeeming leaves, till at laſt it appeared in its former bloom. However, it would again recoil, with a ſurpriſing quickneſs, when my hand came within a ſmall diſtance of it. Having tried the ſame experiment by attempting to touch it with my cane, and a ſmall ſlender rod, the effect was the ſame. But, though I could not by any means contrive to take or pluck one of theſe animals entire, yet I cut off (with a knife which I had held for a long time out of ſight, near the mouth of a hole out of which one of theſe animals appeared) two of theſe ſeeming leaves. Theſe, when out of the water, retain their ſhape and colour; but, being compoſed of a membrane-like ſubſtance, ſurpriſingly thin, they ſoon ſhrivelled up, and decayed." The reproductive power of the Barbadoes animal-flower is prodigious. Many people coming to ſee theſe ſtrange creatures, and occaſioning ſome inconvenience to a perſon through whoſe grounds they were obliged to paſs, he reſolved to deſtroy the object of their curioſity; and, that he might do ſo effectually, he cauſed all the holes out of which they appeared, to be carefully bored and drilled with an iron inſtrument, ſo that we cannot ſuppoſe but their bodies muſt have been entirely cruſhed to a pulp: nevertheleſs they again appeared in a few weeks in ſtill greater abundance, from the very ſame places.

The ſea-carnation, or animal-flower, found among the rocks at Haſtings in Suſſex, is very ſimilar to the animal-flower of Barbadoes. This animal adheres by its tail, or ſucker, to the under part of the projecting rocks oppoſite the town; and, when the tide is out, has the appearance of a long white fig; which is alſo the form of it when put into a glaſs of ſea-water.

Of the POLYPUS.

THE Polype, or Polypus, which ſo long divided naturaliſts in opinion whether it was of vegetable or animal conformation, is a freſh-water animal, of the hydra genus, in the claſs of worms, and order of *zoophytes*, in the Linnæan ſyſtem. It is of a cylindrical figure, but variable, with very long tentacula, or claws. There is ſcarcely an animal in the world more difficult to deſcribe, than this ſurpriſing worm; it varies its whole figure at pleaſure, and is frequently found beſet with young in ſuch a manner as to appear ramoſe and divaricated; theſe young ones adhering to it ſo as to appear parts of its body.

When ſimple and in a moderate ſtate as to contraction or dilation, it is oblong, ſlender, pellucid, and of a pale-reddiſh colour; its body is ſomewhat ſmaller towards the tail, by which it fixes itſelf to ſome ſolid body: and larger towards the other extremity, where it has a larger opening, called its mouth, around which are the tentacula, or claws, which are eight in number, and are uſually extended to about half the length of its body. By means of theſe tentacula, or arms, as they are commonly called, expanded into a circle of more than half a foot diameter, the creature feels every thing that can ſerve it for food; and, ſeizing the prey with one of them, calls in the aſſiſtance of the others, if neceſſary, to conduct it to its mouth.

The production of its young is different from the common courſe of nature in other animals; for the young one iſſues from the ſide of its parent in the form of a ſmall pimple, which, lengthening every hour, becomes, in about two days, a perfect animal, and drops from off its parent to ſhift for itſelf: but, before it does this, it has often another growing from its ſide; and ſometimes a third from that, even before the firſt is ſeparated from its parent; and what is very extraordinary, there has never yet been diſcovered among them any diſtinction of ſex, or appearance of copulation; every individual of the whole ſpecies being prolific, and that as much if kept ſeparate as if ſuffered to live among others; but what is even ſtill more ſurpriſing, is the reproduction of its ſeveral parts when cut off; for, when cut into a number of ſeparate pieces, it becomes in a day or two ſo many diſtinct and ſeparate animals; each piece having the property of producing a head and tail, and the other organs neceſſary for life, and all the animal functions.

There is no diſtinguiſhed place in the body of the polypus, from whence the young are brought forth; for they ſpring out like ſhoots or branches of a tree, from all the exterior parts of their bodies. M. Trembley, who had heard much of this creature, and being determined to convince himſelf, by real experiments, whether it was a vegetable or an animal, cut one in the middle, when, to his utter amazement,

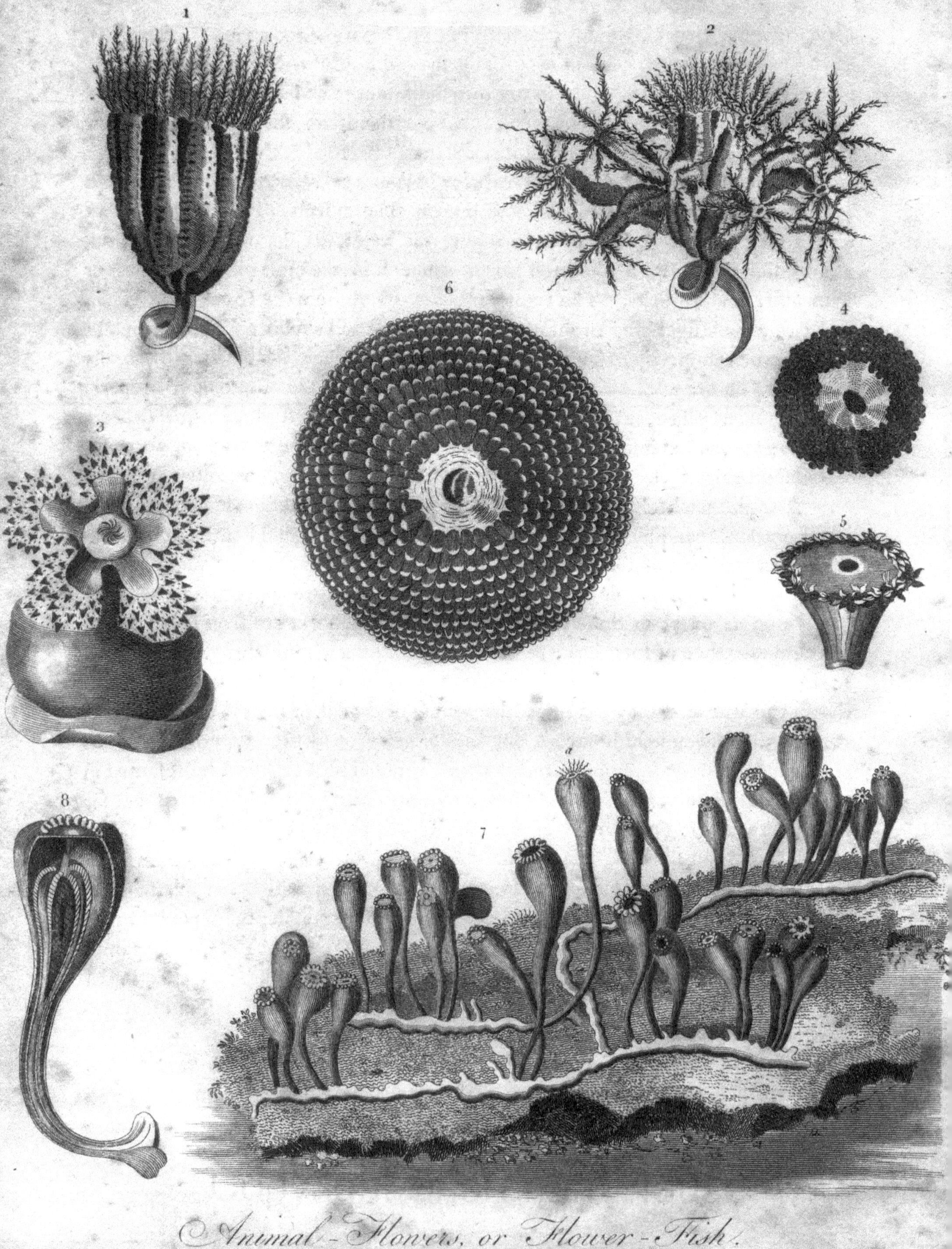

Animal-Flowers, or Flower-Fish.

London Published as the Act directs 7 Jan.y 1809 by J. Wilkes.

J. Pass sculp.

ment, he found that in two days each of thoſe pieces was become a perfect animal, the head-part having ſhot forth a tail, and the tail a head. Numerous trials of a ſimilar nature have been made in my own laboratory; and I have always found that it is of no conſequence how often you cut them, for they ſtill put out new members, and become ſo many diſtinct polypes.

They are always to be found in clear ſlowly-running waters, adhering by the tail to ſticks, ſtones, and water-plants, and live on ſmall inſects. They are eaſily kept alive a long time in glaſſes, often changing the water, keeping the glaſſes clean, and feeding them with a ſmall red worm, common in the mud of the Thames, or with other ſmall inſects. The creature has its name from the Greek πολυς, many, and πους, a foot, ſignifying an animal with many feet; but a more appoſite one might eaſily have been invented, ſince it has in reality no feet at all. What were originally taken for feet, are what have ſince been called its horns, and of late more properly its arms, their office being to catch its prey. With theſe little arms, which are capable of great extenſion, it ſeizes minute worms, and various kinds of water-inſects, and brings them to its mouth: and, like the ſea-anemone, often ſwallows bodies larger than itſelf: having a ſurpriſing property of extending its mouth wider, in proportion, than any other animal. After its food is digeſted in its ſtomach, it returns the remains of the animals upon which it feeds through its mouth again, having no other obſervable emunctory. In a few days there appear ſmall knobs or papillæ on its ſides; as theſe increaſe in length, little fibres are ſeen riſing out of the circumference of their heads, as in the parent animal; which fibres they ſoon begin to uſe for the purpoſe of procuring nouriſhment, &c. When theſe are arrived at mature ſize, they ſend out other young ones on their ſides in the ſame manner; ſo that the animal branches out into a numerous offspring, growing out of one common parent, and united together and diſpoſed in the manner repreſented in the annexed plate. Each of theſe provides nouriſhment not only for itſelf, but for the whole ſociety; an increaſe of the bulk of one polype, by its feeding, tending to an increaſe in the reſt. Thus a polype of the freſh-water kind becomes like a plant branched out, or compoſed of many bodies, each of which has this ſingular characteriſtic, that, if one of them be cut in two in the middle, the ſeparated part becomes a complete animal, and ſoon, adhering to ſome fixed baſe, like the parent from which it was ſeparated, produces a circle of arms; a mouth is formed in the centre; it increaſes in bulk, emits a numerous progeny, and is ſoon, in every reſpect, as perfect an animal as that from which it was ſevered.

The ſeveral ſtrange properties recorded of the polypes and animal flowers, though very ſurpriſing, are not, however, peculiar to them alone. The Surinam toad is well known to produce its young, not in the ordinary way, but in cells upon its back. And, as to the most amazing of their properties, the re-production of their parts, we know the crab and lobster, if a leg be broken off, can always produce a new one.

I have annexed a copper-plate of some excellent drawings of these curious phenomena in nature; in which No. 1 represents the fresh-water polype, with its tentacula or arms extended upwards. No. 2 represents the same animal, with its young branching from it, and putting out their claws for food, which, howsoever taken, goes to the common support of the whole family. No. 3 shows the animal carnation-flower of the rocks near Hastings in Sussex, with its tentacles extended in search of food. No. 4 is an exact representation of the sea-anemone, above described. No. 5 shows the head of the animal-flower of Barbadoes. No. 6 is a brilliant display of the sea sun-flower animal, with its innumerable tentacles expanded to catch its prey, which being allured to it by its elegant appearance, they close instantly upon it, and convey it to the interior concavity or mouth. No. 7 represents a cluster of the animal-flower described by Mr. Ellis, in the islands ceded by France; in which *a* shows one of the animals stretching out its tentacles in search of food. No. 8 is a perpendicular dissection of one of the same animals, in order to show the gullet, intestines, stomach, and the fibres or tendons that move the claws.

Of ANIMALCULES.

THE next most surprising part of animal nature, is that of animalcules, an innumerable tribe of living beings, wholly invisible to the naked eye, and which cannot even be perceived to exist, but by the assistance of microscopes. The smallest living creatures our instruments can show, are those that inhabit the waters; for, though possibly animalcules equally minute, or perhaps more so, may fly in the air, or creep upon the earth, it is scarcely possible to bring such under our examination; but, water being transparent, and confining the creatures in it, we are able, by applying a drop of it to our glasses, to discover, to a certain degree of smallness, all that it contains.—Some of the most curious of these animalcules, which have been described by microscopical observers, are as follow.

1. *The Hair-like Insect.* This is so called on account of its shape; being extremely slender, and frequently an hundred and fifty times as long as broad. The body or middle part, which is nearly straight, appears, in some, composed of such rings as the windpipe of land-animals; but, in others, seems rather scaled, or made up of rings that obliquely cross one another. Its two ends are hooked or bent, pretty nearly in the same degree, but in a direction opposite to one another; and, as no eyes can

can be difcerned, it is difficult to judge which is the head or tail. Its progreffive motion is very fingular, being performed by turning upon one end as a centre, and defcribing almoft a quarter of a circle with the other; its fhape and form may be feen greatly magnified in the following curious Plate of Animalcules, at fig. 1. Its motions are very flow, and require much patience and attention in the obferver. Thefe creatures are fo fmall, that millions of millions of them might be contained in the circle, fig. 2. When viewed fingly, they are exceedingly tranfparent, and of a beautiful green colour; but, when numbers of them are brought together, they become opaque, lofe their green colour, and grow entirely black. The hair-like infect was firft difcovered in a ditch at Norwich, one end of which communicates with the river there, and the other end with a fecond ditch, into which feveral kennels empty themfelves. The length of this ditch was at leaft a hundred yards, and its breadth nine. The bottom, for more than a foot thick, was covered with a blackifh green fubftance in appearance like mud, made up for the moft part of thefe infects; but, fuppofing only half or a quarter part of it to be compofed of them, according to the above dimenfions, their numbers muft exceed all imagination.

2. *Eels in Pafte,* &c. When pafte is allowed to ftand till it becomes four, it is then found to be the habitation of numberlefs animalcules, which may be difcerned by the naked eye; and, though their form cannot be perfectly diftinguifhed, their motion is very perceptible, and the whole pafte will feem to be animated. Fig. 3 reprefents one of thefe anguillæ highly magnified. The moft remarkable property of thefe infects is, that they are viviparous. If one of them is cut through near the middle, feveral oval bodies, of different fizes, will be feen to iffue forth. Thefe are young anguillæ, each of them coiled up and inclofed in its proper membrane, which is fo exquifitely fine, as fcarce to be difcernible by the greateft magnifier, while it inclofes the embryo animal. The largeft and moft forward immediately break through this covering, unfold themfelves, and wriggle about in the water nimbly; others get out, uncoil, and move themfelves about more flowly; and the leaft mature continue entirely without motion. The *uterus,* or veffel that contains all thefe oval bodies, is compofed of many ringlets, not unlike the afpera arteria of land-animals, and feems to be confiderably elaftic; for, as foon as the animalcule is cut in two, the oval bodies are thruft out with fome degree of violence, from the fpringing-back or action of this bowel. An hundred and upwards of the young ones have been feen to iffue from the body of one fingle eel, whereby the prodigious increafe of them may be acounted for; as probably feveral fuch numerous generations are produced in a fhort time. Animalcules of a fimilar kind are likewife found in vinegar; and, like thofe already defcribed, are found to be viviparous. But it is not only in acid matters that fuch appearances are obferved. In

fome

ſome fields of wheat, many grains may be obſerved, that appear blackiſh outwardly, as if ſcorched; but when opened, are found to contain a ſoft white ſubſtance, which, attentively conſidered, appears to be nothing elſe than a congeries of threads or fibres, lying cloſe to each other in a parallel direction, much reſembling the unripe down of ſome thiſtles on cutting open the flower-heads before they begin to blow. This fibrous matter diſcovers not the leaſt ſign of life or motion, unleſs water is applied: but immediately on wetting, provided the grains of wheat have been newly gathered, the ſuppoſed fibres ſeparate, and appear to be living creatures. Their motions at firſt are very languid; but gradually become more active and vigorous, twiſting and wriggling themſelves ſomewhat in the manner of the eels in paſte, but always ſlower and with leſs regularity.

3. *The Proteus,* ſo called on account of its aſſuming a great number of different ſhapes, ſo as ſcarcely to be known as the ſame animal in its various transformations; and indeed, unleſs it be carefully watched while paſſing from one ſhape to another, it will often become ſuddenly inviſible. When water, wherein any ſort of vegetable has been infuſed, or animals preſerved, has ſtood quietly for ſome days or weeks in any glaſs or other veſſel, a ſlimy ſubſtance will be collected about the ſides: ſome of which being taken up with the point of a pen-knife, placed on a ſlip of glaſs in a drop of water, and looked at through the microſcope, will be found to harbour ſeveral kinds of little animals that are ſeldom found ſwimming about at large; among which the proteus is one. Its ſhape is better underſtood from the following plate, than from any deſcription that could be given. Its ſubſtance and colour reſemble that of a ſnail; and its whole ſhape ſeems to bear a conſiderable reſemblance to that of a ſwan. It ſwims to and fro with great vivacity: but will now and then ſtop for a minute or two; during which time its long neck is uſually employed as far as it can reach, forwards, and on every ſide, with a ſomewhat-ſlow but equable motion, like that of a ſnake, frequently extending thrice the length of its body, and ſeemingly in ſearch of food. There are no eyes, nor any opening in the head like a mouth, to be diſcerned: but its actions plainly prove it to be an animal that can ſee; for, though multitudes of different animalcules ſwim about in the ſame water, and its own progreſſive motion is very ſwift, it never ſtrikes againſt any of them, but directs its courſe between them with a dexterity wholly unaccountable ſhould we ſuppoſe it deſtitute of ſight. When it is alarmed, it ſuddenly draws in its neck, repreſented in the Plate at fig. 4 and 5. transforming itſelf into the ſhape repreſented at fig. 6. when it becomes more opaque, and moves about very ſlowly, with the large and foremoſt. When it has continued ſome time in this poſture, it will often, inſtead of the head and neck it had formerly, put forth a new one, with a kind of wheel-machinery,

repreſented

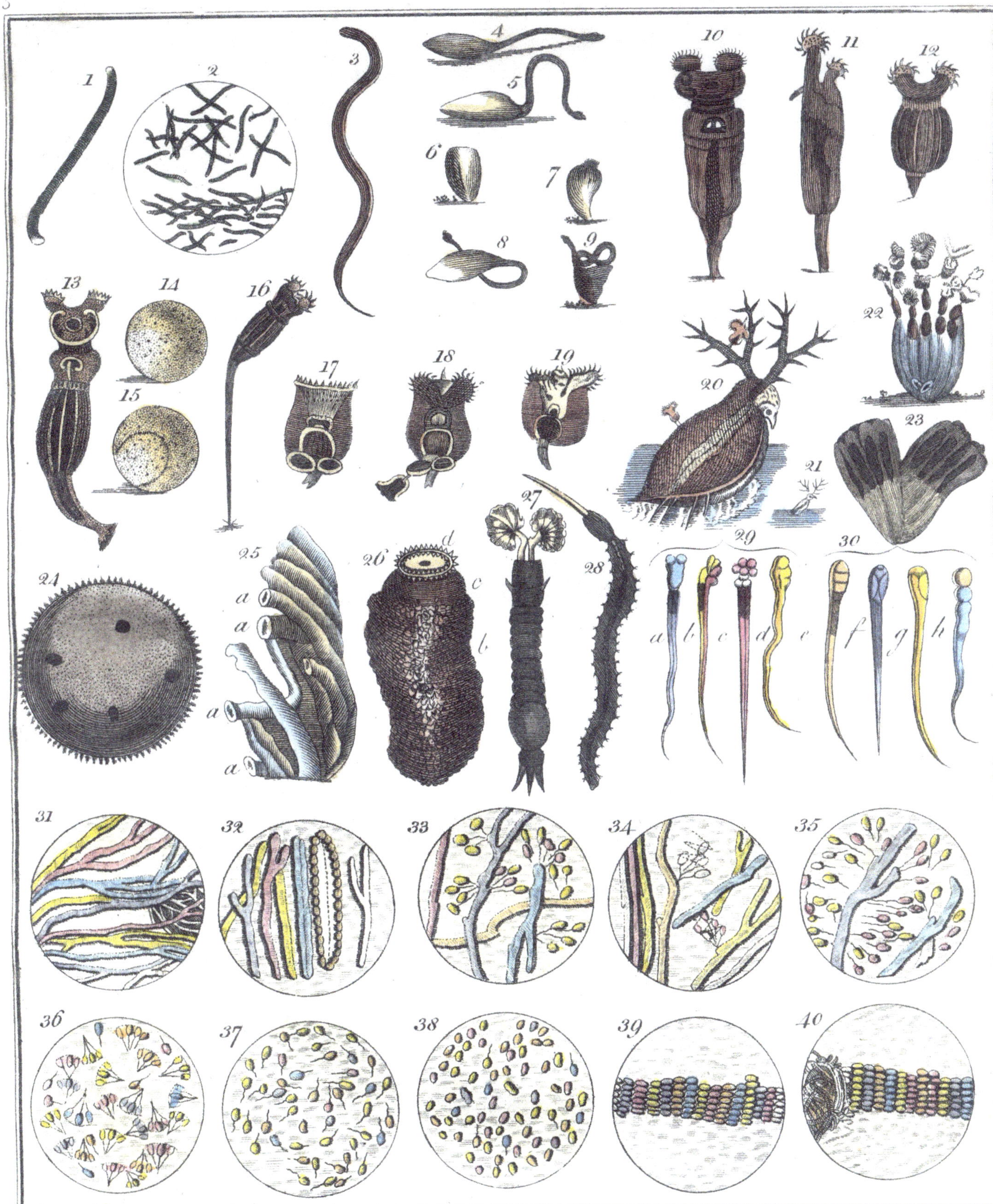

Dodd Delin. Pass Sculp.

Animalcules.

reprefented at fig. 7. the motions of which draw a current of water to it from a confiderable diftance. Having often pulled in and thruft out this fhort head, fometimes with and fometimes without the wheel-work, the creature, as if weary, will remain motionlefs for a while; then its head and long neck will be very flowly protruded, as at fig. 8. and it foon refumes its former agility. Sometimes it difpofes of its neck and head as reprefented at fig. 9.

4. *The Wheel-animal*, or *Vorticella.* This wonderful animalcule is found in rain-water that has ftood fome days in leaden gutters, or in hollows of lead on the tops of houfes, or in the flime or fediment left by fuch water; and perhaps may alfo be found in other places; but, if the water ftanding in gutters of lead, or the fediment left behind it, has any thing of a red colour in it, one may be almoft certain of finding them therein. Though it difcovers no figns of life except when in the water, yet it is capable of continuing alive for many months after it is taken out of the water, and kept in a ftate as dry as duft. In this ftate it is of a gobular fhape, exceeds not the bignefs of a grain of fand, and no figns of life appear; but, being put into water, in the fpace of half an hour a languid motion begins, the globule turns itfelf about, lengthens itfelf by flow degrees, affumes the form of a lively maggot, and moft commonly in a few minutes afterwards puts out its wheels; fwimming vigoroufly through the water, as if in fearch of food; or elfe, fixing itfelf by the tail, works the wheels in fuch a manner as to bring its food to it. Fig. 10, 11, 12, and 13, fhow the different appearances of its wheels; and fig. 14, and 15, fhow its globular form. The moft remarkable part of this animalcule is its wheel-work. This confifts of a couple of femicircular inftruments, round the edges of which many little fibrillæ move themfelves very brifkly, fometimes with a kind of rotation, and fometimes in a trembling or vibrating manner. When in this ftate, it fometimes unfaftens its tail, and fwims along with a great deal of fwiftnefs, feemingly in purfuit of its prey. Sometimes the wheels feem to be entire circles, armed with fmall teeth, like thofe of the balance-wheel of a watch, appearing projected forwards beyond the head, and extending fideways fomewhat wider than its diameter. The teeth or cogs of thefe wheels feem to ftand very regularly at equal diftances; but the figure of them varies according to their pofition, the degree of their protrufion, and perhaps the will of the animal itfelf. All the actions of this creature feem to imply fagacity and quicknefs of fenfation. At the leaft touch or motion in the water, they inftantly draw in their wheels; and their eyes feem to be lodged fomewhere about the wheels; becaufe, while in the maggot ftate, its motions are flow and blundering; but, after the wheels are protruded, they are performed with great regularity, fwiftnefs, and fteadinefs.

Beſides the above, there are found in our waters ſeveral other ſpecies of animals furniſhed with wheels, ſome of which appear to have a rotatory, and others a vibratory, motion. Fig. 16. repreſents a kind found in the ditch at Norwich, where the hair-like inſect is produced. They differ from the foregoing only in having very long tails. Fig. 17, 18, and 19, repreſent a ſpecies of wheel animals, which are alſo covered with ſhells. The body of this ſpecies conſiſts of three parts, in like manner as the other; only the thorax and abdomen, in this, are not ſeparated by any gut, or intermediate veſſel, but are joined immediately together. The heart is plainly perceived, having a regular ſyſtole and diaſtole, at *a a a*, as in the former ſpecies. Theſe creatures occaſionally draw themſelves entirely within their ſhells; and the ſhell then appears terminated by ſix ſhort ſpikes on one ſide and two on the other. The young ones of this ſpecies are carried in oval ſacculi, or integuments, faſtened externally to the lower part of their ſhells ſomewhere about the tail. When a young one is about to burſt its integuments, the parent aſſiſts it greatly, by wagging its tail, and ſtriking the oval bag, ſo that the young one's head becomes as it were forced into the water, though the tail cannot be ſo ſoon diſengaged. In this condition the young one ſets its wheel a-going, and exerts all its endeavours to free itſelf from its confinement. When it has got clear, it ſwims away, wagging its tail as the old one does, and leaving the integument adhering to the ſhell of the parent. Theſe wheel-animals are great tormentors of the water-flea, of which a figure is given in the plate. Fig. 20. ſhows it magnified, with ſome of the wheel-animals adhering to it; fig. 21. ſhows the natural ſize of the flea. Theſe inſects are often found in great numbers in the ſame water; and, when that is the caſe, it is not uncommon to diſcover five or ſix of theſe cruſtaceous wheel-animals faſtened by their tail to the ſhell or horns of the flea: cauſing it, ſeemingly, a vaſt deal of uneaſineſs; nor can they be driven away, or ſhaken off, by all the efforts the flea can uſe for that purpoſe.

5. *The Bell-flower Animal, or Plumed Polype.* Theſe animalcules dwell in colonies together, from ten to fifteen, (ſeldom falling ſhort of the former number, or exceeding the latter,) in a ſlimy kind of mucilaginous or gelatinous caſe; which out of the water has no determined form, appearing like a little lump of ſlime; but, when expanded therein, has ſome reſemblance to the figure of a bell with its mouth upwards; and is uſually about half an inch long and a quarter of an inch in diameter. Theſe bells, or colonies, are to be found adhering to the large leaves of duckweed, and other aquatic plants. They may be moſt eaſily diſcovered by letting a quantity of water, with duckweed in it, ſtand quietly for three or four hours in glaſs veſſels in a window, or other place where a ſtrong light comes; for then,

if

if any are about the duckweed, they will be found, on careful inſpection, extending themſelves out of their caſes, and making an elegant appearance. Beſides the particular and ſeparate motion which each of theſe creatures is able to exert within its own caſe, and independent of the reſt; the whole colony together has a power of altering the poſition of the bell, or even of removing it from one place to another; and hence this bell is ſometimes found ſtanding perfectly upright, as at fig. 22. and ſometimes bending the upper part downwards. As theſe animalcules ſeem not to chooſe to ſtay together in ſocieties whoſe number exceeds fifteen, when the colony happens to increaſe in number, the bell may be obſerved to ſplit gradually, beginning from about the middle of the upper or anterior extremity, and proceeding downwards towards the bottom, as at fig. 23. till they at laſt ſeparate entirely, and become two complete colonies independent of each other, one of which ſometimes removes to another part of the veſſel. The arms of each individual of this colony are ſet round the head, to the number of forty, having each the figure of an Italic *ſ*, one of whoſe hooked ends is faſtened to the head, and all together, when expanded, compoſe a figure ſhaped ſomewhat like a horſe's ſhoe, convex on one ſide next the body, but gradually opening and turning outwards, ſo as to leave a conſiderable area within the outer extremities of the arms. When the arms are thus extended, the creature, by giving them a vibrating motion, can produce a current in the water, which brings the animalcules, or whatever other minute bodies are within the ſphere of its action, with great velocity to its mouth, ſituated between the arms; where they are taken in if liked, or driven away by a contrary motion. Though their eyes cannot be diſcovered, yet they have perception of the light: for, when kept in the dark, they always remain contracted; but, on being expoſed to the light of the ſun or of a candle, they conſtantly extend their arms, and ſhow evident ſigns of being pleaſed.

6. *The Globe-animal.* This animalcule, repreſented at fig. 24. ſeems exactly globular, having no appearance of either head, tail, or fins. It moves in all directions, forwards or backwards, up or down, either rolling over and over like a bowl, ſpinning horizontally like a top, or gliding along ſmoothly without turning itſelf at all. Sometimes its motions are ſlow, at other times very ſwift; and, when it pleaſes, it can turn round, as it were upon an axis, very nimbly, without removing out of its place. The whole body is tranſparent, except where the circular black ſpots are ſhown in the figure. Some of the animals have no ſpots, and others from one to ſeven. The ſurface of the whole body appears, in ſome, as if all over dotted with points; in others, as if granulated like ſhagreen: but their more general appearance is, as if beſet thinly round with ſhort moveable hairs or briſtles, which

probably

probably are the inftruments by which their motions are performed. Thefe animalcules may be feen by the naked eye, but appear only like moving points.

7. *The Pipe-animal.* Thefe creatures are found on the coaft of Norfolk, living in fmall tubes or cafes of fandy matter, in fuch multitudes as to compofe a mafs fometimes of three feet in length. Fig. 25. fhows a piece of fuch a congeries broke off, where *a a a a* reprefent the mouths or openings of the pipes wherein the little animals make their abode. Fig. 26. fhows one fingle pipe, with its inhabitant, feparated from the reft, and magnified nine or ten times in diameter. The pipe or cafe *b* is made of fand, intermixed here and there with minute fhells, and all cemented together by a glutinous flime, probably iffuing from the animal's own body *c*, which is compofed of mufcular ringlets like thofe of a worm, capable of great extenfion or contraction. The anterior end or head, *d*, is exceedingly beautiful, having round it a double row of little arms difpofed in a very regular order, and probably capable of extenfion, in order to catch its food, and bring it to its mouth. Some of thefe tubes are found petrified.

8. *An Infect with net-like arms.* The properties and fhape of this little animal are very extraordinary. It is found only in cafcades, where the water runs very fwift. There thefe infects are found in clufters, ftanding erect on their tails; and refembling, when all together, the combs of bees at the time they are filled with their aureliæ. On being taken out of the water, they fpin threads, by which they hang exactly in the fame manner as the garden-fpider. Fig. 27. fhows one of thefe infects magnified. Its body appears curioufly turned as on a lathe; and at the tail are three fharp fpines, on which it raifes itfelf, and ftands upright in the water: but the moft curious apparatus is about its head, where it is furnifhed with two inftruments like fans or nets, which ferve to provide its food. Thefe it frequently fpreads out and draws in again; and when drawn up they are folded together with the utmoft nicety and exactnefs, fo as to be indifcernible when brought clofe to the body. At the bottom of thefe fans a couple of claws are faftened to the lower part of the head, which, every time the nets are drawn in, conduct to the mouth of the animal whatever is taken in them. Some of thefe creatures being kept with water in a vial, moft of them died in two days; and the reft, having fpun themfelves tranfparent cafes, (which were faftened either to the fides of the glafs, or to pieces of grafs put into it,) feemed to be changed into a kind of chryfalis. None of them lived above three days; and, though frefh water was given them two or three times a-day, yet in a few hours it would ftink to a degree fcarcely conceivable, and that too at feveral yards diftance, though, in proportion to the water, all the included infects were not more than as one to one million one hundred and fifty thoufand.

ſand. This makes it probable, that it is neceſſary for them to live in a rapid ſtream, leſt they ſhould be poiſoned by the effluvia iſſuing from their own bodies, as no doubt they were in the phial.

9. *A curious aquatic worm.* This animalcule is ſhown, magnified, at fig. 28. It is found in ditch-water, and is of various ſizes, from one fortieth to half an inch in length. About the head it has ſomewhat of a yellowiſh colour; but all the reſt of the body is perfectly colourleſs and tranſparent, except the inteſtines, which are conſiderably opaque, and diſpoſed as in the figure. Along its ſides are ſeveral papillæ, with long hairs growing from them: it has two black eyes, and is very nimble. But the moſt remarkable thing in this creature, is a long horn or proboſcis; which, in the large ones, may be ſeen with the naked eye, if the water is clear, and is ſometimes one tenth of an inch in length; this it waves to and fro as it moves in the water, or creeps up the ſide of the glaſs; but it is not known whether it is hollow, or of what uſe it is to the creature itſelf.

10. *Spermatic Animals* and *Animalcula Infuſoria.* The diſcovery of living animalcules in the ſemen of moſt animals, is claimed by Leeuwenhoek, a Dutch natutiraliſt. According to this naturaliſt, theſe animalcules are found in the male ſeed of every kind of animal; but their general appearance is very much the ſame, nor doth their ſize differ in proportion to the bulk of the animal to which they belong. The bodies of all of them ſeem to be of an oblong oval form, with long tapering ſlender tails iſſuing from them: and, as by this ſhape they reſemble tadpoles, they have been frequently called by that name; though the tails of them, in proportion to their bodies, are much longer than the tails of tadpoles are: and it is obſervable, that the animalcules in the ſeed of fiſhes have tails much longer and more ſlender than the tails of thoſe in other animals; inſomuch, that the extremity of them is not to be diſcerned without the beſt glaſſes, and the utmoſt attention. Fig. 29, *a, b, c, d,* repreſent the ſpermatic animalcula of the rabbit; and fig. 30, *e, f, g, h,* thoſe found in the ſeed of a dog. The numbers of theſe animalcula are inconceivable. On viewing with a microſcope the milt or ſeed of a male cod-fiſh, innumerable multitudes of animalcules are found therein, of ſuch a diminutive ſize, that at leaſt ten thouſand of them are capable of being contained in the bulk of a grain of ſand; whence it is concluded, that the milt of this ſingle fiſh contained more living animalcules than there are to be found people living in the whole world. To find the comparative ſize of theſe animalcules, Mr. Leeuwenhoek placed a hair of his head near them; which hair, through his microſcope, appeared an inch in breadth; and he was ſatisfied, that at leaſt ſixty ſuch animalcules could eaſily lie within that diameter; whence, their bodies being ſpherical, it follows, that two hundred and ſix-

teen thousand of them are but equal to a globe whose diameter is the breadth of a hair. He observed, that, when the water wherewith he had diluted the seed of a cod-fish was exhaled, the little bodies of the animalcules burst in pieces; which did not happen to those in the seed of a ram: and this is imputed to the greater firmness and consistency of the latter, as the flesh of a land-animal is more compact than fish. These animalcules appear to be very vigorous and tenacious of life; for they may be observed to move long after the animal from which they are taken is dead. They have this peculiarity also, that they are continually in motion, without the least rest or intermission, provided there is fluid sufficient for them to swim about in. These animalcules are only peculiar to the seed; nothing that has the least token of life being discoverable by the best glasses, either in the blood, spittle, urine, gall, or chyle. Great numbers, however, are to be found in the whitish matter that sticks between our teeth; some of which are of an oval figure, and others resemble eels.

The *Animalcula Infusoria* take their name from their being found in all kinds either of vegetable or animal infusions. Indeed, there is scarcely any kind of water, unless impregnated with some mineral substance, but what will discover living creatures. Leeuwenhoek says, that at first he could discern no living creatures in rain-water; but, after standing some days, he discovered innumerable animalcules, many thousands of times less than a grain of sand, and in proportion to a mite as a bee is to a horse. In other rain-water, which had likewise stood some time, he found the smallest sort he had ever seen; and, in a few days more, met with others eight times as big as these, and almost round. In another quantity of rain-water, that had been exposed like the former, he discovered a kind of animalcules with two little horns, in continual motion. The space between the horns was flat, though the body was roundish, but tapering a little towards the end; where a tail appeared, four times as long as the body, and the thickness of a spider's web. He observed several hundreds of these within the space a grain of sand would occupy. If they happened on the least filament or string, they were entangled in it; and then would extend their bodies into an oblong round, and struggle hard to disengage their tails. He observed a second sort, of an oval figure, and imagined the head to stand at the sharpest end. The body was flat, with several small feet, moving exceeding quick, but not discernible without a great deal of attention. Sometimes they changed their shape into a perfect round, especially when the water began to dry away. He met also with a third sort, twice as long as broad, and eight times smaller than the first: yet in these he discerned little feet, whereby they moved very nimbly. He perceived likewise a fourth sort, a thousand times smaller than a louse's eye, and which exceeded all the rest in briskness: he found these turning themſelves round, as it were upon a point, with the celerity of a top. And he ſays, there were ſeveral other ſorts. The production of *animalcula infuſoria* is very ſurpriſing. In four hours time, an infuſion of cantharides has produced animalcula leſs than even the tails of the ſpermatic animals we have already deſcribed. Neither do they ſeem to be ſubject to the fate of other animals; but, ſeveral kinds of them at leaſt, by dividing themſelves in two, to enjoy a ſort of immortality. Nor do the common methods by which other animals are deſtroyed, ſeem to be effectual for deſtroying their vital principle. Hot mutton-gravy, ſecured in a phial with a cork, and afterwards ſet among hot aſhes, to deſtroy as effectually as poſſible every living creature that could be ſuppoſed to exiſt in it, has nevertheleſs been found ſwarming with animalcules, after ſtanding a few days. In the Philoſophical Tranſactions, vol. lix. we have a very curious account, given us by Mr. Ellis, of animalcules produced from an infuſion of potatoes and of hemp-ſeed.

"On the 25th of May, 1768, Fahrenheit's thermometer ſeventy degrees, I boiled a potatoe in the New-River water, till it was reduced to a mealy conſiſtence. I put part of it, with an equal proportion of the boiling liquor, into a cylindrical glaſs veſſel, that held ſomething leſs than half a wine pint, and covered it cloſe immediately with a glaſs cover. At the ſame time I ſliced an unboiled potatoe, and, as near as I could judge, put the ſame quantity into a glaſs veſſel of the ſame kind; with the ſame proportion of New-River water not boiled; and covered it with a glaſs cover; and placed both veſſels cloſe to each other. On the twenty-ſixth of May, twenty-four hours afterwards, I examined a ſmall drop of each, by the firſt magnifier of Wilſon's microſcope, whoſe focal diſtance is reckoned at the fiftieth part of an inch; and, to my amazement, they were both full of animalcules of a linear ſhape, very diſtinguiſhable, moving to and fro with great celerity; ſo that there appeared to be more particles of animal than vegetable life in each drop. This experiment I have repeatedly tried, and always found it to ſucceed in proportion to the heat of the circumambient air; ſo that even in winter, if the liquors are kept properly warm, at leaſt in two or three days the experiment will ſucceed. What I have obſerved are infinitely ſmaller than ſpermatic animals, and of a very different ſhape: the truth of which every accurate obſerver will ſoon be convinced of, whoſe curioſity may lead him to compare them; and I am perſuaded he will find they are no way akin. At preſent I ſhall paſs over many other curious obſervations, which I have made on two years experiments, in order to proceed to the explaining a hint which I received laſt January from M. de Sauſſure, of Geneva, when he was here; which is, that he found one kind of theſe animalcula infuſoria that increaſe by dividing acroſs into nearly two equal parts. I had often ſeen this

appearance

appearance in various ſpecies a year or two ago, as I found upon looking over the minutes I had taken when I made any new obſervation; but always ſuppoſed the animal, when in this ſtate, to be in coition. Not hearing, till after M. de Sauſſure left this kingdom, from what infuſion he had made his obſervation, his friend Dr. de la Roche, of Geneva, informed me, the latter end of February laſt, that it was from hemp-ſeed. I immediately procured hemp-ſeed from different ſeedsmen, in diſtant parts of the town. Some of it I put into New-River water, ſome into diſtilled water, and ſome I put into very hard pump-water. The reſult was, that in proportion to the heat of the weather, or the warmth in which they were kept, there was an appearance of millions of minute animalcula in all the infuſions; and, ſome time after, ſome oval ones made their appearance. Theſe were much larger than the firſt, which ſtill continued; they wriggled to and fro in an undulatory motion, turning themſelves round very quick all the time that they moved forwards. Nothing more plainly ſhows theſe animals to be zoophytes than this circumſtance; that when, by accident, the extremity of their bodies has been ſhrivelled for want of a ſupply of freſh water, the applying more freſh water has given motion to the part of the animal that was ſtill alive; by which means, this ſhapeleſs figure has continued to live and ſwim to and fro all the time it was ſupplied with freſh water."—Thus we have given as full an account as our limits would admit, of the moſt curious kind of animalcules that have hitherto been obſerved. We cannot, however, diſmiſs this ſubject, without taking notice of the animalcules found in the ſeed of man.

Before the invention of microſcopes, the doctrine of equivocal generation, both with regard to animals and plants of ſome kinds, was univerſally received: but this inſtrument ſoon convinced every intelligent perſon, that thoſe plants which formerly were ſuppoſed to be produced by equivocal generation, aroſe from ſeeds; and the animals, in like manner, from a male and female. But, as the microſcope threw light upon one part of nature, it left another involved in darkneſs: for the origin of the animalcula infuſoria, or of the ſpermatic animals already mentioned, remains as much unknown, as that of many other kinds was, when the doctrine of equivocal generation reigned in full force.

The diſcovery of ſpermatic animalcules was thought to throw ſome light on the myſterious affair of generation itſelf, and theſe minute creatures were imagined to be each of them individuals of the ſame ſpecies with the parent. Here the infinite number of theſe animalcules was an objection, and the difficulty remained as great as before; for, as every one of theſe animalcules behoved to be produced from a male and female, to explain their origin by animalcular generation in the ſame manner,

manner, was only explaining generation by itſelf. This hypotheſis, therefore, having proved unſatisfactory, others have been invented, but which are likewiſe involved in doubt. M. Buffon, however, ſo far as concerns human generation, has given ſuch a particular account of the animalcules in the ſeed of man, that we ſhall ſtate it here, for the information of the curious.

Having procured the private parts of a man who died a violent death, he extracted all the ſeed from them while they were ſtill warm: and having examined a drop of it with a double microſcope, it had the appearance as in the Plate, at fig. 31. Large filaments appeared, which in ſome places ſpread out into branches, and in others intermingled with one another. Theſe filaments clearly appeared to be agitated by an internal undulatory motion, like hollow tubes which contained ſome moving ſubſtance. He ſaw diſtinctly this appearance changed for that at fig. 32. Two of theſe filaments, which were joined longitudinally, gradually ſeparated from each other in the middle, alternately approaching and receding, like two tenſe cords fixed by the ends, and drawn aſunder in the middle. Theſe filaments were compoſed of globules that touched one another, and reſembled a chaplet of beads. After this, he obſerved the filaments ſwelled in ſeveral places, and perceived ſmall globular bodies iſſue from the ſwelled parts, which had a vibratory motion like a pendulum. Theſe ſmall bodies were attached to the filaments by ſmall threads, which gradually lengthened as the bodies moved. At laſt, the ſmall bodies detached themſelves entirely from the filaments, drawing after them the ſmall thread, which looked like a tail. When a drop of the ſeminal liquor was diluted, theſe ſmall bodies moved in all directions very briſkly. The ſeminal matter was at firſt too thick, but gradually became more fluid; and, in proportion as its fluidity increaſed, the filaments diſappeared, but the ſmall bodies became exceedingly numerous. Each of them had a long thread or tail attached to it, from which it evidently endeavoured to get free. Their progreſſive motion was extremely ſlow, during which they vibrated to the right and left; and at each vibration they had a rolling unſteady motion in a vertical direction.

At the end of two or three hours, the ſeminal matter becoming ſtill more fluid, a greater number of theſe moving bodies appeared. They were then more free of incumbrances; their tails were ſhorter; their progreſſive motion was more direct, and their horizontal motion greatly diminiſhed. In five or ſix hours, the ſeed had acquired almoſt all the fluidity it could acquire without being decompoſed. Moſt of the ſmall bodies were now diſengaged from their threads: their figure was oval; they moved forward with conſiderable quickneſs, and, by their irregular motions backward and forward, they had now more than ever the appearance of animals.

Thoſe that had tails adhering to them, ſeemed to have leſs vivacity than the others: and of thoſe that had no tails, ſome altered both their figure and their ſize. In twelve hours, the ſeed had depoſited at the bottom of the vial a kind of aſh-coloured gelatinous ſubſtance, and the fluid at top was almoſt as tranſparent as water. The little bodies, being now entirely freed from their threads, moved with great agility, and ſome of them turned round their centres. They alſo often changed their figures, from oval becoming round, and often breaking into ſmaller ones. Their activity always increaſed as their ſize diminiſhed. In twenty-four hours the ſeed had depoſited a greater quantity of gelatinous matter, which, being with ſome difficulty diluted in water, exhibited an appearance ſomewhat reſembling lace. In the clear ſeed itſelf, only a few ſmall bodies were now ſeen moving; next day, theſe were ſtill farther diminiſhed; and after this nothing was to be ſeen but globules, without the leaſt appearance of motion. All the above-mentioned appearances in the ſeed of man, are ſhown in the Plate, at fig. 33, 34, 35, 36, 37, and 38. Fig. 39 and 40 repreſent an appearance of the globules in another experiment, in which they arranged themſelves in troops, and paſſed very quickly over the field of the microſcope: in this experiment tey were found to proceed from a ſmall quantity of gelatinous mucilage, depoſited by the ſeed.

An objection has, however, been made to the exiſtence of animalcules in the ſeed, or in any other part of animal bodies, from the total excluſion of air, which is found ſo neceſſary to the life of larger animals. Many inſtances, however, have been obſerved of large animals being found in ſuch ſituations as they could not poſſibly have enjoyed the leaſt benefit from the air for a great number of years; and in this ſtate have not only lived, but lived much longer than they would otherwiſe have done.

In Toulon harbour, and the road, are found ſolid hard ſtones, and perfectly entire, containing, in different cells, ſecluded from all communication with the air, ſeveral living ſhell-fiſh, of an exquiſite taſte, called *dactyli*. To come at theſe fiſh, the ſtones are broken with mauls. Alſo, along the coaſt of Anconia, in the Adriatic, are ſtones uſually weighing about fifty pounds, and ſometimes even more; the outſide rugged, and eaſily broken, but the inſide ſo hard as to require a ſtrong arm and an iron maul to break them; within them, and in ſeparate niches, are found ſmall ſhell-fiſh, quite alive, and very palatable, called *ſolenes*, or *cappe lunghe*. Theſe facts are atteſted by Gaſſendi, Blondel, Mayol, the learned biſhop of Sulturara; and more particularly by Aldrovandus, a phyſician of Bologna. The two latter ſpeak of it as a common fact which they themſelves ſaw.

In the volume for the year 1719, of the Memoirs of the Academy of Sciences at Paris, is the following paſſage: "In the foot of an elm, of the bigneſs of a pretty corpulent

corpulent man, three or four feet above the root, and exactly in the centre, has been found a live toad, middle-ſized, but lean, and filling up the whole vacant ſpace: no ſooner was a paſſage opened by ſplitting the wood, that it ſcuttled away very haſtily: a more firm and ſound elm never grew; ſo that the toad cannot be ſuppoſed to have got into it. The egg or ſpawn whence it was formed, muſt, by ſome very ſingular accident, have been lodged in the tree at its firſt growth. There the creature had lived without air, feeding on the ſubſtance of the tree, and growing only as the tree grew. This is atteſted by Mr. Hubert, profeſſor of philoſophy at Caen."

The volume for the year 1731 has a ſimilar obſervation, expreſſed in theſe words: —"In 1719, we gave an account of a fact, which, though improbable, was well atteſted; that a toad had been found living and growing in the trunk of a middling elm, without any way for the creature to come out or to have got in. M. Seigne, of Nantes, lays before the academy a fact juſt of the very ſame nature, except that inſtead of an elm it was an oak, and larger than the elm, which ſtill heightens the wonder. He judges, by the time requiſite for the growth of the oak, that the toad muſt have ſubſiſted in it, without air or any adventitious aliment, during eighty or one hundred years. M. Seigne ſeems to have known nothing of the fact in 1719."

With the two foregoing may be claſſed a narrative of Ambroſe Paré, chief ſurgeon to Henry III. king of France, who, being a very ſenſible writer, relates the following fact, of which he was an eye-witneſs:—"Being (ſays he) at my ſeat, near the village of Meudon, and overlooking a quarry-man whom I had ſet to break ſome very large and hard ſtones; in the middle of one we found a huge toad, full of life, and without any viſible aperture by which it could get there. I began to wonder how it received birth, had grown and lived: but the labourer told me, it was not the firſt time he had met with a toad, and the like creatures, within huge blocks of ſtone, having no viſible opening or fiſſure."

Obſervations of living toads, found in very hard and entire ſtones, occur in ſeveral authors, particularly Baptiſt Fulgoſa, doge of Genoa, the famous phyſicians Agricola and Horſtius, and Lord Verulam; others give very ſpecious accounts of ſnakes, frogs, crabs, and lobſters, being found alive, incloſed within blocks of marble, rocks, and large ſtones.

An inſtance ſimilar to theſe, of the truth of which we have no reaſon to doubt, was obſerved in this country in the year 1773, where a large toad was found in the middle of a piece of coal, having not the leaſt viſible crack or fiſſure.

Thus much as to the faculty of ſome animals to live without air.

Upon the whole, though philoſophers are not yet able to diſcover how very minute creatures are produced; yet, that there really are animals much ſmaller than

what

what we can discern with our naked eye, seems to be indisputable. And the subject evidently requires the utmost attention of philosophers, as well as further improvements in the construction of microscopes, fully to investigate and explain it. —Most naturalists suppose another species or order of invisible animalcules, such as escape the cognizance even of the best microscopes, and give many probable conjectures in relation to them. Reason and analogy give some support to the existence of infinite imperceptible animalcules. The naked eye takes in from the elephant to the mite; but there commences a new order, reserved only for the microscope, which comprehends all those from the mite to those twenty-seven millions of times smaller; and this order cannot be yet said to be exhausted, if the microscope be not arrived at its last and highest perfection.

Animalcules are the cause of various disorders. The itch, from several experiments, is a disorder arising from the irritations of a species of insect found in the pustules of that disorder, (the Acarus exulcerans;) whence the communication of it by contact from one to another is easily conceived; as also the reason of the cure being effected by external applications. Many other cutaneous eruptions, often supposed to originate in the blood, are nothing more than settlements made by colonies of these invisible beings. A swarm of them light upon the skin, and, finding in its pores a comfortable habitation, soon produce a puncture, with scabs and irritation. But this is not the worst. Observation has long convinced me that a variety of internal complaints in the stomach, pancreas, lungs, liver, and intestines, are brought on by swallowing myriads of these, and other imperceptible living creatures, which inhabit raw vegetables and foul water; and, finding the heat and food of the stomach congenial to their growth, they become a new species, of an alarming size, and prey upon the vital parts, to the great detriment of the patient's health, and oftentimes at the expense of his life, before the malady can be known, or even suspected.

A patient of mine, a young man near eighteen years of age, had been a considerable time in a consumptive habit, and disordered in the stomach; and notwithstanding he had the advice of several eminent physicians, and had taken a variety of medicines, he never found the smallest alleviation of his pain. Upon enquiring into the nature of his food for some time before, he told me he came from a village near Bridport, in Dorsetshire, which abounds with water-cresses, and on these he had fed almost daily for some months previous to his coming to London. I gave him three emetics successively, with a view to cleanse the stomach from all slime, phlegm, and undigested food; and immediately after the last had operated, he took a strong dose, undiluted, of my Solar Tincture. In less than ten minutes it brought up an animal of the most hideous form, which at first appeared incapable of motion, being

ing overcome by the ſtrength of the medicine; but, on putting it into a baſon of warm water, it quickly recovered, and ſent forth a ſet of tentacles or claws, which, though greatly enlarged, and diverſely altered, ſoon convinced me it muſt originally have been a ſhoot from the freſh-water polypus; that, on leaving the parent animal, it had attached itſelf to a root or leaf of the water-creſſes which this young man unfortunately ſwallowed. And it appeared further, that theſe tentacles or claws had been ſo ſtrongly affixed to the bowels or coats of the ſtomach, as to have defied the power of all common remedies to remove them. The patient happily found immediate relief, and is now healthy and robuſt.

From ſeveral other patients, apparently in conſumptions, or afflicted with nauſea, or uncommon ſenſations in the ſtomach and bowels, I have brought away living animals that would terrify many people to look upon; and which muſt have come from the ſpawn, or eggs, of minute animals, taken in with the food. For this reaſon I would admoniſh all my readers to have the utmoſt care taken in the waſhing and cleanſing of ſallads, water-creſſes, and all raw vegetables; and particularly to guard againſt the long red worm which almoſt continually lies concealed in the very heart or centre of a head of celery. The ſame caution is neceſſary in eating all kinds of fruit; ſince nothing much more abounds with animalcules, and various living creatures. Cold raw water, particularly when ſtagnant, ought never to be drunk. It is ever the ſafeſt way to boil your water, before it be uſed in the compoſition of any kind of beverage, or even to drink alone.

I might here adduce many other inſtances of perſons having engendered living creatures in their bowels, by ſwallowing the eggs or ſpawn of the parent animal. A young man, ſervant to Lord Stawell, at Holt-park near Farnham, Surry, had eaten voraciouſly of water-creſſes. Some time afterwards he went into a decline, and complained of a continual ſenſation of pain at the pit of his ſtomach, which no medicine could remove. His lordſhip, having a value for the man, ſent him to town for the advice of the moſt able phyſicians; but ſtill to no purpoſe. He was in this ſtate ſent home to his friends, and given over as a loſt and incurable caſe. In this ſtage ſome ſtrong emetics were given him by a country apothecary; and he threw up, to the amazement of all the country round, an incredible number of ſmall tadpoles, which were evidently the production of ſpawn attached to the water-creſſes, eaten without care, and perhaps without waſhing. The patient recovered rapidly, and in leſs than a month was able to reſume his former avocation.

But a ſtill more extraordinary caſe happened in the county of Hants, in the year 1792, of a girl about fourteen years of age, who found a moſt uncommon ſenſation in her ſtomach and bowels, and could plainly feel and diſtinguiſh ſomething alive, and moving within her. The girl's defcription was for fome time treated as a chimera. At laft, however, fhe brought up a living toad! This unqueftionably muft have been taken into her ftomach in that ftate of the fpawn which is juft emerging to tadpoles, and was attributed to her eating water-creffes, which had long been a common food with her. Nothing could have faved her from poifon, but the creature having been bred and nourifhed up as it were in her own body, and had affimilated fo much with the nature thereof as to have thus long proved harmlefs. It is however certain, that, had it not been thus timely brought away, fhe muft very foon have died.

Animalcules are the moft common caufes of foul and rotten teeth. They attack the roots below the enamel, which they perforate, and in a fhort time form cruftations or fcales round the teeth, as hard as ftones; but which are nothing more than a congeries or cruftaceous fhell, which thefe little animals inhabit, and are probably formed of the fine particles falling from the teeth during their perforations, cemented together by a glutinous flimy matter iffuing from their own bodies, which are compofed of ringlets like a worm. Hence too we difcover the true caufe of fœtid or ftinking breath; for, when thefe little eels have made their way to the marrow, or internal part, of the tooth, the whole crown foon becomes rotten, and the marrow fends forth a putrid effluvia, fomewhat fimilar, but much more offenfive, than the animalcules in ftinking cheefe. Thefe circumftances feem to be but little known to the generality of dentifts and operators on the teeth; otherwife I am perfuaded their mode of practice would be widely different. Inftead of applying powders and dentifrices calculated to deftroy thefe little worms, they prepare fuch as multiply and nourifh them; of which any perfon may be convinced, who will take the pains to make a few fimple experiments. Let the roots of the teeth be fcraped, and the matter collected from them put into a few drops of any dentifrice or tooth-tincture, particularly of the aromatic kind. If viewed with a microfcope, it will be feen that the animalcules or eels found in this matter will move about with great celerity, as if delighted with the liquor; and, in proportion as it evaporates or dries away, the animals appear diffatisfied and become very uneafy.

Happening to have a patient who had a very bad fet of teeth, he fuffered me to make fome experiments upon them. I took off a few of the fcales, and emerged them in a fmall quantity of fpring water. It was quickly filled with the little eels or animalcules; but imparted no ill fmell. I examined the fcales with the microfcope, and found them full of pores, out of which thefe invifible animals were iffuing. I then took out as much of the foul matter from the cavities of his hollow teeth as I could conveniently get at; and, the moment I put it in the water, it became fœtid, and fent forth an offenfive fmell. Viewing it with the microfcope, the

animals

animals appeared in the ſame ſhape as the former, but quite opaque, and the inteſtinal canal much fuller, and more diſtended. I poured into the water a few drops of my Solar Tincture, and in leſs than five minutes all motion ceaſed, and they were quite dead. This induced me to perſuade the man to waſh his teeth and gums well with the Solar Tincture. He did ſo; and I then took off more of the ſcales, and collected all the matter I could from the rotten teeth; but very few living animalcules could be found therein, and the fœtid ſmell was conſiderably abated. He continued to waſh his mouth with the Tincture every other day for a week, and then uſed the following preparation: Chalk finely powdered, burnt hartſhorn levigated, Florentine orris-root, and myrrh, of each two drams; ſpirit of ſalt, ſix drops; the whole mixed into a fine powder. With this he rubbed his teeth every third day, with a ſoft bruſh, and in leſs than three weeks his black ſet of teeth became beautifully white; his breath ſweet; and his gums hard and firm; and he has ever ſince continued them ſo by the ſimple means above deſcribed. I am no dentiſt; what I have ſtated was matter of mere accident: but I would caution all my readers againſt too free a uſe of thoſe numerous powders and preparations continually recommended for the teeth. Inſtead of preſerving them, they have too often the direct contrary tendency, by deſtroying the whole ſet. The world however is fond of tooth-powders; and a moderate uſe of ſome of them may be of ſervice; but the daily rubbing does more injury to the teeth than wholly neglecting them. Fine levigated powders may be prudently uſed once in ſix or ſeven days, to keep the teeth white and ſplendid. But the generality of powders prepared for this purpoſe are much too hard, and wear away the gums, as well as the enamel of the teeth. Yet, notwithſtanding the danger of theſe preparations, it is a very deſirable thing to be enabled to preſerve the beauty and ſoundneſs of the teeth, from infancy to old age. For this purpoſe I would adviſe, that children ſhould be accuſtomed to waſh their teeth every morning with common water, and a ſoft tooth-bruſh; and after meals to rince the mouth, and rub the teeth with their fingers, where a bruſh cannot conveniently be uſed. Thoſe who conſtantly purſue this method, may expect to be free from rotten teeth, putrid ſlough, external diſcolourations, flaccid gums, and pain and looſeneſs of the teeth, ariſing from the animalcules which prey upon them. When teeth have been neglected for a time, and ſcales and cruſtations are affixed to them, theſe ſhould be removed by inſtruments, and the teeth and gums well waſhed with a powerful inſinuating tincture, to kill the animalcules; then the teeth ſhould be rubbed with a fine teſtaceous powder, in order to remove the diſcolouration. When the enamel is become perfectly clean, white, and poliſhed, even this fine powder ſhould be uſed very ſparingly, and at diſtant intervals. The colour

and

and ſweetneſs are only to be preſerved by frequent bruſhing and waſhing. The common trick of dentiſts is to give a waſh that inſtantly cleanſes and whitens the teeth, the ſudden effects of which are apt to ſurpriſe and pleaſe people; but their conſequences are very pernicious. All the acid ſpirits will do this; but they deſtroy the enamel, and rot the teeth. The ſafeſt liquid to take off black, green, and yellow, diſcolorations, is the following: Take plantane water, an ounce; honey of roſes, two drams; ſpirit of ſalt, ten drops: mix the whole together, and rub the teeth well with a linen rag dipped in the mixture, every day till they are perfectly white. If the roots of the teeth are very foul, inclined to grow rotten, and ſurrounded with ſcales and cruſtations, I ſhould by all means recommend them to be well waſhed with the Solar Tincture, which will ſtop them from further decay, entirely remove the ſcurvy from the gums, and perfectly ſweeten the breath.

The form, diſpoſition, and order, of the teeth, are admirable; and furniſh us with a noble inſtance of the wiſdom and goodneſs of the Creator; the foremoſt are weak, and far from the centre, as being only preparers to the reſt; the others, being to grind and maſticate, are accordingly ſtronger, and placed near the centre of motion. Their peculiar hardneſs is very remarkable, conſidering the tender ſubſtance they are formed of. Again, their various forms, in various creatures, are no leſs conſiderable, being all curiouſly adapted to the peculiar food and occaſions of the different ſpecies of animals. Thus in the rapacious, they are fitted for the catching, holding, and tearing, the prey; in herbaceous, for the gathering and comminution of vegetables; and in ſuch as have no teeth, as birds, the bill ſupplies the defect. Add to this, that the temporary defect of them is no leſs obſervable in ſome: that children, for inſtance, ſhould have none while they are not able to uſe them, but to hurt themſelves or the mother; and that, at the very age when they can take in the more ſubſtantial food, and live without the breaſt, and begin to need teeth for the help of ſpeech, that then their teeth ſhould begin to appear, and gradually grow, as they more and more ſtand in need of them; and that, when this firſt crop are worn out or decayed, they ſhould be ſucceeded by a new ſet, more firm and durable than the former. Nature, indeed, ſometimes deviates from the ordinary rule; according to the conſtruction of the elementary influx then operating, as is ſhown at large in my Illuſtration of the Occult Sciences; whence we have inſtances of perſons born with all their teeth, as Marcus Curius Dentatus and Cneius Papirius Carbo: others have only had one continued tooth, reaching the whole length of the jaw, as Pyrrhus king of Epirus, and Pruſias ſon of the king of Bithynia. A German phyſician, named Mentrelius, aſſures us, that he ſaw an old man at Cleves, in 1666, aged a hundred and twenty years, who had a new ſet of teeth only two years before, which were cut with great pain; and he alſo ſaw an Engliſhman at the Hague, who cut a new ſet of teeth in his hundred and eighteenth year. To the ſame purpoſe Dr. Slare mentions a relation, who had all his teeth at eighty years of age, and afterwards ſhed them, and had a new ſet all round. See Phil. Tranſ. Abr. vol. v. p. 353.

Of INSTINCT.

INSTINCT is an occult power or diſpoſition of the mind, by which animals are unerringly directed to do ſpontaneouſly whatever is neceſſary for the preſervation of the individual, or the continuation of the ſpecies. From this cauſe, all the actions of brutes, or inferior animals, are ſaid to be directed by *inſtinct*; but thoſe of man by *reaſon*. Philoſophers, however, have greatly differed in their opinions concerning this ſubject; and modern authors are extremely at a loſs where to draw the line. Some maintain that man is endowed with a greater number of inſtincts than any ſpecies of brutes whatever. Others inſiſt, that in human nature there is not any power or propenſity at all, which can properly be called inſtinctive. Some contend that brutes are guided wholly by an invariable inſtinct, without the ſmalleſt power of memory, or of any intellectual faculty; whilſt others inſiſt, that they poſſeſs a vegetative ſoul, directed by a certain inſtinct, capable of reaſon, of memory, and of experience.

With reſpect to man, nothing can be more apparent, than that, as being the microcoſm, or epitome of all created nature, he muſt of neceſſity partake of all its eſſential properties; of which *reaſon* and *inſtinct* rank amongſt the foremoſt.

Upon the ſlighteſt reflection, it will be obvious to every reader, that *reaſon* can never be exerciſed but from experience; conſequently, until man is arrived to a certain degree of maturity, he muſt be directed, in moſt of the propenſities of nature, by mere inſtinct. Thus an infant, a few moments after its birth, is directed by an inſtinctive impulſe for its preſervation, to ſeek the breaſt, and to ſuck it; and to the ſame cauſe, in the earlier ſtages of life, and in all ſavage uneducated countries, are to be attributed the firſt ſenſations or deſires of copulation, not from the pleaſures of enjoyment, for they are then unknown; but from an impulſive inſtinct, for the propagation of the ſpecies. It has been inſiſted, that the firſt commerce of the ſexes amongſt human beings is directed by reaſon; and the arguments aſſigned for it are theſe; that, as ſoon as the organs of generation in either ſex become ſufficiently ripe for the purpoſe intended by nature, they ſympathiſe with the ſenſes, and are affected with vibrations in the nerves, which riſe into pleaſure above the power of controul, and are heightened by youth, health, grateful aliment, imagination, ambition, ſympathy, and various other involuntary ſenſations, which, under ſuch circumſtances, pervade the whole ſyſtem. And as theſe organs are endued with a greater degree of ſenſibility than the other parts, both from their make, and the peculiar ſtructure and diſpoſition of their nerves; from the great diſtention of the muſcular ſyſtem and ſeed-veſſels in males; as well as from the extenſion of the clitoris and ſinuſes of the uterus in females, which never fail to take place about the time of puberty, the genital organs in both ſexes become ſo extremely irritable, that *reaſon*, being thereby *awakened*, directs and impels to that act, by which alone the human ſpecies can poſſibly be continued, and the works of an Omnipotent Creator carried on and conducted to the ends intended.

In the above ſtatement, I am perſuaded every rational mind will agree, that the word *inſtinct* ought to have been ſubſtituted where that of *reaſon* is uſed; becauſe in civilized ſocieties we are taught by *reaſon* to overcome thoſe inſtinctive paſſions, inſtead of having our reaſon *awakened* by them; but we too often find that theſe inſtinctive paſſions are proof againſt both reaſon and reſolution, even in the moſt virtuous families, in all countries, and in the beſt-regulated ſocieties. What ſhall we ſay then of that part of the human race which yet remains in a ſtate of nature, uncultivated, and unenlightened by any precepts of morality or ſcience? They are ſubject to the primary command, "*Increaſe and multiply;*" and they obey it. A couple of young ſavages go together *for the firſt time*, without any view to offspring, without any knowledge of the pleaſure to be derived from it, and without any determinate idea at all; and, as we ſee theſe means invariably purſued by all animals, as well rational as irrational, without experience, and without inſtruction, we muſt refer the mutual deſire of the ſexes to a much higher principle than can poſſibly ariſe from human motives; and that principle can be nothing but inſtinct. But as I ſhall have occaſion to ſpeak more at large on this ſubject under the article LOVE, when I come to treat of the affections and paſſions of the mind, and of the nature and perfections of Man, I ſhall in the interim proceed to ſhow, that the inferior animals are directed by inſtinct to performances of the moſt ſurpriſing kind; and are, within certain limits, endued with memory, and a reaſoning intellect.

The moſt remarkable inſtance of the power of inſtinct is obſerved in the conſtruction of a honey-comb. Bees, it is well known, conſtruct their combs with ſmall cells on both ſides, fit both for holding their ſtore of honey, and for rearing their young. There are only three poſſible figures of the cells, which can make them all equal and ſimilar, without any uſeleſs interſtices. Theſe are the equilateral triangle, the ſquare, and the regular hexagon. Of the three, the hexagon is the moſt

moſt proper, both for convenience and ſtrength. Bees, as if they knew this, make their cells regular hexagons. As the combs have cells on both ſides, the cells may either be exactly oppoſite, having partition againſt partition, or the bottom of a cell may reſt upon the partitions between the cells on the other ſide, which will ſerve as a buttreſs to ſtrengthen it. The laſt way is the beſt for ſtrength; accordingly the bottom of each cell reſts againſt the point where three partitions meet on the other ſide, which gives it all the ſtrength poſſible. The bottom of a cell may either be one plane, perpendicular to the ſide-partitions; or it may be compoſed of ſeveral planes, meeting in a ſolid angle in the middle point. It is only in one of theſe two ways that all the cells can be ſimilar without loſing room. And for the ſame intention, the planes, of which the bottom is compoſed, if there be more than one, muſt be three in number, and neither more nor fewer. It has been demonſtrated, that, by making the bottoms of the cells to conſiſt of three planes meeting in a point, there is a ſaving of material and labour no-way inconſiderable. The bees, as if acquainted with theſe principles of ſolid geometry, follow them moſt accurately; the bottom of each cell being compoſed of three planes, which make obtuſe angles with the ſide partitions and with one another, and meet in a point in the middle of the bottom; the three angles of this bottom being ſupported by three partitions on the other ſide of the comb, and the point of it by the common interſection of theſe three partitions. One inſtance more of the mathematical ſkill diſplayed in the ſtructure of a honey-comb deſerves to be mentioned. It is a curious mathematical problem, at what preciſe angle the three planes which compoſe the bottom of a cell ought to meet, in order to make the greateſt poſſible ſaving of material and labour. This is one of thoſe problems belonging to the higher parts of the mathematics, which are called problems of maxima and minima. The celebrated Maclaurin reſolved it by a fluxionary calculation, which is to be found in the Tranſactions of the Royal Society of London, and determined preciſely the angle required. Upon the moſt exact menſuration which the ſubject could admit, he afterwards found, that it is the very angle in which the three planes in the bottom of the cell of a honey-comb do actually meet. If a honey-comb were a work of human art, every man of common ſenſe would conclude, without heſitation, that he who invented the conſtruction muſt have underſtood the principles on which it was conſtructed. We need not ſay that bees know none of theſe things. They work moſt geometrically without any knowledge of geometry; ſomewhat like a child, who by turning the handle of an organ makes good muſic without any knowledge of muſic. The art is not in the child, but in him who made the organ. In like manner, when a bee makes its comb ſo geometrically,

cally, the geometry is not in the bee, but in that great Geometrician who made the bee, and made all things in number, weight, and meaſure. This places in a moſt ſtriking point of view the difference betwixt inſtinct and reaſon. There are no improvements made by man, but what we ſee carried ſtill further by ſucceeding generations; but in bees, and in all inferior animals, we ſee preciſely the ſame economy and contrivance now, in conſtructing their cells, building their neſts, laying up proviſions, &c. as at the beginning; and that in all ages, and in all generations, they have neither improved, nor departed from, that fixed ſyſtem aſſigned to them by nature, for their preſervation and guidance; whereas men, acting by reaſon and ſcience, improve from the labours and inventions of each other. Were we to attribute reaſon inſtead of inſtinct to bees, in the conſtruction of their combs, we ſhould at the ſame time admit them to be rational creatures, endued with thinking and reaſoning faculties far ſuperior to men; for the principle upon which the honey-comb is conſtructed, is founded on thoſe high departments of the mathematics, which were altogether unknown to the human race till the beginning of the preſent century, and which at this moment are beyond the comprehenſion of nine-tenths of mankind in the moſt enlightened nations on earth. Hence it is plain that the contrivance is not in the bees, but in the Creator of the bees, who directs them, and all brute creatures, to act by an inſtinct for their own immediate benefit, without knowing the principles upon which they act. And this is by no means contrary to reaſon; for we daily ſee men, working under the direction of others of ſuperior underſtanding, to effect purpoſes, and accompliſh ends, without having themſelves any idea of either; and, if we look through the endleſs variety of human avocations, we ſhall find that the greater part of mankind ſeem deſtined by God and nature to be governed in this way. But to proceed—

Caterpillars, when ſhaken off a tree in every direction, inſtantly turn round towards the trunk, and climb up, though they had never formerly been on the ſurface of the ground. This is a ſtriking inſtance of inſtinct. On the tree, and not upon the ground, the caterpillar finds its food. If therefore it did not turn and climb up the trunk, it would inevitably periſh. The ſolitary waſp digs holes in the ſand, in each of which ſhe depoſits an egg: ſhe collects a few ſmall green worms, which ſhe rolls up in a circular form, and fixes in the hole in ſuch a manner that they cannot move. When the waſp-worm is hatched, it is amply ſtored with the food which nature has deſtined for its ſupport. The green worms are devoured in ſucceſſion; and the number depoſited is exactly proportioned to the time neceſſary for the growth and transformation of the waſp-worm into a fly; then it iſſues from the hole, and is capable of procuring its own nouriſhment. This in-

ſtinct

moſt proper, both for convenience and ſtrength. Bees, as if they knew this, make their cells regular hexagons. As the combs have cells on both ſides, the cells may either be exactly oppoſite, having partition againſt partition, or the bottom of a cell may reſt upon the partitions between the cells on the other ſide, which will ſerve as a buttreſs to ſtrengthen it. The laſt way is the beſt for ſtrength; accordingly the bottom of each cell reſts againſt the point where three partitions meet on the other ſide, which gives it all the ſtrength poſſible. The bottom of a cell may either be one plane, perpendicular to the ſide-partitions; or it may be compoſed of ſeveral planes, meeting in a ſolid angle in the middle point. It is only in one of theſe two ways that all the cells can be ſimilar without loſing room. And for the ſame intention, the planes, of which the bottom is compoſed, if there be more than one, muſt be three in number, and neither more nor fewer. It has been demonſtrated, that, by making the bottoms of the cells to conſiſt of three planes meeting in a point, there is a ſaving of material and labour no-way inconſiderable. The bees, as if acquainted with theſe principles of ſolid geometry, follow them moſt accurately; the bottom of each cell being compoſed of three planes, which make obtuſe angles with the ſide partitions and with one another, and meet in a point in the middle of the bottom; the three angles of this bottom being ſupported by three partitions on the other ſide of the comb, and the point of it by the common interſection of theſe three partitions. One inſtance more of the mathematical ſkill diſplayed in the ſtructure of a honey-comb deſerves to be mentioned. It is a curious mathematical problem, at what preciſe angle the three planes which compoſe the bottom of a cell ought to meet, in order to make the greateſt poſſible ſaving of material and labour. This is one of thoſe problems belonging to the higher parts of the mathematics, which are called problems of maxima and minima. The celebrated Maclaurin reſolved it by a fluxionary calculation, which is to be found in the Tranſactions of the Royal Society of London, and determined preciſely the angle required. Upon the moſt exact menſuration which the ſubject could admit, he afterwards found, that it is the very angle in which the three planes in the bottom of the cell of a honey-comb do actually meet. If a honey-comb were a work of human art, every man of common ſenſe would conclude, without heſitation, that he who invented the conſtruction muſt have underſtood the principles on which it was conſtructed. We need not ſay that bees know none of theſe things. They work moſt geometrically without any knowledge of geometry; ſomewhat like a child, who by turning the handle of an organ makes good muſic without any knowledge of muſic. The art is not in the child, but in him who made the organ. In like manner, when a bee makes its comb ſo geometri-

cally,

cally, the geometry is not in the bee, but in that great Geometrician who made the bee, and made all things in number, weight, and meafure. This places in a moft ftriking point of view the difference betwixt inftinct and reafon. There are no improvements made by man, but what we fee carried ftill further by fucceeding generations; but in bees, and in all inferior animals, we fee precifely the fame economy and contrivance now, in conftructing their cells, building their nefts, laying up provifions, &c. as at the beginning; and that in all ages, and in all generations, they have neither improved, nor departed from, that fixed fyftem affigned to them by nature, for their prefervation and guidance; whereas men, acting by reafon and fcience, improve from the labours and inventions of each other. Were we to attribute reafon inftead of inftinct to bees, in the conftruction of their combs, we fhould at the fame time admit them to be rational creatures, endued with thinking and reafoning faculties far fuperior to men; for the principle upon which the honey-comb is conftructed, is founded on thofe high departments of the mathematics, which were altogether unknown to the human race till the beginning of the prefent century, and which at this moment are beyond the comprehenfion of nine-tenths of mankind in the moft enlightened nations on earth. Hence it is plain that the contrivance is not in the bees, but in the Creator of the bees, who directs them, and all brute creatures, to act by an inftinct for their own immediate benefit, without knowing the principles upon which they act. And this is by no means contrary to reafon; for we daily fee men, working under the direction of others of fuperior underftanding, to effect purpofes, and accomplifh ends, without having themfelves any idea of either; and, if we look through the endlefs variety of human avocations, we fhall find that the greater part of mankind feem deftined by God and nature to be governed in this way. But to proceed—

Caterpillars, when fhaken off a tree in every direction, inftantly turn round towards the trunk, and climb up, though they had never formerly been on the furface of the ground. This is a ftriking inftance of inftinct. On the tree, and not upon the ground, the caterpillar finds its food. If therefore it did not turn and climb up the trunk, it would inevitably perifh. The folitary wafp digs holes in the fand, in each of which fhe depofits an egg: fhe collects a few fmall green worms, which fhe rolls up in a circular form, and fixes in the hole in fuch a manner that they cannot move. When the wafp-worm is hatched, it is amply ftored with the food which nature has deftined for its fupport. The green worms are devoured in fucceffion; and the number depofited is exactly proportioned to the time neceffary for the growth and transformation of the wafp-worm into a fly; then it iffues from the hole, and is capable of procuring its own nourifhment. This in-ſeem to be earneſtly contending for her favours. From the time of her appearance till after the middle of ſummer, the neſts of the birds ſelected to receive her egg are to be found in great abundance; but, like the other migrating birds, ſhe does not begin to lay till ſome weeks after her arrival. It is on all hands allowed, that the cuckow does not hatch its own eggs. The hedge-ſparrow, the water-wagtail, the titlark, the red-breaſt, the yellow-hammer, the green-linnet, or the winchat, is generally the nurſe of the young cuckow. It may be ſuppoſed, that the female cuckow lays her egg in the abſence of the bird in whoſe neſt ſhe intends to depoſit; as it has been known, that, on ſight of one of theſe, a red-breaſt and its mate jointly attacked her on approaching the neſt, putting her to flight; and ſo effectually drove her away, that ſhe did not dare to return. Among the birds above mentioned, it generally ſelects the three firſt, but ſhows a much greater partiality to the hedge-ſparrow. This laſt commonly takes up four or five days in laying her eggs. During this time (generally after ſhe has laid one or two), the cuckow contrives to depoſit her egg among the reſt, leaving the future care of it entirely to the hedge-ſparrow. When the hedge-ſparrow has ſat her uſual time, and diſengaged the young cuckow and ſome of her own offspring from the ſhell, her own young ones, and any of her eggs that remain unhatched, are ſoon turned out, the young cuckow remaining poſſeſſor of the neſt, and ſole object of her future care. The young birds are not previouſly killed, nor are the eggs demoliſhed; but all are left to periſh together, either entangled about the buſh which contains the neſt, or lying on the ground under it. The early fate of the young hedge-ſparrow (ſays Mr. Jenner, who made theſe experiments) is a circumſtance that has been noticed by others, but attributed to wrong cauſes; but the true cauſe we ſhall preſently explain. A variety of conjectures have been formed upon it. A cuckow laid her egg in a water-wagtail's neſt in the thatch of an old cottage. The wagtail ſat her uſual time, and then hatched all the eggs but one; which, with all the young ones except the cuckow, was turned out of the neſt. The young birds, conſiſting of five, were found upon the rafter that projected from under the thatch, and with them was an egg not in the leaſt injured. The cuckow was reared by the wagtails till it was nearly capable of flying, when it was killed by an accident.

A hedge-ſparrow built her neſt in a hawthorn-buſh in a timber-yard. After ſhe had laid two eggs, a cuckow dropped in a third. The ſparrow continued laying as if nothing had happened, till ſhe had laid five, her uſual number, and then ſat. On inſpecting the neſt, June 20, 1786, (ſays Mr. Jenner,) I found that the bird had hatched this morning, and that every thing but the young cuckow was thrown out. Under the neſt I found one of the young hedge-ſparrows dead, and one egg by the ſide of the neſt entangled with the coarſe woody materials that formed its out-

ſide

fide covering. On examining the egg, I found one end of the fhell a little cracked, and could fee that the fparrow it yet contained was alive. It was then reftored to the neft, but in a few minutes was thrown out. The egg, being again fufpended by the outfide of the neft, was faved a fecond time from breaking. To fee what would happen if the cuckow was removed, I took out the cuckow, and placed the egg containing the hedge-fparrow in the neft in its ftead. The old birds, during this time, flew about the fpot, fhowing figns of great anxiety; but, when I withdrew, they quickly came to the neft again. On looking into it in a quarter of an hour afterwards, I found the young one completely hatched, warm, and lively. The hedge-fparrows were fuffered to remain undifturbed with their new charge for three hours (during which time they paid every attention to it), when the cuckow was again put into the neft; and, on examining it again in a few minutes, I found the young fparrow was tumbled out. It was a fecond time reftored, but again experienced the fame fate. From thefe experiments, and fuppofing, from the feeble appearance of the young cuckow juft difengaged from the fhell, that it was utterly incapable of difplacing either the egg or the young fparrows, I was induced to believe that the old fparrows were the only agents in this feemingly-unnatural bufinefs. But I afterwards clearly perceived the caufe of this ftrange phenomenon, by difcovering the young cuckow in the act of difplacing its fellow-neftlings, as the following relation will fully evince. June 18, 1787, I examined the neft of a hedge-fparrow, which then contained a cuckow's and three hedge-fparrow's eggs. On infpecting it the day following, I found the bird had hatched, but that the neft now contained only a young cuckow and one young hedge-fparrow. The neft was placed fo near the extremity of a hedge, that I could diftinctly fee what was going forward in it; and, to my aftonifhment, faw the young cuckow, though fo newly hatched, in the act of turning out the young hedge-fparrow. The mode of accomplifhing this was very curious. The little animal, with the affiftance of its rump and wings, contrived to get the bird upon its back; and making a lodgement for the burden by elevating its elbows, clambered backward with it up the fide of the neft, till it reached the top; where refting for a moment, it threw off its load with a jerk, and quite difengaged it from the neft. It remained in this fituation a fhort time, feeling about with the extremities of its wings, as if to be convinced whether the bufinefs was properly executed, and then dropped into the neft again. With thefe (the extremities of its wings) I have often feen it examine, as it were, an egg and neftling before it began its operations; and the nice fenfibility which thefe parts appeared to poffefs, feemed fufficiently to compenfate the want of fight, which as yet it was deftitute of. I afterwards put in an egg; and this, by a fimilar procefs, was conveyed to the edge of the neft and thrown out. Thefe experiments ment I have ſince repeated ſeveral times in different neſts, and have always found the young cuckow diſpoſed to act in the ſame manner. In climbing up the neſt, it ſometimes drops its burden, and thus is foiled in its endeavours; but, after a little reſpite, the work is reſumed, and goes on almoſt inceſſantly till it is effected. It is wonderful to ſee the extraordinary exertions of the young cuckow, when it is two or three days old, if a bird be put into the neſt with it that is too weighty for it to lift out. In this ſtate it ſeems ever reſtleſs and uneaſy. The ſingularity of its ſhape is well adapted to theſe purpoſes; for, different from other newly-hatched birds, its back, from the ſcapulæ downward, is very broad, with a conſiderable depreſſion in the middle. This depreſſion ſeems formed by nature for the deſign of giving a more ſecure lodgement to the egg of the hedge-ſparrow, or its young one, when the young cuckow is employed in removing either of them from the neſt. When it is about twelve days old, this cavity is quite filled up, and then the back aſſumes the ſhape of neſtling birds in general.

It appears a little extraordinary that two cuckows eggs ſhould ever be depoſited in the ſame neſt, as the young one produced from one of them muſt inevitably periſh; yet two inſtances of this kind fell under our author's obſervation, one of which he thus relates: Two cuckows and a hedge-ſparrow were hatched in the ſame neſt this morning, (June 27, 1787;) one hedge-ſparrow's egg remained unhatched. In a few hours after, a conteſt began between the cuckows for the poſſeſſion of the neſt, which continued undetermined till the next afternoon, when one of them, which was ſomewhat ſuperior in ſize, turned out the other, together with the young hedge-ſparrow and the unhatched egg. This conteſt was very remarkable. The combatants alternately appeared to have the advantage, as each carried the other ſeveral times nearly to the top of the neſt, and then ſunk down again, oppreſſed by the weight of its burden; till at length, after various efforts, the ſtrongeſt prevailed, and was afterwards brought up by the hedge-ſparrows.

But the principal circumſtance that has agitated the mind of naturaliſts reſpecting the cuckow is, why, like other birds, it ſhould not build a neſt, incubate its eggs, and rear its own young? The moſt probable ſuggeſtion is, the ſhort reſidence this bird is allowed to make in the country where it is deſtined to propagate its ſpecies; and the call that nature has upon it, during that reſidence, to produce a numerous progeny. The cuckow's firſt appearance here is about the middle of April, commonly on the 17th. Its egg is not ready for incubation till ſome weeks after its arrival, ſeldom before the middle of May. A fortnight is taken up by the ſitting bird in hatching the egg. The young bird generally continues three weeks in the neſt before it flies, and the foſter-parents feed it more than five weeks after this period; ſo that, if a cuckow ſhould be ready with an egg much ſooner than the time pointed out, not a ſingle neſtling, even one of the earlieſt, would be fit to provide for itſelf before its parent would be inſtinctively directed to ſeek a new reſidence, and be thus compelled to abandon its young one; for old cuckows take their final leave of this country the firſt week in July. Among the many peculiarities of the young cuckow, there is one that ſhows itſelf very early. Long before it leaves the neſt, it frequently, when irritated, aſſumes the manner of a bird of prey, looks ferocious, throws itſelf back, and pecks at any thing preſented to it, with great vehemence, often at the ſame time making a chuckling noiſe like a young hawk. Hence probably the vulgar opinion, that this bird changes into a hawk, and devours its nurſe on quitting its neſt; whence the French proverb, *Ingrat comme un coucou,* "As ungrateful as a cuckow." Sometimes, when diſturbed in a ſmaller degree, it makes a kind of hiſſing noiſe, accompanied with a heaving motion of the whole body. From what has been ſaid, it becomes evident, that the ſame inſtinctive impulſe which directs the cuckow to depoſit her eggs in the neſts of other birds, directs her young one to throw out the eggs and young of the owner of the neſt. The ſcheme of nature would be incomplete without it; for it would be extremely difficult, if not impoſſible, for the little birds deſtined to find ſuccour for the cuckow, to find it alſo for their own young ones after a certain period; nor would there be room for them to inhabit the neſt. Cuckows may be, and often are, brought up tame, ſo as to become familiar. They will eat in this ſtate bread and milk, fruits, inſects, eggs, and fleſh, either cooked or raw; but in a ſtate of nature, they chiefly live on caterpillars. When fat, they are ſaid to be as good eating as a land-rail: the French and Italians eat them to this day. The ancient Romans admired them greatly as food: Pliny ſays that there is no bird which can be compared to them for delicacy. In migrating, the major part of theſe birds are ſuppoſed to go into Africa; ſince they are obſerved to viſit the iſland of Malta twice in a year, in their paſſage backwards and forwards, as is ſuppoſed, to that part of the world.

The inſtinct which has been diſcovered in ants, beavers, &c. is too well known and admired, to need any mention in this place: and we ſee in a great variety of birds, inſects, and quadrupeds, a ſimilar economy in laying up ſtores of proviſion in time of plenty, that they might have acceſs to it in time of need. The common daw has a peculiar knack of this ſort; and, in houſes where they have been brought up tame, have frequently been known to hide, with their meat, money, rings, ſeals, lockets, and other ſmall trinkets, thereby occaſioning injurious ſuſpicions of theſt in ſervants or others, who are perfectly innocent.

We have a remarkable anecdote given by the Rev. Mr. Robinſon of Ouſby in Weſtmoreland, relative to an inſtinct in crows, by which they are made the natural

tural planters of all ſorts of wood and trees. They diſſeminate the kernels upon the earth, which like nurſeries bring them forth till they grow up to their natural ſtrength and perfection. He ſays, "About twenty-five years ago, coming from Roſecaſtle early in the morning, I obſerved a great number of crows very buſy at their work upon a declining ground of a moſſy ſurface; I went out of my way on purpoſe to view their labour, and I found they were planting a grove of oaks. The manner of their planting was thus: they firſt made little holes in the earth with their bills, going about and about till the hole was deep enough; and then they dropped in the acorn, and covered it with earth and moſs. The ſeaſon was at the latter end of autumn, when all ſeeds are full ripe." Mr. Robinſon ſeems to think that Providence had given the crows this inſtinct ſolely for the propagation of trees; but I imagine it was given them principally for their own preſervation, by hiding proviſion in time of plenty, in order to ſupply them in a time of ſcarcity; ſo that ſuch an inſtinct in theſe birds may anſwer a double purpoſe; both their own ſupport in times of need, and the propagation of the trees they plant: for, wherever they hide a great number of nuts or grains in the earth, we cannot ſuppoſe they find them all again; but that as many will remain in the plot of ground they make uſe of, as can well grow by one another.

A wonderful ſpirit of ſociality in the brute creation, independent of ſexual attachment, has been frequently remarked. Many horſes, though quiet with company, will not ſtay one minute in a field by themſelves: the ſtrongeſt fences cannot reſtrain them. A horſe has been known to leap out at a ſtable-window through which dung was thrown, after company; and yet in other reſpects to be remarkably quiet. Oxen and cows will not fatten by themſelves; but will neglect the fineſt paſture that is not recommended by ſociety. It would be needleſs to inſtance in ſheep, which conſtantly flock together. But this propenſity ſeems not to be confined to animals of the ſame claſs or ſpecies. Even great diſparity of kind and ſize does not always prevent ſocial advances and mutual fellowſhip. Of this the following remarkable inſtance is given: A gentleman who kept but one horſe, happened alſo on a time to have but one ſolitary hen: theſe two incongruous animals ſpent much of their time together in a lonely orchard, where they ſaw no creature but each other: by degrees an apparent regard began to take place between theſe two ſequeſtered individuals: the fowl would approach the quadruped with notes of complacency, rubbing herſelf gently againſt his legs; while the horſe would look down with ſatisfaction, and move with the greateſt caution and circumſpection, leſt he ſhould trample on his diminutive companion; and thus by mutual good offices each ſeemed to conſole the vacant hours of the other.

In

In the Gentleman's Magazine for March 1788, we have the following anecdotes of a raven. The raven alluded to "lives, or did live three years ſince, at the Red Lion at Hungerford; his name, I think, is Rafe. You muſt know then, that, coming into that inn, my chaiſe ran over and bruiſed the leg of my Newfoundland dog; and, while we were examining the injury done to the dog's foot, Rafe was evidently a concerned ſpectator; for, the minute the dog was tied up under the manger with my horſe, Rafe not only viſited, but fetched him bones, and attended upon him with particular and repeated marks of kindneſs. The bird's notice of the dog was ſo marked, that I obſerved it to the hoſtler; for I had not heard a word before of the hiſtory of this benevolent creature. John then told me, that he had been bred from his pin-feather in intimacy with a dog; that the affection between them was mutual; and that all the neighbourhood had often been witneſſes of the innumerable acts of kindneſs they had conferred upon each other. Rafe's poor dog, after a while, unfortunately broke his leg; and during the long time he was confined, Rafe waited upon him conſtantly, carried him his proviſions daily, and never ſcarcely left him alone! One night by accident the hoſtler had ſhut the ſtable door, and Rafe was deprived of the company of his friend the whole night; but the hoſtler found in the morning the bottom of the door ſo pecked away, that, had it not been opened, Rafe would in another hour have made his own entrance-port. I then enquired of my landlady, and heard what I have related confirmed by her, with ſeveral other ſingular traits of the kindneſſes this bird ſhows to all dogs in general, but particularly to maimed or wounded ones. I hope and believe, however, the bird is ſtill living; and the traveller will find I have not over-rated his merit."

To theſe inſtances of attachment between incongruous animals from a ſpirit of ſociality or the feelings of ſympathy, may be added the following inſtance of fondneſs from a different motive, recounted by Mr. White, in his Hiſtory of Selborne: "My friend had a little helpleſs leveret brought to him, which the ſervants fed with milk in a ſpoon; and about the ſame time his cat kittened, and the young were diſpatched and buried. The hare was ſoon loſt, and ſuppoſed to be gone the way of moſt foundlings, or killed by ſome dog or cat. However, in a fortnight after, as the maſter was ſitting in his garden in the duſk of the evening, he obſerved his cat, with tail erect, trotting towards him, and calling with little ſhort inward notes of complacency, ſuch as they uſe towards their kittens, and ſomething gamboling after, which proved to be the leveret, which the cat had ſupported with her milk, and continued to ſupport with great affection. Thus was a graminivorous animal nurtured by a carnivorous and predacious one! Why ſo cruel and ſanguinary a beaſt as a cat, of the ferocious genus of felis, the murium leo, as Linnæus calls it, ſhould be affected with any tenderneſs towards an animal which is its natural prey, is not ſo eaſy to determine. This ſtrange affection probably was occaſioned by that deſiderium, thoſe tender maternal feelings, which the loſs of her kittens had awakened in her breaſt; and by the complacency and eaſe ſhe derived to herſelf from procuring her teats to be drawn, which were too much diſtended with milk, till from habit ſhe became as much delighted with this foundling as if it had been her real offspring. This incident is no bad ſolution of that ſtrange circumſtance which grave hiſtorians as well as poets aſſert, of expoſed children being ſometimes nurtured by female wild beaſts that had probably loſt their young. For it is not one whit more marvellous that Romulus and Remus, in their infant ſtate, ſhould be nurſed by a ſhe-wolf, than that a poor little ſucking leveret ſhould be foſtered and cheriſhed by a bloody grimalkin.

That brute animals poſſeſs reflection and ſentiment, and are ſuſceptible of the kindly as well as the iraſcible paſſions, independently of ſexual attachment and natural affection; and that they have a great ſhare of fidelity, of pride, and even a ſenſe of glory; may be demonſtrated from the elephant, the horſe, and the dog. Elephants, even in a ſavage ſtate, are peaceable and gentle creatures. They never uſe their weapons but in defence of themſelves or companions. Their ſocial diſpoſitions are ſo ſtrong, that they are ſeldom found alone, but march always in large troops: the oldeſt and moſt experienced lead the van; the younger or lame ones keep in the middle; and thoſe of a ſecond rate, as to age, walk in the rear. The females carry their young on their tuſks, embracing them at the ſame time with their trunk. They ſeldom march in this regular order but when they reckon the journey dangerous, ſuch as an expedition to cultivated lands, where they expect to meet with reſiſtance. On other occaſions they are leſs cautious; ſome of them falling behind or ſeparating from the reſt, but ſeldom ſo far as to be without the reach of aſſiſtance by alarming and aſſembling their companions. It is dangerous to offer them the leaſt injury; for they run ſtraight upon the offender; and, although the weight of their body be great, their ſteps are ſo large, that they eaſily outrun the ſwifteſt man, whom they either pierce with their tuſks, or ſeize with their trunk, dart him in the air like a ſtone, and then trample him under their feet. But they never attack any perſon unleſs when provoked. However, as they are extremely ſenſible and delicate with regard to injuries, it is always prudent to keep out of their way. Travellers who frequent thoſe countries kindle large fires, and beat drums, during the night, in order to prevent their approach. After being once attacked by men, or falling into any ambuſh, they are ſaid never to forget the

the injury, but ſearch for every opportunity of getting revenge. As they are endowed perhaps with a more exquiſite ſenſation of ſmell than any other animal, owing to the great extent of their noſe, they can ſcent a man at a very great diſtance, and trace him by his footſteps.

The elephant, when tamed, is the moſt friendly and obedient of all animals; he is entirely attached to the perſon who feeds and takes care of him. In a ſhort time he underſtands ſigns, and the ſound of his maſter's voice. He diſtinguiſhes the language of paſſion, of command, of ſatisfaction; and acts accordingly. He receives his orders with attention; and executes them with prudence and alacrity, but without precipitation. He eaſily learns to bow his knees and lower his body, for the convenience of thoſe who mount him. He careſſes his friends with his trunk. He lifts burdens with his trunk, and aſſiſts thoſe who are loading him in laying them on his back. He delights in ſhining harneſs and trappings. When yoked in a cart or waggon, he pulls equally and cheerfully, unleſs he be abuſed by injudicious chaſtiſement. His guide is generally mounted on his neck, with a ſmall rod of iron, ſharp at the point, in his hand; he directs his motion by pricking him on the ears and head; but, for the moſt part, a word is ſufficient. A tame elephant will do more labour than ſix horſes; but then he requires a proportional quantity of food. They are the principal beaſts of burden in many parts of Africa and the Eaſt Indies. They carry ſacks and bundles of all kinds on their neck, back, and tuſks. They never loſe or damage any thing committed to their care: they will ſtand on the edge of a river, take bundles off their necks and tuſks, lay them carefully in a boat, whenever they are deſired, and try with their trunk whether they are properly ſituated; if they be loaded with caſks, they go in queſt of ſtones to prop them and prevent them from rolling. The elephant is not only the moſt tractable, but the moſt intelligent, of animals; ſenſible of benefits, and reſentful of injuries. In India, they were once employed in the launching of ſhips: one was directed to force a very large veſſel into the water; the work proved ſuperior to his ſtrength; his maſter, with a ſarcaſtic tone, bade the keeper take away this lazy beaſt and bring another: the poor animal inſtantly repeated his efforts, fractured his ſkull, and died on the ſpot. In Delhi, an elephant, paſſing along the ſtreets, put his trunk into a tailor's ſhop where ſeveral people were at work; one of them pricked the end with his needle: the beaſt paſſed on; but at the next dirty puddle filled his trunk with water, returned to the ſhop, and, ſpirting every drop among the people who had offended him, ſpoiled their work. An elephant in Adſmeer, which often paſſed through the bazar, or market, as he went by a certain herb-woman, always received from her a mouthful of greens: at

6 length

length he was ſeized with one of his periodical fits of rage, broke his fetters, and, running through the market, put the crowd to flight; among others, this woman, who in haſte forgot a little child ſhe had brought with her. The animal, recollecting the ſpot where his benefactreſs was wont to ſit, took up the infant gently in his trunk, and placed it in ſafety on a ſtall before a neighbouring houſe. Another in his madneſs, killed his cornac, or governor: the wife, ſeeing the misfortune, took her two children and flung them before the elephant, ſaying, "Now you have deſtroyed their father, you may as well put an end to their lives and mine." It inſtantly ſtopped, relented, took the greateſt of the children, placed him on its neck, adopted him for his cornac, and never afterwards would permit any body elſe to mount it. A ſoldier at Pondicherry, who was accuſtomed, whenever he received the portion that came to his ſhare, to carry a certain quantity of it to one of theſe animals, having one day drank rather too freely, and finding himſelf purſued by the guards, who were going to take him to priſon, took refuge under the elephant's body, and fell aſleep. In vain did the guard try to force him from this aſylum, as the elephant protected him with his trunk. The next morning the ſoldier, recovering from his drunken fit, ſhuddered with horror to find himſelf ſtretched under the belly of this huge animal. The elephant, who without doubt perceived the man's embarraſſment, careſſed him with his trunk, in order to inſpire him with courage and make him underſtand that he might now depart in ſafety. A painter was deſirous of drawing the elephant which was kept in the menagerie at Verſailles in an uncommon attitude, which was that of holding his trunk raiſed up in the air with his mouth open. The painter's boy, in order to keep the animal in this poſture, threw fruit into his mouth; but as the lad frequently deceived him, and made an offer only of throwing him the fruit, he grew angry; and, as if he had known that the painter's intention of drawing him was the cauſe of the affront that was offered him, inſtead of revenging himſelf on the lad, he turned his reſentment on the maſter, and, taking up a quantity of water in his trunk, threw it on the paper on which the painter was drawing, and ſpoiled it. At the Cape of Good Hope, it is cuſtomary to kill thoſe animals for the ſake of their teeth, by the chaſe. Three horſemen, well mounted and armed with lances, attack the elephant alternately, each relieving the other as they ſee their companion preſſed, till the beaſt is ſubdued. Three Dutchmen (brothers), who had made large fortunes by this buſineſs, determined to retire to Europe, and enjoy the fruits of their labours; but reſolved, before they went, to have a laſt chace by way of amuſement: they met with their game, and began the attack in the uſual manner; but unfortunately one of their horſes

fell

fell down and flung its rider: the enraged animal inftantly feized the unhappy man with its trunk, flung him up to a vaft height in the air, and received him on one of his tufks; then turning towards the two other brethren, as if it were with an afpect of revenge and infult, held out to them the impaled wretch writhing on the bloody tooth.

When the elephant is properly managed, he lives very long even in a ftate of flavery and labour: that fome have lived in this ftate an hundred and thirty years, is pretty well authenticated. In a natural ftate they often exceed two hundred years, and propagate their fpecies till they are an hundred and twenty: it is thirty years before they come to their full growth. The mode of connection between the male and female is now afcertained beyond the poffibility of doubt; as Mr. Buller, Lieut. Hawkins, and many others, faw a male copulate with a female, after they were fecured, in the Eaft-Indies, in a manner exactly fimilar to the conjunction of the horfe with a mare. This fact entirely overturns what has been fo often related concerning the fuppofed delicacy of this ufeful animal, and a variety of other hypothefes, which are equally void of foundation. The time an elephant goes with young, has been afcertained to be fomewhat lefs than two years, as an elephant brought forth a young one twenty-one months and three days after fhe was taken. She was obferved to be with young in April or May 1788, and fhe was only taken in January preceding; fo that it is very likely fhe muft have had connection with the male fome months before fhe was fecured, otherwife they could not difcover that fhe was with young, as a fœtus of lefs than three months cannot well be fuppofed to make any alteration in the fize or fhape of fo large an animal. The young one, a male, was produced October 16, 1789, and appeared in every refpect to have arrived at its full time. He was thirty-five inches high at his birth, and grew four inches in nearly as many months. Elephants are always meafured at the fhoulder; for the arch or curve of the back, of young ones particularly, is confiderably higher than any other part, and it is a fure fign of old age whenever this curve is found flattened or confiderably depreffed, after an elephant has once attained its full growth. The young elephants fuck conftantly with their mouths, and never with their trunks, as Buffon has afferted; a conclufion he made merely from conjecture, and the great and various ufes to which the trunk is adapted and applied by every elephant.—The approach of the rutting-feafon is eafily known; for fome days before it happens, an oily liquor flows from a fmall hole on each fide of the head. The domeftic female on thefe occafions fometimes makes her efcape, and joins the wild males in the woods. Some days afterwards her cornac goes in queft of her, and calls her by her name till fhe comes. She fubmits to him with complacence, and allows herfelf to be conducted home, and fhut ſhut up in the ſtable. They bring forth but one at a time, though the female has two dugs, one on each ſide the breaſt. The young one, as ſoon as it comes into the world, is as large as a wild boar, and is furniſhed with teeth: however, the large tuſks do not make their appearance till ſome time after, and at the age of ſix months they are ſeveral inches long. Elephants of this age are as large as an ox when in a natural ſtate.

The intrepidity and ſagacity of the HORSE have been regarded with admiration by all ranks of men, and in all ages of the world. Even in a domeſtic ſtate he is bold and fiery; and, equally undaunted as his maſter, faces danger and death with ardour and magnanimity. He delights in the noiſe and tumult of arms, and ſeems to feel the glory of victory: he exults in the chace; his eyes ſparkle with emulation in the courſe. But, though bold and intrepid, he is docile and tractable: he knows how to govern and check the natural vivacity and fire of his temper. He not only yields to the hand, but ſeems to conſult the inclination of his rider. Conſtantly obedient to the impreſſions he receives, his motions are entirely regulated by the will of his maſter. He in ſome meaſure reſigns his very exiſtence to the pleaſure of man. He delivers up his whole powers, he reſerves nothing; he will rather die than diſobey. Who could endure to ſee a character ſo noble, abuſed? who could be guilty of ſuch groſs barbarity? none but wretches moſt relentleſs and unfeeling! We need go no further than the horſe, to prove how ſtrongly nature has endowed brute animals with memory; for whatever roads, paſtures, inns, or ſtables, a horſe has been accuſtomed to, though removed for years to a diſtant part of the country, he never forgets them; but, if ever he returns, or paſſes by them again, he gives evident tokens that he has been accuſtomed to the place. The celebrated equeſtrians, Hughes and Aſtley, could, I doubt not, furniſh a thouſand curious anecdotes to illuſtrate this fact; but the following, I think, being incontrovertible, will be ſufficient for the purpoſe.

Mr. James George, a gentleman of Southampton, in the county of Hants, loſt his horſe from off the common on which he had been accuſtomed to be turned out. About twelve months after, the horſe was ſeen and recognized, grazing on a common near Burſledon. Soon as this was made known, Mr. George ſent his ſervant with a bridle, and ordered the horſe to be caught, and brought home. In a few days after, a Mr. Langtree of Burſledon came to Mr. George, to demand the horſe, inſiſting it was his property, and had been ſo for years. The horſe, however, being poſitively withheld, Mr. Langtree brought an action for his recovery, which was tried at Wincheſter aſſiſes, in March 1789. The plaintiff, on the poſitive affirmations of ſix or ſeven witneſſes, traced the ſale of the horſe from one dealer's hands to another, as far back as the year 1784, when he was purchased of one farmer Moses, in the county of Sussex, who bred him; and the horse was identified by them all. On the part of the defendant, the most positive evidence was adduced, to show that this horse was not the horse sworn to have been so bought and sold by the witnesses on the other side, but was, and had been, the real property of Mr. George, from the year 1786. In this situation, with the horse equally sworn to and identified by the witnesses on both sides, who had all been to inspect him but the day before the trial, the judge by cross-examining the last witness, discovered, that, when the defendant's servant brought the horse back, he turned him into his old pasturage on Southampton common, where he had not been many minutes, before he set off, of his own accord, towards Southampton; and in his way crossed a number of lanes, and turnings, passed by eight or ten stables, until he came to the stable of Mr. George, where he instantly stopped, and neighed at the stable-door, as much as to say he was come back, and begged to be taken in. This circumstance decided the verdict. The learned judge remarked, that there could be no collusion in the evidence derived from the *memory of the horse;* and directed the jury to find for the defendant.

This character, though natural to the animal, is improved by habit and education. His education commences with the loss of liberty, and is finished by constraint. In the vast deserts of America, they roam at large without any restraint. M. de Salle relates, that he saw, in the year 1685, horses feeding in the meadows of North America, near the bay of St. Louis, which were so ferocious that nobody durst come near them. Oexmelin says, that he has seen large troops of them in St. Domingo running in the valleys; that, when any person approached, they all stopped; and one of them would advance till within a certain distance, then snort with his nose, take to his heels, and the whole troop after him. These relations sufficiently prove, that the horse, when at full liberty, has no inclination to associate with mankind: that all the softness and ductility of his temper proceed entirely from the culture and polish he receives in his domestic education, which in some measure commences as soon as he is brought forth.—The horse has not only a grandeur in his general appearance, but there is the greatest symmetry and proportion in the different parts of his body. The regularity and proportion of the different parts of the head give him an air of lightness, which is well supported by the strength and beauty of his chest. He erects his head, as if willing to exalt himself above the condition of other quadrupeds; his eyes are open and lively; his ears are handsome, and of a proper height; his mane adorns his neck, and gives him the appearance of strength and boldness. At the age of two years, or two years and an half, the horse is in a condition to propagate; and the mare, like most other females, is ready to receive him still sooner. But the foals produced

duced by ſuch early embraces are generally ill-made and weakly. The horſe ſhould never be admitted to the mare till he is four or four and a half; this is only meant with regard to draught-horſes. Fine horſes ſhould not be admitted to the mare before they are ſix years old; and Spaniſh ſtallions not till ſeven. The mares are generally in ſeaſon from the beginning of April to the end of June; but their chief ardour for the horſe laſts only about fifteen or twenty days, and this critical ſeaſon ſhould always be embraced. The ſtallion ought to be ſound, well-made, vigorous, and of a good breed. For fine ſaddle-horſes, foreign ſtallions, as Arabians, Turks, Barbs, and Andaluſians, are preferable to all others. Next to theſe, Britiſh ſtallions are the beſt; becauſe they originally ſprang from thoſe above-mentioned, and are very little degenerated. The ſtallions of Italy, and eſpecially the Neapolitans, are very good. The beſt ſtallions for draught or carriage horſes, are thoſe of Naples, Denmark, Holſtein, and Friezeland. The ſtallions for ſaddle-horſes ſhould be from fourteen to fifteen hands high, and for draught-horſes at leaſt fifteen hands. Neither ought the colour of ſtallions to be overlooked; as a fine black, grey, bay, ſorrel, cheſnut, &c. Beſides theſe external qualities, a ſtallion ought to have courage, tractability, ſpirit, agility, a ſenſible mouth, ſure limbs, &c. Theſe precautions in the choice of a ſtallion are the more neceſſary, becauſe he has been found by experience to communicate to his offspring almoſt all his good or bad qualities, whether natural or acquired.

To ſhow, more obviouſly, the reaſoning-faculty of brutes, and to diſtinguiſh the operations of intellect from thoſe of inſtinct, we need only contemplate the actions and diſpoſition of the DOG. In a ſavage ſtate, it muſt be allowed, that he is fierce, cruel, and voracious; but, when civilized and accuſtomed to live with men, he is poſſeſſed of every amiable quality. He ſeems to have no other deſire than to pleaſe and protect his maſter. He is gentle, obedient, ſubmiſſive, and faithful. Theſe diſpoſitions, joined to his almoſt unbounded ſagacity, juſtly claim the eſteem of mankind. Accordingly no animal is ſo much careſſed or reſpected: he is ſo ductile, and ſo much formed to pleaſe, that he aſſumes the very air and temper of the family in which he reſides. An animal endowed with ſuch uncommon qualities, muſt anſwer many uſeful purpoſes. His fidelity and vigilance are daily employed to protect our perſons, our flocks, or our goods. The acuteneſs of his ſmell gains him employment in hunting: he is frequently employed as a turnſpit: at Bruſſels and in Holland he is trained to draw little carts to the herb-market; and in the northern regions draws a ſledge with his maſter in it, or loaden with proviſions. It is a remarkable inſtinct in the dog, that, when oppreſſed with ſickneſs, to which he is very ſubject, eſpecially in the beginning of ſummer, and before ill weather, he eats the leaves of the quicken graſs, the bearded wheat-grafs, or the rough cock's-foot grafs, which gives him immediate relief by making him vomit. He does not throw out his excrements promifcuoufly upon every thing that happens to be in the way, but carefully feeks ftones, trunks of trees, or barren places. This is a wife inftitution of nature; for the excrements of a dog deftroy almoft every vegetable or animal fubftance. They are of fuch a putrid nature, that, if a man's fhoe touches them when recently expelled, that particular part will rot in a few days. He obferves the fame method in making his urine, which he throws out at a fide. It is remarkable, that a dog will not pafs a ftone or a wall againft which any other dog has urined, without following his example, although an hundred fhould occur in a few minutes; infomuch that it is aftonifhing how fuch a quantity can be fecreted in fo fhort a time. The principal objection to dogs, in the fhocking circumftance of their going mad, and of communicating the diforder to whatever perfon, or animal, they may chance to bite; and of which the cure has ever been confidered precarious and uncertain. From a minute inveftigation of the poifonous qualities of the hydrophobia, and the effect it has on the blood, as well as from a confideration of what the blood and juices undergo by emerging the body in the fea, I am bold to affirm that my Solar Tincture, adminiftered in the way I have directed, is a certain and infallible cure for this deplorable malady; at leaft as far as human certainty can go with refpect to medicine. I would not however be underftood to encourage a negligence in thofe who keep dogs, to watch well their actions, and, on the fmalleft fufpicion that fuch a misfortune is near, to have them inftantly difpatched, as they may be eafily replaced, and much anxiety and diftrefs prevented.—With regard to the propagation of dogs, the females admit the males before they are twelve months old. They remain in feafon ten, twelve, or even fifteen, days, during which time they will admit a variety of males. They come in feafon generally twice in the year, and more frequently in the cold than in the hot months. The male difcovers the condition of the female by the fmell; but fhe feldom admits him the firft fix or feven days. One coition will make her conceive a great number of young; but, when not reftrained, fhe will admit feveral dogs every day; fhe feems to have no choice or predilection, except in favour of large dogs; from this circumftance it fometimes happens, that a fmall female, who has admitted a maftiff, perifhes in bringing forth her young. During the time of copulation, thefe animals cannot feparate themfelves, but remain united fo long as the erection fubfifts. This is owing to the ftructure of the parts. The dog has not only a bone in his penis, but in the middle of the corpus cavernofum there is a large hollow, which is blown up in the time of erection to a confiderable bulk. The female, on the other hand, has a larger clitoris than perhaps any other animal: befides, a large firm protuberance

 rifes

rises in the time of copulation, and remains perhaps longer than that of the male, and prevents him from retiring till it subsides; accordingly, after the act of penetrating is effected, the male turns about in order to rest himself on his legs, and remains in that position till the parts turn flaccid. The female goes with young about nine weeks. They generally bring forth from six to twelve puppies. Those of a small size bring forth five, four, and sometimes but two. They continue to copulate and bring forth during life, which lasts generally about fourteen or fifteen years. The whelps are commonly blind, and cannot open their eyes till the tenth or twelfth day; the males are like the dog, the females like the bitch.—The dog, the wolf, and the fox, are certainly derived from one original parent; and all dogs whatsoever, from the terrible boar-dog to Pompey the little, were all one in the first creation. All the variety we behold in them, is either produced by change of climate, or the accidental effect of soil, food, or situation; or from the issue of human care, experiment, or caprice. Every huntsman knows what a vast alteration may be made in dogs, by industriously improving the breed for twenty or thirty years. Nature wisely tends to render every kind of creature fit for the country where it is to inhabit, or be employed; which is the reason why hounds, and all other animals, degenerate, by being removed into contrary climates. This is manifest from the following experiment: if a couple of right southern hounds be removed to the north, and suffered to propagate without art or mixture, they will, by sensible degrees, decline into lighter bodies and shriller accents; and in the same way are all dogs varied, by being carried from one country to another. But the utmost efforts of human industry and contrivance, whether assisted by change of climate, or mixture of breed, could never add one new species to the works of the creation. Nature is still uniform as to the main, nor suffers the Almighty Creator to be imitated by short-sighted mortals. In spite of art, our mules are always barren; nor can the most curious projector produce one amphigeneous animal that will increase and multiply. There appears a distinct specific difference in all living creatures; the horse, the dog, the bear, the goat, however diversified by art, by copulation, or by climate, either in size, shape, or figure, will ever discover something that approximates to the character of their species. Above all, the peculiar instinct and appetite for generation, will prompt them to own and indicate their relation. Animals of different species will never copulate together. This is one of the most undeniable arguments that wolves, foxes, and dogs, are originally the same species, because in coition they are not only all held together in the same manner, but we have some instances of litters of puppies produced from the dog and fox, and also from the dog and wolf. Mr. Brooke, animal-merchant in Holborn, turned a wolf to a Pomeranian bitch in heat; the

the congreſs was immediate, and as uſual between dog and bitch: ſhe produced ten puppies. Mr. Pennant ſaw one of them at Gordon Caſtle, that had very much the reſemblance of a wolf, and alſo much of its nature; being ſlipped at a weak deer, it inſtantly caught at the animal's throat and killed it. I could not learn (ſays Mr. Pennant) whether this mongrel continued its ſpecies; but another of the ſame kind did, and ſtocked the neighbourhood of Fochabers, in the county of Moray, (where it was kept,) with a multitude of curs of a moſt wolfiſh aſpect. There was lately living a mongrel offspring of this kind. It greatly reſembled its wolf parent. It was firſt the property of Sir Wolſtein Dixey; afterwards of Sir Willoughby Aſton. During day it was very tame; but at night ſometimes relapſed into ferocity. It never barked, but rather howled; when it came into fields where ſheep were, it would feign lameneſs, but if no one was preſent would inſtantly attack them. It had been ſeen in copulation with a bitch, which afterwards pupped: the breed was imagined to reſemble in many reſpects the ſuppoſed ſire. It died between the age of five and ſix.—The woodman of the manor of Mongewell, in Oxfordſhire, has a bitch, which conſtantly follows him, the offspring a tame dog-fox by a ſhepherd's cur; and ſhe again has had puppies by a dog. Since there are ſuch authentic proofs of the further continuance of the breed, we may ſurely add the wolf and fox to the other ſuppoſed ſtocks of theſe faithful domeſtics, particularly as moſt naturaliſts ſuppoſe the dog to have been originally the production of one or other of theſe animals, tamed and civilized.

Many and wonderful are the inſtances of ſagacity, fidelity, and attention, and even of foreſight, which theſe faithful animals have evinced towards their maſters. Some ſuch will doubtleſs occur to the minds of my readers, as falling under their own obſervation; I ſhall therefore only recite two or three ſuch inſtances, of unqueſtionable authenticity.—In the year 1791, a perſon went to a houſe in Deptford, to take lodgings, under pretence than he was juſt arrived from the Weſt-Indies; and, after having agreed on terms, ſaid he ſhould ſend in his trunk that night, and come himſelf the next day. About nine o'clock in the evening, the trunk was brought in by two porters, and was carried into his bed-room. Juſt as the family were going to bed, their little houſe-dog, deſerting his uſual ſtation in the ſhop, placed himſelf cloſe to the door of the chamber where the cheſt was depoſited, and kept up an inceſſant barking. The moment the chamber door was opened, the dog flew to the cheſt, againſt which it barked and ſcratched with redoubled vehemence and fury. At firſt they tried to get the dog out of the room; but in vain. Calling in ſome neighbours, and making them eye-witneſſes of the circumſtance, they began to move the trunk about, when they quickly diſcovered that it contained ſomething alive. Suſpicion

falling

falling very ſtrong, they were induced to open it, when, to their utter aſtoniſhment, who ſhould preſent himſelf but their new lodger, who had been thus conveyed in, to rob the houſe!—In the ſummer of the year 1792, a gentleman went down to Portſmouth for the benefit of ſea-bathing. He went to one of Mr. Bradley's machines, to be conducted into the water. Being unacquainted with the depth, and no ſwimmer, he found himſelf, the inſtant he quitted the machine, nearly out of his depth. Fright increaſed the peril of his ſituation, and, unnoticed by the perſon who attends the machines, he had ſunk for the laſt time in the agonies of drowning. A large Newfoundland dog, ſtanding by accident on the ſhore, and ſeeing the diſtreſs of this ſtranger, plunged in after him; and, ſeizing him by the hair of the head, conducted him ſafely on-ſhore, though it was ſome time before he recovered. The gentleman afterwards purchaſed the dog at a high price, but values him equally with the ſum total of his fortune.—At the ſeat of the late Earl of Litchfield, three miles from Blenheim, there is a portrait in the dining-room of Sir Henry Lee, by Johnſton, with that of a maſtiff dog which ſaved his life. It ſeems a ſervant had formed the deſign of aſſaſſinating his maſter and robbing the houſe; but the night he had fixed on, the dog, which had never been much noticed by Sir Henry, for the firſt time followed him up-ſtairs, got under his bed, and could not be driven thence by either maſter or man; in the dead of night, the ſame ſervant entered the room to execute his horrid deſign, but was inſtantly ſeized by the dog, and, being ſecured, confeſſed his intentions. There are ten quaint lines in one corner of the picture, which conclude thus:

> But in my dog, whereof I made no ſtore,
> I find more love than thoſe I truſted more.

Upon what hypotheſis can we account for a degree of foreſight and penetration ſuch as this? Or will it be ſuggeſted, as a ſolution of the difficulty, that a dog may poſſibly become capable in a great meaſure of underſtanding human diſcourſe, and of reaſoning and acting accordingly; and that, in the preſent inſtance, the villain had either uttered his deſign in ſoliloquy, or imparted it to an accomplice, in the hearing of the animal? It has been much diſputed whether brutes have any language whereby they can expreſs their minds to each other; or whether all the noiſe they make conſiſts only of cries inarticulate, and unintelligible even to themſelves. We may indeed, from analogy, conclude, with great reaſon, that ſome of the cries of beaſts are really expreſſions of their ſentiments; but whether one beaſt is capable of forming a deſign, and communicating that deſign by any kind of language to others, is what I ſhall leave to the judgment of the reader,

 after

after submitting to his consideration the following instance.—A sparrow, finding a nest that a martin had just built, standing very conveniently for him, possessed himself of it. The martin, seeing the usurper in her house, called for help to expel him. A thousand martins came full speed, and attacked the sparrow; but the latter, being covered on every side, and presenting only his large beak at the entrance of the nest, was invulnerable, and made the boldest of them who durst approach him repent of their temerity. After a quarter of an hour's combat, all the martins disappeared. The sparrow thought he had got the better, and the spectators judged that the martins had abandoned their undertaking. Not in the least. Immediately they returned to the charge; and, each of them having procured a little of that tempered earth with which they make their nests, they all at once fell upon the sparrow, and inclosed him in the nest to perish there, though they could not drive him thence. Can it be imagined that the martins could have been able to hatch and concert this design all of them together, without speaking to each other, or without some medium of communication equivalent to language?

From all these extraordinary endowments, manifested by brute animals of different countries and kinds, some philosophers have maintained that brutes are gifted with a soul, though essentially inferior to that of men; and to this soul they have allowed immortality. Father Bougeant, a Jesuit, published, about the year 1740, a treatise expressly on this subject, entitled, *A Philosophical Amusement on the Language of Brutes*, in which he affirms that they are animated by evil spirits, or devils. The strangeness of this doctrine has induced me to give the outline of his argument, since it cannot fail to prove entertaining to the reader. "Reason (says he) naturally inclines us to believe that beasts have a spiritual soul; and the only thing that opposes this sentiment is, the consequences that might be inferred from it. If brutes have a soul, that soul must be either matter or spirit; it must be one of the two, and yet you dare affirm neither. You dare not say it is a matter, because you must then necessarily suppose matter to be capable of thinking; nor will you say that it is spirit, this opinion bringing with it consequences contrary to the principles of religion; and this, among others, that man would differ from beasts only by the degrees of plus and minus; which would demolish the very foundation of all religion. Therefore, if I can elude all these consequences; if I can assign to beasts a spiritual soul, without striking at the doctrines of religion; it is evident, that my system, being moreover the most agreeable to reason, is the only warrantable hypothesis. Now I shall, and can do it, with the greatest ease imaginable. I even have means, by the same method, to explain many very obscure passages in the Holy Scriptures, and to resolve some very great difficulties which are not well confuted. This I shall unfold in a more particular manner. Religion teaches us, us, that the devils, from the very moment they had finned, were reprobate, and that they were doomed to burn for ever in hell; but the church has not yet determined whether they do actually endure the torments to which they are condemned. It may then be thought that they do not yet fuffer them, and that the execution of the verdict brought againft them is referved for the day of final judgement.—Now what I pretend to infer from hence is, that, till doomfday comes, God, in order not to fuffer fo many legions of reprobate fpirits to be of no ufe, has diftributed them through the feveral fpaces of the world, to ferve the defigns of his Providence and make his omnipotence to appear. Some, continuing in their natural ftate, bufy themfelves in tempting men, in feducing and tormenting them; either immediately, as Job's devil, and thofe that lay hold of human bodies; or by the miniftry of forcerers or phantoms. Thefe wicked fpirits are thofe whom the fcripture calls the *powers of darknefs*, or the *powers of the air*. God, with the others, makes millions of beafts of all kinds, which ferve for the ufes of men, which fill the univerfe, and caufe the wifdom and omnipotence of the Creator to be admired. By that means I can eafily conceive, on the one hand, how the devils can tempt us; and, on the other, how beafts can think, know, have fentiments, and a fpiritual foul, without any way ftriking at the doctrines of religion. I am no longer furprifed to fee them have forecaft, memory, and judgment. I fhould rather have occafion to wonder at their having no more, fince their foul very likely is more perfect than ours. But I difcover the reafon of this: it is becaufe, in beafts as well as in ourfelves, the operations of the mind are dependent on the material organs of the machine to which it is united; and, thofe organs being groffer and lefs perfect than in us, it follows, that the knowledge, the thoughts, and the other fpiritual operations, of beafts, muft of courfe be lefs perfect than ours: and, if thefe proud fpirits know their own difmal ftate, what an humiliation muft it be to them thus to fee themfelves reduced to the condition of beafts! But whether they know it or no, fo fhameful a degradation is ftill, with regard to them, the primary effect of the divine vengeance I juft mentioned; it is an anticipated hell."—Having mentioned the prejudices againft this hypothefis, fuch particularly as the pleafure which people of fenfe and religion take in beafts and birds, efpecially all forts of domeftic animals; he proceeds, "Do we love beafts for their own fakes? No. As they are altogether ftrangers to human fociety, they can have no other appointment but that of being ufeful and amufing. And what care we whether it be a devil or any other creature that amufes us? The thought of it, far from fhocking, pleafes me mightily. I with gratitude admire the goodnefs of the Creator, who gave me fo many little devils to ferve and amufe me. If I am told that thefe poor devils are doomed to fuffer eternal tortures, I admire God's decrees, but I have no manner of ſhare in that dreadful ſentence; I leave the execution of it to the Sovereign Judge; and, notwithſtanding this, I live with my little devils as I do with a multitude of people of whom religion informs me that a great number ſhall be damned. But the cure of a prejudice is not to be effected in a moment; it is done by time and reflection: give me leave then lightly to touch upon this difficulty, in order to obſerve a very important thing to you. Perſuaded as we are that beaſts have intelligence, have we not all of us a thouſand times pitied them for the exceſſive evils which the majority of them are expoſed to, and in reality ſuffer? How unhappy is the condition of horſes! we are apt to ſay, upon ſeeing a horſe whom an unmerciful carman is murdering with blows. How miſerable is a dog whom they are breaking for hunting! How diſmal is the fate of beaſts living in woods; they are perpetually expoſed to the injuries of the weather, always ſeized with apprehenſions of becoming the prey of hunters, or of ſome wilder animal; for ever obliged, after long fatigue, to look out for ſome poor inſipid food; often ſuffering cruel hunger; and ſubject, moreover, to illneſs and death! If men are ſubject to a multitude of miſeries that overwhelm them, religion acquaints us with the reaſon of it; viz. the being born ſinners. But what crimes can beaſts have committed by birth to be ſubject to evils ſo very cruel? What are we, then, to think of the horrible exceſſes of miſeries undergone by beaſts? miſeries, indeed, far greater than thoſe endured by men. This is, in any other ſyſtem, an incomprehenſible myſtery; whereas nothing is more eaſy to be conceived from the ſyſtem I propoſe. The rebellious ſpirits deſerve a puniſhment ſtill more rigorous, and happy is it for them that their puniſhment is deferred. In a word, God's goodneſs is vindicated, and man himſelf is juſtified: for what right can we have, without neceſſity, and often in the way of mere diverſion, to take away the life of millions of beaſts, if God had not authoriſed us ſo to do? And, beaſts being as ſenſible as ourſelves of pain and death, how could a juſt and merciful God have given man that privilege, if they were not ſo many guilty victims of the divine vengeance?—But hear ſtill ſomething more convincing, and of greater conſequence: beaſts, by nature, are extremely vicious. We know well that they never ſin, becauſe they are not free; but this is the only condition wanting to make them ſinners. The voracious birds and beaſts of prey are cruel. Many inſects of one and the ſame ſpecies devour one another. Cats are perfidious and ungrateful; monkeys are miſchievous; and dogs envious. All beaſts in general are jealous and revengeful to exceſs; not to mention many other vices we obſerve in them: and at the ſame time that they are by nature ſo very vicious, they have, ſay we, neither the liberty nor any helps to reſiſt the bias that hurries them into ſo many bad actions. They are, according to the ſchools, neceſſitated to do evil,

to

to disconcert the general order, to commit whatever is most contrary to the notion we have of natural justice and to the principles of virtue. What monsters are these in a world originally created for order and justice to reign in? This is, in good part, what formerly persuaded the Manicheans, that there were of necessity two orders of things, one good, and the other bad; and that the beasts were not the work of the good principle: a monstrous error! But how then shall we believe that beasts came out of the hands of their Creator with qualities so very strange! If man is so very wicked and corrupt, it is because he has himself through sin perverted the happy nature God had given him at his creation. Of two things, then, we must say one: either that God has taken delight in making beasts so vicious as they are, and of giving us in them models of what is most shameful in the world; or that they have, like man, original sin, which has perverted their primitive nature.—The first of these propositions finds very difficult access to the mind, and is an express contradiction to the Holy Scriptures; which say, that whatever came out of God's hands, at the time of the creation of the world, was good, yea very good. What good can there be in a monkey's being so very mischievous, a dog so full of envy, a cat so malicious? But then many authors have pretended, that beasts, before man's fall, were different from what they are now; and that it was in order to punish man that they became so wicked. But this opinion is a mere supposition, of which there is not the least footstep in holy scripture. It is a pitiful subterfuge to elude a real difficulty: this at most might be said of the beasts with whom man has a sort of correspondence; but not at all of the birds, fishes, and insects, which have no manner of relation to him. We must then have recourse to the second proposition, that the nature of beasts has, like that of man, been corrupted by some original sin: another hypothesis, void of foundation, and equally inconsistent with reason and religion, in all the systems which have been hitherto espoused concerning the souls of beasts. What side are we to take? Why, admit of my system, and all is explained. The souls of beasts are refractory spirits which have made themselves guilty towards God. The sin in beasts is no original sin; it is a personal crime, which has corrupted and perverted their nature in its whole substance; hence all the vices and corruption we observe in them, though they can be no longer criminal, because God, by irrecoverably reprobating them, has at the same time divested them of their liberty."

These quotations contain the strength of Father Bougeant's hypothesis, which also hath had its followers; but the reply to it is obvious. Beasts, though remarkably mischievous, are not completely so; they are in many instances capable of gratitude and love, which devils cannot possibly be. The very same passions that are in the brutes exist in the human nature; and, if we chose to argue from the existence of those passions, and the ascendency they have over mankind at some times, we may say, with as great justice, that the souls of men are devils, as that the souls of brutes are. All that can be reasonably inferred from the greater prevalency of the malignant passions among the brutes than among men, is, that the former have less rationality than men: and accordingly it is found, that among savages, who exercise their reason less than other men, every species of barbarity is practised, without being deemed a crime.—Upon the whole, it is impossible to deduce this variety of action, in animals, from a general and uniform instinct only. For they accommodate their operations to times and circumstances. They combine; they choose the favourable moment; they avail themselves of the occasion, and seem to receive instruction by experience. Many of their operations announce reflection: the bird repairs a shattered nest, instead of constructing instinctively a new one: the hen, who has been robbed of her eggs, changes her place in order to lay the remainder with more security: the cat discovers both care and artifice in concealing her kittens. Again it is evident, that, on many occasions, animals know their faults and mistakes, and correct them; they sometimes contrive the most ingenious methods of obtaining their ends, and when one method fails, have recourse to another; and they have, without doubt, a kind of language for the mutual communication of their ideas. How is all this to be accounted for, unless we suppose them endowed with the powers of perceiving, thinking, remembering, comparing, and judging? They certainly have these powers, in a degree inferior to the human species, and form classes below them in the graduated scale of intelligent beings; but, their actions not being directed to moral ends, are consequently not accountable and proper subjects for reward or punishment in a future world.

After all, it does not appear upon what principle of reason and justice it is, that mankind have founded their right over the *life* of every creature that is placed in a subordinate rank of being to themselves. Whatever claim they may have in right of food and self-defence, did they extend their privilege no farther, numberless beings might enjoy their lives in peace, who are now hurried out of them by the most wanton and unnecessary cruelties. It is surely difficult to discover why it should be thought less inhuman to crush to death a harmless insect, whose single offence is that he eats that food which nature has prepared for his sustenance, than it would be were we to kill any bulky creature for the same reason. There are few tempers so hardened against the impressions of humanity, as not to shudder at the thought of the latter; and yet the former is universally practised, without the least check of compassion. This seems to arise from the gross error of supposing, that every creature is really in itself contemptible, which happens to be clothed

with

with a body infinitely diſproportionate to our own, not conſidering that great and little are merely relative terms. But the inimitable Shakeſpeare would teach us, that

——the poor beetle that we tread upon,
In corp'ral ſuff'rance feels a pang as great
As when a giant dies.——

And, indeed, there is every reaſon to believe, that the ſenſations of many inſects are as requiſite as thoſe of creatures of far more enlarged dimenſions, perhaps even more ſo. The millepede, for inſtance, rolls itſelf round upon the ſlighteſt touch; and the ſnail gathers in her horns upon the leaſt approach of our hand. Are not theſe the ſtrongeſt indications of their ſenſibility? and is it any evidence of ours, that we are not therefore induced to treat them with a more ſympathiſing tenderneſs?

I cannot conclude theſe obſervations on the inſtinct and œconomy of brute animals, without reciting the following moſt remarkable account of the land-crab, which inhabits the Bahama Iſlands, as well as moſt parts between the tropics, and feeds upon vegetables. Theſe creatures live not only in a kind of orderly ſociety in their retreats in the mountains, but regularly, once a-year, march down to the ſea-ſide in a body of ſome millions at a time. As they multiply in great numbers, they chooſe the month of April or May to begin their expedition; and then ſally out by thouſands from the ſtumps of hollow trees, from the clefts of rocks, and from the holes which they dig for themſelves under the ſurface of the earth. At that time the whole ground is covered with this band of adventurers; there is no ſetting down one's foot without treading upon them. The ſea is their place of deſtination, and to that they direct their march with right-lined preciſion. No geometrician could ſend them to their deſtined ſtation by a ſhorter courſe; they turn neither to the right nor to the left, whatever obſtacles intervene; and, even if they meet with a houſe, they will attempt to ſcale the walls to keep the unbroken tenor of their way. But, though this be the general order of their route, they upon other occaſions are obliged to conform to the face of the country; and, if it is interſected with rivers, they are then ſeen to wind along the courſe of the ſtream. The proceſſion ſets forward from the mountains with the regularity of an army under the guidance of an experienced commander. They are commonly divided into three battalions; of which the firſt conſiſts of the ſtrongeſt and boldeſt males, that, like pioneers, march forward to clear the way, and face the greateſt dangers. Theſe are often obliged to halt for want of rain, and to go into the moſt convenient encampment till the weather changes. The main body of the army is composed of females, which never leave the mountains till the rain has set in for some time, and then descend in regular battalia, being formed into two columns of fifty paces broad, and three miles deep, and so close that they almost cover the ground. Three or four days after this, the rear guard follows, a straggling undisciplined tribe, consisting of males and females, but neither so robust nor so vigorous as the former. The night is their chief time of proceeding; but, if it rains by day, they do not fail to profit by the occasion; and they continue to move forward in their slow uniform manner. When the sun shines and is hot upon the surface of the ground, they then make an universal halt, and wait till the cool of the evening. When they are terrified, they march back in a confused disorderly manner, holding up their nippers, with which they sometimes tear off a piece of the skin, and then leave the weapon where they inflicted the wound. They even try to intimidate their enemies; for they often clatter their nippers together, as if it were to threaten those that come to disturb them. But, though they thus strive to be formidable to man, they are much more so to each other; for they are possessed of one most unsocial property, which is, that, if any of them by accident is maimed in such a manner as to be incapable of proceeding, the rest fall upon and devour it on the spot, and then pursue their journey.—When, after a fatiguing march, and escaping a thousand dangers, (for they are sometimes three months in getting to the shore,) they have arrived at their destined port, they prepare to cast their spawn. The pease are as yet within their bodies, and not excluded, as is usual in animals of this kind, under the tail; for the creature waits for the benefit of sea-water to help the delivery. For this purpose the crab has no sooner reached the shore, than it eagerly goes to the edge of the water, and lets the waves wash over its body two or three times. This seems only a preparation for bringing their spawn to maturity; for, without further delay, they withdraw to seek a lodging upon land. In the mean time the spawn grows larger, is excluded out of the body, and sticks to the barbs under the flap, or more properly the tail. This bunch is seen as big as a hen's egg, and exactly resembling the roes of herrings. In this state of pregnancy they once more seek the shore for the last time; and, shaking off their spawn into the water, leave accident to bring it to maturity. At this time whole shoals of hungry fish are at the shore in expectation of this annual supply; the sea to a great distance seems black with them; and about two thirds of the crabs' eggs are immediately devoured by these rapacious invaders. The eggs that escape are hatched under the sand; and, soon after, millions at a time of the little crabs are seen quitting the shore,

ſhore, and ſlowly travelling up to the mountains. The old ones, however, are not ſo active to return; they have become ſo feeble and lean, that they can hardly creep along, and their fleſh at that time changes its colour. The moſt of them, therefore, are obliged to continue in the flat parts of the country till they recover, making holes in the earth, which they cover at the mouth with leaves and dirt, ſo that no air may enter. There they throw off their old ſhells, which they leave, as it were, quite whole; the place where they opened on the belly being unſeen. At that time they are quite naked, and almoſt without motion for ſix days together, when they become ſo fat as to be delicious food. They have then under their ſtomachs four large white ſtones, which gradually decreaſe in proportion as the ſhell hardens, and, when they come to perfection, are not to be found. It is at that time that the animal is ſeen ſlowly making its way back; and all this is moſt commonly performed in the ſpace of ſix weeks.—This animal, when poſſeſſed of its retreats in the mountains, is impregnable; for only ſubſiſting on vegetables, it ſeldom ventures out; and, its habitation being in the moſt inacceſſible places, it remains for a great part of the ſeaſon in perfect ſecurity. It is only when impelled by the deſire of bringing forth its young, and when compelled to deſcend into the flat country, that it is taken. At that time the natives wait for its deſcent in eager expectation, and deſtroy thouſands; but, diſregarding their bodies, they only ſeek for that ſmall ſpawn which lies on each ſide the ſtomach within the ſhell, of about the thickneſs of a man's thumb. They are much more valuable upon their return after they have caſt their ſhell; for, being covered with a ſkin reſembling ſoft parchment, almoſt every part except the ſtomach may be eaten. They are taken in the holes by feeling for them with an inſtrument; they are ſought after by night, when on their journey, by flambeaux. The inſtant the animal perceives itſelf attacked, it throws itſelf on its back, and with its claws pinches moſt terribly whatever it happens to faſten on. But the dexterous crab-catcher takes them by the hinder legs in ſuch a manner that the nippers cannot touch them, and thus he throws them into his bag. Sometimes alſo they are caught, when they take refuge in the bottoms of holes in rocks by the ſea-ſide, by clapping a ſtick to the mouth of the hole, which prevents their getting out; and then ſoon after, the tide coming, enters the hole, and the animal is found, upon its retiring, drowned in its retreat.—Theſe crabs are of various ſizes, the largeſt about ſix inches wide; they walk ſide-ways like the ſea-crab, and are ſhaped like it: ſome are black, ſome are yellow, ſome red, and others variegated with red, white, and yellow, mixed. Some of theſe are poiſonous; and ſeveral people have died of eating of the crabs, particularly of the black kind.

The light-coloured are reckoned beſt; and, when full in fleſh, are very well taſted. In ſome of the ſugar-iſlands they are eaten without danger; and are no ſmall help to the negro-ſlaves, who, on many of theſe iſlands, would fare very hard without them.

OF SCENT.

NOTHING more eminently demonſtrates the doctrine of atoms, than *ſcent.* It is an effluvium continually ariſing from the corpuſcles that iſſue out of all bodies; and, being impregnated with the peculiar ſtate and quality of the blood and juices of that particular animal from which they flow, occaſions the vaſt variety of ſmells or ſcents cognizable by the olfactory nerves, or organs of ſmelling. Hence the reaſon why one perſon differs from another in ſcent, and why a dog will trace the footſteps of his maſter for a hundred miles together, follow him into any houſe, church, or other building, and diſtinguiſh him from every other perſon, though ſurrounded by ten thouſand; and when the faithful animal has thus diligently ſought out and recogniſed his maſter, he is ſeldom willing even to truſt the evidence of his own eyes, until, with erected creſt, he has taken a few cordial ſniffs, to convince himſelf he is right. Hence alſo we perceive how a pack of hounds are enabled to purſue the hare, fox, ſtag, or any other animal they are trained to hunt, acroſs the ſcent, and amidſt the ſociety of others of the ſame ſpecies, without being diverted from the purſuit of that ſelf-ſame animal they had *firſt* on foot. And hence too we diſcover how it is poſſible for birds and beaſts of prey to be directed to their food at ſuch vaſt diſtances; for theſe corpuſcles, iſſuing from putrid bodies, and floating in the air, are carried by the wind to different quarters; and, ſtriking the olfactory nerves of whatever animals they meet in their way,* immediately conduct them to the ſpot. It matters not how much the effluvia may be gone off, ſo as enough remains to irritate the olfactory organ; for, whether it be bird or beaſt, they try the ſcent in all directions, till at length they diſcover that which is ſtronger and ſtronger in proportion as they proceed; and this nature has taught them to know is the direct and obvious road to their prey, and prevents them from following the contrary courſe, which is naturally weaker and weaker, and what in hunting is termed *heel.* This obſervation is confirmed by the increaſed eagerneſs we perceive in all animals, the nearer they approach the object of purſuit; as we ſee hounds and ſpaniels in hunting and

* It is by this means the ſmall-pox, meaſles, putrid fevers, and all epidemic complaints, are communicated, and the plague and peſtilence conveyed from one place to another.

ſhooting,

ſhooting, are the more earneſt, in proportion as the ſcent is recent, and as they draw nearer to the game. The ſame thing, amongſt quadrupeds, whether wild or domeſtic, directs the male to the female that is in ſeaſon for love; and hence we ſee the dog, the boar, the bull, and the ſtallion, when turned looſe, apply their noſtrils to the ambient air, and proceed accordingly. By the ſame medium the vermin which infeſt our dwellings kuow how to direct their operations, whether to undermine walls, eat through ſolid boards, croſs rivers, or climb ſpouts; which ſhows how much ſtronger the faculty of ſmelling is poſſeſſed by the brute ſpecies than by the human; wiſely ordained by nature, to enable them to ſeek their food, and propagate their ſpecies, but for which they would often periſh, or have long ſince become extinct.

There are wonderful inſtances of ſome animal carcaſes, which, though ſlaked with lime, and buried ten feet under ground, have ſent forth ſo ſtrong a ſcent as to have attracted dogs to the ſpot, that eagerly endeavoured to dig away the earth to get at them. And an inſtance happened only a few years ſince at Petersfield in Hampſhire, where an unfortunate female, having privately delivered herſelf of two children, went and buried them in a deep hole in an adjoining field; but within three days ſome dogs were attracted to the ſpot by the ſcent, dug them up, and partly devoured them before the ſhocking circumſtance was diſcovered. No wonder then a pack of hounds, which have caught the ſcent of a polecat or weaſel, will purſue it into the thickeſt foreſt, and aſſemble round the very tree up the trunk of which it had run for ſhelter; or that blood-hounds, as in times of old, ſhould trace out fugitives and robbers in ſubterraneous caverns, in trees, caves, or foreſts, or in clefts of inacceſſible rocks, of which inſtances are given by the moſt reputable hiſtorians. It is however to be remarked, that, as all animals hunt for and purſue their prey by its ſcent, ſo they ſeem inſtinctively to know, that they themſelves are hunted and purſued by means of the ſcent iſſuing from their own bodies. But, as this ſubject admits of ſo much curious and occult ſpeculation, I ſhall give a few inſtances of the effects of ſcent upon different animals, and the ſenſe and ſagacity they diſplay in the management of it.—And firſt, of the hare.

The hare is naturally a timid animal, but emanates a very ſtrong ſcent. He ſleeps in his form or ſeat during the day; and feeds, copulates, &c. in the night. In a moon-light evening, a number of them are ſometimes ſeen ſporting together, leaping and purſuing each other: but the leaſt motion, the falling of a leaf, alarms them; and then they all run off ſeparately, each taking a different route. They are extremely ſwift in their motion, which is a kind of gallop, or a ſucceſſion of quick leaps. When purſued, they always take to the higher grounds: as their fore feet are much ſhorter than the hind ones, they run with more eaſe up-hill than down-hill. The hare is endowed with all thoſe inſtincts which are neceſſary for his own preſervation. In winter he chooſes a form expoſed to the ſouth, and in ſummer to the north; and has a thouſand contrivances to elude the vigilance of his purſuers, and to cut off his ſcent from the hounds. If it be rainy, the hare uſually takes to the highways; and if ſhe come to the ſide of a young grove, or ſpring, ſhe ſeldom enters, but ſquats down till the hounds have over-ſhot her; and then ſhe will return the very way ſhe came, for fear of the wet and dew that hang on the boughs.—When ſhe comes near brook-ſides and plaſhes, ſhe will make all her croſſings, doublings, and workings. Some hares have been ſo crafty, that, as ſoon as they have heard the ſound of a horn, they would inſtantly ſtart out of their form, though it was at the diſtance of a quarter of a mile, and go and ſwim in ſome pool, and reſt upon ſome ruſh-bed in the midſt of it; and would not ſtir from thence till they have heard the ſound of the horn again, and then have ſtarted out, ſwimming to land, and have ſtood up before the hounds for hours before they could kill them, ſwimming and uſing all ſubtilties and croſſings in the water. Nay, ſuch is the natural craft and ſubtilty of the hare, that ſometimes after ſhe has been hunted three hours, ſhe will drive up a freſh hare, and ſquat in the ſame form herſelf. Others, having been hunted a conſiderable time, will creep under the door of a ſheep-cot, and hide themſelves among the ſheep; or, when they have been hard hunted, will run in among a flock of ſheep, and will by no means be gotten out from among them till the hounds are coupled up, and the ſheep driven into their pens. Some of them will take the ground like a rabbit, and run up a wall, and hide in the graſs on the top of it. Some hares will go up one ſide of the hedge and come down the other, the thickneſs of the hedge being the only diſtance between the courſes. A hare that has been ſorely hunted, has got upon a quickſet hedge, and run a good way upon the top thereof, and then leapt off upon the ground. And they will frequently betake themſelves to furze-buſhes, and will leap from one to the other, to cut off the ſcent, whereby the hounds are frequently at fault.—In the ſpring-time or ſummer, a hare will not ſit in buſhes, becauſe they are frequently infeſted with piſmires, ſnakes, and adders; but will ſit in corn-fields and open places. In the winter-time they ſit near towns and villages, in tufts of thorns and brambles, eſpecially when the wind is northerly or ſoutherly.—It is remarkable that the hare, although ever ſo frequently purſued by the dogs, ſeldom leaves the place where ſhe was brought forth, or even the form in which ſhe uſually ſits. It is common to find them in the ſame place next day, after being long and keenly chaſed the day before. The females

 are

are more groſs than the males, and have leſs ſtrength and agility; they are likewiſe more timid, and never allow the dogs to approach ſo near their form before riſing as the males. They likewiſe practiſe more arts, and double more frequently, than the males. The hare is diffuſed almoſt over every climate; and, notwithſtanding they are every-where hunted, their ſpecies never diminiſhes. They are in a condition of propagating the firſt year of their lives; the females go with young about thirty days, and produce four or five at a time; and, as ſoon as they have brought forth, they again admit the embraces of the male; ſo that they may be ſaid to be always pregnant. The eyes of the young are open at birth; the mother ſuckles them about twenty days, after which they ſeparate from her, and procure their own food. The young never go far from the place where they were brought forth; but ſtill they live ſolitary, and make forms about thirty paces diſtant from each other: thus, if a young hare be found any where, you may almoſt be certain of finding ſeveral others within a very ſmall diſtance.

The fecundity of the rabbit is ſtill greater than that of the hare. They will breed ſeven times in the year, and the female ſometimes brings eight young ones at a time. Suppoſing this to happen regularly for four years, the number of rabbits from a ſingle pair will amount to one million two hundred ſeventy-four thouſand eight hundred and forty.—They are in a condition for generating when ſix months old; and, like the hare, the female is almoſt conſtantly in ſeaſon; ſhe goes with young about thirty days, and brings forth from four to eight at a litter. A few days before littering, ſhe digs a hole in the earth, not in a ſtraight line, but in a zig-zag form: the bottom of this hole ſhe enlarges every way, and then pulls off a great quantity of hair from her belly, of which ſhe makes a kind of bed for her young. During the two firſt days after birth, ſhe never leaves them but when preſſed with hunger, and then ſhe eats quickly and returns; and in this manner ſhe ſuckles and attends her young for ſix weeks. All this time both the hole and the young are concealed from the male: ſometimes, when the female goes out, ſhe, in order to deceive the male, fills up the mouth of the hole with earth, mixed with her own urine. But when the young ones begin to come to the mouth of the hole, and to eat ſuch herbs as the mother brings to them, the father ſeems to know them; he takes them betwixt his paws, ſmooths their hair, and careſſes them with great fondneſs.

The fox is eſteemed to be the moſt ſagacious and moſt crafty of all beaſts of prey. The former quality he ſhows in his method of providing himſelf with an aſylum, where he retires from preſſing dangers, where he dwells, and where he brings up his young: and his craftineſs is chiefly diſcovered by the ſchemes he falls upon in order to catch lambs, geeſe, hens, and all kinds of ſmall birds. The

fox

fox fixes his abode on the border of a wood, in the neighbourhood of cottages: he liſtens to the crowing of the cock, and the cries of the poultry. He ſcents them at a diſtance; he chooſes his time with judgment; he conceals his road as well as his deſign; he ſlips forward with caution, ſometimes even trailing his body; and ſeldom makes a fruitleſs expedition. In this manner he has been ſeen, on a moon-light night, to enter a paſture where ſeveral hares were feeding, when lying down, and taking his tail in his mouth, he has trailed along like a rolling-ſtone, unſuſpected by his prey, till he had got too near for them all to eſcape. If he can leap the wall, or get in underneath, he ravages the court-yard, puts all to death, and then retires ſoftly with his prey, which he either hides under the herbage, or carries off to his kennel. He returns in a few minutes for another, which he carries off, or conceals in the ſame manner, but in a different place. In this way he proceeds till the progreſs of the ſun, or ſome movements perceived in the houſe, advertiſe him that it is time to ſuſpend his operations, and to retire to his den. He plays the ſame game with the catchers of thruſhes, woodcocks, &c. He viſits the nets and birdlime very early in the morning, carries off ſucceſſively the birds which are entangled, and lays them in different places, eſpecially near the ſides of highways, in the furrows, under the herbage or bruſhwood, where they ſometimes lie two or three days; but he knows perfectly where to find them when he is in need. He hunts the young hares in the plains, ſeizes old ones in their ſeats, never miſſes thoſe which are wounded, digs out the rabbits in the warrens, diſcovers the neſts of partridges and quails, ſeizes the mothers on the eggs, and deſtroys a vaſt quantity of game. The fox is exceedingly voracious; beſides fleſh of all kinds, he eats, with equal avidity, eggs, milk, cheeſe, fruits, and particularly grapes. When the young hares and partridges fail him, he makes war againſt rats, field-mice, ſerpents, lizards, toads, &c. Of theſe he deſtroys vaſt numbers; and this is the only ſervice he does to mankind. He is ſo fond of honey, that he attacks the wild bees, waſps, and hornets. They at firſt put him to flight by a thouſand ſtings; but he retires only for the purpoſe of rolling himſelf on the ground to cruſh them; and he returns ſo often to the charge, that he obliges them to abandon the hive, which he ſoon uncovers, and devours both the honey and wax. He will alſo devour fiſhes, lobſters, graſs-hoppers, &c. Foxes produce but once a-year; and the litter commonly conſiſts of four or five, ſeldom ſix, and never leſs than three. When the female is full, ſhe retires, and ſeldom goes out of her hole, where ſhe prepares a bed for her young. When ſhe perceives that her retreat is diſcovered, and that her young have been diſturbed, ſhe carries them off one by one, and goes in ſearch of another habitation. The fox, as well as the congenerous wolf, will produce with the dog kind, as noticed before. The fox ſleeps ſound, and may be eaſily approached without awakening: he ſleeps in a round form, like the dog; but, when he only

repoſes

repofes himfelf, he extends his hind legs, and lies on his belly. It is in this fituation that he fpies the birds along the hedges, and meditates fchemes for catching them. The fox flies when he hears the explofion of a gun, or fmells gunpowder. Being exceedingly fond of grapes, he does much mifchief in vineyards.—When purfued by the hounds, he feldom fails to deceive and fatigue them, becaufe he purpofely paffes through the thickeft parts of the foreft or places of the moft difficult accefs, where the dogs are hardly able to follow him; and, when he takes to the plains, he runs ftraight out, without ftopping or doubling.—He is a great admirer of his own tail, with which he frequently amufes and exercifes himfelf, by running in circles to catch it: and, in cold weather, wraps it round his nofe. The fmell of this animal is in general very ftrong, but that of the urine is remarkably fetid. This feems fo offenfive even to itfelf, that it will take the trouble of digging a hole in the ground, ftretching its body at full length over it; and there, after depofiting its water, covers it over with the earth, as the cat does its dung. The fmell is fo obnoxious, that it has often proved the means of the fox's efcape from the dogs; who have fo ftrong an averfion at the filthy effluvia, as to avoid encountering the animal it came from. It is faid the fox makes ufe of its urine as an expedient to force the cleanly badger from its habitation: whether that be the means, is rather doubtful; but that the fox makes ufe of the badger's hole is certain; not through want of ability to form its own retreat, but to fave itfelf fome trouble; for, after the expulfion of the firft inhabitant, the fox improves as well as enlarges it confiderably, adding feveral chambers, and providently making feveral entrances to fecure a retreat from every quarter. In warm weather, it will quit its habitation for the fake of bafking in the fun, or to enjoy the free air; but then it rarely lies expofed, but choofes fome thick brake, that it may reft fecure from furprife. Crows, magpies, and other birds, who confider the fox as their common enemy, will often, by their notes of anger, point out his retreat.

The ftag or buck is the moft crafty of all the fpecies of deer. He conceals himfelf with great addrefs, is moft difficult to trace, and derives fuperior refources from inftinct; for, though he has the misfortune to leave behind him a ftrong fcent, which redoubles the ardour and appetite of the hounds, he knows how to withdraw himfelf from their purfuit, by the rapidity with which he begins his flight, and by his numerous doublings. He delays not his arts of defence till his ftrength fails him; but, as foon as he finds that the firft efforts of a rapid chace have been unfuccefsful, he repeatedly retraces his former fteps; and after confounding, by thefe oppofite movements, the direction he has taken, after intermixing the prefent with the paft fcent from his body, he rifes from the earth by a great bound, and, retiring to a fide, he lies down flat on his belly, and in this immoveable ſituation he allows the whole troop of his deceived enemies to paſs very near him. His laſt refuge, when ſorely hunted, is the ſoil, keeping the middle, fearing leſt, by touching a bough, or a ſhrub, he may give greater ſcent to the hounds. He always ſwims againſt the ſtream, and will often cover himſelf under water, ſo as to ſhow nothing but his noſe. Where opportunity of water fails, he will fly into herds of cattle, as cows, ſheep, &c. and will ſometimes leap on an ox, cow, or the like, that he may leave no ſcent on the ground. What is ſtill more remarkable, it is related by the principal huntſman of Louis XII. that a buck which they had hunted for a long time, being at laſt hard preſſed, leaped into the middle of a very large white-thorn, in order to cut off its ſcent; and there ſtood aloft till he was run through by the huntſman, rather than ſtir from the place to be worried by the dogs.—Their ſeaſon of love commences about the end of Auguſt or beginning of September, when they leave the coppice, return to the foreſts, and ſearch for the hinds. They cry with a loud voice; their neck and throat ſwell; they become extremely reſtleſs, and traverſe in open day the fields and the fallow grounds: they ſtrike their horns againſt the trees and hedges; in a word, they ſeem to be tranſported with fury, and run from country to country, till they find the hinds or females, whom they purſue and compel into compliance; for the female at firſt avoids and flies from the male, and never ſubmits to his embraces till ſhe is fatigued with the purſuit. The old hinds likewiſe come in ſeaſon before the younger ones. When two bucks approach the ſame hind, they muſt fight before they enjoy. If nearly equal in ſtrength, they threaten, paw the ground, ſet up terrible cries, and attack each other with ſuch fury, that they often inflict mortal wounds with the ſtrokes of their horns. The combat never terminates but in the defeat or flight of one of the rivals. The conqueror loſes not a moment in enjoying his victory, unleſs another rival approaches, whom he is again obliged to attack and repel. The oldeſt ſtags are always maſters of the field; becauſe they are ſtronger and more furious than the young ones, who muſt wait patiently till their ſuperiors tire, and quit their miſtreſſes. Sometimes, however, the young ſtags accompliſh their purpoſe while the old ones are fighting, and, after a haſty gratification, fly off. The hinds prefer the old ſtags, not becauſe they are moſt courageous, but becauſe they are much more ardent. It has been alleged, that, attracted by the ſcent of the hinds, the ſtags in the rutting-ſeaſon throw themſelves into the ſea, and paſs from one iſland to another at the diſtance of ſeveral leagues. They leap ſtill more nimbly than they ſwim; for, when purſued, they eaſily clear a hedge or fence of ſix or ſeven feet high; and on all preſſing occaſions ſhow aſtoniſhing ſenſe and ſagacity.

The ſenſes of the wolf are likewiſe excellent, particularly his ſenſe of ſmelling, which often extends farther than his eye. The odour of carrion ſtrikes him at the diſtance

of

of more than a league. He likewife fcents live animals very far, and hunts them a long time by following their track. When he iffues from the wood, he never lofes the wind. He ftops on the borders of the foreft, fmells on all fides, and receives the corpufcles of living or dead animals brought to him from a diftance by the wind. Though he prefers living to dead animals, yet he devours the moft putrid carcafes. He is fond of human flefh; and, if ftronger, he would perhaps eat no other. Wolves have been known to follow armies, and to come in troops to the field of battle, where bodies are carelefsly interred, to tear them up, and to devour them with an infatiable avidity; and, when once accuftomed to human flefh, they ever after attack men, prefer the fhepherd to the flock, devour women, and carry off children. The wolf, unlike the dog, is an enemy to all fociety, and keeps no company even with thofe of his own fpecies. When feveral wolves unite together, it is not a fociety of peace, but of war; it is attended with tumult and dreadful howlings, and indicates an attack upon fome large animal, as a ftag, an ox, or a formidable maftiff. This military expedition is no fooner finifhed, than they feparate, and each returns in filence to his folitude. There is even little intercourfe between the males and females: they feel the mutual attractions of love but once a-year, and never remain long together. The females come in feafon in winter: many males follow the fame female; and this affociation is more bloody than the former; for they growl, chace, fight, and tear one another, and often facrifice him that is preferred by the female. The female commonly flies a long time, fatigues her admirers, and retires, while they fleep, with the moft alert or favourite male. The wolves copulate like dogs, and have an offeous penis, furrounded with a ring, which fwells and hinders them from feparating. When the females are about to bring forth, they fearch for a concealed place in the inmoft receffes of the foreft. The puppies come into the world blind, like dogs; the mother fuckles them fome weeks, and foon teaches them to eat flefh, which fhe prepares for them by tearing it into fmall pieces. Some time after fhe brings them field-mice, young hares, partridges, and live fowls. The young wolves begin by playing with thefe animals, and at laft worry them; then the mother pulls off the feathers, tears them in pieces, and gives a part to each of her young. They never leave their den till the end of fix weeks or two months. They then follow their mother, who leads them to drink in the hollow trunk of a tree, or in fome neighbouring pool. She conducts them back to the den, or, when any danger is apprehended, obliges them to conceal themfelves elfewhere. Though, like other females, the fhe-wolf is naturally more timid than the male, yet, when her young are attacked, fhe defends them with intrepidity; fhe lofes all fenfe of danger, and becomes perfectly furious. She never leaves them till they are fo ftrong as to need no affiftance or protection, and till

they

they have acquired talents fit for rapine. The wolf has great ſtrength, eſpecially in the anterior parts of the body, in the muſcles of the neck, and jaws. He carries a ſheep in his mouth, and at the ſame time outruns the ſhepherds; ſo that he can only be ſtopped or deprived of his prey by dogs. His bite is cruel, and always more obſtinate in proportion to the ſmallneſs of the reſiſtance; for, when an animal can defend itſelf, he is cautious and circumſpect. He never fights but from neceſſity, and not from motives of courage. When wounded with a ball, he cries; and yet, when diſpatching him with bludgeons, he complains not. When he falls into a ſnare, he is ſo overcome with terror, that he may be either killed or taken alive without reſiſtance; he allows himſelf to be chained, muzzled, and led where you pleaſe, without exhibiting the leaſt ſymptom of reſentment or diſcontent. Wolves are now ſo rare in the populated part of America, that the inhabitants leave their ſheep the whole night unguarded; yet the government of Pennſylvania and New Jerſey did ſome years ago allow a reward of twenty ſhillings, and the laſt even thirty ſhillings, for the killing of a wolf. Tradition informed them what a ſcourge thoſe animals had been to the colonies; ſo they wiſely determined to prevent the like evil. In their infant ſtate, wolves came down in multitudes from the mountains, often attracted by the ſmell of the corpſes of hundreds of Indians who died of the ſmall-pox, brought among them by the Europeans: but the animals did not confine their inſults to the dead, but even devoured in their huts the ſick and dying ſavages.—Britain, a few centuries ago, was much infeſted by them. They were, as appears by Hollingſhead, very noxious to the flocks in Scotland in 1577; nor were they entirely extirpated till about 1680, when the laſt wolf fell by the hands of the famous Sir Owen Cameron. Edward I. iſſued out his royal mandate to Peter Corbet to ſuperintend and aſſiſt in the deſtruction of them in the ſeveral counties of Glouceſter, Worceſter, Hereford, Salop, and Stafford; and in the adjacent county of Derby, certain perſons at Wormhill held their lands by the duty of hunting and taking the wolves that infeſted the country, whence they were ſtiled *wolve-hunt*. To look back into the Saxon times, we find that in Athelſtan's reign wolves abounded ſo in Yorkſhire, that a retreat was built at Flixton in that county, "to defend paſſengers from the wolves, that they ſhould not be devoured by them;" and ſuch ravages did theſe animals make during winter, particularly in January, when the cold was ſevereſt, that the Saxons diſtinguiſhed that month by the name of the *wolf-month*.—At the Cape of Good Hope, there is a ſpecies called the tiger-wolf, which is poſſeſſed of the peculiar gift of being enabled, in ſome meaſure, to imitate the cries of other animals; by which means this arch-deceiver is ſometimes enabled to beguile and attract calves, foals, lambs, and other animals. Near ſome of the larger farms, where there is a great deal of cattle, this ravenous

beaſt

beaſt is to be found almoſt every night; and at the ſame time frequently from one hour to another betraying itſelf by its howlings, gives the dogs the alarm. In this caſe the cunning of the wolves is ſo great, that a party of them, half flying and half defending themſelves, will decoy the whole pack of dogs to follow them to the diſtance of a gun-ſhot or more from the farm, with a view to give an opportunity to the reſt of the wolves to come out from their ambuſcade, and, without meeting with the leaſt reſiſtance, carry off booty ſufficient for themſelves and their fugitive brethren. The tiger-wolf, though a much larger and ſtronger animal, does not venture, without being driven to the utmoſt neceſſity, to meaſure its ſtrength with the common dog, which is certainly an evident proof of its cowardice. Notwithſtanding this, the Hottentots inform us, that it is ſtill within the memory of man, that the tiger-wolf was bold enough to ſteal upon them and moleſt them in their huts, particularly by carrying off their children. This, however, is now no longer the caſe; a circumſtance, perhaps, proceeding from the introduction of fire-arms into the country, an invention which, in theſe latter times, has cauſed this as well as other wild beaſts to ſtand in greater awe of man than formerly. I have heard the following ſtory of the tiger-wolf mentioned, which is laughable enough, though perhaps not quite ſo probable: "At a feaſt near the Cape one night, a trumpeter who had got his fill was carried out of doors, in order that he might cool himſelf, and get ſober again. The ſcent of him ſoon drew thither a tiger-wolf, which threw him on his back, and dragged him along as a corpſe, up towards Table-mountain. During this, however, the drunken muſician waked, enough in his ſenſes to know the danger of his ſituation, and to ſound the alarm with his trumpet, which he carried faſtened to his ſide. The wild beaſt, as may eaſily be ſuppoſed, was not leſs frightened in his turn." Any other beſides a trumpeter would in ſuch circumſtances have been undoubtedly no better than wolf's meat.

The jackal appears to have the gift of ſcent equal to a dog, of which it ſeems to be a wild ſpecies. They go in packs of forty, fifty, and even two hundred; and hunt like hounds in full cry from evening to morning. They deſtroy flocks and poultry, but in a leſs degree than the wolf or fox: ravage the ſtreets of villages and gardens near towns, and will even deſtroy children, if left unprotected. They will enter ſtables and outhouſes, and devour ſkins, or any thing made of leather. They will familiarly enter a tent, and ſteal whatſoever they find from the ſleeping traveller. In default of living prey, they will feed on roots and fruits; and even on the moſt infected carrion: they will greedily diſinter the dead, and devour putrid carcaſes. They attend caravans, and follow armies, in hopes that death will provide them a banquet. Their voice naturally is a howl. Barking, Mr. Pennant obſerves, is latently inherent, and in their ſtate of nature ſeldom exerted; but its different modifications are adventitious, and expreffive of the new paffions and affections gained by a domeftic ftate. Their howlings and clamours in the night are dreadful, and fo loud that people can fcarcely hear one another fpeak. Dellon fays, their voice is like the cries of a great many children of different ages mixed together: when one begins to howl, the whole pack join in the cry. This animal is vulgarly called the *lion's provider*, from an opinion that it roufes the prey for that quadruped. The fact is, every creature in the foreft is fet in motion by the fearful cries of the jackals; the lion, and other beafts of rapine, by a fort of inftinct, attend to the chace, and lie in wait, to feize fuch timid animals as betake themfelves to flight at the noife of this nightly pack.

From what has been ftated, as well as from the contemplation of nature in general, it will appear, that there is an occult inftinctive principle infufed into the whole race of animal beings, whereby they are unerringly led on to the propagation and prefervation of their fpecies; yet fo as that no one fhall become too numerous for the exiftence of another, upon which they prey, or with which they live in a continual ftate of warfare. We may likewife remark, that the more fimilarity we difcover among brutes, the more amicable we find them towards each other, becaufe the corpufcles of their bodies have an agreement pleafing to their fenfitive faculty, without exciting the appetite—but for which the fame fpecies would inceffantly devour each other, and the purpofes of creation would be annihilated by the operation of its own works. Contrary however to fuch a violation of order, we find the beafts of the foreft, and brute animals in general, prey by *antipathy* upon thofe which are oppofite or inimical in fcent and fpecies to themfelves; and affociate by *fympathy* with thofe of fimilar and concordant qualities; but the moft powerful effect of fympathy is to be found between the male and female of one and the fame clafs of beings; as we fhall demonftrate more fatisfactorily and pleafingly, in our confiderations

On MAN.

MAN is placed at the head of the animal creation; and is a being who feels, reflects, thinks, contrives, and acts; who has power of changing his place upon the earth at pleafure; who poffeffes the faculty of communicating his thoughts by means of fpeech; and who has dominion over all other creatures on the face of the globe. Animated and enlightened by a ray from the Divinity, he furpaffes in dignity every material being. He fpends lefs of his time in folitude than in fociety, or in obedience to thofe laws which he himfelf has framed.

The hiftory of man is an object of attention highly interefting, whether we confider him in the different periods of his life, or take a view of the varieties of his

 fpecies,

ſpecies, or examine the wonderful ſymmetry and conſtruction of his parts in the womb, or the more mature completion and organization of his body in perfect manhood.—I ſhall therefore attempt firſt to give a ſhort ſketch of him in theſe different points of view; and then, by conſidering the actions and paſſions of his mind, the infirmities of his nature, the affections of his heart, the objects of his purſuits, and the impreſſion of the celeſtial, elementary, and atmoſpherical influx; of light, heat, colour, motion, magnetiſm, electricity, and the univerſal ſpirit of nature which acts upon his conſtitution; deduce thoſe obvious and inevitable cauſes that reſult from them, and which it ſhould be the care of every man to know, who would wiſh to diſcover the golden KEY to the occult operations of Nature, and to the ſecret of preſerving HEALTH and LONG LIFE.

Noſce teipſum, "Know thyſelf," is a precept worthy of the lawgiver of Athens: it has been called the firſt ſtep to wiſdom, and was formerly written in letters of gold over the portal of the Temple of Diana. In purſuit of this important information, MAN may be contemplated in the following reſpects:

PHYSIOLOGICALLY,—as a frail machine, chiefly compoſed of nerves and fibres interwoven with each other. His moſt perfect ſtate is during youth; and he is endowed with faculties more numerous, and in higher perfection, than thoſe of all other animals. "Man, intended for exerciſing dominion over the whole animal creation, is ſent by Nature into the world naked, forlorn, and bewailing his lot; he is then unable to uſe his hands or feet, and is incapable of acquiring any kind of knowledge without inſtruction; he can neither ſpeak, nor walk, nor eat, nor perform any action whatever by natural inſtinct:" *Pliny.*—"We may judge what kind of life is allotted to us by Nature, ſince it is ordained, as an omen, that we ſhould come weeping into the world:" *Seneca.*—"It is humiliating to the pride of man, to conſider the pitiable origin of this moſt arrogant of all the animals:" *Pliny.*

DIETETICALLY.—*Cura valetudinem.* Bodily health and tranquillity of mind are more to be deſired than all the riches, pomp, or glory, of a Crœſus, a Solomon, or an Alexander. Health is to be preſerved by moderation; it is deſtroyed by abſtinence, injured by a variety of delicacies, weakened by unuſual things, and ſtrengthened by the uſe of proper and accuſtomed fare. Man, learned in the pernicious art of cookery, is fond of many diſhes, rendered palatable by the injurious effects of fire, and by the baneful addition of wine. "Hunger is ſatisfied with a ſmall quantity of food, while luxury demands over-abundance. Imagination requires vaſt ſupplies; while nature is contented with a moderate quantity of ordinary food, and is burthened by ſuperfluity:" *Seneca.*—According as thou liveſt, ſo ſhall thy life be enjoyed.

PATHO-

Pathologically.—*Memento mori!* The life of man resembles a bubble ready to burst; his fate is suspended by a hair, and is dependent on the uncertain lapse of time. "The earth contains nothing more frail than man:" *Homer.*—"Nothing is weaker than human life: to what dangers, and to how many diseases, is it not exposed! Hence the whole period of a man's life is but a span: half of it is necessarily spent in a state resembling death; without including the years of infancy, wherein there is no judgment; or the period of old age, fertile in sufferings, during which the senses are blunted, the limbs become stiff, and the faculties of sight and hearing, the powers of walking, and the teeth, the instruments of nourishment, fail before the rest of the body:" *Pliny.*—"Thus a considerable part of death is suffered during life; and death possesses all that belonged to the times which are past. Finally, nature will speedily recal and destroy all the beings which thou seest, and all that thy imagination can suppose to exist hereafter; for death calls equally upon all, whether they be good or whether they be evil:" *Seneca.*

Naturally.—*Innocui vivite, Numen adest!* Man, the prince of animated beings, who is a miracle of nature, and for whom all things on this earth were created, is a mimic animal, weeping, laughing, singing, speaking; tractable, judicious, inquisitive, and most wise; he is weak and naked, unprovided with natural weapons, exposed to all the injuries of fortune, needful of assistance from others, of an anxious mind, solicitous of protection, continually complaining, changeable in temper, obstinate in hope, and slow in the acquisition of wisdom. He despises the time which is past, abuses that which is present, and sets his affections on the uncertain future; thus continually neglecting winged time, which, though infinitely precious, can never be recalled: for thus the best and readiest time, in every age, flies on with miserable mortals; some it summons to attend their daily and burthensome labours; some it confines to luxurious inaction, pampered even to suffocation with superfluities; some it solicits in the ever-restless paths of ambition; some it renders anxious for the acquisition of wealth, and distresses by the possession of the thing desired; some it condemns to solitude, and others to have their doors continually crouded with visitors; here one bewails the conduct of his children, there one grieves their loss. Tears will sooner fail us than their causes, which only oblivion can remove. "On every hand our evils overbalance our advantages; we are surrounded with dangers; we rush forward into untried situations; we are enraged without having received provocation; like wild beasts, we destroy those we do not hate; we wish for favourable gales, which lead us only to destruction; the earth yawns wide, ready for our death:" *Seneca.*—"Other animals unite together against enemies of a kind different from their own, while man suffers most injuries from his own species:" *Pliny.*

Poli-

Politically.—*Esto antiqua virtute et fide!* Man, instead of following that which is right, is subjected to the guidance of manifest error; this envelopes all his faculties under the thick veil of custom, as soon as he is born; according to its dictates he is fed, educated, brought up, and directed in all things; and by its arbitrary rules his honesty, fortitude, wisdom, morality, and religion, are judged of; thus, governed by opinion, he lives conformably to custom, instead of being guided by reason. Though sent into the world a perishable being, (for all are evidently born to suffer,) instead of endeavouring to secure those things which are most advantageous and truly beneficial, he, infatuated by the smiles of fortune, anxiously collects her gaudy trifles for future enjoyment, and neglects her real benefits; he is driven to madness by envious snarlers; he persecutes with hatred the truly religious for differing from himself in speculative opinions; he excites numberless broils, not that he may do good, but for a purpose that even himself is ignorant of. He wastes his precious and irrecoverable time in trifles; he thinks lightly of immortal and eternal concerns, while regulating the succession of his posterity; and perpetually entering on new projects, forgetful of his real condition, he builds palaces instead of preparing his grave; till at length, in the midst of his schemes, death seizes him; and then, first opening his eyes, he perceives, O man! that all is delusion. "Thus we live as if immortal, and first learn in death that we have to die:" *Seneca.*

Morally.—*Benefac et lætare!* Man is composed of an animated medullary substance, which prompts him to that which is right; and of a bodily frame liable to impressions which instigate him to the enjoyment of pleasure. In his natural state he is foolish, wanton, an inconsiderate follower of example, ambitious, profuse, dissatisfied, cunning, peevish, invidious, malicious, and covetous; by the influence of just morals he is transformed to be attentive, chaste, considerate, modest, temperate, quiet, sincere, mild, beneficent, grateful, and contented. "Sorrow, luxury, ambition, avarice, the desire of life, and anxiety for the future, are common to all animals:" *Pliny.*

Theologically.—*Memento Creatoris tui!* Man, the ultimate purpose of creation, and masterpiece of the works of Omnipotence, was placed on earth that he might contemplate its perfections; he was endowed with sapient reason, and made capable of forming conclusions from the impressions of his senses, that, from a consideration of created objects he might know their Creator as the Almighty, the Infinite, the Omniscient, the Eternal, God: that we may live morally under his governing care, it is requisite that we have a thorough conviction of his existence, and must have it ever in remembrance. "There are two things which lead to a knowledge of God; creation and revelation:" *Augustine.*—"God, therefore, may be found out by the light of nature, but is only to be known by the aſſiſtance of doctrine:" *Tertullian.*—"Man alone has the ineſtimable privilege of contemplating the perfections of God, who is the author both of nature and of revelation:" *Ibid.*—"Learn that God has both ordered you to exiſt, and that you ſhould ſtudy to act that part properly which is alloted for you in life:" *Perſ.* Sat. iii. 71.

In the Syſtema Naturæ, MAN *(Homo)* is ranked as a diſtinct genus of the order *Primates* or "Chiefs," belonging to the *Mammalia* claſs of animals, or thoſe which nouriſh their young by means of lactiferous teats or paps. Of this genus he is the only ſpecies: and denominated *Sapiens*, as being endowed with wiſdom far ſuperior to, or rather in excluſion of, all other animals.—He *varies*, from climate, education, and habits; and the following varieties, excluſive of *wild men*, are enumerated by Linnæus.

Americans. "Of copper-coloured complexion, choleric conſtitution, and remarkably erect."—Their hair is black, lank, and coarſe; their noſtrils are wide; their features harſh, and the chin is ſcantily ſupplied with beard. Are obſtinate in their tempers, free and ſatisfied with their condition; and are regulated in all their proceedings by traditional cuſtoms.—Paint their ſkin with red ſtreaks.

Europeans. "Of fair complexion, ſanguine temperament, and brawny form." The hair is flowing, and of various ſhades of brown; the eyes are moſtly blue.—They are of gentle manners, acute in judgment, of quick invention, and governed by fixed laws.—Dreſs in cloſe veſtments.

Aſiatics. "Of ſooty complexion, melancholic temperament, and rigid fibre."—The hair is ſtrong, black, and lank; the eyes are dark brown. They are of grave, haughty, and covetous, manners; and are governed by opinions.—Dreſs in looſe garments.

Africans. "Of black complexion, phlegmatic temperament, and relaxed fibre." The hair is black and frizzly; the ſkin ſoft and ſilky; the noſe flat: the lips are thick; and the female has a natural apron, and long lax breaſts.—They are of crafty, indolent, and careleſs, diſpoſitions, and governed in their actions by caprice. —Anoint the ſkin with greaſe.

The following arrangement of the varieties in the human ſpecies is offered by Dr. Gmelin as more convenient than that of Linnæus:

a, White, *(Hom. albus.)* Formed by the rules of ſymmetrical elegance and beauty; or at leaſt what we conſider as ſuch.—This diviſion includes almoſt all the inhabitants of Europe; thoſe of Aſia on this ſide of the Oby, the Caſpian, Mount Imaus, and the Ganges; likewiſe the natives of the north of Africa, of Greenland, and the Eſquimaux.

b, Brown:

b, Brown: *(Hom. badius.)* Of a yellowiſh brown colour; has ſcanty hair, flat features, and ſmall eyes.—This variety takes in the whole inhabitants of Aſia not included in the preceding diviſion.

c, Black: *(Hom. niger.)* Of black complexion; frizzly hair, flat noſe, and thick lips.—The whole inhabitants of Africa, excepting thoſe of its more northern parts.

d, Copper-coloured: *(Hom. cupreus.)* The complexion of the ſkin reſembles the colour of copper not burniſhed.—The whole inhabitants of America, except the Greenlanders and Eſquimaux.

e, Tawny: *(Hom. fuſcus.)* Chiefly of a dark blackiſh-brown colour; having a broad noſe, and harſh coarſe ſtraight hair.—The inhabitants of the ſouthern iſlands, and of moſt of the Indian iſlands.

Monſters. Of theſe there are ſeveral varieties; the firſt and ſecond of which, in the following liſt, are occaſioned by peculiarity of climate, while the reſt are produced by artificial management. 1. *Alpini:* The inhabitants of the northern mountains: they are ſmall in ſtature, active, and timid in their diſpoſitions. 2. *Patagonici:* The Patagonians of South America; of vaſt ſize, and indolent in their manners. 3. *Monorchides:* The Hottentots; having one teſticle extirpated. 4. *Imberbes:* Moſt of the American nations; who eradicate their beards and the hair from every part of the body except the ſcalp. 5. *Macrocephali:* The Chineſe; who have their heads artificially forced into a conical form. 6. *Plagiocephali:* The Canadian Indians, who have the fore part of their heads flattened, when young, by compreſſion.

We have likewiſe the following account of monſters: *Homines feri;* deſcribed as walking on all-fours, as being dumb, and as covered with hair.—1. A youth found in Lithuania, in 1761, reſembling a bear. 2. A youth found in Heſſe, in 1544, reſembling a wolf. 3. A youth in Ireland reſembling a ſheep, *(Tulp. Obſ.* iv. 9.) 4. A youth in Bamberg reſembling an ox, *(Camerarius.)* 5. A wild youth found in 1724 in Hanover. 6. Wild boys found in 1719 in the Pyrenees. 7. A wild girl found in 1717 in Overyſel. 8. A wild girl found in 1631 in Champagne. 9. A wild lad found near Leyden, *(Boerhaave.)*—Theſe and other inſtances of wild men, their ſimilitudes, extraction, and generation, being foreign to the preſent ſubject, I ſhall treat largely of them in a future work on NATURAL HISTORY.*

Thoſe characters in the form of man by which he is diſtinguiſhed from brute animals, are reducible to two heads. The firſt is the ſtrength of the muſcles of the legs, by which the body is ſupported in a vertical poſition above them: the ſecond conſiſts in the articulation of the head with the neck by the middle of its baſe. We ſtand upright, bend our body, and walk, without thinking on the power by which we

* Since publiſhed at No. 17, Ave-Maria-Lane, St Paul's, in 14 vols. 8vo. See vol. i. of that work.

we are ſupported in theſe ſeveral poſitions. This power reſides chiefly in the muſcles which conſtitute the principal part of the calf of the leg. Their exertion is felt, and their motion is viſible externally, when we ſtand upright and bend our body backwards and forwards. This power is no leſs great when we walk even on an horizontal plane. In aſcending a height, the weight of the body is more ſenſibly felt than in deſcending. All theſe motions are natural to man. Other animals, on the contrary, when placed on their hind legs, are either incapable of performing them at all, or do it partially, with great difficulty, and for a very ſhort time. The *gibbon*, and the *jocko* or *ourang-outang*, are the animals moſt reſembling man in their conſtruction: they can ſtand upright with much leſs difficulty than other brutes; but the reſtraint they are under in this attitude plainly ſhows that it is not natural to them. The reaſon is, that the muſcles in the back part of the leg in the gibbon and the jocko are not, as in man, ſufficiently large to form a calf, and conſequently not ſufficiently ſtrong to ſupport the thighs and body in a vertical line, and to preſerve them in that poſture.—The attitudes proper to man, and to the animals, are pointed out by the different manners in which the head is articulated with the neck. The two points, by which the oſſeous part of the head is connected with the firſt vertebra of the neck, and on which every movement of the head is made with the greateſt facility, are placed at the edge of the great foramen of the occipital bone, which in man is ſituated near the centre of the baſe of the cranium, affords a paſſage for the medullary ſubſtance into the vertebræ, and determines the place of the articulation of the head with the neck. The body and neck being, according to the natural attitude, in a vertical direction, the head muſt be placed in equilibrium upon the vertebræ as upon a pivot or point of ſupport. The face is on a vertical line, almoſt parallel to that of the body and neck. The jaws, which are very ſhort compared with thoſe of moſt other animals, extend very little farther forwards than the forehead.—No animal has, like man, its hind legs as long as the body, neck, and head, taken together, meaſuring from the top of the head to the os pubis.—In the frame of the human body the principal parts are nearly the ſame with thoſe of other animals; but in the connection and form of the bones, there is as great a difference as in the attitudes proper to each. Were a man to aſſume the natural poſture of quadrupeds, and try to walk by the help of his hands and feet, he would find himſelf in a very unnatural ſituation; he could not move his feet and head but with the greateſt difficulty and pain; and, let him make what exertions he pleaſed, he would find it impoſſible to attain a ſteady and continued pace. The principal obſtacles he would meet with would ariſe from the ſtructure of the pelvis, the hands, the feet, and the head.—The plane of the great occipital foramen, which in man is almoſt horizontal, puts the head in a kind of equilibrium upon

 the

the neck when we ſtand erect in our natural attitude; but, when we are in the attitude of quadrupeds, it prevents us from raiſing our head ſo as to look forwards, becauſe the movement of the head is ſtopped by the protuberance of the occiput, which then approaches too near the vertebræ of the neck.—In moſt animals, the foramen magnum of the occipital bone is ſituated at the back part of the head; the jaws are very long; the occiput has no protuberance beyond the aperture, the plane of which is in a vertical direction, or inclined a little forwards or backwards; ſo that the head is pendant, and joined to the neck by its poſterior part. This poſition of the head enables quadrupeds, though their bodies are in a horizontal direction, to preſent their muzzle forwards, and to raiſe it ſo as to reach above them, or to touch the earth with the extremity of their jaws when they bring their neck and head down to their feet. In the attitude of quadrupeds, man could touch the earth only with the fore part or the top of the head.—When man is ſtanding, his heel reſts upon the earth as well as the other parts of his foot; when he walks, it is the firſt part which touches the ground; man can ſtand on one foot: theſe are peculiarities in ſtructure and in the manner of moving which are not to be found in other animals. We may therefore conclude that man cannot be ranked in the claſs of quadrupeds. We may add, that in man the brain is much larger, and the jaws much ſhorter, than in any other animal. The brain, by its great extent, forms the protuberance of the occipital bone, the forehead, and all that part of the head which is above the ears. In animals, the brain is ſo ſmall, that moſt of them have no occiput, or the front is either wanting or little raiſed. In animals which have large foreheads, ſuch as the horſe, the ox, the elephant, &c. they are placed as low, and even lower, than the ears. Theſe animals likewiſe want the occiput, and the top of the head is of very ſmall extent. The jaws, which form the greateſt portion of the muzzle, are large in proportion to the ſmallneſs of the brain. The length of the muzzle varies in different animals: in ſolipede animals it is very long; it is ſhort in the ourang-outang; and in man it does not exiſt at all. No beard grows on the muzzle.

Anatomiſts have employed much pains in the ſtudy of the material part of man, and of that organization which determines his place in the animal creation. From tracing and combining his different external parts; from obſerving that his body is in ſome places covered with hair; that he can walk upon his hands and his feet at the ſame time, in the manner of quadrupeds; that, like certain animals which hold their food in their paws, he has two clavicles; that the female brings forth her young alive, and that her breaſts are ſupplied with milk; from theſe circumſtances we might be led to aſſign man a place in the claſs of viviparous quadrupeds. But, in truth, ſuch an arrangement would be defective, arbitrary, and abſurd. Man is not a quadruped: of all the animals, he alone can support himself continually, and without restraint, in an erect posture; (that is, with his head and body in a vertical line upon his legs.) In this majestic and dignified attitude, he can change his place, survey this earth which he inhabits, and turn his eyes towards the vault of heaven. By a noble and easy gait, he preserves an equilibrium in the several parts of his body, and transports himself from one place to another with different degrees of celerity. To man alone nature has denied a covering; but still he is her masterpiece, the last work which came from the hands of the Almighty Artist, the sovereign and the chief of animals, a world in miniature, the centre which connects the universe together. The form of his body, the organs whereof are constructed in such a manner as to produce a much greater effect than those of other animals, announces his power. Every thing demonstrates the excellence of his nature, and the immense distance placed by the bounty of the Creator between man and beast. Man is a reasonable being; brute animals are deprived of that noble faculty. The weakest and most stupid of the human race is able to manage the most sagacious quadruped; he commands it, and makes it subservient to his use. The operations of brutes are purely the effect of mechanical impulse, and continue always the same; human works are varied without end, and infinitely diversified in the manner of execution. The soul of man is free, independent, and immortal. He is fitted for the study of science, and the cultivation of art; he has the exclusive privilege of examining every thing which has existence, and of holding communication with his fellow-creatures by language, by particular motions of the body, and by marks and characters mutually agreed upon. Hence arises that physical pre-eminence which he enjoys over all animals; and hence that power which he possesses over the elements, and (so to speak) over nature itself. Man, therefore, is unequalled in his kind; but the individuals thereof differ greatly from one another in figure, stature, colour, manners, and dispositions. The globe which man inhabits is covered with the productions of his industry and the works of his hands: it is his labour, in short, which gives a value to the whole terrestrial mass.

Nothing (says M. Buffon) exhibits such a striking picture of our weakness as the condition of an infant immediately after birth. Incapable of employing its organs, it needs assistance of every kind. In the first moments of our existence, we present an image of pain and misery, and are more weak and helpless than the young of any other animal. At birth, the infant passes from one element to another: when it leaves the gentle warmth of the tranquil fluid by which it was completely surrounded in the womb of the mother, it becomes exposed to the impressions of the air, and instantly feels the effects of that active element. The air acting upon the olfactory nerves, and upon the organs of respiration, produces a ſhock ſomething like ſneezing, by which the breaſt is expanded, and air is admitted into the lungs. In the mean time, the agitation of the diaphragm preſſes upon the viſcera of the abdomen, and the excrements are thus for the firſt time diſcharged from the inteſtines, and the urine from the bladder. The air dilates the veſicles of the lungs, and, after being rarefied to a certain degree, is expelled by the ſpring of the dilated fibres re-acting upon this rarefied fluid. The infant now reſpires; and articulates ſounds, or cries.—Moſt animals are blind for ſome days after birth: infants open their eyes to the light the moment they come into the world; but they are dull, fixed, and commonly blue. The new-born child cannot diſtinguiſh objects, becauſe he is incapable of fixing his eyes upon them. The organ of viſion is yet imperfect; the cornea is wrinkled; and perhaps the retina is too ſoft for receiving the images of external objects, and for communicating the ſenſation of diſtinct viſion. At the end of forty days, the infant begins to hear and to ſmile. About the ſame time it begins to look at bright objects, and frequently to turn its eyes towards the window, a candle, or any light. Now likewiſe it begins to weep; for its former cries and groans were not accompanied with tears. Smiles and tears are the effect of two internal ſenſations, both of which depend on the action of the mind. Thus they are peculiar to the human race, and ſerve to expreſs mental pain or pleaſure; while the cries, motions, and other marks of bodily pain and pleaſure, are common to man and moſt of the other animals. Conſidering the ſubject as metaphyſicians, we ſhall find that pain and pleaſure are the univerſal power which ſets all our paſſions in motion.

The ſize of an infant born at the full time is commonly twenty-one inches; and that *fœtus*, which nine months before was an imperceptible bubble, now weighs ten or twelve pounds, and ſometimes more. The head is large in proportion to the body; and this diſproportion, which is ſtill greater in the firſt ſtage of the fœtus, continues during the period of infancy. The ſkin of a new-born child is of a reddiſh colour, becauſe it is ſo fine and tranſparent as to allow a ſlight tint of the colour of the blood to ſhine through. The form of the body and members is by no means perfect in a child ſoon after birth; all the parts appear to be ſwollen. At the end of three days, a kind of jaundice generally comes on, and at the ſame time milk is to be found in the breaſts of the infant, which may be ſqueezed out by the fingers. The ſwelling decreaſes as the child grows up.

The liquor contained in the amnios leaves a viſcid whitiſh matter upon the body of the child. In this country we have the precaution to waſh the new-born infant only with warm water; but it is the cuſtom with whole nations, inhabiting the coldeſt climates, to plunge their infants into cold water as ſoon as they are born, without their receiving the leaſt injury. It is even ſaid that the Laplanders leave

their

their children in the ſnow till the cold has almoſt ſtopped their reſpiration, and then plunge them into a warm bath. Among theſe people, the children are alſo waſhed thrice a-day during the firſt year of their life. The inhabitants of northern countries are perſuaded that the cold bath tends to make men ſtronger and more robuſt, and on that account accuſtom their children to the uſe of it from their infancy. The truth is, that we are totally ignorant of the power of habit, or how far it can make our bodies capable of ſuffering, of acquiring, or of loſing.

The child is not allowed to ſuck as ſoon as it is born; but time is given for diſcharging the liquor and ſlime from the ſtomach, and the *meconium* or excrement, which is of a black colour, from the inteſtines. As theſe ſubſtances might ſour the milk, a little diluted wine mixed with ſugar is firſt given to the infant; and the breaſt is not preſented to it before ten or twelve hours have elapſed.

The young of quadrupeds can of themſelves find the way to the teat of the mother: it is not ſo with man; the mother, in order to ſuckle her child, muſt raiſe it to her breaſt; and, at this feeble period of life, the infant can expreſs its wants only by its cries.

New-born children have need of frequent nouriſhment. During the day, the breaſt ought to be given to them every two hours, and during the night as often as they wake. At firſt they ſleep almoſt continually; and they ſeem never to wake but when preſſed by hunger or pain. Sleep is uſeful and refreſhing to them; and it ſometimes becomes neceſſary to employ narcotic doſes, proportioned to the age and conſtitution of the child, for the purpoſe of procuring them repoſe. The common way of appeaſing the cries of children is by rocking them in the cradle; but this agitation ſhould be very gentle, otherwiſe a great riſk is run of confuſing the infant's brain, and of producing a total derangement. It is neceſſary to their being in good health, that their ſleep be long and natural. It is poſſible, however, that they may ſleep too much, and thereby endanger their conſtitution. In that caſe, it would be proper to take them out of the cradle, and awaken them by a gentle motion, or by preſenting ſome bright object to their eyes. At this age we receive the firſt impreſſions from the ſenſes, which, without doubt, are more important during the reſt of life than is generally imagined. Great care ought to be taken to place the cradle in ſuch a manner that the child ſhall be directly oppoſite to the light: for the eyes are always directed towards that part of the room where the light is ſtrongeſt: and, if the cradle be placed ſideways, one of them, by turning towards the light, will acquire greater ſtrength than the other, and the child will ſquint. For the two firſt months, no other food ſhould be given to the child but the milk of the nurſe; and, when it is of a weak and delicate conſtitution, this nouriſhment alone ſhould be continued during the third or fourth month. A child,

however

however robuſt and healthful, may be expoſed to great danger and inconvenience, if any other aliment is adminiſtered before the end of the firſt month. In Holland, Italy, Turkey, and the whole Levant, the food of children is limited to the milk of the nurſe for a whole year. The ſavages of Canada give their children ſuck for four, five, and ſometimes even ſeven, years. In this country, as nurſes generally have not a ſufficient quantity of milk to ſatisfy the appetite of their children, they commonly ſupply the want of it by panada, or other light preparations.

The teeth uſually begin to appear about the age of ſeven months. The cutting of theſe, although a natural operation, does not follow the common laws of nature, which act continually on the human body without occaſioning the ſmalleſt pain, or even producing any ſenſation. Here a violent and painful effort is made, accompanied with cries and tears. Children at firſt loſe their ſprightlineſs and gaiety; they become ſad, reſtleſs, and fretful. The gums are red, and ſwelled; but they afterwards become white, when the preſſure of the teeth is ſo great as to ſtop the circulation of the blood. Children apply their fingers to their mouth, that they may remove the irritation which they feel there. Some relief is given, by putting into their hands a bit of ivory or of coral, or of ſome other hard and ſmooth body, with which they rub the gums at the affected part. This preſſure, being oppoſed to that of the teeth, calms the pain for a moment, contributes to make the membrane of the gum thinner, and facilitates its rupture. Nature here acts in oppoſition to herſelf; and an inciſion of the gum muſt ſometimes take place, to allow a paſſage to the tooth.

When children are allowed to cry too long and too often, ruptures are ſometimes occaſioned by the efforts they make. Theſe may eaſily be cured by the ſpeedy application of bandages; but, if this remedy has been too long delayed, the diſeaſe may continue through life. Children are very much ſubject to worms. Some of the bad effects occaſioned by theſe animals might be prevented by giving them a little wine now and then, for fermented liquors have a tendency to prevent their generation.

Though the body is very delicate in the ſtate of infancy, it is then leſs ſenſible of cold than at any other part of life. The internal heat appears to be greater: the pulſe in children is much greater than in adults; from which we are certainly intitled to infer, that the internal heat is greater in the ſame proportion. For the ſame reaſon, it is evident that ſmall animals have more heat than large ones; for the beating of the heart and of the arteries is always quicker in proportion to the ſmallneſs of the animal. The ſtrokes of the heart in a ſparrow ſucceed one another ſo rapidly, that they can ſcarcely be counted.

Till three years of age, the life of a child is very precarious. In the two or three following years, it becomes more certain; and at ſix or ſeven years of age, a child has a better chance of living than at any other period of life. From the bills of mortality publiſhed in London, it appears, that, of a certain number of children born at the ſame time, one half of them die the three firſt years: according to which, one half of the human race are cut off before they are three years of age. But the mortality among children is not near ſo great every-where as in London. M. Dupré de Saint Maur, from a great number of obſervations made in France, has ſhown that half of the children born at the ſame time are not extinct till ſeven or eight years have elapſed.

The period of infancy is followed by that of adoleſcence. This begins, together with puberty, at the age of twelve or fourteen, and commonly ends in girls at ſixteen, and in boys at eighteen, but ſometimes not till twenty-one, twenty-three, or twenty-five, years of age. According to its etymology (being derived from the Latin word *adoleſcentia,*) it is completed when the body has attained its full height. Thus, puberty accompanies adoleſcence, and precedes youth. This is the ſpring of life; this is the ſeaſon of pleaſures, of loves, and of graces: but alas! this ſmiling ſeaſon is of ſhort duration. Hitherto nature ſeems to have had nothing in view but the preſervation and increaſe of her work: ſhe has made no proviſion for the infant except what is neceſſary to its life and growth. It has lived, or rather enjoyed a kind of vegetable exiſtence, which was ſhut up within itſelf, and which it was incapable of communicating. In this firſt ſtage of life, reaſon is ſtill aſleep: but the principles of life ſoon multiply; and man has not only what is neceſſary to his own exiſtence, but what enables him to give exiſtence to others. This redundancy of life, this ſource of health and vigour, can no longer be confined, but endeavours to diffuſe and expand itſelf.

The age of puberty is announced by ſeveral marks. The firſt ſymptom is a kind of numbneſs and ſtiffneſs in the groins, accompanied with a new and peculiar ſenſation in thoſe parts which diſtinguiſh the ſexes. There, as well as in the arm-pits, ſmall protuberances of a whitiſh colour appear, which are the germs of a new production, of a kind of hair by which theſe parts are afterwards to be veiled. The voice, for a conſiderable time, is rough and unequal; after which it becomes fuller, ſtronger, and graver, than it was before. This change may eaſily be diſtinguiſhed in boys; but leſs ſo in girls, becauſe their voices are naturally ſharper. Theſe marks of puberty are common to both ſexes: but there are marks peculiar to each, ſuch as the diſcharge of the menſes and growth of the breaſts, in girls; the beard, and the emiſſion of ſemen, in boys: in ſhort, the feeling of venereal deſire, and the appetite which unites the ſexes. Among all races of mankind, the

females

females arrive at puberty ſooner than the males; but the age of puberty is different in different nations, and ſeems partly to depend on the temperature of the climate and the quality of the food. In all the ſouthern countries of Europe, and in cities, the greateſt part of girls arrive at puberty about twelve, and boys about fourteen, years of age. But in the northern parts, and in the country, girls ſcarcely arrive at puberty till they are fourteen or fifteen, and boys not till they are ſixteen or ſeventeen. In our climate, girls, for the greateſt part, have attained complete maturity at eighteen, and boys at twenty, years of age.

At the age of adoleſcence, and of puberty, the body commonly attains its full height. About that time, young people ſhoot out ſeveral inches almoſt at once. But there is no part of the human body which increaſes more quickly and more perceptibly than the organs of generation in both ſexes. In males, this growth is nothing but an unfolding of the parts, an augmentation in ſize; but in females, it often occaſions a ſhrinking and contraction, which have received different names from thoſe who have treated of the ſigns of virginity.

Marriage is a ſtate ſuitable to man, wherein he muſt make uſe of thoſe new faculties which he has acquired by puberty. At this period of life, the deſire of producing a being like himſelf is ſtrongly felt. The external form and the correſpondence of the organs of ſex occaſion without doubt that irreſiſtible attraction which unites the ſexes and perpetuates the race. By connecting pleaſure with the propagation of the ſpecies, nature has provided moſt effectually for the continuance of her work. *Increaſe and multiply* is the expreſs command of the Creator, and one of the natural functions of life. We may add, that at the age of puberty a thouſand impreſſions act upon the nervous ſyſtem, and reduce man to ſuch a ſituation, that he feels his exiſtence only in that voluptuous ſenſe, which then appears to become the ſeat of his ſoul, which engroſſes the whole ſenſibility of which he is ſuſceptible, and which at length proceeds to ſuch a height, that its attacks cannot long be ſupported without a general derangement of the whole machine. The continuance of ſuch a feeling may ſometimes indeed prove fatal to thoſe who indulge in exceſſive enjoyment; but it is equally dangerous to thoſe who obſtinately perſiſt in celibacy, eſpecially when ſtrongly ſolicited by nature to the contrary. The ſemen, being too long confined in the ſeminal veſſels, may, by its ſtimulant property, occaſion diſeaſes in both ſexes, and excite irritations ſo violent as to reduce man to a level with the brutes, which, when acted upon by ſuch impreſſions, are perfectly furious and ungovernable. When this irritation proceeds to extremity, it produces what is called the *furor uterinus* in women. The oppoſite habit, however, is infinitely more common, eſpecially in the temperate, and above all in the frozen, zones. After all, exceſs is much more to be dreaded than continency.

nency. The number of diſſolute and intemperate men afford us plenty of examples. Some have loſt their memory, ſome have been deprived of ſight, many have become early bald, and ſome have died, through mere weakneſs. In ſuch a caſe, bleeding is well known to be fatal. Young men cannot be too often warned of the irreparable injury they may do to their health; and parents, to whoſe care they are entruſted, ought to employ all the means in their power to turn them from ſuch dangerous exceſſes. But at the age of puberty, young men know not of how great importance it is to prolong this ſmiling ſeaſon of their days, whereon the happineſs or miſery of their future life ſo much depends. Then they look not forward to futurity, nor reflect on what is paſt, nor enjoy preſent pleaſures with moderation. How many ceaſe to be men, or at leaſt to have the faculties of men, before the age of thirty? Nature muſt not be forced: like a true mother, her object is the ſober and diſcreet union of the ſexes. It is ſufficient to obey when ſhe commands, and to anſwer when ſhe calls. Neither muſt we forget here to mention and condemn an outrage committed againſt nature, the ſhameful practice of which endangers the loſs of health, and the total ruin of the conſtitution; I mean that ſolitary libertiniſm, ſo extenſively explained in the Medical Part of my edition of Culpeper, by which a man or woman, deceiving nature as it were, endeavours to procure thoſe enjoyments which religion has forbidden except when connected with the happineſs of being a parent. Such then is the phyſical order which the Author of Nature, the great preſerver of the ſpecies as well as the individual, has appointed to induce man, by the attraction of pleaſure, to propagate and continue his race.

According to the ordinary courſe of nature, women are not fit for conception till after the firſt appearance of the menſes. When theſe ſtop, which generally happens about forty or fifty years of age, they are barren ever after. Their breaſts then ſhrink and decay, and the voice becomes feebler. Some, however, have become mothers before they have experienced any menſtrual diſcharge; and others have conceived at the age of ſixty, and ſometimes at a more advanced age. Such examples, though not unfrequent, muſt be conſidered as exceptions to the general rule; but they are ſufficient to ſhow that the menſtrual diſcharge is not abſolutely eſſential to generation. The age at which man acquires the faculty of procreating is not ſo diſtinctly marked. In order to the production of ſemen, the body muſt have attained a certain growth, which generally happens between twelve and eighteen years of age. At ſixty or ſeventy, when the body begins to be enervated by old age, the voice becomes weaker, the ſemen is ſecreted in ſmaller quantities, and it is often unprolific. There are inſtances, however, of old men who have procreated at the age of eighty or ninety. Boys have been found who had the faculty of generating

at

at nine, ten, or eleven, years of age; and young girls who have become pregnant at the age of ſeven, eight, or nine. But ſuch facts, which are very rare, ought to be conſidered as extraordinary phænomena in the courſe of nature.

At the age of puberty, or a few years after, the body attains its full ſtature. Some young men grow no taller after fifteen or ſixteen, and others continue to grow till the age of twenty or twenty-three. At this period they are very ſlender; but by degrees the members ſwell and begin to aſſume their proper ſhape; and, before the age of thirty, the body in men has attained its greateſt perfection with regard to ſtrength, conſiſtence, and ſymmetry Adoleſcence ends at the age of twenty or twenty-five; and at this period youth (according to the diviſion which has been made of the years of man's life into different ages) begins. It continues till the age of thirty or thirty-five.

The common ſtature of man is about five feet and three, four, five, ſix, or ſeven, inches; and of women about five feet and two, three, and four, inches. Men below five feet are of a ſmall ſtature. The Laplanders do not exceed four feet and a half; and the natives of ſome other countries are ſtill ſmaller. Women attain their full height ſooner than men. Haller computes, that, in the temperate climates of Europe, the medium ſtature of men is about five feet and five or ſix inches. It is obſerved by the ſame author, that in Swiſſerland the inhabitants of the plains are taller than thoſe of the mountains. It is difficult to aſcertain with preciſion the actual limits of the human ſtature. In ſurveying the inhabited earth, we find greater differences in the ſtatures of individuals than in thoſe of nations. In the ſame climate, among the ſame people, and ſometimes in the ſame family, there are men whoſe ſtature is either too tall or too diminutive.—The body, having acquired its full height during the period of adoleſcence, and its full dimenſions in youth, remains for ſome years in the ſame ſtate before it begins to decay. This is the period of manhood, which extends from the age of thirty or thirty-five to that of forty or forty-five years. During this ſtage, the powers of the body continue in full vigour, and the principal change which takes place in the human figure ariſes from the formation of fat in different parts. Exceſſive fatneſs disfigures the body, and becomes a very cumberſome and inconvenient load.

The body of a well-ſhaped man ought to be ſquare, the muſcles ought to be ſtrongly marked, the contour of the members boldly delineated, and the features of the face well defined. In women, all the parts are more rounded and ſofter, the features are more delicate, and the complexion brighter. To man belong ſtrength and majeſty; gracefulneſs and beauty are the portion of the other ſex.—Every thing in both ſexes points them out as the ſovereigns of the earth; even the external appearance of man declares his ſuperiority to other living creatures. His body is erect; his attitude is that of command; his august countenance, which is turned towards heaven, bears the impression of his dignity. The image of his soul is painted in his face; the excellence of his nature pierces through the material organs, and gives a fire and animation to the features of his countenance. His majestic deportment, his firm and emboldened gait, announce the noblenefs of his rank. He touches the earth only with his extremity; he views it only at a distance, and seems to despise it. It has been justly observed, that the countenance of man is the mirror of his mind. In the looks of no animal are the expressions of passion painted with such energy and rapidity, and with such gentle gradations and shades, as in those of man. We know, that in certain emotions of the mind, the blood rises to the face, and produces blushing; and that in others the countenance turns pale. These two symptoms, the appearance of which depends on the structure and transparency of the reticulum, especially rednefs, constitute a peculiar beauty. In our climates, the natural colour of the face of a man in good health is white, with a lively red suffused upon the cheeks. Palenefs of the countenance is always a suspicious symptom. That colour which is shaded with black is a sign of melancholy and of vitiated bile; and constant and universal rednefs is a proof that the blood is carried with too great impetuosity to the brain. A livid colour is a morbid and dangerous symptom; and that which has a tint of yellow is a sign of jaundice or repletion of bile. The colour of the skin is frequently altered by want of sleep or of nourishment, or by loosenefs and diarrhœa.

Notwithstanding the general similitude of countenance in nations and families, there is a wonderful diversity of features. No one, however, is at a lofs to recollect the person to whom he intends to speak, provided he has once fully seen him. One man has livelinefs and gaiety painted in his countenance, and announces beforehand, by the cheerfulnefs of his appearance, the character which he is to support in society. The tears which bedew the cheeks of another man would excite compassion in the most unfeeling heart. Thus the face of man is the rendezvous of the symptoms both of his moral and physical affections; tranquillity, anger, threatening, joy, smiles, laughter, malice, love, envy, jealousy, pride, contempt, disdain or indignation, irony, arrogance, tears, terror, astonishment, horror, fear, shame or humiliation, sorrow and affliction, compassion, meditation, particular convulsions, sleep, death, &c. &c. The difference of these characters is of sufficient importance to form a principal article in the natural history of man.

When the mind is at ease, all the features of the face are in a state of profound tranquillity. Their proportion, harmony, and union, point out the serenity of the thoughts. But, when the soul is agitated, the human face becomes a living canvass, whereon the passions are represented with equal delicacy and energy, where every

7 emotion

emotion of the ſoul is expreſſed by ſome feature, and every action by ſome mark; the lively impreſſion of which anticipates the will, and reveals by pathetic ſigns our ſecret agitation, and thoſe intentions which we are anxious to conceal. It is particularly in the eyes that the ſoul is painted in the ſtrongeſt colours and with the moſt delicate ſhades.

The different colours of the eyes are, dark hazel, light hazel, green, blue, grey, and whitiſh-grey. The moſt common of theſe colours are hazel and blue, both of which are often found in the ſame eye. Eyes which are commonly called black are only dark hazel; they appear black in conſequence of being contraſted with the white of the eye. Wherever there is a tint of blue, however ſlight, it becomes the prevailing colour, and outſhines the hazel, with which it is intermixed, to ſuch a degree, that the mixture cannot be perceived without a very narrow examination. The moſt beautiful eyes are thoſe which appear black or blue. In the former, there is more expreſſion and vivacity; in the latter, more ſweetneſs and perhaps delicacy. Next to the eyes, the parts of the face by which the phyſiognomy is moſt ſtrongly marked are the eye-brows. Being of a different nature from the other parts, their effect is increaſed by contraſt. They are like a ſhade in a picture, which gives relief to the other colours and forms.

The forehead is one of the largeſt parts of the face, and contributes moſt to its beauty. Every body knows of how great importance the hair is in the phyſiognomy, and that baldneſs is a very great defect. When old age begins to make its approaches, the hair which firſt falls off is that which covers the crown of the head and the parts above the temples. We ſeldom ſee the hair of the lower part of the temples, or of the back of the head, completely fall of. Baldneſs is peculiar to men; women do not naturally loſe their hair, though it becomes white, as well as that of men, at the approach of old age.

The noſe is the moſt prominent feature of the face; but, as it has very little motion, and that only in the moſt violent paſſions, it contributes leſs to the expreſſion than to the beauty of the countenance. The noſe is ſeldom perpendicular to the middle of the face, but for the moſt part is turned to one ſide or the other. The cauſe of this irregularity, which, according to the painters, is perfectly conſiſtent with beauty, and of which even the want would be a deformity, appears to be frequent preſſure on one ſide of the cartilage of the child's noſe againſt the breaſt of the mother when it receives ſuck. At this early period of life, the cartilages and bones have acquired very little ſolidity, and are eaſily bent, as may be obſerved in the legs and thighs of ſome individuals, who have been injured by the bandages of the ſwaddling clothes.

Next

Next to the eyes, the mouth and lips have the greateſt motion and expreſſion. Theſe motions are under the influence of the paſſions. The mouth, which is ſet off by the vermilion of the lips and the enamel of the teeth, marks, by the various forms which it aſſumes, their different characters. The organ of the voice likewiſe gives animation to this feature, and communicates to it more life and expreſſion than is poſſeſſed by any of the reſt. The cheeks are uniform f eatures, and have no motion or expreſſion excepting from that involuntary redneſs or paleneſs with which they are covered in different paſſions; ſuch as ſhame, anger, pride, and joy, on the one hand; and fear, terror, and ſorrow, on the other.

In different paſſions, the whole head aſſumes different poſitions, and is affected with different motions. It hangs forward during ſhame, humility, and ſorrow; it inclines to one ſide in languor and compaſſion; it is elevated in pride; erect and fixed in obſtinacy and ſelf-conceit; in aſtoniſhment it is thrown backwards; and it moves from ſide to ſide in contempt, ridicule, anger, and indignation.—In grief, joy, love, ſhame, and compaſſion, the eyes ſwell and the tears flow. The effuſion of tears is always accompanied with an extenſion of the muſcles of the face, which opens the mouth.—In ſorrow, the corners of the mouth are depreſſed, the under lip riſes, the eye-lids fall down, the pupil of the eye is raiſed and half concealed by the eye-lid. The other muſcles of the face are relaxed, ſo that the diſtance between the eyes and the mouth is greater than ordinary; and conſequently the countenance appears to be lengthened.—In fear, terror, conſternation, and horror, the forehead is wrinkled, the eye-brows are raiſed, the eye-lids are opened as wide as poſſible, the upper lid uncovers a part of the white above the pupil, which is depreſſed and partly concealed by the under lid. At the ſame time, the mouth opens wide, the lips recede from each other, and diſcover the teeth both above and below.—In contempt and deriſion, the upper lip is raiſed at one ſide and expoſes the teeth, while the other ſide of the lip moves a little, and wears the appearance of a ſmile. The noſtril on the elevated ſide of the lip ſhrivels up, and the corner of the mouth falls down. The eye on the ſame ſide is almoſt ſhut, while the other is open as uſual; but the pupils of both are depreſſed, as when one looks down from a height.—In jealouſy, envy, and malice, the eye-brows fall down and are wrinkled; the eye-lids are elevated, and the pupils are depreſſed. The upper lip is elevated on both ſides, while the corners of the mouth are a little depreſſed, and the under lip riſes to join the middle of the upper.—In laughter, the corners of the mouth are drawn back and a little elevated; the upper part of the cheeks riſe; the eyes are more or leſs cloſed; the upper lip riſes, and the under one falls down; the mouth opens; and, in caſes of immoderate laughter, the ſkin of the noſe wrinkles. That gentler and more gracious kind of laughter which is called *ſmiling*, is ſeated

wholly

wholly in the parts of the mouth. The under lip rifes; the angles of the mouth are drawn back; the cheeks are puffed up; the eye-lids approach one another; and a fmall twinkling is obferved in the eyes. It is very extraordinary, that laughter may be excited either by a moral caufe without the immediate action of external objects, or by a particular irritation of the nerves without any feeling of joy. Thus an involuntary laugh is excited by a flight tickling of the lips, of the palm of the hand, of the fole of the foot, of the arm-pits, and below the middle of the ribs. We laugh when two diffimilar ideas, the union of which was unexpected, are prefented to the mind at the fame time; and when one or both of thefe ideas, or their union, include fome abfurdity which excites an emotion of difdain mingled with joy. In general, ftriking contrafts never fail to produce laughter.—A change is produced in the features of the countenance by weeping as well as by laughing. When we weep, the under lip is feparated from the teeth, the forehead is wrinkled, the eye-brows are depreffed, the dimple, which gives a gracefulnefs to laughter, forfakes the cheek; the eyes are more compreffed, and almoft conftantly bathed in tears, which in laughter flow more feldom and lefs copioufly.

The arms, hands, and every part of the body, contribute to the expreffion of the paffions. In joy, for inftance, all the members of the body are agitated with quick and various motions. In languor and forrow, the arms hang down, and the whole body remains fixed and immoveable. In admiration and furprife, this total fufpenfion of motion is likewife obferved. In love, defire, and hope, the head and eyes are raifed to heaven, and feem to folicit the wifhed-for good; the body leans forward, as if to approach it; the arms are ftretched out, and feem to feize beforehand the beloved object. On the contrary, in fear, hatred, and horror, the arms feem to pufh backward and repel the object of our averfion; we turn away our head and eyes as if to avoid the fight of it; we recoil in order to fhun it.

Although the human body is externally much more delicate than that of any other animal, yet it is very nervous, and perhaps ftronger in proportion to its fize than that of the ftrongeft animals. We are affured that the porters at Conftantinople carry burdens of nine hundred pounds weight, yet drink nothing but water. A thoufand wonderful ftories are related of the Hottentots and other favages, concerning their agility in running. Civilized man knows not the full extent of his powers, nor how much he lofes by that effeminacy and inactivity by which they are weakened and deftroyed. He is contented even to be ignorant of the ftrength and vigour which his members are capable of acquiring by motion, and by being accuftomed to fevere exercifes, as is obferved in runners, tumblers, and rope-dancers. The conclufion is, therefore, founded on the moft juft and indifputable induction and analogy.

analogy.—The attitude of walking is lefs fatiguing to man than that in which he is placed when he is ftopped in running. Every time he fets his foot upon the ground, he paffes over a more confiderable fpace; the body leans forward, and the arms follow the fame direction; the refpiration increafes, and breathing becomes difficult. Leaping begins with great inflexions of the members; the body is then much fhortened, but immediately ftretches itfelf out with a great effort. The motions which accompany leaping make it very fatiguing.

It is obferved that a ceffation from exercife is not alone fufficient to reftore the powers of the body when they are exhaufted by fatigue. The fprings, though not in action, are ftill wound up while we are awake, even when every movement is fufpended. In *fleep* nature finds that repofe which is fuited to her wants; and the different organs enjoy a falutary relaxation. This is that wonderful ftate in which man, unconfcious of his own exiftence, and funk in apparent death, repairs the lofs which his faculties have fuftained, and feems to affume a new exiftence. In this ftate of drowfinefs and repofe, the fenfes ceafe to act, the functions of the body are fufpended, and it feems abandoned to itfelf. The external fymptoms of fleep, which alone are the objects of our attention, are eafily diftinguifhed. At the approach of fleep, the eyes begin to wink, the eye-lids fall down, the head nods and hangs down: its fall aftonifhes the fleeper; he ftarts up, and makes an effort to drive away fleep, but in vain; a new inclination, ftronger than the former, deprives him of the power of raifing his head; his chin refts upon his breaft, and in this pofition he enjoys a tranquil fleep.

The age of decline extends from forty or forty-five to fixty or fixty-five years of age. At this time of life, the diminution of the fat is the caufe of thofe wrinkles which begin to appear in the face and fome other parts of the body. The fkin, not being fupported by the fame quantity of fat, and being incapable, from want of elafticity, of contracting, finks down and forms folds. In the decline of life, a remarkable change takes place alfo in vifion. In the vigour of our days, the cryftalline lens, being thicker and more diaphanous than the humours of the eye, enables us to read letters of a fmall character at the diftance of eight or ten inches. But, when the age of decline comes on, the quantity of the humours of the eye diminifhes, they lofe their clearnefs, and the tranfparent cornea becomes lefs convex. To remedy this inconvenience, we place what we wifh to read at a greater diftance from the eye: but vifion is thereby very little improved, becaufe the image of the object becomes fmaller and more obfcure. Another mark of the decline of life is a weaknefs of the ftomach, and indigeftion, in moft people who do not take fufficient exercife in proportion to the quantity and the quality of their food.—At fixty, fixty-three, or fixty-five, years of age, the figns of decline become more and more

vifible,

visible, and indicate *old age*. This period commonly extends to the age of seventy, sometimes to seventy-five, but seldom to eighty. When the body is extenuated and bent by old age, man then becomes *crazy*. Craziness therefore is nothing but an *infirm old age*. The eyes and stomach then become weaker and weaker; leanness increases the number of the wrinkles; the beard and the hair become white; the strength and the memory begin to fail.—After seventy, or at most eighty years of age, the life of man is nothing but labour and sorrow: such was the language of David near three thousand years ago. Some men of strong constitutions, and in good health, enjoy old age for a long time without decrepitude; but such instances are not very common. The infirmities of decrepitude continually increase, and at length death concludes the whole. This fatal term is uncertain. The only conclusions which we can form concerning the duration of life, must be derived from observations made on a great number of men who were born at the same time, and who died at different ages.

The signs of *decrepitude* form a striking picture of weakness, and announce the approaching dissolution of the body. The memory totally fails; the nerves become hard and blunted; deafness and blindness take place; the senses of smell, of touch, and of taste, are destroyed; the appetite fails; the necessity of eating, and more frequently that of drinking, are alone felt; after the teeth fall out, mastication is imperfectly performed, and digestion is very bad; the lips fall inwards; the edges of the jaws can no longer approach one another; the muscles of the lower jaw become so weak, that they are unable to raise and support it; the body sinks down; the spine is bent outward; and the vertebræ grow together at the anterior part; the body becomes extremely lean: the strength fails; the decrepid wretch is unable to support himself; he is obliged to remain on a seat, or stretched in his bed: the bladder becomes paralytic; the intestines lose their spring; the circulation of the blood becomes slower; the strokes of the pulse no longer amount to the number of eighty in a minute, as in the vigour of life, but are reduced to twenty-four, and sometimes fewer: respiration is slower; the body loses its heat; the circulation of the blood ceases; death follows; and the dream of life is no more.

Man, however, has no right to complain of the shortness of life. Throughout the whole of living beings, there are few who unite in a greater degree all the internal causes which tend to prolong its different periods. The term of gestation is very considerable; the rudiments of the teeth are very late in unfolding; his growth is slow, and is not completed before about twenty years have elapsed.—The age of puberty, also, is much later in man than in any other animal. In short, the parts of his body, being composed of a softer and more flexible substance, are not

so

so soon hardened as those of inferior animals. Man, therefore, seems to receive at his birth the seeds of a long life: if he reaches not the distant period which nature seemed to promise him, it must be owing to accidental or acquired causes, foreign to himself. Instead of saying that he has finished his life, we ought rather to say that he has not completed it.—The natural and total duration of life is in some measure proportioned to the period of growth. A tree or an animal, which soon acquires its full size, decays much sooner than another which continues to grow for a longer time. If it be true that the life of animals is eight times longer than the period of their growth, we might conclude that the boundaries of human life may be extended to a century and a half.

It does not appear that the life of man becomes shorter in proportion to the length of time the world has existed. In the days of the Psalmist, the ordinary limits of human life did not exceed seventy or eighty years. No king of Judah lived beyond that period. When the Romans, however, were numbered by Vespasian, there were found in the empire, in that age of effeminacy, ten men aged an hundred and twenty and upwards. Among the princes of modern times, Frederic the Great of Prussia lived to the age of 74. George II. of Britain lived to that of 77. Louis XIV. lived to the same age. Stanislaus King of Poland and Duke of Lorrain exceeded that age. Pope Clement XII. lived to the age of 80. George I. of Britain attained the age of 83. William Lecomte, a shepherd, died suddenly in 1776, in the county of Caux in Normandy, at the age of 110. Cramers, physician to the emperor, saw at Temeswar two brothers, the one aged 110 and the other 112, both of whom were fathers at that age. Saint Paul the hermit was 113 at his death. The Sieur Iswan-Horwaths, knight of the order of St. Louis, died at Sar-Albe in Lorraine, in 1775, aged almost 111; he was a great hunter; he undertook a long journey a short time before his death, and performed it on horseback. Rosine Iwiwarouska died at Minsk in Lithuania at the age of 113. Fockjel Johannes died at Oldeborn in Friesland, aged 113 years and 16 days. Marsk Jones died in the year 1775 at Villejac in Hungary, aged 119. John Niethen of Bakler in Zealand lived to the age of 120. Eleonora Spicer died in 1773, at Accomack in Virginia, aged 121. John Argus was born in the village of Lastua in Turkey, and died the 6th of March 1779, at the age of 123; having six sons and three daughters, by whom he had posterity to the fifth generation; they amounted to the number of 160 souls, and all lived in the same village: his father died at the age of 120. In December 1777, there lived in Devonshire a farmer named John Brookey, who was 134 years of age, and had been fifteen times married. The Philosophical Transactions mention an Englishman, of the name of Eccleston, who lived to the age of 143. Another Englishman, of the name of Effingham, died in 1757, at the

the age of 144. Niels Jukens, of Hammerſet in Denmark, died in 1764, aged 146. Chriſtian Jacob Drakemberg died in 1770 at Archuſen, in the 146th year of his age: this old man of the north was born at Stavangar in Norway in 1624, and at the age of 130 married a widow of 60. In Norway ſome men have lived to the age of 150. John Rovin, who was born at Szatlova-Carantz-Betcher, in the bannat of Temeſwar, lived to the age of 172, and his wife to that of 164, having been married to him during the ſpace of 147 years: when Rovin died, their youngeſt ſon was 99 years of age. Peter Zoten, a peaſant, and a countryman of John Rovin, died in 1724, at the age of 185: his youngeſt ſon was then 97 years of age. The hiſtory and whole-length pictures of John Rovin, Henry Jenkins, and Peter Zoten, are to be ſeen in the library of H. R. H. Prince Charles at Bruſſels. Hanovins, profeſſor at Dantzic, mentions in his Nomenclature an old man who died at the age of 184; and another ſtill alive in Wallachia, whoſe age, according to this author, amounted to 186. Thomas Parr, of Shropſhire, died November 16, 1635, aged 152. Henry Jenkins, of Yorkſhire, died December 8, 1670, aged 169. Robert Montgomery, of Yorkſhire, died in 1670, aged 196. James Sands, of Staffordſhire, aged 140, and his wife, 120. The Counteſs of Deſmond, of Ireland, aged 140. J. Sagar, of Lancaſhire, died in 1668, aged 112. —— Laurence, of Scotland, aged 140. Simon Sack, of Trionia, died May 30, 1764, aged 141. Col. Thomas Winſlow, of Ireland, died Auguſt 26, 1766, aged 146. Francis Conſiſt, of Yorkſhire, died in January 1768, aged 150. Margaret Forſter, aged 136, and her daughter, aged 104, of Cumberland, were both living in 1771. Francis Bons, of France, died Feb. 6, 1769, aged 121. James Bowels, of Killingworth, aged 152. John Tice, of Worceſterſhire, died March 1774, aged 125. John Mount, of Scotland, died Feb. 27, 1766, aged 136. A. Goldſmith, of France, died in June 1776, aged 140. Mary Yates, of Shropſhire, died in 1776, aged 128. John Bales, of Northampton, died April 5, 1766, aged 126. William Ellis, of Liverpool, died Auguſt 16, 1780, aged 130. Louiſa Truxo, a negreſs of Tucomea, South America, was living October 5, 1780, aged 175. Margaret Patten, of Lockneugh near Paiſley, aged 138. Janet Taylor, of Fintray, Scotland, died October 10, 1780, aged 108. Richard Lloyd, of Montgomery, aged 133. Suſannah Hilliar, of Piddington, Northamptonſhire, died Feb. 19, 1781, aged 110. Ann Cockbolt, of Stoke-Bruerne, Northamptonſhire, died April 5, 1775, aged 105. James Ayley, of Middlewich, Cheſhire, died March 17, 1781, aged 112. William Walker, who was a ſoldier at the battle of Edgehill, lived to the age of 112. Hippocrates, phyſician, of the iſland of Cos, aged 104. Democritus, philoſopher, of Abdera, aged 109. Galen, phyſician, of Pergamus, aged 140. Albuna Marc, of Ethiopia, aged 150. Dumitur Raduly, of Haromſzeck,

Tranſylvania, died Jan. 12, 1782, aged 140. Titus Fullonius, of Bononia, aged 150. Abraham Paiba, of Charleſtown, South Carolina, aged 142. L. Tertulia, of Arminium, aged 137. Lewis Cornaro, of Venice, aged 100. Robert Blakeney, Eſq. of Armagh, Ireland, aged 114. Margaret Scott, of Dalkeith, Scotland, aged 125. W. Gulſtone, of Ireland, aged 140. J. Bright, of Ludlow, aged 105. William Poſtell, of France, aged 120. Jane Reeves, of Eſſex, aged 103. W. Paulet, Marquis of Wincheſter, of Hampſhire, aged 106. John Wilſon, of Suffolk, aged 116. Patrick Wian, of Leſbury, Northumberland, aged 115. M. Laurence of Arcades, aged 140. Evan Williams, of Caermarthen work-houſe, was alive in October 1782, aged 145. John Jacobs, of Mount Jura, aged 121. This man, in 1789, at the age of 120, quitted his native hills, and from the ſummit of Mount Jura undertook a journey to Verſailles, to behold and return thanks to the national aſſembly for the vote which had freed him and his poor countrymen from the feudal yoke. In the early part of his life, he was a ſervant in the family of the Prince de Beaufremont. His memory continued good to the laſt day of his life; and the principal inconveniences which he felt from his great age were, that his ſight was weakened, and the natural heat of his body was ſo diminiſhed, that he ſhivered with cold in the middle of the dog-days if he was not ſitting by a good fire. This old man was received in the body of the houſe by the national aſſembly, indulged with a chair, and directed to keep on his hat leſt he ſhould catch cold if he were to ſit uncovered. A collection was made for him by the members, which exceeded 500l. ſterling; but he lived not to return to Mount Jura. He was buried on Saturday the 31ſt of January 1790, with great funeral pomp, in the pariſh-church of St. Euſtace, at Paris. Mathew Tait, of Auchinleck, Airſhire, died Feb. 19, 1792, aged 123: he ſerved as a private at the taking of Gibraltar in 1704. Donald Macleod, of the Iſle of Sky, was living in May 1793, aged 105. There was living in Portſmouth poor-houſe, in May 1793, one Elizabeth Bennett, aged 104 years.

Before we proceed to aſſign the common cauſes of longevity, it is proper to inquire into the manner of life and the ſituation of thoſe by whom it has been enjoyed. We find, then, that thoſe who have lived to the greateſt age have been ſuch as did not attain their full growth till a very advanced period of life, and who have kept their appetites and paſſions under the moſt complete ſubjection. In a word, thoſe who have exceeded 100 years, have in general been robuſt, laborious, ſober, and careful to obſerve the ſtricteſt regimen. Enjoying a good conſtitution from nature, they have ſeldom or never been ſubject to diſeaſe. They have even enjoyed the greateſt health and vigour, and retained the uſe of their ſenſes, to the laſt moment of their lives.

Among

Among thoſe who have led a life of contemplation and ſtudy, many have reached a very advanced age. Longevity is frequent among the different orders of religious, who by their ſtatutes are confined to a moderate diet, and obliged to abſtain from wine and the uſe of meat. Some celebrated anchorets have lived to a great age while they fed upon nothing but the wild roots and fruits which they found in the deſert whither they had retired. The philoſopher Xenophilus, who lived to the age of 106, was of the Pythagorean ſect. It is well known, that thoſe philoſophers who held the tranſmigration of ſouls denied themſelves the uſe of meat, becauſe they imagined that killing an animal would be to aſſaſſinate another ſelf. A country life has produced many ſound and vigorous old men. It is ſuppoſed that a happy old age is attained with greater difficulty in towns than in the country. Sir Hans Sloane, Duverney, and Fontenelle, however, are inſtances of men whoſe lives have been ſpent in cities, and yet extended to a very great length. It has been obſerved, that men deprived of reaſon live very long; which is to be imputed to their being exempt from thoſe inquietudes which are the moſt deadly poiſon. Perſons poſſeſſing a ſufficiently good underſtanding, but deſtitute of ambition, have been found to enjoy very long life. Men who are devoid of pretenſions, who are free from thoſe cares which a deſire of ſhining by a diſplay of talents, or of acquiring dignity and power, neceſſarily brings in its train, who feel no regret for the paſt nor anxiety about the future, are ſtrangers to thoſe torments of the mind which waſte and conſume the body. To that tranquillity of ſoul, which is ſo excellent a prerogative of infancy, they add that of being long young by phyſical conſtitution, on which the moral has a ſtriking and powerful influence.

Premature wiſdom, and early talents, are often fitter to excite aſtoniſhment than expectation. The rapid unfolding of the moral faculties, by ſhortening the period of youth, ſeems to diminiſh in proportion the total duration of life. We have known a young lady of ſeventeen, who could ſpeak very correctly ſeven languages: ſhe tranſlated and wrote Latin, Greek, Italian, Spaniſh, German, Engliſh, and French; but ſhe died at the age of eighteen. The young man by whom ſhe was aſked in marriage, having been informed that he could not obtain her hand till he had made himſelf worthy of her by the ſame degree of talents and information, died the ſame year, and at the ſame age.

From the preceding obſervations, Haller has attempted to deduce the cauſes why a few men are longer exempted than others from the common fate.—The circumſtances which oppoſe their influence are independent of our will; ſuch as the ravages of epidemic diſtempers; trouble and anxiety of mind, which create diſeaſes in the body: or the torments of ambition. It is neceſſary to live in a ſalubrious climate; to enjoy a fortune ſufficiently eaſy to exclude thoſe uneaſy deſires which create a feeling of want and privation; to be deſcended from healthy parents, to avoid drinking wine in youth; to drink water; and to eat little meat, and a great deal of vegetables. It is neceſſary alſo to be temperate in meals; moderate in pleaſures, ſtudy, and exerciſe; to be naturally inclined to cheerfulneſs; and to allot a due time to ſleep and repoſe.—Long life is certainly very rare; but, as has been already obſerved, we muſt diſtinguiſh between what is natural to the conſtitution of man and that which is the conſequence of his condition. By the former he is made to be long-lived: but nature is arreſted in her courſe by local and accidental cauſes, which it is not always in our power to avoid.

Let us take a retroſpective view of man's life from his infancy, and enumerate the chief of theſe different cauſes. Of a thouſand infants, extracted from the London bills of mortality, twenty-three died almoſt as ſoon as they came into the world; teething carried off fifty, and convulſions two hundred and ſeventy-ſeven: eighty died of the ſmall-pox, and ſeven of the meaſles. Among the adult females, eight at leaſt died in child-bed: conſumption and aſthma, diſeaſes more frequent in England than in France, carried off a hundred and ninety-one of the ſame ſex, and almoſt a fifth part of the full-grown men. A hundred and fifty died of fevers. At a more advanced age, twelve died of apoplexy, and forty-one of dropſy, without mentioning thoſe to whom diſeaſes of little importance in themſelves became mortal. There only remained ſeventy-eight whoſe death could be aſcribed to old age; and of theſe twenty-ſeven lived to the age of eighty and upwards. Among the different diſeaſes of which we have juſt now ſeen the fatal effects, and which carry off more than nine-tenths of mankind, not one, it muſt be allowed, is natural to the conſtitution. The inhabitants of this iſland are in general but little ſubject to diſeaſes, excepting the ſmall-pox and the meaſles; and many of them enjoy uninterrupted health to old age.—And here it may be proper to mention what are the moſt prevalent diſeaſes in other countries, which prove equally fatal to the duration of human life. In northern climates, ſcurvy, the cholic of the Laplanders, and the diſeaſes of the lungs, moſt frequently occaſion death. In temperate climates, dropſy carries off a great many at the beginning of old age, which is the boundary of life in the greateſt part of both ſexes, when they have eſcaped the acute diſeaſes, ſuch as putrid fever, &c. Acute diſeaſes are moſt common in warm countries. In ſome places, the rays of the ſun kill in a few hours thoſe who are expoſed to its burning heat. The air of Egypt and of Aſia Minor engenders the plague, by which one half of their inhabitants are carried off. Between the tropics men are ſubject to dyſenteries and violent fevers. The cold of the night, in warm climates, occaſions ſometimes violent diſeaſes, ſuch as palſy, quinſey, and a ſwelling of the head. Damp and marſhy places give riſe to fevers

of

of a different kind, but alſo very dangerous. The life of ſailors has a great tendency to produce ſcurvy. How many profeſſions prove fatal to the health, and haſten that period which nature would have brought on by ſlow degrees! Miners, ſtone-cutters, gilders, perſons employed in emptying privies, &c. are ſubject to diſeaſes of the lungs, and become paralytic. Other profeſſions of life bring on other accidents, of which it would carry us too far to give a particular account. What has been ſaid is ſufficient to ſhow, that it is the dangers with which we are ſurrounded that ſhorten the period of human exiſtence.

By examining the liſt of thoſe who have attained a great age, it will be found that mankind are longer lived in northern than in ſouthern countries. It has been obſerved, that there are more old men in mountainous and elevated ſituations than in plains and low countries. We repeat it, if the duration of life among the inhabitants of ſouthern climates be compared with the duration of life in northern nations, it will be allowed, that the latter enjoy both longer life and better health than the former. Their growth being retarded by the rigour of the climate, their decay muſt alſo be ſlower, becauſe of the proportion which exiſts between the growth of animals and the length of their lives. Among ten perſons who have lived to the age of an hundred, eight or nine will be found to have lived in the north.

It appears from the bills of mortality, that in the country more boys are born than girls; in cities, on the contrary, the number of females is commonly greateſt. Obſervations made with great care prove, that in moſt countries there are fewer men alive than women, and that more males die, chiefly at the firſt and laſt periods of life. In Sweden, the whole number of females, in 1763, was to that of males in the proportion of ten to nine. The number of old women who exceeded eighty years of age was to that of old men of the ſame age in proportion of thirty-three to nineteen; and there were more women than men who had attained the age of eighty-ſix, in proportion of almoſt two to one.

Dr. Price made obſervations, after Dr. Percival, on the difference of longevity, and the duration of human life, in towns, country-pariſhes, and villages; of which the following is the reſult: a greater number in proportion die in great towns than in ſmall ones, and a greater number in the latter than in villages. The cauſe of this difference, which is found to be very great, muſt be, in the firſt place, the luxury and diſſipation which prevail in towns; and, ſecondly, the badneſs of the air. In the town of Mancheſter, according to obſervation, 1-28th of the inhabitants die annually; whereas in the neighbouring country, the number of deaths does not exceed 1-46th of the whole inhabitants. It may be laid down as a general principle, that in great towns, the number of deaths annually is from one in nineteen to one in twenty-two or twenty-three; in middling towns, from one in twenty-four to one in twenty-eight; and in country pariſhes and villages ſeldom more than one in forty or fifty. In 1763, the number of inhabitants in Stockholm amounted to 72979. The average number of deaths for the ſix years preceding had been 3802, which makes one in nineteen annually; while throughout all Sweden, including the towns and the country, not more than one in thirty-five die annually. At Rome the inhabitants are numbered every year. In 1771 they were found to amount to 159675: the average number of deaths for ten years was 7367; which makes one in twenty-three and a half annually. In London not leſs than one in twenty three-fourths of the inhabitants die every year.

M. Daubenton has given, in the Encyclopédie Méthodique, a table of the probabilities of the duration of life, conſtructed from that which is to be found in the ſeventh volume of the Supplémens à l'Hiſtoire Naturelle de M. de Buffon. The following is an abridgement of it:

Of twenty-three thouſand nine hundred and ninety-four children, born at the ſame time, there will probably die,

In one year - - - -	7998
Remaining 2-3ds, or 15996.	
In eight years - - - -	11997
Remaining 1-half, or 11997.	
In thirty-eight years - - - -	15996
Remaining 1-3d, or 7998.	
In fifty years - - - -	17994
Remaining 1-4th, or 5998	
In ſixty-one years - - - -	19995
Remaining 1-6th, or 3999.	
In ſeventy years - - - -	21595
Remaining 1-10th, or 2399.	
In eighty years - - - -	22395
Remaining 1-40th, or 599.	
In ninety years - - - -	23914
Remaining 1-300th, or 79.	
In one hundred years - - -	23992
Remaining 1-10000th, or 2.	

It thus appears, that a very ſmall number of men indeed paſs through all the periods of life, and arrive at the goal marked out by nature. Innumerable cauſes accelerate our diſſolution. The life of man, we have obſerved, conſiſts in the activity and

and exercife of his organs, which grow up and acquire ftrength during infancy, adolefcence, and youth. No fooner has the body attained its utmoft perfection, than it begins to decline. Its decay is at firft imperceptible; but in the progrefs of time the membranes become cartilaginous, the cartilages acquire the confiftence of bone; the bones become more folid, and all the fibres are hardened. Almoft all the fat waftes away; the fkin becomes withered and fcaly; wrinkles are gradually formed; the hair grows white; the teeth fall out; the face lofes its fhape; the body is bent; and the colour and confiftence of the cryftalline humour become more perceptible. The firft traces of this decay begin to be perceived at the age of forty, and fometimes fooner; this is the *age of decline.* They increafe by flow degrees till fixty, which is the *period of old age.* They increafe more rapidly till the age of feventy or feventy-five. At this period *crazinefs* begins, and continues always to increafe. Next fucceeds *decrepitude,* when the memory is gone, the ufe of the fenfes loft, the ftrength totally annihilated, the organs worn out, and the functions of the body almoft deftroyed. Little now remains to be loft; and, before the age of ninety or a hundred, death terminates at once decrepitude and life.

The body then dies by little and little; its motion gradually diminifhes; life is extinguifhed by fucceffive gradations, and death is only the laft term in the fucceffion. When the motion of the heart, which continues longeft, ceafes, man has then breathed his laft; he has paffed from the ftate of life to the ftate of death; and, as at his birth a breath opened to him the career of life, fo with a breath he finifhes his courfe.

This natural caufe of death is common to all animals, and even to vegetables. We may obferve that the centre of an oak firft perifhes and falls into the duft, becaufe thefe parts, having become harder and more compact, can receive no further nourifhment. The caufes of our diffolution, therefore, are as neceffary as death is inevitable; and it is no more in our power to retard this fatal term than to alter the eftablifhed laws of the univerfe. Hence the following maxim has been univerfally adopted; *Contra vim mortis, nullum medicamentum in hortis.* In whatever manner death happens, the time and circumftances thereof are unknown. It is confidered, however, as at all times terrible, and the very thoughts of it fill the mind with fear and trouble. It is notwithftanding our duty frequently to direct our thoughts to that event, which muft inevitably happen, and by a life of virtue and innocence to prepare againft thofe confequences which we fo much dread.

As in women the bones, the cartilages, the mufcles, and every other part of the body, are fofter and lefs folid than thofe of men, they muft require more time in hardening to that degree which occafions death.—Women of courfe ought to live longer than men. This reafoning is confirmed by experience; for, by confulting

4 the

the bills of mortality, it appears, that, after women have paſſed a certain age, they live much longer than men who have arrived at the ſame age.—In like manner, it is found by experience, that in women the age of youth is ſhorter and happier than in men, but that the period of old age is longer, and attended with more trouble. *Citius pubeſcunt, citius ſeneſcunt.*

After death, the organization of the body begins to be diſſolved, and all the parts relax, corrupt, and ſeparate. This is produced by an inteſtine fermentation, which occaſions putrefaction, and reduces the body to volatile alkali, fetid oil, and earth.

The deſire of ſelf-preſervation, and of protracting the ſhort ſpan of life, is ſo intimately interwoven with our conſtitution, that it is juſtly eſteemed one of the firſt principles of our nature, and, in ſpite even of pain and miſery, ſeldom quits us to the laſt moments of our exiſtence. It ſeems, therefore, to be no leſs our duty than our intereſt to examine minutely into the various means that have been conſidered as conducive to health and long life; and, if poſſible, to diſtinguiſh ſuch circumſtances as are eſſential to that great end, from thoſe which are merely accidental.

It has long been known that freſh air is more immediately neceſſary to life than food; for a man may live two or three days without the latter, but not many minutes without the former. The vivifying principle contained in the atmoſphere, ſo eſſential to the ſupport of flame, as well as animal life, concerning which authors have propoſed ſo many conjectures, is nothing elſe but the pure oxygenated fluid diſcovered by that ingenious philoſopher Lavoiſier. The common atmoſphere may well be ſuppoſed to be more or leſs healthy in proportion as it abounds with this animating principle. As this exhales in copious ſtreams from the green leaves of all kinds of vegetables, even from thoſe of the moſt poiſonous kind, may we not, in ſome meaſure, account why inſtances of longevity are ſo much more frequent in the country than in large cities; where the air, inſtead of partaking ſo largely of this ſalutary impregnation, is daily contaminated with noxious animal effluvia.

With reſpect to climate, various obſervations conſpire to prove, that thoſe regions which lie within the temperate zones are beſt calculated to promote long life. Hence, perhaps, may be explained, why Italy has produced ſo many long livers, and why iſlands in general are more ſalutary than continents; of which Bermudas and ſome others afford examples. And it is a pleaſing circumſtance that our own iſland appears to contain far more inſtances of longevity than could well be imagined. The ingenious Mr. Whitehurſt aſſures us, from certain facts, that Engliſhmen are in general longer lived than North Americans; and that a Britiſh conſtitution will laſt longer, even in that climate, than a native one. But it muſt be allowed

allowed in general, that the human conſtitution is adapted to the peculiar ſtate and temperature of each reſpective climate, ſo that no part of the habitable globe can be pronounced too hot or too cold for its inhabitants. Yet, in order to promote a friendly intercourſe between the moſt remote regions, the Author of nature has wiſely enabled the inhabitants to endure great and ſurpriſing changes of temperature with impunity.

Though foods and drink of the moſt ſimple kinds are allowed to be the beſt calculated for ſupporting the body in health, yet it can hardly be doubted but variety may be ſafely indulged occaſionally, provided men would reſtrain their appetites within the bounds of temperance; for bountiful Nature cannot be ſuppoſed to have poured forth ſuch a rich profuſion of proviſions, merely to tantalize the human ſpecies, without attributing to her the part of a cruel ſtep-dame, inſtead of that of a kind and indulgent parent. Beſides, we find, that, by the wonderful powers of the digeſtive organs, a variety of animal and vegetable ſubſtances, of very diſcordant principles, are happily aſſimilated into one bland homogeneous chyle; therefore it ſeems natural to diſtruſt thoſe cynical writers, who would rigidly confine mankind to one ſimple diſh, and their drink to the mere water of the brook. Nature, it is true, has pointed out that mild inſipid fluid as the univerſal diluent, and therefore moſt admirably adapted for our daily beverage: but experience has equally proved, that vinous and ſpirituous liquors, on certain occaſions, are no leſs ſalutary and beneficial, whether it be to ſupport ſtrength againſt ſickneſs or bodily fatigue, or to exhilarate the mind under the preſſure of heavy misfortunes. But, alas! what Nature meant for innocent and uſeful cordials, to be uſed only occaſionally, and according to the direction of reaſon, cuſtom and caprice have, by degrees, rendered habitual to the human frame, and liable to the moſt enormous and deſtructive abuſes. Hence it may be juſtly doubted, whether gluttony and intemperance have not depopulated the world more than even the ſword, peſtilence, and famine. True, therefore, is the old maxim, *Modus utendi ex veneno facit medicamentum, ex medicamento venenum.*

It is allowed on all hands, that alternate motion and reſt, and ſleep and watching, are neceſſary conditions to health and longevity; and that they ought to be adapted to age, temperament, conſtitution, temperature of the climate, &c. but the errors which mankind daily commit in theſe reſpects become a fruitful ſource of diſeaſes. While ſome are bloated and relaxed with eaſe and indolence, others are emaciated and become rigid through hard labour, watching, and fatigue.—Where the animal functions are duly performed, the ſecretions go on regularly; and the different evacuations ſo exactly correſpond to the quantity of aliment taken in, in a given time, that the body is found to return daily to nearly the ſame weight. If any particular

particular evacuation happen to be preternaturally diminiſhed, ſome other evacuation is proportionally augmented, and the equilibrium is commonly preſerved; but continued irregularities, in theſe important functions, cannot but terminate in diſeaſe.—The due regulation of the paſſions, perhaps, contributes more to health and longevity than that of any other of the non-naturals. The animating paſſions, ſuch as joy, hope, love, &c. when kept within proper bounds, gently excite the nervous influence, promote an equable circulation, and are highly conducive to health; while the depreſſing affections, ſuch as fear, grief and deſpair, produce the contrary effect, and lay the foundation of the moſt formidable diſeaſes.

From the light which hiſtory affords us, as well as from the foregoing liſt of long lives, there is great reaſon to believe, that longevity is in ſome meaſure hereditary; and that healthy long-lived parents would commonly tranſmit the ſame to their children, were it not for intemperance, and the frequent errors in medical advice, which ſo evidently tend to the abbreviation of human life.—Where is it, but from theſe cauſes, and the unnatural modes of living, that, of all the children which are born in the capital cities of Europe, nearly one half die in early infancy? To what elſe can we attribute this extraordinary mortality? Such an amazing proportion of premature deaths is a circumſtance unheard-of among ſavage nations, or among the young of other animals! In the earlieſt ages, we are informed, that human life was protracted to a very extraordinary length; yet how few perſons, in theſe latter times, arrive at that period which nature ſeems to have deſigned! Man is by nature a field-animal, and ſeems deſtined to riſe with the ſun, and to ſpend a large portion of his time in the open air, to inure his body to robuſt exerciſes and the inclemency of the ſeaſons, and to make a plain homely repaſt only when hunger dictates. But art has ſtudiouſly defeated the kind intentions of nature; and, by enſlaving him to all the blandiſhments of ſenſe, has left him, alas! an eaſy victim to folly and caprice. Let the conſideration of the following ſubjects direct every one, who values health and long life, to purſue the means nature has pointed out for their preſervation and ſuſtenance.

Of NUTRITION.

NUTRITION, in the animal œconomy, is the acceſſion of new parts to the body, either for its augmentation, or for the reparation of ſuch as are worn off, or exhaled through the pores and perſpiring veſſels, whereby the fluids are diminiſhed, and the body falls away. So that, to preſerve life, it is neceſſary that a reſtitution be made to the juices and ſolids of the body, at leaſt equal to what is loſt by thoſe motions; which is what we call the *action of nutrition.* Now the loſt juices are eaſily and quickly ſupplied by aliment, air, &c. but the nutrition of the ſolid parts is much more obſcure. This, indeed, has proved a ſubject of infinite doubts and differences among authors; nor had we any rational or ſatisfactory account of the ſame, till that of the accurate Boerhaave, whoſe doctrine is as follows.

Every ſolid part of the body conſiſts of other ſmaller ones, in all reſpects like the larger; veſſels, of veſicles, and thoſe of others ſtill ſmaller; bones, of oſſicles, &c. Which ſtructure goes beyond all limits of ſenſe, however aſſiſted by art; as appears by the experiments and obſervations of Malpighi, Ruyſch, Leeuwenhoek, and Hooke. Yet it is ſcarcely poſſible this diviſion and ſubdiviſion ſhould be infinite, as thoſe of foods and juices are. Again, it appears from microſcopes, injections, ſmall wounds, exſiccations, &c. that the ſolid parts of the body are very ſmall, compared with the fluids; and it is alſo demonſtrable, from conſidering the riſe and generation of the veſſels, and the reſolution of the greater veſſels into their ſmaller conſtituent ones, that all the ſolid maſs of the body is conſtructed of mere nerves, as its elements. And, in effect, all this maſs, an incredible ſmall particle only excepted, at firſt aroſe out of what was a very ſmall colliquament, much like the nervous juice itſelf; as is abundantly ſhown by the great Malphighi, in his two treatiſes on incubated eggs. For neither does the white of the egg nouriſh, till, by means of the incubation, it has paſſed innumerable degrees of fluidity, from its firſt thickneſs, to that exceeding ſubtilty wherein it terminates. But, even then, the liquor, thus given to the embryo, is exceedingly thick, in compariſon with what it is to be when converted into its veſſels and viſcera. Now, the firſt tender ſolids, ariſing from this ſubtle humour, do again paſs infinite intermediate degrees, before they arrive at their utmoſt ſtate and conſiſtence; as is ſhown by Malpighi in eggs, and by Ruyſch in embryoes and fœtuſes. Hence, therefore, it follows, that the ſolids, in their firſt formation out of the liquids whence they ariſe, only differ from them in reſt, coheſion, and figure. Therefore ſuch a particle, now in its fluid ſtate, will become a part of the ſolid to be formed out of it, as ſoon as there happens to be a power to effect its coheſion with the other ſolid parts, howſoever that coheſion be effected.

This coheſion is eaſily produced in a fibre already formed, if there happen to be a proper cavity in the ſolid, left open by ſome loſt particle; and, at the ſame time, a particle in the fluid, anſwerable thereto in bulk, figure, and nature; and, laſtly, if there be a power wherewithal to intrude it into that place, or accomodate it thereto. Thus will ariſe a real nutrition of the ſolids in the minute veſſels, by whoſe union the large ones are formed; that is, in the nerves, or in veſſels ſimilar thereto. Which being impracticable by any other liquid than that brought into theſe veſſels, it appears very evident, that the nervous juice, at leaſt a juice perfectly like

like it, is the immediate matter of nutrition: whence nutrition appears one of the laſt and moſt perfect actions of the body; ſince, to have this laudable, all the precedent actions muſt of neceſſity have been ſo. The chyle, therefore, which ſome make the immediate matter of nutrition, is, indeed, fitted to fill the larger veſſels; but it cannot nouriſh or reſtore them. This, when attenuated, changed, more intimately mixed in the lungs by means of reſpiration, and thus fitted for the paſſage of certain veſſels, is indeed rendered fitter, yet far from being quite fit, to be the matter of nutrition. But, by the repeated action of the lungs, the viſcera, veſſels, &c. there is formed, out of this humour, a ſoft, tenacious, plaſtic, inſipid, ſerum, which, thickening by the fire, becomes perfectly like the white of an egg. This fluid, therefore, has in it all the conditions found in that, from whence, by ſure experience, we know all the ſolid parts of an animal ariſe by mere incubation. It is, therefore, a ſtep nearer; but is not yet quite diſpoſed for nutriment; much leſs is the cruor, or red globular part of the blood, ſo. Neither are yet fitted to enter the veſſels; yet both the one and the other are, by different authors, made the nutritive juice. But, as the heat of the incubation, ſo the action of the viſcera and veſſels on the ſerum, introduces various changes therein, till at length a part of it be rendered ſubtile enough for the purpoſe required. This, when exhauſted, is inſtantly repaired: and thus we have the true immediate matter of nutrition.

The matter of nutrition thus aſcertained, the manner wherein, and the cauſe whereby, it is effected, are as follows: A juice being driven directly through a full, conic or cylindric, elaſtic or rigid, canal; if its courſe be from a wider to a narrower part, or if it have any thing to oppoſe its motion, will endeavour to ſtretch the ſides of its canal, according to the axis of its length. This muſt be the caſe every-where in the body, except, perhaps, in the veins and receptacles. By this *niſus*, or endeavour, how weak ſoever, continually repeated, the veſſels will be inſenſibly lengthened out; and, in lengthening, they will be made more and more ſlender. Hence the laſt extremities of the veſſels, which in man are extremely ſmall, are continually ſtretched, and rendered leſs and leſs coherent, i. e. ſtill nearer and nearer to a diſſolution; and thus at length will they cohere ſo weakly, as ſcarcely to differ from fluids. While ſuch motion goes on, therefore, and the propulſion is continued, there will, of neceſſity, happen theſe two things: Firſt, the outmoſt particles of the minuteſt tubes being torn off, will again be converted into a kind of humour, what part of the body ſoever they ſtick in. Secondly, the ſmalleſt particles, which, by their union, compoſed the ſlendereſt fibrillæ, will be ſo ſeparated from each other, as to leave open interſtices in thoſe places, where, before, they cohered. Both theſe effects will be produced at all times, and in all parts of the body, ſo long as life continues, eſpecially where nature is ſtrong, and the

the actions of the body violent. But the ſame humour whereby theſe effects are produced, containing abundance of particles ſimilar to thoſe thus ſeparated and loſt, conveys and applies them to thoſe interſtices, by that very impetus whereby it endeavours to diſtend the canals; and, thus intercepted, at length it forms, adapts, and faſtens, them, ſo as to adhere in the ſame manner as the former. The matter, preparation, application, energy of motion, ſtill remaining the ſame; what, from time to time, is loſt, is thus preſently reſtored; and the ſolids continue in the ſame ſtate as before, that is, they are perpetually nouriſhed, and ſupplied, and preſerved.

In this the Creator's wiſdom is very conſpicuous; in that the ſame power which inevitably deſtroys does repair again at the ſame time, and by the ſame action; and that, the greater the loſs is, the more copious the ſupply; and, laſtly, that thoſe parts firſt ſpent in the action of the body are the firſt reſtored. Farther, it is evident, that the newer, the more tender, and the nearer to the moving cauſe, theſe veſſels are, the more eaſily will they be lengthened, diſtended, deſtroyed, and repaired: our bodies, therefore, the nearer to the origin, the more do they grow. For, the action ſtill continuing, the greater veſſels become more extended by their fluid; and at the ſame time the ſmaller, whereof the membranes or coats, of the larger ſort are compoſed, are compreſſed, dried, and at laſt concreted, and grow up; whence ariſes a firmneſs, indeed, of the fibres, but a loſs of the veſicles. Thus what were formerly veſſels commence mere hard ligaments; and thus, the fluids being once fixed, the ſeveral veſſels coaleſce; from the concurrence of theſe cauſes ariſe the ſtrength, hardneſs, rigidity, and thickneſs, of the ſolid parts. Hence the number of veſſels is greateſt in embryoes, and, as age comes on, it ſenſibly diminiſhes; and hence it is, that their weakneſs conſtantly declines, and their ſtrength and firmneſs increaſe. In young people, therefore, the quantity of humours is redundant, and greatly exceeds the ſolids: in old men, the ſolids exceed the fluids. And hence we ſee the reaſon, manner, and appearance, of growth, ſtate, declenſion, and, at length, of death from pure old age.

A perſon who conſiders this account, and compares it with what is actually obſervable in the body, will find every circumſtance to obtain: thus the whole cuticula is every-where, and at all times, conſtantly deſquamating, peeling off, and again renewing; and thus the hair, nails, teeth, continually rubbed, torn, and worn off, come again: parts taken off from the veſſels and the bones ſoon grow again: and the ſordes, or filth, rubbed off from the extremities of the veſſels, when examined by a microſcope, or diluted, and viewed in water, appear plainly to conſiſt both of ſolid and fluid parts; and thoſe carried off by waſhing, ſhaving, &c. are the ſame. Hence too, we ſee, that a general increaſe of the bulk of the body, with regard to habit, as in fat, fleſhy, brawny, perſons, does not ariſe from any increaſe of the ſolids, but by their extenſion into larger cavities, crowded with ſtagnant humours. And hence fatneſs becomes hurtful, as it loads, weakens, and ſuffocates. Whence ariſes a very conſiderable diſtinction between nutrition and repletion, to which a phyſician ſhould have ſpecial regard; the one ſtrengthening and condenſing the veſſels; the other weakening, looſening, and extending, the ſame. Hence, laſtly, we ſee why the fabric of the ſolids is not deſtroyed by the contained fluids; how our machine comes to ſubſiſt ſo long; why, when a nerve is corrupted, the nutrition of that part it belongs to ceaſes; and why the ſame obtains in an artery; why in an embryo there are no ſolids, in a fœtus very few, in old men a great deal; and why even the nerves, tendons, arteries, and receptacles, become firſt cartilaginous, and then bony. Dr. Prieſtley concludes, from ſome experiments undertaken with a view of diſcovering the principle of nutrition in vegetable and animal ſubſtances, that this principle is phlogiſton, in ſuch a ſtate as to be capable of becoming, by putrefaction, a true inflammable air, (hydrogenous gas;) but not generally ſuch as to burn with exploſion, but rather with a blue and lambent flame, mixed with a certain proportion of fixed air. This principle in nutrition is immediately held in ſolution by the gaſtric juice, and in the chyle formed by it; and, when it has entered into the circulation with the chyle, and anſwered the purpoſe in the animal œconomy for which it is deſigned, it is thrown out again by means of the blood in the lungs, and communicated to the air, which is phlogiſticated with it.

Of FOOD, or ALIMENT.

FROM aliment, or food, by the proceſs of digeſtion, is prepared a very mild, ſweet, and whitiſh, liquor, reſembling milk, and diſtinguiſhed by the name of *chyle;* which, being abſorbed by the lacteal veins, by them conveyed into the circulation, and there aſſimilated into the nature of blood, affords that ſupply of nutrition, which, as we have ſeen above, the continual waſte of the body is found to require. Food is the moſt neceſſary thing for the preſervation of our bodies: and, as on the choice thereof our health greatly depends, it is of much importance to underſtand, in general, what is the propereſt for our nouriſhment; and, in particular deviations from health, what is the beſt adapted to reſtore us. Our blood and juices naturally incline to become putrid and acrimonious: freſh chyle, duly received, prevents this deſtructive tendency, and preſerves in them that mild ſtate which alone conſiſts with health. An animal diet affords the moſt of this bland nutritious mucilage; watery fluids dilute the too groſs parts, and carry off what is become unfit for uſe. It is only the ſmall portion of jelly which is ſeparated from the farinaceous parts of vegetables, that, after being much elaborated, is converted

verted into the animal nature; yet the uſe of vegetables prevents both repletion and a too great tendency to a putreſcent acrimony of the blood. In hot climates, as well as againſt the conſtitutional heat of particular perſons, vegetables are demanded in the largeſt portion; animal ſubſtances afford the higheſt reliſh while our appetite continues; but will ſate the appetite before the ſtomach is duly filled. Vegetables may be eaten after either fleſh or fiſh: few herbs or fruits ſatiate ſo much as that the ſtomach may not be filled with them when it is already ſatiſfied with fleſh or fiſh; whence it may be obſerved, that no diet which is very nouriſhing can be eaten to fulneſs, becauſe its nutritious parts are oily and ſatiating. Health depends almoſt wholly on a proper craſis of the blood; and to preſerve this, a mixture of vegetables in ſome degree is always required, for a loathing is ſoon the conſequence of animal food alone: hot acrid habits, too, receive from milk and vegetables the needful for correcting their exceſſes; but in cold, pituitous, and nervous, habits, which want moſt nouriſhment from leaſt digeſtion, and from the ſmalleſt quantity of food, animal diet is to be uſed more freely.

As the blood, the nutritive juice, and in general all the parts of the body, are made up of three elements, viz. of one which is ſulphureous, oily, and inflammable; of one of an earthy, ſubtile, alkaline, nature; and of one of an aqueous nature: ſo the ſeveral kinds and virtues of food may be moſt commodiouſly reduced to theſe three claſſes; and aliments of theſe three ſeveral qualities, duly mixed with one another, afford a proper nouriſhment for the human body.—The fleſh of animals, eſpecially when roaſted, affords the body its principal ſupply of the ſulphureous part; but it is to be obſerved, that wild animals are preferable in this reſpect to the tame and domeſtic kind, becauſe their oils and ſalts are exalted by habitual exerciſe. Among the aliments which furniſh the blood with its humid parts, of animals, fiſh; and of vegetables, pot-herbs, the milder roots, and ſome ſummer-fruits; are reckoned the principal. To the third claſs, which ſupplies the blood with its fixed and earthy parts, belong all kinds of grain, as the ſeveral ſorts of bread, rice, peaſe, beans, lentils, cheſnuts, almonds, cocoa, cheeſe, &c. From what has been ſaid, it will appear that all ſuch aliments as are of a mild quality, and reſemble the chyle and blood, are fit for nouriſhment; that all ſuch food as either recedes from, or is quite oppoſite to, the nature of the chyle and blood, is unfit for nouriſhing the parts; that all food in which there is too much of an acid, is improper for nouriſhment, becauſe milk and blood will not mix with an acid, which is quite oppoſite to their natures, and induces a coagulation of the circulating juices; that all ſalts, and all foods too highly ſalted, muſt be unfit for nouriſhment, becauſe no ſalt whatever can be mixed with the blood, chyle, and milk;

and

and laſtly, that the free uſe of ſpirits muſt be very detrimental both to health and nouriſhment, becauſe blood and chyle never incorporate with ſpirituous liquors, but rather ſeparate from them.

Thus much being obvious as general principles with reſpect to the matter and quality of our aliment, the valetudinarian may eaſily regulate his diet with ſome advantage to himſelf by an attention to the few enſuing particulars. In winter, eat freely, but drink ſparingly: roaſt meat is to be preferred, and what is drunk ſhould be ſtronger than at other ſeaſons. In ſummer, let thirſt determine the quantity to be drunk; cold ſtomachs never require much: boiled meats and vegetables, if not otherwiſe contradicted, may now be more freely uſed. Lax habits require the winter's diet to be continued all the year, and rigid ones ſhould be confined to that of ſummer. Occaſional faſting will prevent the neceſſity of periodical bleeding, &c. Thoſe who are troubled with eructations occaſioned by their food ſhould drink but little, and uſe ſome unaccuſtomed exerciſe. The thirſty ſhould drink freely, but eat ſparingly. In general, let moderation be obſerved; and, though no dinner hath been had, a light ſupper is at all times to be preferred. After very high ſeaſoned meats, a glaſs of water acidulated with the acid elixir of vitriol, or in very weak ſtomachs the ſweet elixir of vitriol, is far more aſſiſtant to the work of digeſtion than the common method of taking brandy.

As to common drink, water alone is ſufficient and effectual for all the purpoſes of nature. Strong liquors were never deſigned for common uſe. They were formerly kept here in England, as other medicines are, in apothecaries' ſhops, and preſcribed by phyſicians, as they do diaſcordium, and Venice treacle, to refreſh the weary, ſtrengthen the weak, and raiſe the low-ſpirited. The effect of the ordinary uſe of wine and ſpirituous liquors, as natural cauſes will always produce their effects, is to inflame the body into gout, ſtone, and rheumatiſm, fevers, pleuriſies, ſmall-pox, &c. to dry up the juices, and ſcorch and ſhrivel the ſolids. Thoſe whoſe appetite and digeſtion are good and entire, never want ſtrong liquors to ſupply them with ſpirits; ſuch ſpirits are too volatile and fugitive for any ſolid or uſeful purpoſes of life. Two ounces of fleſh-meat, well digeſted, beget a greater ſtock of more durable and uſeful ſpirits than ten times as much ſtrong liquor.

All ſtrong liquors are as hard to digeſt, and require as much labour of the concoctive powers, as ſtrong food itſelf. Water is the only univerſal diſſolvent, or menſtruum, and the moſt certain diluter of all bodies proper for food. There are a great many ſpirituous liquors, which not only will not diſſolve, but which will harden, and make more indigeſtible, certain parts, eſpecially the ſalts of bodies, wherein their active qualities, that is, thoſe which can do moſt harm to human conſtitutions,

ſtitutions, conſiſt. And we have known perſons of tender conſtitutions, who could neither eat nor digeſt upon drinking wine, but who, by drinking at meals common water, warmed, have recovered their appetites and digeſtion, and have thriven and grown plump. It is true, ſtrong liquors, by their heat, and ſtimulation on the organs of concoction, by increaſing the velocity of the motion of the fluids, and thereby quickening the other animal functions, will carry off the load that lies upon the ſtomach, with more *preſent* cheerfulneſs. But then, beſides the *future* damage of ſuch a quantity of wine to the ſtomach and fluids, by its heat and inflammation, the food is hurried into the habit unconcocted, and lays a foundation for a fever, a fit of the cholic, or ſome chronical diſeaſe. With reſpect to fermented liquors, which are commonly uſed, it may be obſerved, that thoſe which are too ſtrong hurt digeſtion, and are ſo far from ſtrengthening the body, that they weaken and relax it. They keep up a conſtant fever, which exhauſts the ſpirits, heats and inflames the blood, diſpoſes to numberleſs diſeaſes, and occaſions a premature old age. But fermented liquors may be too weak, as well as too ſtrong: theſe muſt either be drunk new, before the fermentation is over, and in this caſe will generate air in the bowels, and occaſion flatulencies; or they ſoon become ſtale, ſour the ſtomach, and injure digeſtion. On this account all malt liquors, cider, &c. ſhould be ſufficiently ſtrong to keep till they are ripe, and then they ſhould be uſed; and neither ſooner nor later. Liquors that are adulterated with a mixture of ingredients of the opiate kind, which are poiſonous in their quality, as they moſtly are by thoſe who make them for ſale, hurt the nerves, relax and weaken the ſtomach, and ſpoil its digeſtive powers.

A due regulation of the quantity and quality of our meat and drink, and a nice adjuſtment thereof to the concoctive powers, is of the utmoſt conſequence to health and long life. What we expend in motion, excretion, effluvia, &c. is but a determinate quantity; and the ſupply ſhould only keep pace with the expence; a juſt proportion of the two would, probably, preſerve us from acute diſtempers, as it certainly would from chronical ones; moſt or all of which proceed from repletion, as appears from their being cured by evacuation.

Phyſicians have attempted to determine the healthful quantity of food for a human body. Some ſay, that in winter, when the perſpiration of an unexerciſed perſon is only equal to the urine, the diet for twenty-four hours ought not to exceed four pounds, or four pounds and a half. In ſummer, the diet may be ſix pounds and a half, which may be carried off without the help of exerciſe, when the air is hot and dry. If the quantity of food be ſuch as to make the perſpiratien and urine of a natural day always nearly equal, and the morning weight of the body always nearly the ſame, that quantity is the truly healthful quantity of food for grown bodies

bodies which uſe but little exerciſe. The quantity of food neceſſary to keep a grown body in health, will be better and more eaſily digeſted, when it is ſo divided as to make the meals equal, than when they are very unequal. The diſtance between one meal and another ſhould bear ſome proportion to the largeneſs of the preceding meal. Good and conſtant health conſiſts in a juſt quantity of food, and a juſt proportion of the meat to the drink; and, in order to be freed from chronical diſorders contracted by intemperance, the quantity of food ought to be leſſened, and the proportion of the meat to the drink increaſed, more or leſs, according to the greatneſs of the diſorders; and both the quantity of food, and the proportion of meat to the drink, ought to be ſuch as ſhall make perſpiration and urine nearly equal at all ſeaſons of the year.

The quantity of animal food conſumed by the Engliſh is generally pernicious, becauſe it produces but little of that air which is antiſeptic: hence they are ſubject to the ſcurvy, and its numerous train of conſequences, indigeſtion, low ſpirits, hypochondriaciſm, &c. whereas, if vegetables and milk, whoſe antiſeptic quality, ariſing from the gas or air which they plentifully afford, were more uſed as food, we ſhould have leſs ſcurvy, and likewiſe fewer putrid and inflammatory fevers. One great reaſon why leproſies, hot ſcurvies, dyſenteries, plagues, peſtilential fevers, and the like diſtempers, formerly ſo frequent in London, are now ſo rare, is the change that has been made in the food of the inhabitants. Hopped beer, and wine, coming into general uſe, have been a great means of ſuppreſſing putrid diſeaſes; greens and fruit are likewiſe more univerſally eaten, and ſalted meats make a much leſs part of our food than formerly: to which may be added the more general conſumption of tea and ſugar.

Vegetable food is moſt proper for ſcorbutic and hectical perſons, and does very well with people who have much exerciſe; but in other circumſtances, a mixed diet of vegetable and animal ſubſtances, ſuch as is commonly uſed, ſeems beſt calculated to nouriſh and preſerve the body from decay.

The ſofter and milder kinds of aliment are proper for children, and for youth the ſtronger. Old people ought to leſſen the quantity of their food, and increaſe that of their drink: but yet ſome allowance is to be made for cuſtom, eſpecially in cold climates; for, as in theſe the appetite is keener, ſo is the digeſtion ſtronger and better performed.

Different ſexes alſo require a different food and regimen. Women are weaker than men, and for that reaſon require a food and regimen peculiar to themſelves; they are of a ſpungy and lax habit, and for the moſt part addicted to indolence and pleaſure, drink little, have bodies of a highly delicate and ſenſible nature, much inclined to ſpaſms and convulſive motions, and diſpoſed to generate a redundance of of blood. Befides, at certain ftated times they have a regular evacuation by the veins of the uterus; and in confequences of thefe circumftances it is neceffary that women, rather than men, fhould obferve a regimen and method of living peculiarly and accurately adapted to their habit and conftitution.

Hence it is obvious, that the phyfician acts a prepofterous and unaccountable part who to every one prefcribes the fame method of living; or thinks, that what contributes to the health of one will without diftinction or referve prove falutary to all. For we are fufficiently taught by daily experience, that all fubftances are not equally adapted to all patients; and that what one may bear without being fenfible of any bad effects, may to another prove prejudicial, and even fatal. Time itfelf has a confiderable influence in determining the falutary or noxious effects of aliments; fince fome fubftances may fafely, and without any bad confequence, be ufed at one feafon, which at another may contribute not a little to the deftruction of health.

As to the effects of food on the mind, it is plain, that delicacy of feeling, livelinefs of imagination, quicknefs of apprehenfion, and acutenefs of judgment, very frequently accompany a weak ftate of the body. True it is, indeed, that the fame ftate is liable to timidity, fluctuation, and doubt; while the ftrong have that fteadinefs of judgment, and firmnefs of purpofe, which are proper for the higher and more active fcenes of life. The moft valuable ftate of the mind, however, appears to refide in fomewhat lefs firmnefs and vigour of body. Vegetable aliment, as never over-diftending the veffels or loading the fyftem, never interrupts the ftronger motions of the mind; while the heat, fulnefs, and weight, of animal food, are an enemy to its vigorous efforts. Temperance, then, does not fo much confift in the quantity, for that always will be regulated by our appetite, as in the quality, viz. a large proportion of vegetable aliment.

Of Air.

IT is no eafy tafk to afcertain the nature and origin of air, as being a fluid imperceptible to all our fenfes, except that of feeling. Indeed, from the refiftance and impreffion it makes, we know that there is fuch a body, which every-where furrounds our earth, and is of the utmoft importance, not only to mankind, in promoting many ufeful arts, but abfolutely neceffary for the prefervation of health and life.

The wholefomenefs or unwholefomenefs of air is certainly owing to the different effluvia with which it abounds, and ought to be particularly attended to by the valetudinarian. The beft air is to be met with in open champaign countries; where the foil is dry, not parched or fandy, and fpontaneoufly produces wild

 thyme,

thyme, wild marjoram, and the like ſweet-ſcented plants. That near rivers is rather prejudicial, unleſs they are ſmall, clear, and have a gravelly channel. The morning air is deemed more refreſhing than that of the evening, and air agitated with breezes than that which is ſerene and ſtill. As good air contributes greatly to health, ſo that which is bad is no leſs prejudicial to it. Stagnating air is productive of putrid and malignant diſorders, as dyſenteries, bilious fevers, &c. and that which is too moiſt, of inflammatory ones, as coughs, rheumatiſms, &c. Moiſt and rainy ſeaſons, however, differ widely in this reſpect; ſince, in marſhy countries, intenſe and continued heats occaſion the greateſt moiſture in the air: whereas frequent ſhowers, during the hot ſeaſon, cool it, check the exceſs of vapour, dilute and refreſh the corrupted ſtagnating water, and precipitate all noxious and putrid effluvia.

To the preſſure of air we are to attribute the coherence of the parts of bodies. Breathing too, on which depends animal life, is owing to the preſſure and ſpring of the air; and to the ſame cauſe may be attributed the production of fire and flame, as appears from the ſudden extinction of a coal or candle in the exhauſted receiver. It is likewiſe neceſſary for the exiſtence and preparation of ſounds, for the germination and growth of plants, for conveying all the variety of ſmells, and for tranſmitting the rays and influence of the celeſtial bodies. In ſhort, ſuch is the generating and vivifying power of air, that ſome of the ancient philoſophers conſidered it as the firſt principle of all things. Air not only acts upon all bodies by its common properties of weight and elaſticity, but by the peculiar virtues of the ingredients whereof it is compoſed. By means of a corroding acid, it diſſolves iron and copper, unleſs well defended by oil. Even gold, in the chemiſt's laboratory, when the air is impregnated with the effluvia of aqua regia, contracts a ruſt like other bodies. It fixes volatile bodies, and volatilizes thoſe which are fixed. From the different effluvia, diffuſed through the air, proceed a variety of effects. Near mines of copper, it will diſcolour ſilver and braſs; and in London, the air of which abounds with acid and corroſive particles, metalline utenſils ruſt ſooner than in the country. It is very difficult to obtain oil of ſulphur in a clear dry air, as its parts are then more ready to evaporate; whereas in a moiſt cloudy air it may be obtained in abundance. All ſalts melt moſt readily in cloudy weather; and ſeparations ſucceed beſt in the ſame ſtate of the air. If pure wine be carried into a place where the air is full of the fumes of wine then fermenting, it will begin to ferment afreſh.

Wherever air ſtagnates long, it becomes unwholeſome. Hence the unhappy perſons confined in gaols not only contract malignant fevers themſelves, but often communicate them to others. Nor are many of the holes, for we cannot call them houſes, houfes, poffeffed by the poor in great towns, much better than gaols. These low dirty habitations are the very lurking-places of bad air and contagious difeafes. Such as live in them feldom enjoy good health; and their children commonly die young. In the choice of a houfe, thofe who have it in their power ought always to pay the greateft attention to open free air. The various methods which luxury has invented to make houfes clofe and warm, contribute not a little to render them unwholefome. No houfe can be wholefome unlefs the air has a free paffage through it. For which reafon houfes ought daily to be ventilated, by opening oppofite windows, and admitting a current of frefh air into every room. Beds, inftead of being made up as foon as people rife out of them, ought to be turned down, and expofed to the frefh air from the open windows through the day. This would expel any noxious vapour, and could not fail to promote the health of the inhabitants. In hofpitals, gaols, fhips, &c. where that cannot be conveniently done, ventilators fhould be ufed. The method of expelling foul, and introducing frefh, air, by means of ventilators, is a moft falutary invention, and is indeed the moft ufeful of all our modern medical improvements. It is capable of univerfal application, and is fraught with numerous advantages, both to thofe in health and in ficknefs. In all places where numbers of people are crowded together, ventilation becomes abfolutely neceffary. * Air which ftagnates in mines, wells, cellars, &c. is extremely noxious. That kind of air is to be avoided as the moft deadly poifon. It often kills almoft as quickly as lightning. For this reafon, people fhould be very cautious in opening cellars that have been long fhut, or going down into deep wells, or pits, efpecially if they have been kept clofe covered. We have daily accounts of perfons who lofe their lives by going down into deep wells and other places where the air ftagnates. All thefe accidents might be prevented by only letting down a lighted candle before them, and ftopping when they perceive it go out: yet this precaution, fimple as it is, is feldom ufed.

If frefh air be neceffary for thofe in health, it is ftill more fo for the fick, who often lofe their lives for want of it. The notion that fick people muft be kept very hot, is fo common, that one can hardly enter the chamber where a patient lies, without being ready to faint, by reafon of the hot fuffocating fmell. How this muft affect the fick any one may judge. No medicine is fo beneficial to the fick as frefh air. It is the moft reviving of all cordials, if it be adminiftered with prudence. We are not, however, to throw open doors and windows at random upon the fick. Frefh air is to be let into the chamber gradually, and, if poffible, by opening the windows of fome other apartment.

There are many kinds of air, produced by accidental or artificial caufes, of which the following are the moft material:

Dephlogiſticated air, the oxygenous gas, or vital air, of the new chemiſtry, is an elaſtic fluid naturally extricated in the proceſs of vegetation; but artificially procured from nitre, minium, magneſia, water, &c. This is eminently capable of ſupporting flame and animal life, and is one of the component parts of our atmoſphere.

Phlogiſticated air, or azotic gas, is produced in great quantities during putrefaction and fermentation; and is alſo obtained in the calcination of metals and other phlogiſtic proceſſes. It deſtroys animal life, and extinguiſhes flame; but is very friendly to vegetation, and is another of the component parts of our atmoſphere.

Fixed air, or carbonic acid gas, derives its name from the property of adhering to certain bodies, and fixing itſelf in them. It conſiſts of dephlogiſticated air united to charcoal; this is obtained by fermentation, and in all phlogiſtic proceſſes, and manifeſts the properties of an acid. It extinguiſhes flame, and deſtroys animal life.

Inflammable air, (hydrogenous gas,) conſiſts wholly of charcoal and water rarefied by heat; and is remarkable for being the lighteſt of all gravitating ſubſtances. It is produced naturally from all putrid waters, and may be artificially procured from certain metallic ſolutions, by paſſing the ſteam of water over red-hot iron, and by diſtilling wood, pit-coal, &c. with a ſtrong heat, or by oppoſing charcoal to the heat of a burning lens in vacuo. It extinguiſhes flame, unleſs it be mixed with a certain proportion of atmoſpherical, or dephlogiſticated, air; in which caſe, it explodes violently. It deſtroys animal life, but is friendly to vegetation.

Nitrous air, or nitrous gas, is procured artificially by diſſolving metallic or other ſubſtances in nitrous acid. Being mixed with dephlogiſticated air, both the fluids loſe their elaſticity, and a ſmall quantity of nitrous acid is produced. It inſtantly kills animals and extinguiſhes flame. By union with ſome metals it is converted into volatile alkali. In ſome caſes it may be made to ſupport flame, and even animal life. Its property of condenſing with dephlogiſticated air, renders it a teſt of the ſalubrity of the atmoſphere.

Marine acid air, the muriatic acid gas of the new chemiſtry, is the ſame as marine acid reduced into vapour, and deprived of moſt of its waters.

Dephlogiſticated marine acid air, or oxygenated muriatic acid gas, is ſuppoſed by ſome to be the marine acid deprived of its phlogiſton; by others, to be the ſame acid with an addition of pure air. It deſtroys many kinds of colours, and, with inflammable air, regenerates common marine acid.

Alkaline air, or ammoniacal gas, is the ſame with pure volatile alkali, and is formed by an union of phlogiſticated and inflammable air.

Hepatic air, or ſulphureous acid gas, is produced from the decompoſition of liver of ſulphur by acids; and in the common atmoſphere it is inflammable, but does not burn with exploſion.

Atmoſpherical

Atmoſpherical air, is compoſed of dephlogiſticated and phlogiſticated air, and thus ſupports and ſuſtains both animal life and vegetation.

The exterior part of our habitable world is the air or atmoſphere, a ſpringy body, that encompaſſes the ſolid earth on all ſides, and is near a thouſand times lighter than water; and the higher it is, the leſs it is compreſſed by the ſuperior incumbent air; and ſo conſequently, it being a ſpringy body, the thinner it is. And, as a pillar of air of any diameter is equal in weight to a pillar of quickſilver of the ſame diameter of between twenty-nine and thirty inches high, we may infer that the top of the atmoſphere is not very near the ſurface of the ſolid earth. Now, as quickſilver is near fourteen times heavier than water, the atmoſphere will ſuſtain a column of water about fourteen times higher than the column of quickſilver, that is, about thirty-four feet; and, if we conſider that air is a thouſand times lighter than water, then a pillar of air, equal in weight to a pillar of quickſilver of thirty inches high, will be thirty-four thouſand feet; whereby we come to know, that the air or atmoſphere is at leaſt 34,000 feet, that is, about ſix miles, high, but probably much more. And if we conſider that the air is a ſpringy body, and that that which is neareſt the earth is compreſſed by the weight of all the atmoſphere above it, we ſhall find that the air near the ſurface of the earth is much denſer and thicker than it is in the upper regions. On this theory it may be accounted for why great cities are not ſo healthful to reſide in as ſmall towns and country villages; and why London is much more prejudicial to health, owing to the many works containing noxious effluvia poſſeſſed of the component parts mentioned in the different kinds of air, and conſequently forms an air to breathe in that is not congenial to the life of Man.

Of EXERCISE.

EXERCISE may be ſaid to be either active or paſſive. The active is walking, hunting, dancing, playing at bowls, and the like; as alſo ſpeaking, and other labour of the body and mind. The paſſive is riding in a coach, on horſeback, or in any other manner. Exerciſe may be continued to a beginning of wearineſs, and ought to be uſed before dinner in a pure light air; for which reaſon, journeys, and going into the country, contribute greatly to preſerve and re-eſtabliſh health. Exerciſe increaſes the circulation of the blood, attenuates and divides the fluids, and promotes a regular perſpiration, as well as a due ſecretion of all the humours; for it accelerates the animal ſpirits, and facilitates their diſtribution into all the fibres of the body, ſtrengthens the parts, creates an appetite, and helps digeſtion. Whence it ariſes, that thoſe who accuſtom themſelves to exerciſe are generally very robuſt, and ſeldom ſubject to diſeaſes.

Boerhaave

Boerhaave recommends bodily exercise in diseases of a weak and lax fibre. By riding on horseback, the pendulous viscera of the abdomen are shaken every moment, and gently rubbed as it were one against another, while in the mean time the pure air acts on the lungs with greater force. But it is to be observed, that a weak man should not ride with a full stomach, but either before dinner or after the digestion is near finished; for, when the stomach is distended, weak people do not bear these concussions of the horse without difficulty; but, when the primæ viæ are near empty, the remaining fæces are discharged by this concussion. Sailing in a ship is also an exercise of great use to weak people. If the vessel moves with an even motion, by increasing perspiration it usually excites a wonderful alacrity, creates an appetite, and promotes digestion. These exercises are more especially serviceable to weak people; but, in order to strengthen the body by muscular motion, running and bodily exercises are to be used. In these we should begin with the most gentle, such as walking, and increase it by degrees till we come to running. Those exercises of the body are more especially serviceable which give delight to the mind at the same time, as tennis, fencing, &c. for which reason, the wisdom of antiquity appointed rewards for those who excelled in these gymnastic exercises, that by this means the bodies of their youth might be hardened for warlike toils.

As nothing is more conducive to health than moderate exercise, so violent exercise dissipates the spirits, weakens the body, destroys the elasticity of the fibres, and exhausts the fluid parts of the blood. No wonder, then, that acute and mortal fevers often arise from too violent exercise of the body; for the motion of the venous blood towards the heart being quickened by the contraction of the muscles, and the veins being thus depleted, the arteries more easily propel their contained humours through the smallest extremities into the now less-resisting veins; and therefore the velocity of the circulation will be increased through all the vessels. But this cannot be performed without applying the humours oftener, or in a greater quantity, to the secretory organs in the same time, whence the more fluid parts of the blood will be dissipated, and what remains will be inspissated; and, by the greater action of the vessels upon their contained fluids, and of the re-acting fluids upon the vessels, the blood acquires an inflammatory density. Add to this, that by the violent attrition of the solids and fluids, together with the heat thence arising, all the humours will incline to a greater acrimony, and the salts and oils of the blood will become more acrid and volatile. Hence those fevers which arise from too much exercise or motion, are cured by the rest of body and mind, with such aliments and medicines as moisten, dilute, and soften or allay acrimony.

The exercise of a soldier in camp, considered as conducive to health, Dr. Pringle distinguishes into three heads; the first relating to his duty, the second to his living more commodiously, and the third to his diversions. The first, consisting chiefly in the exercise of his arms, will be no less the means of preserving health than of making him expert in his duty: and frequent returns of this, early, and before the sun grows hot, will be more advantageous than repeating it seldom, and staying out long at a time; for, a camp affording little convenience for refreshment, all unnecessary fatigue is to be avoided. As to the second article, cutting boughs for shading the tents, making trenches round them for carrying off the water, airing the straw, cleaning their clothes and accoutrements, and assisting in the business of the mess, ought to be no disagreeable exercise to the men for some part of the day. Lastly, as to diversions, the men must be encouraged to them either by the example of their officers, or by small premiums to those who shall excel in such kind of sports as shall be judged most conducive to health: but herein great caution is necessary, not to allow them to fatigue themselves too much, especially in hot weather or sickly times; but above all, that their clothes be kept dry, wet clothes being frequent causes of diseases and death.

Exercise, above all, is peculiarly necessary to the philosopher, the student, and young gentlemen at school. How useful, how agreeable soever, study may be to the mind, it is very far from being equally salutary to the body. Every one observes, that the Creator has formed an intimate connection between the body and the mind; a perpetual action and re-action, by which the body instantly feels the disorders of the mind, and the mind those of the body. The delicate springs of our frail machines lose their activity and become enervated, and the vessels are choaked by obstructions, when we totally desist from exercise; and the consequences necessarily affect the brain: a life entirely studious and sedentary is therefore equally prejudicial to body and mind. The limbs likewise become stiff; we contract an aukward constrained manner; a certain disgustful air attends all our actions, and we are very near being as disagreeable to ourselves as to others. An inclination to study is highly commendable; but it ought not, however, to inspire us with an aversion to society. The natural lot of man is to live among his fellows: and, whatever may be the condition of our birth or our situation in life, there are a thousand occasions where a man must naturally desire to render himself agreeable; to be active and adroit; to dance with a grace; to command the fiery steed; to defend himself against a brutal enemy; to preserve his life by dexterity, as by leaping, swimming, &c. Many rational causes have therefore given rise to the practice of particular exercises; and the most sagacious and benevolent legislators have instituted, in their academies and universities, proper methods of enabling youth, who devote themselves to study, to become expert also in laudable athletic exercises.

Whoever conſiders the ſtructure of the human body will ſoon be convinced of the neceſſity of exerciſe for the health of children. The body is compoſed of an infinite number of veſſels, whoſe fluids cannot be puſhed on without the action and preſſure of the muſcles. But, if the fluids remain inactive, obſtructions muſt happen, and the humours will, of courſe, be vitiated, which cannot fail to occaſion diſeaſes. Nature has furniſhed both the veſſels which carry the blood and lymph with numerous valves, in order that the action of every muſcle might puſh forward their contents; but, without action, this admirable contrivance can have no effect. This part of the animal œconomy proves to a demonſtration the neceſſity of exerciſe for the preſervation of health. Without exerciſe, the circulation of the blood cannot be properly carried on, nor the different ſecretions duly performed; without exerciſe, the humours cannot be properly prepared, nor the ſolids rendered ſtrong or firm. The action of the heart, the motion of the lungs, and all the vital functions, are greatly aſſiſted by exerciſe. But to point out the manner in which theſe effects are produced, would lead us farther into the œconomy of the human body than moſt of thoſe for whom this treatiſe is intended would be able to follow. We ſhall therefore only add, that, where exerciſe is neglected, none of the animal functions can be duly performed; and, when that is the caſe, the whole conſtitution muſt go to wreck.

The love of activity ſhows itſelf very early in man. So ſtrong is this principle, that a healthy youth cannot be reſtrained from exerciſe, even by the fear of puniſhment. Our love of motion is ſurely a ſtrong proof of its utility. Nature implants no diſpoſition in vain. It ſeems to be a catholic law throughout the whole animal creation, that no creature, without exerciſe, ſhould enjoy health, or be able to find ſubſiſtence. Every creature, except man, takes as much of it as is neceſſary. He alone, and ſuch animals as are under his direction, deviate from this original law; and they ſuffer accordingly. Inactivity never fails to induce an univerſal relaxation of the ſolids, which diſpoſes the body to innumerable diſeaſes. When the ſolids are relaxed, neither the digeſtion nor any of the ſecretions can be duly performed. In this caſe, the worſt conſequences muſt enſue. How can perſons who loll all day in eaſy chairs, and ſleep all night on beds of down, fail to be relaxed? Nor do ſuch greatly mend the matter, who never ſtir abroad but in a coach, ſedan, or ſuch like. Theſe elegant pieces of luxury are become ſo common, that the inhabitants of great towns ſeem to be in ſome danger of loſing the uſe of their limbs altogether. It is now below any one to walk, who can afford to be carried. How ridiculous would it ſeem, to a perſon unacquainted with modern luxury, to behold the young and healthy ſwinging along on the ſhoulders of their fellow-creatures! or to ſee a fat carcaſe, over-run with diſeaſes occaſioned by inactivity, dragged through the ſtreets by half a dozen horſes!

Glandular obſtructions, now ſo common, generally proceed from inactivity. Theſe are the moſt obſtinate of maladies. So long as the liver, kidneys, and other glands, duly perform their functions, health is ſeldom impaired; but, when they fail, nothing can reſtore it. Exerciſe is almoſt the only cure we know for glandular obſtructions; indeed, it does not always ſucceed as a *remedy;* but there is reaſon to believe that it would ſeldom fail to *prevent* theſe complaints, were it uſed in due time. One thing is certain, that, amongſt thoſe who take ſufficient exerciſe, glandular diſeaſes are very little known; whereas the indolent and inactive are ſeldom free from them. Weak nerves are the conſtant companions of inactivity. Nothing but exerciſe and open air can brace and ſtrengthen the nerves, or prevent the endleſs train of diſeaſes which proceed from a relaxed ſtate of theſe organs. We ſeldom hear the active or laborious complain of nervous diſeaſes; theſe are reſerved for the ſons of eaſe and affluence. Many have been completely cured of theſe diſorders by being reduced from a ſtate of opulence to labour for their daily bread. This plainly points out the ſources from whence nervous diſeaſes flow, and the means by which they may be prevented. It is abſolutely impoſſible to enjoy health, where the perſpiration is not duly carried on; but that can never be the caſe where exerciſe is neglected. When the matter which ought to be thrown off by perſpiration is retained in the body, it vitiates the humours, and occaſions the gout, fevers, rheumatiſm, &c. Exerciſe alone would prevent many of thoſe diſeaſes which cannot be cured, and would remove others where medicine proves ineffectual.

No piece of indolence hurts the health more than the modern cuſtom of lying a-bed too long in a morning. This is the general practice in great towns. The inhabitants of cities ſeldom riſe before nine or ten o'clock; but the morning is undoubtedly the beſt time for exerciſe, while the ſtomach is empty, and the body refreſhed with ſleep. Beſides, the morning air braces and ſtrengthens the nerves, and, in ſome meaſure, anſwers the purpoſe of a cold bath. Let any one who has been accuſtomed to lie a-bed late, riſe by ſix or ſeven, ſpend a couple of hours in walking, riding, or any active diverſion without doors, and he will find his ſpirits cheerful and ſerene through the day, his appetite keen, and his body braced and ſtrengthened. Cuſtom ſoon renders early riſing agreeable, and nothing contributes more to the preſervation of health. The inactive are continually complaining of pains of the ſtomach, flatulencies, indigeſtions, &c. Theſe complaints, which pave the way to many others, are not to be removed by medicine: they can only be cured by a vigorous courſe of exerciſe, to which they ſeldom fail to yield.

Exerciſe, if poſſible, ought always to be taken in the open air. When that cannot be done, various methods may be contrived for exerciſing the body within doors.

doors. It is not neceſſary to adhere ſtrictly to any particular kind of exerciſe. The beſt way is to take them by turns, and to uſe that longeſt which is moſt ſuitable to the ſtrength and conſtitution. Thoſe kinds of exerciſe which give action to moſt of the bodily organs, are always to be preferred, as walking, running, riding, digging, ſwimming, and ſuch like. It is much to be regretted, that active and manly diverſions are now ſo little practiſed. Diverſions make people take more exerciſe than they otherwiſe would do, and are of the greateſt ſervice to ſuch as are not under the neceſſity of labouring for their bread. As active diverſions loſe ground, thoſe of a ſedentary kind ſeem to prevail. Sedentary diverſions are of no other uſe but to conſume time. Inſtead of relieving the mind, they often require more thought than either ſtudy or buſineſs. Every thing that induces people to ſit ſtill, unleſs it be ſome neceſſary employment, ought to be avoided.

The diverſions which afford the beſt exerciſe are hunting, ſhooting, playing at cricket, bowls, &c. Theſe exerciſe the limbs, promote perſpiration, and the other ſecretions. They likewiſe ſtrengthen the lungs, and give firmneſs and agility to the whole body. Such as can, ought to ſpend two or three hours a-day on horſeback; thoſe who cannot ride, ſhould employ the ſame time in walking. Exerciſe ſhould never be continued too long. Over-fatigue prevents the benefit of exerciſe, and inſtead of ſtrengthening the body tends to weaken it. Every man ſhould lay himſelf under ſome ſort of neceſſity to take exerciſe. Indolence, like other vices, when indulged, gains ground, and at length becomes agreeable. Hence many who were fond of exerciſe in the early part of life, become quite averſe from it afterwards. This is the caſe of moſt hypochondriac and gouty people, which renders their diſeaſes in a great meaſure incurable. Indolence not only occaſions diſeaſes, and renders men uſeleſs to ſociety, but promotes all manner of vice. To ſay a man is lazy, is little better than calling him vicious. The mind, if not engaged in ſome uſeful purſuit, is conſtantly in queſt of ideal pleaſures, or impreſſed with the apprehenſion of ſome imaginary evil. From theſe ſources proceed moſt of the miſeries of mankind. Certainly man was never intended to be idle. Inactivity fruſtrates the very deſign of his creation: whereas an active life is the beſt guardian of virtue, and the greateſt preſervative of health.

Of SLEEP.

SLEEP, as well as food, ought to be duly regulated. Too little ſleep weakens the nerves, exhauſts the ſpirits, and occaſions diſeaſes; and too much renders the mind dull, the body groſs, and diſpoſes to apoplexies, lethargies, and other complaints of a ſimilar nature. A medium ought therefore to be obſerved; but this is not eaſy to fix. Children require more ſleep than grown perſons, the laborious

than

than the idle, and ſuch as eat and drink freely than thoſe who live abſtemiouſly. Beſides, the real quantity of ſleep cannot be meaſured by time, as one perſon will be more refreſhed by five or ſix hours ſleep than another by eight or ten. Children may always be allowed to take as much ſleep as they pleaſe; but, for adults, ſix or ſeven hours are certainly ſufficient, and no one ought to exceed eight. Thoſe who lie a-bed more than eight hours may ſlumber, but they can be hardly ſaid to ſleep; ſuch generally toſs and dream away the fore part of the night, ſink to reſt towards morning, and doze till noon. The beſt way to make ſleep ſound and refreſhing is to riſe betimes. The cuſtom of lying a-bed for nine or ten hours, not only makes the ſleep leſs refreſhing, but weakens the conſtitution. Nature points out night as the proper ſeaſon for ſleep. Nothing more certainly deſtroys the conſtitution than night-watching. It is a great pity that a practice ſo deſtructive to health ſhould be ſo much in faſhion. How quickly the want of reſt in due ſeaſon will blaſt the moſt blooming complexion, or ruin the beſt conſtitution, is evident from the ghaſtly countenances of thoſe who, as the phraſe is, turn day into night, and night into day. To make ſleep refreſhing, the following things are requiſite: Firſt, to take ſufficient exerciſe in the open air; to avoid ſtrong tea or coffee; next, to eat a light ſupper; and laſtly, to lie down with a mind as cheerful and ſerene as poſſible.

It is certain that too much exerciſe will prevent ſleep, as well as too little. We ſeldom, however, hear the active and laborious complain of reſtleſs nights. It is the indolent and ſlothful who generally have theſe complaints. Is it any wonder that a bed of down ſhould not be refreſhing to a perſon who ſits all day in an eaſy chair? A great part of the pleaſure of life conſiſts in alternate reſt and motion; but they who neglect the latter can never reliſh the former. The labourer enjoys more true luxury in plain food and ſound ſleep, than is to be found in ſumptuous tables and downy pillows, where exerciſe is wanting. That light ſuppers cauſe ſound ſleep, is true even to a proverb. Many perſons, if they exceed the leaſt at that meal, are ſure to have uneaſy nights; and, if they fall aſleep, the load and oppreſſion on their ſtomach and ſpirits occaſion frightful dreams, broken and diſturbed repoſe, the night-mare, &c. Were the ſame perſons to go to bed with a light ſupper, or ſit up till that meal was pretty well digeſted, they would enjoy ſound ſleep, and riſe refreſhed and cheerful. There are indeed ſome people who cannot ſleep unleſs they have ſome ſolid food at night; but this does not imply the neceſſity of a heavy ſupper.

Nothing more certainly diſturbs our repoſe than anxiety. When the mind is not at eaſe, one ſeldom enjoys ſound ſleep. That greateſt of human bleſſings flies the wretched, and viſits the happy, the cheerful, and the gay. This is a ſufficient reaſon why every man ſhould endeavour to be as eaſy in his mind as poſſible when he goes to reſt. Many, by indulging grief and anxious thought, have baniſhed ſound ſleep ſo long, that they could never afterwards enjoy it. Sleep, when taken in the fore part of the night, is generally reckoned moſt refreſhing. Whether this be the effect of habit or not is hard to ſay; but, as moſt people are accuſtomed to go early to bed when young, it may be preſumed that ſleep, at this ſeaſon, will prove moſt refreſhing to them ever after. Whether the fore part of the night be beſt for ſleep or not, ſurely the fore part of the day is fitteſt both for buſineſs and amuſement; and we hardly ever find an early-riſer who does not enjoy a good ſtate of health.

Experience proves that, the more a perſon ſleeps, the more he is inclined to ſleep: if in the morning we ſleep an hour beyond our cuſtom, the conſequence is, that we ſhall be dull and heavy all the day; and, as to theſe facts, there are ſome very remarkable.—A youth in Germany, of immenſe wealth, was ſummoned by his prince to take up a title of nobility; on which occaſion he drank to ſuch an exceſs, that the prince, in order to cure him of ſuch a ſcandalous vice, had him carried into a dark and remote place, where he ſlept three days and three nights; for, whenever he awoke, believing it to be the middle of the night, he betook himſelf to ſleep again.—The Memoirs of the Academy of Sciences at Paris mention a ſleep of two months, cauſed by a catalepſy, a diſeaſe by which the patient is inſtantly rendered as immoveable as a ſtatue.—Samuel Chilton, a labourer in Somerſetſhire, fell, and without any viſible cauſe, into a profound ſleep, out of which no means could recover him, till, after a month's time, he aroſe of himſelf. His mother, fearing he would be ſtarved in that ſullen humour, as ſhe called it, put bread and cheeſe and ſmall beer by him, and it was daily ſpent. On the ninth of April, 1696, he was ſeized with a like ſleepy fit, which laſted till the ſeventh of Auguſt, when he awoke, without knowing he had ſlept above a night. He occaſionally uſed the food ſet by him, and had evacuations, till, about the tenth week, his jaws ſeemed to be ſet, and his teeth clinched ſo cloſe that his mouth could not be opened; and all the nouriſhment he received, during theſe ſeven weeks, was about three pints of tent inſinuated through a cavity in one of his teeth. He had made water but once, and never had a ſtool all the time. On the 17th of Auguſt, 1697, his fit returned; and Dr. Oliver, the author of the memoir, in order to try whether there might not be ſome impoſture in this extraordinary phenomenon, went to the houſe; he put his mouth to his ear, and called him ſeveral times, by his name, as loud as he could; pulled him by the ſhoulders, pinched his noſe, ſtopped his mouth and noſe at the ſame time; lifted up his eye-lids, when he found the balls drawn up under the brows; he, farther, held a phial of ſpirit of ſal ammoniac

niac under one nostril; that producing no effect, he poured up his nose near a half-ounce bottle, and the spirit, he says, being drawn from quicklime, was almost as hot as fire itself. Not satisfied with this, he crammed the same nostril with powder of white hellebore: all these experiments producing no other effect than to make his eyes shiver a little. Dr. Oliver left him, convinced that he was really asleep. A few days after, an apothecary drew some ounces of blood from his arm, and bound up the orifice, without his making the least motion: likewise, a gentleman, though somewhat indiscreetly, ran a pin into his arm, up to the very bone; and in this state of insensible sleep he continued till the nineteenth of November; during all this time he ate and evacuated, but never fouled his bed. The above instance of sleep is to be seen at large in Jones's Abridgement of the Philosophical Transactions, vol. v.

Of Dreams.

SCARCELY any part of nature is less open to our observation than the human mind in this state. The dreamer himself cannot well observe the manner in which dreams arise or disappear to him. When he awakes, he cannot recollect the circumstances of his dreams with sufficient accuracy. Were we to watch over him with the most vigilant attention, we could not perceive with certainty what emotions are excited in his mind, or what thoughts pass through it, during his sleep. But, even though we could ascertain these phenomena, many other difficulties would still remain. What parts of a human being are active, what dormant, when he dreams? Why does he not always dream while asleep? Or why dreams he at all? Do any circumstances in our constitution, situation, and peculiar character, determine the nature of our dreams? We may lay before the reader such facts as have been ascertained concerning dreaming, and the most plausible conjectures that have been offered to explain those particulars, about which we can only conjecture, or have at least hitherto obtained nothing more certain than conjecture.

In dreaming, we are not conscious of being asleep. This is well known from a thousand circumstances. When awake, we often recollect our dreams; and we remember on such occasions, that, while those dreams were passing through our minds, it never occurred to us that we were separated by sleep from the active world. We are often observed to act and talk in dreaming as if we were busily engaged in the intercourse of social life. In dreaming we do not consider ourselves as witnessing or bearing a part in a fictitious scene; we seem not to be in a similar situation with the actors in a dramatic performance, or the spectators before whom they exhibit, but engaged in the business of real life. All the varieties of thought that pass through our minds when awake may also occur in dreams; all the images which imagination presents in the former state she is also able to call up in the latter all the same emotions may be excited, and we are often actuated by equal violence of passion; none of the transactions in which we are capable of engaging while awake is impossible in dreams: in short, our range of action and observation is equally wide in one state as in the other; and, while dreaming, we are not sensible of any distinction between our dreams and the events and transactions in which we are actually concerned in our intercourse with the world.

Though in dreams imagination appears to be free from all restraint, and indulges in the most wanton freaks, yet it is generally agreed, that the imaginary transactions of the dreamer bear always some relation to his particular character in the world, his habits of action, and the circumstances of his life. The lover, we know, dreams of his mistress, and the miser of his money; the philosopher renews his researches in sleep often with the same pain and fatigue as when awake; and even the merchant, at times, returns to balance his books, and compute the profits of an adventure, when slumbering on his pillow. And not only do the more general circumstances of a person's life influence his dreams; his passions and habits are nearly the same when asleep as when awake. *A person whose habits of life are virtuous, does not in his dreams plunge into a series of crimes; nor are the vicious reformed when they pass into this imaginary world.* The choleric man finds himself offended by slight provocations as well in his dreams as in his ordinary intercourse with the world, and a mild temper continues pacific in sleep. The character of a person's dreams is influenced by his circumstances when awake in a still more unaccountable manner. Certain dreams usually arise in the mind after a person has been in certain situations. Dr. Beattie relates, that he once, after riding thirty miles in a high wind, passed a part of the succeeding night in dreams beyond description terrible. The state of a person's health, and the manner in which the vital functions are carried on, have a considerable influence in determining the character of dreams. After too full a meal, or after eating of an unusual sort of food, a person has always dreams of a certain nature. In dreaming, the mind for the most part carries on no intercourse through the senses with surrounding objects. Touch a person gently who is asleep, he feels not the impression. You may awake him by a smart blow; but, when the stroke is not sufficiently violent to effect that, he remains insensible of it. We speak softly beside a person asleep without fearing that he overhears us. His eye-lids are shut; and, even though light should fall upon the eye-ball, yet still his powers of vision are not awakened to active exertion, unless the light be so strong as to rouse him from sleep. He is insensible both to sweet and disagreeable smells. It is not easy to try whether his organs of taste retain their activity, without awakening him; yet from analogy it may be presumed that

that theſe too are inactive. With reſpect to the circumſtances here enumerated, it is indifferent whether a perſon be dreaming or buried in ſleep.

Yet there is one remarkable fact concerning dreaming which may seem to contradict what has been here aſſerted. In dreams, we are liable not only to ſpeak aloud in conſequence of the ſuggeſtions of imagination, but even to get up, and walk about, and engage in little enterpriſes, without awaking. Now, as we are in this inſtance ſo active, it ſeems that we cannot be then inſenſible of the preſence of ſurrounding objects. The ſleep-walker is really ſenſible, in a certain degree, of the preſence of the objects around him; but he does not attend to them with all their circumſtances, nor do they excite in him the ſame emotions as if he were awake. He feels no terror on the top of a houſe, or the brink of a precipice; and, in conſequence of being free from fear, he is alſo without danger in ſuch a ſituation, unleſs ſuddenly awaked. This is one of the moſt inexplicable phenomena of dreaming. There is alſo another fact not quite conſonant with what has been above advanced. It is ſaid, that in ſleep a perſon will continue to hear the noiſe of a cataract in the neighbourhood, or regular ſtrokes with a hammer, or any ſimilar ſound ſufficiently loud, and continued uninterruptedly from before the time of his falling aſleep. We know not whether he awakes on the ſudden ceſſation of the noiſe. This fact is aſſerted on ſufficient evidence: it is curious. Even when awake, if very deeply intent on any piece of ſtudy, or cloſely occupied in buſineſs, the ſound of a clock ſtriking in the neighbourhood, or the beating of a drum, will eſcape us unnoticed: and it is therefore the more ſurpriſing that we ſhould thus continue ſenſible to ſounds when aſleep. Not only do a perſon's general character, habits of life, and ſtate of health, influence his dreams; but thoſe concerns in which he has been moſt deeply intereſted during the preceding day, and the views which have ariſen moſt frequently to his imagination, very often afford the ſubjects of his dreams. When I look forward with anxious expectation towards any future event, I am likely to dream either of the diſappointment or the gratification of my wiſhes. Have I been engaged through the day, either in buſineſs or amuſements which I have found exceedingly agreeable, or in a way in which I have been extremely unhappy, either my happineſs or my miſery is likely to be renewed in my dreams. Though dreams have been regarded among almoſt all nations through the world, at leaſt in ſome periods of their hiſtory, as prophetic of future events, yet it does not appear that this popular opinion has been eſtabliſhed on good grounds. Chriſtianity, indeed, teaches us to believe, that the Supreme Being may, and actually does, operate on our minds, and influence at times the determinations of our will, without making us ſenſible of the reſtraint to which we are thus ſubjected. And, in the ſame manner, no doubt, the ſuggeſtions which ariſe to us in dreams may be produced.

produced. The imaginary tranfactions in which we are then engaged, may be fuch as are actually to occupy us in life; the ftrange and feemingly incoherent appearances which are then prefented to the mind's eye, may allude to fome events which are to befal ourfelves or others. It is, therefore, by no means impoffible, or inconfiftent with the general analogy of nature, that dreams fhould have a refpect to futurity. We have no reafon to regard the dreams which are related in the Holy Scriptures to have been prophetic of future events as not infpired by heaven, or to laugh at the idea of a prophetic dream as abfurd or ridiculous.

We know of no other facts that have been fully afcertained concerning dreaming. But we are by no means fufficiently acquainted with this important phenomenon in the hiftory of the mind. We cannot tell by what laws of our conftitution we are thus liable to be fo frequently engaged in imaginary tranfactions, nor what are the particular means by which the delufion is accomplifhed. The delufion is indeed remarkably ftrong. One will fometimes have a book prefented to him in a dream, and fancy that he reads, and actually enter into the nature of the compofition before him, and even remember, after he awakes, what he knows that he only fancied himfelf reading. Can this be delufion? If delufion, how or for what purpofes is it produced? The mind, it would appear, does not, in fleep, become inactive like the body; or at leaft is not always inactive while we are afleep. When we do not dream, the mind muft either be inactive, or the connection between the mind and the body muft be confidered as in fome manner fufpended: and, when we dream, the mind, though it probably acts in concert with the body, yet does not act in the fame manner as when we are awake. It feems to be clouded or bewildered, in confequence of being deprived for a time of the fervice of the fenfes. Imagination becomes more active and more capricious: and all the other powers, efpecially judgment and memory, become difordered and irregular in their operation.

Various theories have been propofed to explain what appears here moft inexplicable. The ingenious Mr. Baxter, in his treatife on the Immateriality of the Human Soul, endeavours to prove that dreams are produced by the agency of fome fpiritual beings, who either amufe or employ themfelves ferioufly in engaging mankind in all thofe imaginary tranfactions with which they are employed in dreaming. This theory, however, is far from being plaufible. It leads us entirely beyond the limits of our knowledge. It requires us to believe without evidence. It is unfupported by any analogy. It creates difficulties ftill more inexplicable than thofe which it has been propofed to remove. Till it appear that our dreams cannot poffibly be produced without the interference of other fpiritual agents, poffeffing fuch influence over our minds as to deceive us with fancied joys, and involve us in imaginary afflictions, we cannot reafonably refer them to fuch a caufe. Befides, from the facts which have been ftated as well known concerning dreams, it appears that their nature depends both on the ftate of the human body and on that of the mind. But, were they owing to the agency of other fpiritual beings, how could they be influenced by the ftate of the body? Thofe muft be a curious fet of fpiritual beings who depend in fuch a manner on the ftate of our corporeal frame. Better not to allow them exiftence at all, than to place them in fuch a dependance.

Wolfius, and after him M. Formey, have fuppofed, that dreams never arife in the mind except in confequence of fome of the organs of fenfation having been previoufly excited. Either the ear or the eye, or the organs of touching, tafting, or fmelling, communicate information, fomehow, in a tacit, fecret, manner; and thus partly roufe its faculties from the lethargy in which they are buried in fleep, and engage them in a feries of confufed and imperfect exertions. But what paffes in dreams is fo very different from all that we do when awake, that it is impoffible for the dreamer himfelf to diftinguifh whether his powers of fenfation perform any part on the occafion. It is not neceffary that imagination be always excited by fenfation. Fancy, even when we are awake, often wanders from the prefent fcene. Abfence of mind is incident to the ftudious; the poet and the mathematician many times forget what they are. We cannot difcover, from any thing that a perfon in dreaming difplays to the obfervation of others, that his organs of fenfation take a part in the imaginary tranfactions in which he is employed. In thofe inftances, indeed, in which perfons afleep are faid to hear founds; the founds which they hear are faid alfo to influence, in fome manner, the nature of their dreams. But fuch inftances are fingular. Since then it appears that the perfon who dreams is himfelf incapable of diftinguifhing either during his dreams, or by recollection when awake, whether any new impreffions are communicated to him in that ftate by his organs of fenfation; that even by watching over him, and comparing our obfervations of his circumftances and emotions, in his dreams, with what he recollects of them after awaking, we cannot, except in one or two fingular inftances, afcertain this fact; and that the mind is not incapable of acting while the organs of fenfation are at reft, and on many occafions refufes to liften to the information which they convey; we may, without hefitation, conclude, that the theory of Wolfius and Formey has been too haftily and incautioufly advanced.

Other phyfiologifts tell us, that the mind, when we dream, is in a ftate of delirium. Sleep, they fay, is attended with what is called a collapfe of the brain; during which either the whole or a part of the nerves of which it confifts, are in a ftate in which they cannot carry on the ufual intercourfe between the mind and the organs of fenfation. When the whole of the brain is in this ftate, we become entirely unconſcious of exiſtence, and the mind ſinks into inactivity : when only a part of the brain is collapſed, as they term it, we are then neither aſleep nor awake, but in a ſort of delirium between the two. This theory, like the laſt-mentioned, ſuppoſes the mind incapable of acting without the help of ſenſation : it ſuppoſes that we know the nature of a ſtate of which we cannot aſcertain the phenomena ; it alſo contradicts a known fact, in repreſenting dreams as confuſed images of things around us, not fanciful combinations of things not exiſting together in nature or in human life. We muſt conſider it likewiſe, therefore, as a baſeleſs fabric.

Inſtead of the attendant ſpirits watching over our bodies, and inciting us to good or evil in our dreams, may we not more rationally ſuppoſe, that theſe incitements, or rather exertions towards real and ſenſible actions, are produced by the ſoul or ſpirit within us, which, being immortal, never ſleeps ; but which rather, during that paſſive ſtate of the body, aſſumes an endeavour to act without it, or to eſcape from it, as from a priſon, wherein it is reſtrained to certain limits, and obliged to act under the will of its keeper. This ſurely will beſt explain the facts attending ſleep-walkers, who, in the darkeſt nights viſit different apartments, walk up and down ſtairs, lock and unlock doors, open windows, and crawl over the roofs of houſes, with the utmoſt eaſe and celerity; which, if the perſons were awake, would be impoſſible. May we not likewiſe attribute to the ſame ſource, thoſe elevated ideas, and ſublime compoſitions, which Milton and other celebrated authors have confeſſed, were communicated nightly in their dreams?

It lately happened, that a young gentlemen, about fifteen years of age, from one of the public ſchools, ſlept in the ſame room with me. He choſe to go to bed early; and, when I came into the ſame apartment about two hours after, he appeared remarkably intent upon his ſtudies, though faſt locked in the arms of ſleep. I ſtood ſome time at his bed-ſide, and heard him repeat ſeveral lines from Homer and Virgil. After this he repeated, with a bold and nervous accent, the whole of the Hebrew alphabet: then, turning, ſeemed to fall into a more compoſed ſleep. The next morning at breakfaſt I related this circumſtance to the company, in the preſence of the young gentleman; and all began to commend the great progreſs he muſt have made in his ſtudies. The young man inſtantly declared, that, however converſant he might be with Virgil and Homer, he had never heard the Hebrew alphabet repeated, nor did he ever know the name of any one of its characters.—The nature of the connection by which the ſoul and body are united, ſeems to be almoſt beyond our comprehenſion; and, till we can apply experiment and obſervation in a better manner to this branch of phyſiology,

physiology, it must undoubtedly remain unknown. To something mysterious in the nature of that connection, the delusion produced in dreams is in all probability owing.

Amid this uncertainty with respect to the manner in which our powers of mind and body perform their functions in dreaming, it is pleasing to find that we can, however, apply to useful purposes the imperfect knowledge which we have been able to acquire concerning the series of phenomena. Our dreams are affected by the state of our health, by the manner in which we have passed the preceding day, by our general habits of life, by the hopes which we most fondly indulge, and the fears which prevail most over our fortitude when we are awake. From recollecting our dreams, therefore, we may learn to correct many improprieties in our conduct; to refrain from bodily exercises, or from meats and drinks, that have unfavorable effects on our constitution; to resist in due time evil habits that are stealing upon us; and to guard against hopes and fears which detach us from our proper concerns, and unfit us for the duties of life. Instead of thinking what our dreams may forebode, we may, with much better reason, reflect by what they have been occasioned, and look back to those circumstances in our past life to which they are owing. The sleep of innocence and health is sound and refreshing; their dreams delightful and pleasing. A distempered body, and a polluted or perturbed mind, are haunted in sleep with frightful, impure, and unpleasing, dreams.

Of INTEMPERANCE.

It is universally agreed, that temperance and exercise are the two best physicians in the world, and that, if these were duly regarded, there would be little occasion for any other. *Temperance may justly be called the parent of health;* yet numbers of mankind act as if they thought diseases and death too slow in their progress, and, by intemperance and debauch, seem, as it were, to solicit their approach. The danger of intemperance appears from the very construction of the human body. Health depends on the state of the solids and fluids which fits them for the due performance of the vital functions; and, while these go regularly on, we are sound and well; but whatever disturbs them necessarily impairs health. Intemperance never fails to disorder the whole animal œconomy; it hurts the digestion, relaxes the nerves, renders the differeut secretions irregular, vitiates the humours, and occasions numberless diseases.

The analogy between the nourishment of plants and animals affords a striking proof of the danger of intemperance. Moisture and manure greatly promote vegetation; yet an over-quantity of either will entirely destroy it. The best things become hurtful, nay destructive, when carried to excess. Hence we learn, that the higheſt degree of human wiſdom conſiſts in regulating our appetites and paſſions ſo as to avoid all extremes. It is this chiefly which entitles us to the character of rational beings. The ſlave of appetite will ever be the diſgrace of human nature.—The Supreme Being hath endued us with various paſſions for the propagation of the ſpecies, the preſervation of the individual, &c. Intemperance is the abuſe of theſe paſſions; and moderation conſiſts in the proper regulation of them. Men, not contented with ſatisfying the ſimple calls of Nature, create artificial wants, and are perpetually in ſearch of ſomething that may gratify them; but imaginary wants can never be gratified. *Nature is content with little; but luxury knows no bounds.* Hence the epicure, the drunkard, and the debauchee, ſeldom ſtop their career, till their money, or their conſtitution, fails: then, indeed, they generally ſee their error when too late.

It is impoſſible to lay down fixed rules with regard to diet, on account of the different conſtitutions of mankind. The moſt ignorant perſon, however, certainly knows what is meant by exceſs; and it is in the power of every man, if he chooſes, to avoid it. The great rule of diet is to ſtudy ſimplicity. Nature delights in the moſt plain and ſimple food; and every animal, except man, follows her dictates. Man alone riots at large, and ranſacks the whole creation in queſt of luxuries, to his own deſtruction. An elegant writer of the laſt age ſpeaks thus of intemperance in diet: "For my part, when I behold a faſhionable table ſet out in all its magnificence, I fancy that I ſee gouts and dropſies, fevers and lethargies, with other innumerable diſtempers, lying in ambuſcade among the diſhes." Nor is intemperance in other things leſs deſtructive than in diet. How quickly does the immoderate purſuit of carnal pleaſures, or the abuſe of intoxicating liquors, ruin the beſt conſtitution! Indeed theſe vices generally go hand in hand. Hence it is that we ſo often behold the votaries of Bacchus and Venus, even before they have arrive at the prime of life, worn out with diſeaſe, and haſting with ſwift pace to an untimely grave. Did men reflect on the painful diſeaſes, and premature deaths, which are daily occaſioned by intemperance, it would be ſufficient to make them ſhrink back with horror from the indulgence even of their darling pleaſures.

Intemperance does not hurt its votaries alone; the innocent too often feel the direful effects of it. How often do we behold the miſerable mother, with her helpleſs infants, pining in want, while the cruel father is indulging his inſatiate appetites? Families are not only reduced to miſery, but even extirpated, by intemperance. *Nothing tends ſo much to prevent propagation, and to ſhorten the lives of children, as the intemperance of parents.* The poor man who labours all day, and at night lies down contented with his humble fare, can boaſt a numerous offspring; while his pampered lord, ſunk in eaſe and luxury, often languiſhes

without

without an heir to his ample fortunes. Even ſtates and empires feel the influence of intemperance, and riſe or fall as it prevails. Inſtead of mentioning the different kinds of intemperance, and pointing out their influence upon health, we ſhall only, by way of example, make a few obſervations on one particular ſpecies of that vice, viz. the abuſe of intoxicating liquors.

Every act of intoxication puts Nature to the expenſe of a fever, in order to diſcharge the poiſonous draught. When this is repeated almoſt every day, it is eaſy to foreſee the conſequences. That conſtitution muſt be ſtrong indeed, which is able long to hold out under a daily fever! *But fevers occaſioned by drinking do not always go off in a day; they frequently end in an inflammation of the breaſt, liver, or brain, and produce fatal effects.* Though the drunkard ſhould not fall by an acute diſeaſe, he ſeldom eſcapes thoſe of a chronic kind. Intoxicating liquors, when uſed to exceſs, weaken the bowels, and ſpoil the digeſtion; they deſtroy the power of the nerves, and occaſion paralytic and convulſive diſorders: they likewiſe heat and inflame the blood, deſtroy its balſamic quality, render it unfit for circulation, and for the nouriſhment of the body. Hence obſtructions, atrophies, dropſies, and conſumptions of the lungs. Theſe are the common ways in which drunkards make their exit. Diſeaſes of this kind, when brought on by hard drinking, ſeldom admit of a cure. *Many people injure their health by drinking, who ſeldom get drunk.* The continual habit of ſoaking, as it is called, though its effects be not ſo violent, is not leſs pernicious. When the veſſels are kept conſtantly full and upon the ſtretch, the different digeſtions can neither be duly performed, nor the humours properly prepared. Hence moſt people of this character are afflicted with the gout, the gravel, ulcerous ſores in the legs, &c. If theſe diſorders do not appear, they are ſeized with low ſpirits, hypochondriacal affections, and other ſymptoms of indigeſtion.

Conſumptions are now ſo common, that it is thought one-tenth of the inhabitants of great towns die of that diſeaſe. Hard drinking is no doubt one of the cauſes to which we muſt impute the increaſe of conſumptions. The great quantities of viſcid malt-liquor drunk by the common people of England cannot fail to render the blood ſizy and unfit for circulation; from whence proceed obſtructions and inflammations of the lungs. There are few great ale-drinkers who are not phthiſical; nor is that to be wondered at, conſidering the glutinous and almoſt indigeſtible nature of ſtrong ale. Thoſe who drink ardent ſpirits or ſtrong wines run ſtill greater hazard; theſe liquors heat and inflame the blood, and tear the tender veſſels of the lungs in pieces; yet, ſo great is the conſumption of them in this country, that one would almoſt be induced to think the inhabitants lived upon them. The habit of drinking proceeds frequently from misfortunes in life. The miſera-

ble

ble fly to it for relief. It affords them indeed a temporary eaſe. But alas! this ſolace is ſhort-lived; and, when it is over, the ſpirits ſink as much below their uſual tone as they had before been raiſed above it. Hence a repetition of the doſe becomes neceſſary; and every freſh doſe makes way for another, till the unhappy wretch becomes a ſlave to the bottle, and at length falls a ſacrifice to what at firſt perhaps was taken only as a medicine. No man is ſo dejected as the drunkard when his debauch is gone off. Hence it is, that thoſe who have the greateſt flow of ſpirits while the glaſs circulates freely, are of all others the moſt melancholy when ſober, and often put an end to their own miſerable exiſtence in a fit of ſpleen or ill humour.

Drunkenneſs not only proves deſtructive to health, but likewiſe to the faculties of the mind. It is ſtrange that creatures who value themſelves on account of a ſuperior degree of reaſon to that of brutes ſhould take pleaſure in ſinking ſo far below them. Were ſuch as voluntarily deprive themſelves of the uſe of reaſon to continue ever after in that condition, it would ſeem but a juſt puniſhment; and, though this be not the conſequence of one act of intoxication, it ſeldom fails to ſucceed a courſe of it. By a habit of drinking, the greateſt genius is often reduced to a mere idiot. *Intoxication is peculiarly hurtful to young perſons.* It heats their blood, impairs their ſtrength, and obſtructs their growth; beſides, the frequent uſe of ſtrong liquors in the early part of life deſtroys any benefit that might ariſe from them afterwards. *Thoſe who make a practice of drinking generous liquors when young, cannot expect to reap any benefit from them as a cordial in the decline of life.* Drunkenneſs is not only in itſelf a moſt abominable vice, but is an inducement to many others. There is hardly any crime ſo horrid that the drunkard will not perpetrate for the love of liquor. We have known mothers ſell their children's clothes, the food that they ſhould have eaten, and afterwards even the infants themſelves, in order to purchaſe the accurſed draught.

It is amazing that our improvements in arts, learning, and politeneſs, have not put the barbarous cuſtom of drinking to exceſs out of faſhion. It is indeed leſs common in South Britain than it was formerly: but it ſtill prevails very much in the North, where this relic of barbarity is miſtaken for hoſpitality. In Ireland, no man is ſuppoſed to entertain his gueſts well, who does not make them drunk. Forcing people to drink, is certainly the greateſt piece of rudeneſs that any man can be guilty of. Bravado, complaiſance, or mere good-nature, may induce a man to take his glaſs, if urged to it, at a time when he might as well take poiſon. The cuſtom of drinking to exceſs has long been out of faſhion in France; and, as it begins to loſe ground among the politer part of the Engliſh, we hope it will ſoon be baniſhed from every part of the united kingdom.

Of the PASSIONS.

THE paſſions have great influence both in the cauſe and cure of diſeaſes. How the mind affects the body, will, in all probability, ever remain a ſecret. It is ſufficient for us to know, that there is eſtabliſhed a reciprocal influence betwixt the mental and corporeal parts, and that whatever injures the one diſorders the other.

Of Anger.—The paſſion of anger ruffles the mind, diſtorts the countenance, hurries on the circulation of the blood, and diſorders the whole vital and animal functions. It often occaſions fevers, and other acute diſeaſes, and ſometimes even ſudden death. This paſſion is peculiarly hurtful to the delicate, and thoſe of weak nerves. I have known ſuch perſons frequently loſe their lives by a violent fit of anger, and would adviſe them to guard againſt the exceſs of this paſſion with the utmoſt care. It is not indeed always in our power to prevent being angry; but we may ſurely avoid harbouring reſentment in our breaſt. Reſentment preys upon the mind, and occaſions the moſt obſtinate chronical diſorders, which gradually waſte the conſtitution. Nothing ſhows true greatneſs of mind more than to forgive injuries: it promotes the peace of ſociety, and greatly conduces to our own eaſe, health, and felicity. Such as value health ſhould avoid violent guſts of anger as they would the moſt deadly poiſon. Neither ought they to indulge reſentment, but endeavour at all times to keep their minds calm and ſerene. Nothing tends ſo much to the health of the body as a conſtant tranquillity of mind. Add to this, the indecency of extravagent anger; how it renders us, whilſt it laſts, the ſcorn and ſport of all about us, of which it leaves us, when it ceaſes, ſenſible and aſhamed; the inconveniences and irretrievable miſconduct into which our irraſcibility has ſometimes betrayed us; the friendſhips it has loſt us; the diſtreſſes and embarraſſments in which we have been involved by it, and the ſore repentance which on one account or other it always coſts us.

Phyſicians and naturaliſts afford inſtances of very extraordinary effects of this paſſion.—Borrichius cured a woman of an inveterate tertian ague, which had baffled the art of phyſic, by putting the patient in a furious fit of anger. Valeriola made uſe of the ſame means, with the like ſucceſs, in a quartan ague.—The ſame paſſion has been equally ſalutary to paralytic, gouty, and even dumb, perſons; to which laſt it has ſometimes given the uſe of ſpeech. Etmuller gives divers inſtances of very ſingular cures wrought by anger; among others, he mentions a perſon laid up in the gout, who, being provoked by his phyſician, flew upon him, and was cured. It is true, the remedy is ſomewhat dangerous in the application, when a patient does not happen to uſe it with moderation. We meet with ſeveral inſtances of princes to whom it has proved mortal; for example, Valentinian the First, Wenceſlas, Matthias Corvinus king of Hungary, and others. There are alſo inſtances wherein it has produced the epilepſy, jaundice, cholera-morbus, diarrhœa, &c. In fact, this paſſion is of ſuch a nature, that it quickly throws the whole nervous ſyſtem into preternatural commotions, by a violent ſtricture of the nervous and muſcular parts; and ſurpriſingly augments not only the ſyſtole of the heart and of its contiguous veſſels, but alſo the tone of the fibrous parts in the whole body. It is alſo certain that this paſſion, by the ſpaſmodic ſtricture it produces in the parts, exerts its power principally on the ſtomach and inteſtines, which are highly nervous and membranous parts; whence the ſymptoms are more dangerous, in proportion to the greater conſent of the ſtomach and inteſtines with the other nervous parts, and almoſt with the whole body. The unhappy influence of anger, likewiſe, on the biliary and hepatic ducts, is very ſurpriſing; ſince, by an intenſe conſtriction of theſe, the liver is not only rendered ſcirrhous, but ſtones alſo are often generated in the gall-bladder and biliary ducts; theſe accidents have ſcarcely any other origin than an obſtruction of the free motion and efflux of the bile, by means of this violent ſtricture. From ſuch a ſtricture of theſe ducts likewiſe proceeds the jaundice, which in proceſs of time lays a foundation for calculous concretions in the gall-bladder. Laſtly, by increaſing the motion of the fluids, or the ſpaſms of the fibrous parts, by means of anger, a larger quantity of blood is propelled with an impetus to certain parts; whence it happens that they are too much diſtended, and the orifices of the veins diſtributed there are too much opened. It is evident from experience, that anger has a great tendency to excite hæmorrhages, either from the noſe, the aperture of the pulmonary artery, the veins of the anus; or, in women, from the uterus, eſpecially in thoſe previouſly accuſtomed and diſpoſed to ſuch evacuations.

Of Fear.—The influence of fear, both in occaſioning and aggravating diſeaſes, is very great. No man ought to be blamed for a decent concern about life; but too great a deſire to preſerve it, is often the cauſe of loſing it. Fear and anxiety, by depreſſing the ſpirits, not only diſpoſe us to diſeaſes, but often render thoſe diſeaſes fatal which an undaunted mind would overcome. Sudden fear has generally violent effects. Epileptic fits, and other convulſive diſorders, are often occaſioned by it. Hence the danger of that practice, ſo common among young people, of frightening one another. Many have loſt their lives, and others have been rendered miſerable, by frolics of this kind. It is dangerous to tamper with the human paſſions. The mind may eaſily be thrown into ſuch diſorder as never again to act with regularity.

But

But the gradual effects of fear prove most hurtful. The constant dread of some future evil, by dwelling upon the mind, often occasions the very evil itself. Hence it comes to pass, that so many die of those very diseases of which they long had a dread, or which had been impressed on their minds by some accident or foolish prediction. This, for example, is often the case with women in child-bed. Many of those who die in that situation are impressed with the notion of their death a long time before it happens; and there is reason to believe, that this impression is often the cause of it. The methods taken to impress the minds of women with the apprehensions of the great pain and peril of child-birth, are very hurtful. Few women die in labour, though many lose their lives after it; which may be thus accounted for. A woman after delivery, finding herself weak and exhausted, immediately apprehends she is in danger; and this fear seldom fails to ostruct the necessary evacuations upon which her recovery depends. Thus the sex often fall a sacrifice to their own imaginations, when there would be no danger, did they apprehend none. It seldom happens, when two or three women, in a great town, die in child-bed, but their death is followed by many others. Every woman of their acquaintance who is with child dreads the same fate, and the disease becomes epidemical by the mere force of imagination. This should induce pregnant women to despise fear, and by all means to avoid those tattling gossips who are continually buzzing in their ears the misfortunes of others. Every thing that may in the least alarm a patient, or a child-bed woman, ought with the greatest care to be guarded against.

In general, the effects of terror are a contraction of the small vessels and a repulsion of the blood in the large and internal ones; hence proceed a suppression of perspiration, a general oppression, trembling, and anguish of the heart and lungs, overcharged with blood. Frights often occasion incurable diseases, as epilepsy, stupor, madness, &c. In acute diseases, they have evidently killed many, by the agitation into which they have thrown the spirits, already too much disordered. We have also accounts of persons absolutely killed by terrors when in perfect health at the time of receiving the shock from them: people ordered to be executed, but with private orders for a reprieve, have expired at the block without a wound.

Out of many instances of the fatal effects of fear recorded in writers, the following is selected as one of the most singular. "George Grochantzy, a Polander, who had inlisted as a soldier in the service of the King of Prussia, deserted during the war. A small party was sent in pursuit of him; and, when he least expected it, they surprised him singing and dancing among a company of peasants, who were got together at an inn and were making merry. This event, so sudden and unforeseen, and at the same time so dreadful in its consequences, struck him in such a manner,

manner, that, giving a great cry, he became at once altogether ſtupid and inſenſible, and was ſeized without the leaſt reſiſtance. They carried him away to Glogau, where he was brought before the council of war, and received ſentence as a deſerter. He ſuffered himſelf to be led and diſpoſed of at the will of thoſe about him, without uttering a word, or giving the leaſt ſign that he knew what had happened or would happen to him. He remained immoveable as a ſtatue wherever he was placed, and was wholly paſſive with reſpect to all that was done to him or about him. During all the time that he was in cuſtody, he neither ate, nor drank, nor ſlept, nor had any evacuation. Some of his comrades were ſent to ſee him; after that he was viſited by ſome officers of his corps and by ſome prieſts; but he ſtill continued in the ſame ſtate, without diſcovering the leaſt ſigns of ſenſibility. Promiſes, intreaties, and threatenings, were equally ineffectual. The phyſicians who were conſulted upon his caſe were of opinion, that he was in a ſtate of hopeleſs idiotcy. It was at firſt ſuſpected, that thoſe appearances were feigned; but theſe ſuſpicions neceſſarily gave way when it was known that he took no ſuſtenance, and that the involuntary functions of nature were in great meaſure ſuſpended. After ſome time they knocked off his fetters, and left him at liberty to go whither he would. He received his liberty with the ſame inſenſibility that he had ſhown on other occaſions: he remained fixed and immoveable; his eyes turned wildly here and there without taking cognizance of any object, and the muſcles of his face were fallen and fixed like thoſe of a dead body. Being left to himſelf, he paſſed twenty days in this condition, without eating, or having any evacuation, and died on the twentieth day. He had been ſometimes heard to fetch deep ſighs; and once he ruſhed with great violence on a ſoldier, who had a mug of liquor in his hand, forced the mug from him, and, having drank the liquor with great eagerneſs, let the mug drop to the ground."

When a perſon is affected with terror, the principal endeavour ſhould be to reſtore the circulation to its due order, to promote perſpiration, and to allay the agitation of the patient. For theſe purpoſes he may drink a little warm liquor, as camomile-tea, &c. the feet and legs may be put into warm water, the legs rubbed, and the camomile-tea repeated every ſix or eight minutes; and when the ſkin is warm, and there is a tendency to perſpiration, ſleep may be promoted by a gentle opiate. But frights have been known not only to cauſe, but alſo to cure, diſeaſes. Mr. Boyle mentions agues, gout, and ſciatica, cured by this means.

To turn from the ſerious to the ludicrous effects of fear, the following inſtance of the latter ſort, quoted from a French author by Mr. Andrews in his volume of Anecdotes, ſhows upon what ſlight occaſions this paſſion may be ſometimes excited in a very high degree, even in perſons the moſt unlikely to entertain ſuch a

 gueſt,

guest. "Charles Gustavus (the successor of Christina of Sweden) was besieging Prague, when a boor of most extraordinary visage desired admittance to his tent; and, being allowed entrance, offered, by way of amusing the king, to devour a whole hog of one hundred weight in his presence. The old general Koningsmark, who stood by the king's side, and who, soldier as he was, had not got rid of the prejudices of his childhood, hinted to his royal master that the peasant ought to be burnt as a sorcerer. 'Sir,' said the fellow, irritated at the remark, 'if your majesty will but make that old gentleman take off his sword and his spurs, I will eat him immediately, before I begin the hog.' General Koningsmark (who had, at the head of a body of Swedes, performed wonders against the Austrians, and who was looked upon as one of the bravest men of the age) could not stand this proposal, especially as it was accompanied by a most hideous and preternatural expansion of the frightful peasant's jaws. Without uttering a word, the veteran suddenly turned round, ran out of the tent, nor thought himself safe until he had arrived at his own quarters, where he remained above twenty-four hours locked up securely, before he got rid of the panic which had so severely affected him."

Fear should not rise higher than to make us attentive and cautious; when it gains an ascendency in the mind, it becomes an insupportable tyranny, and renders life a burden. The object of fear is evil; and to be exempt from fear, or at least not enslaved to it, gives dignity to our nature, and invigorates all our faculties. Yet there are evils which we ought to fear. Those that arise from ourselves, or which it is in our power to prevent, it would be madness to despise, and audacity not to guard against. External evils, which we cannot prevent, or could not avoid without a breach of duty, it is manly and honourable to bear with fortitude. Insensibility to danger is not fortitude, any more than the incapacity of feeling pain can be called patience; and to expose ourselves unnecessarily to evil is worse than folly, and very blameable presumption.

Of IMPOTENCY occasioned by FEAR.

IT has been proved by Dr. Hunter, that impotency is frequently the result of fear. He observes, that as the "parts of generation are not necessary for the existence or support of the individual, but have a reference to something else in which the mind has a principal concern; so a complete action in those parts cannot take place without a perfect harmony of body and of mind: that is, there must be both a power of body and disposition of mind; for the mind is subject to a thousand alarms which affect the actions of these parts. Copulation is an act of the body, the spring of which is in the mind; but it is not volition: and according to the ſtate of the mind, ſo is the act performed. To perform this act well, the body ſhould be in health, and the mind perfectly confident of the powers of the body: the mind ſhould be in a ſtate entirely diſengaged from every thing elſe; it ſhould have no difficulties, no fears, no apprehenſions, not even an anxiety to perform the act well; for even this anxiety is a ſtate of mind different from what ſhould prevail; there ſhould not be even a fear that the mind itſelf may find a difficulty at the time the act ſhould be performed. Perhaps no function of the machine depends ſo much upon the ſtate of the mind as this.—The will and reaſoning faculty have nothing to do with this power; they are only employed in the act, ſo far as voluntary parts are made uſe of; and if they ever interfere, which they ſometimes do, it often produces another ſtate of mind which deſtroys that which is proper for the performance of the act; it produces a deſire, a wiſh, a hope, which are all only diffidence and uncertainty, and create in the mind the idea of a poſſibility of the want of ſucceſs, which deſtroys the proper ſtate of mind, or neceſſary confidence.—There is perhaps no act in which a man feels himſelf more intereſted, or is more anxious to perform well; his pride being engaged in ſome degree, which, if within certain bounds, would produce a degree of perfection in an act depending upon the will, or an act in voluntary parts; but, when it produces a ſtate of mind contrary to that ſtate on which the perfection of the act depends, a failure muſt be the conſequence.—The body is not only rendered incapable of performing this act by the mind being under the above influence of fear, but alſo by the mind being, though perfectly confident of its power, yet conſcious of an impropriety in performing it; this, in many caſes, produces a ſtate of mind which ſhall take away all power. The ſtate of a man's mind reſpecting his ſiſter takes away all power. A conſcientious man has been known to loſe his powers on finding the woman he was going to be connected with unexpectedly a virgin.—Shedding tears ariſes entirely from the ſtate of the mind, although not ſo much a compound action as the act in queſtion; for none are ſo weak in body that they cannot ſhed tears; it is not ſo much a compound action of the mind and ſtrength of body joined, as the other act is; yet, if we are afraid of ſhedding tears, or are deſirous of doing it, and that anxiety is kept up through the whole of an affecting ſcene, we certainly ſhall not ſhed tears, or at leaſt not ſo freely as would have happened from our natural feelings."

From this account of the neceſſity of having the mind independent reſpecting the act, we muſt ſee that it may very often happen that the ſtate of mind will be ſuch as not to allow the animal to exert his natural powers; and every failure increaſes the evil. We muſt alſo ſee from this ſtate of the caſe, that this act muſt be often interrupted; and, the true cauſe of this interruption not being known, it will be

 laid

laid to the charge of the body, or want of powers. As these cases do not arise from real inability, they are to be carefully distinguished from such as do; and perhaps the only way to distinguish them is, to examine into the state of mind respecting this act. So trifling often is the circumstance which shall produce this inability depending on fear, that the very desire to please shall have that effect, as in making the woman the sole object to be gratified.

Cases of this kind we see every day; one of which I shall relate as an illustration of this subject, and also of the method of cure.—A gentleman told me, that he had lost his virility. After above an hour's investigation of the case, I made out the following facts: that he had at unnecessary times strong erections, which showed that he had naturally this power; that the erections were accompanied with desire, which are all the natural powers wanted; but that there was still a defect somewhere, which I supposed to be from the mind. I inquired if all women were alike to him? His answer was, No; some women he could have connection with as well as ever. This brought the defect, whatever it was, into a smaller compass: and it appeared there was but one woman that produced this inability, and that it arose from a desire to perform the act with this woman well; which desire produced in the mind a doubt or fear of the want of success, which was the cause of the inability of performing the act. As this arose entirely from the state of the mind produced by a particular circumstance, the mind was to be applied to for the cure; and I told him that he might be cured, if he could perfectly rely on his own power of self-denial. When I had explained what I meant, he told me that he could depend upon every act of his will or resolution. I then told him that, if he had a perfect confidence in himself in that respect, he was to go to bed to this woman, but first promise to himself that he would not have any connection with her for six nights, let his inclinations and powers be what they would; which he engaged to do, and also to let me know the result. About a fortnight after, he told me, that this resolution had produced such a total alteration in the state of his mind, that the power soon took place; for, instead of going to bed with the fear of inability, he went with fears that he should be possessed with too much desire, too much power, so as to become uneasy to him; which really happened; for he would have been happy to have shortened the time: and, when he had once broke the spell, the mind and powers went on together, and his mind never returned to its former state.

Impotency also happens from a want of proper correspondence between the action of the testicles and penis: for we find that an irregularity in the actions of these parts sometimes happens in men, producing impotence; and something similar probably may be one cause of barrenness in women. In men the parts subservient

to

to generation may be divided into two; the essential and the accessory. The testicles are the essential; the penis, &c. the accessory. As this division arises from their uses or actions in health, which exactly correspond with one another, a want of exactness in the correspondence or susceptibility of those actions may also be divided into two: where the actions are reversed, the accessory taking place without the first or essential, as in the erections of the penis, where neither the mind nor the testicles are stimulated to action; and the second is where the testicles perform the action of secretion too readily for the penis, which has not a corresponding erection. The first is called *priapism;* and the second is what ought to be called *seminal weakness.*—The mind has considerable effect on the correspondence of the actions of these two parts: but it would appear in many instances, that erections of the penis depend more on the state of the mind than the secretion of the semen does; for many have the secretion, but not the erection; but in such, the want of erection appears to be owing to fears of the mind only.

Priapism often arises spontaneously; and often from visible irritation of the penis, as in the venereal gonorrhœa, especially when violent. The sensation of such erections is rather uneasy than pleasant; nor is the sensation of the glands at the time similar to that arising from the erections of desire, but more like to the sensation of the parts immediately after coition. Such as arise spontaneously are of more serious consequence than those from inflammation, as they proceed probably from causes not curable in themselves or by any known methods. The priapism arising from inflammation of the parts, as in a gonorrhœa, is attended with nearly the same symptoms; but generally the sensation is that of pain, proceeding from the inflammation of the parts. It may be observed, that what is said of priapism is only applicable to it when a disease in itself, and not when a symptom of other diseases, which is frequently the case.

Seminal weakness, or a secretion and emission of the semen without erection, is the reverse of a priapism, and is by much the worst disease of the two. There is great variety in the degrees of this disease, there being all the gradations from the exact correspondence of the actions of all the parts to the testicles acting alone; in every case of the disease, there is too quick a secretion and evacuation of the semen. Like to the priapism, it does not arise from desires and abilities; although when mild it is attended with both, but not in a due proportion; a very slight desire often producing the full effect. The secretion of the semen shall be so quick, that simple thought, or even toying, shall make it flow.—Dreams have produced this evacuation repeatedly in the same night; and even when the dreams have been so slight, that there has been no consciousness of them when the sleep has been broken by the act of emission. I have known cases where the testicles have been

so

ſo ready to ſecrete, that the leaſt friction on the glans has produced an emiſſion: I have known the ſimple action of walking or riding produce this effect, and that repeatedly in a very ſhort ſpace of time. A young man, about four or five and twenty years of age, not ſo much given to venery as moſt young men, had theſe laſt-mentioned complaints upon him. Three or four times in the night he would emit; and if he walked faſt, or rode on horſeback, the ſame thing would happen. He could ſcarcely have connection with a woman before he emitted, and in the emiſſion there was hardly any ſpaſm. He tried every ſuppoſed ſtrengthening medicine, as alſo the cold bath and ſea-bathing, but with no effect. By taking twenty drops of laudanum on going to bed, he prevented the night-emiſſions; and, by taking the ſame quantity in the morning, he could walk or ride without the before-mentioned inconvenience. I directed this practice to be continued for ſome time, although the diſeaſe did not return, that the parts might be accuſtomed to this healthy ſtate of action; and I have reaſon to believe the gentleman is now well. It was found neceſſary, as the conſtitution became more habituated to the opiate, to increaſe the doſe of it.—The ſpaſms, upon the evacuation of the ſemen in ſuch caſes, are extremely ſlight, and a repetition of them ſoon takes place; the firſt emiſſion not preventing a ſecond; the conſtitution being all the time but little affected. When the teſticles act alone, without the acceſſory parts taking up the neceſſary and natural conſequent action, it is ſtill a more melancholy diſeaſe: for the ſecretion ariſes from no viſible or ſenſible cauſe, and does not give any viſible or ſenſible effect, but runs off ſimilar to involuntary ſtools or urine. It has been obſerved that the ſemen is more fluid than natural in ſome of theſe caſes.

There is great variety in the diſeaſed action of theſe parts; of which the following caſe may be conſidered as an example. A gentleman has had a ſtricture in the urethra for many years, for which he has frequently uſed a bougie, but of late has neglected it. He has had no connection with women for a conſiderable time, being afraid of the conſequences. He has often in his ſleep involuntary emiſſions, which generally awake him at the paroxyſm; but what ſurpriſes him moſt is, that often he has ſuch without any ſemen paſſing forwards through the penis, which makes him think that at thoſe times it goes backwards into the bladder. This is not always the caſe, for at other times the ſemen paſſes forwards. At the time the ſemen ſeems to paſs into the bladder, he has the erection, the dream; and is awaked with the ſame mode of action, the ſame ſenſation, and the ſame pleaſure, as when it paſſes through the urethra, whether dreaming or waking. My opinion is, that the ſame irritation takes place in the bulb of the urethra without the ſemen, that takes place there when the ſemen enters, in conſequence of all the natural preparatory ſteps, whereby the very ſame actions are excited as if it came into the paſſage; from which one would ſuppoſe, that either the ſemen is not ſecreted, or, if it be, that a retrograde motion takes place in the action of the acceleratores urinæ. But, if the firſt be the caſe, then we may ſuppoſe, that in the natural ſtate the action of thoſe muſcles does not ariſe ſimply from the ſtimulus of the ſemen in the part, but from their action being a termination of a preceding one, making part of a ſeries of actions. Thus they may depend upon the friction, or the imagination of a friction, on the penis: the teſticles not doing their part, and the ſpaſm in ſuch caſes ariſing from the friction and not from the ſecretion. In many of thoſe caſes of irregularity, when the erection is not ſtrong, it ſhall go off without the emiſſion; and at other times an emiſſion ſhall happen almoſt without an erection; but theſe ariſe not from debility, but affections of the mind. In many of the preceding caſes, waſhing the penis, ſcrotum, and perinæum, with cold water, is often of ſervice; and, to render it colder than we find it in ſome ſeaſons of the year, common ſalt may be added to it, and the parts waſhed when the ſalt is almoſt diſſolved.

Of GRIEF.

GRIEF is the moſt deſtructive of all the paſſions. Its effects are permanent, and when it ſinks deep into the mind, it generally proves fatal. Anger and fear, being of a more violent nature, ſeldom laſt long; but grief often changes into a fixed melancholy, which preys upon the ſpirits, and waſtes the conſtitution. This paſſion ought not to be indulged. It may generally be conquered at the beginning; but, when it has gained ſtrength, all attempts to remove it are vain. No perſon can prevent misfortunes in life; but it ſhows true greatneſs of mind to bear them with ſerenity. Many perſons make a merit of indulging grief; and, when misfortunes happen, they obſtinately refuſe all conſolation, till the mind, overwhelmed with melancholy, ſinks under the load. Such conduct is not only deſtructive to health, but inconſiſtent with reaſon, religion, and common ſenſe.

Change of ideas is as neceſſary for health as change of poſture. When the mind dwells long upon one ſubject, eſpecially of a diſagreeable nature, it hurts the whole functions of the body. Hence grief indulged ſpoils the digeſtion and deſtroys the appetite; by which means the ſpirits are depreſſed, the nerves relaxed, the bowels inflated with wind, and the humours, for want of freſh ſupplies of chyle, vitiated. Thus many an excellent conſtitution has been ruined by a family-miſfortune, or any thing that occaſions exceſſive grief. It is utterly impoſſible, that any perſon of a dejected mind ſhould enjoy health. Life may indeed be dragged out for a few years: but whoever would live to a good old age muſt be good-

humoured

humoured and cheerful. This indeed is not altogether in our own power; yet our temper of mind, as well as our actions, depend greatly upon ourſelves. We can either aſſociate with cheerful or melancholy companions, mingle in the amuſements and offices of life, or ſit ſtill and brood over our calamities, as we chooſe.

Theſe, and many ſuch things, are certainly in our power; and from theſe the mind generally takes its caſt. The variety of ſcenes which preſent themſelves to the ſenſes, were certainly deſigned to prevent our attention from being too long fixed upon any one object. Nature abounds with variety; and the mind, unleſs fixed down by habit, delights in contemplating new objects. This at once points out the method of relieving the mind in diſtreſs. Turn the attention frequently to new objects: examine them for ſome time: when the mind begins to recoil, ſhift the ſcene: by this means a conſtant ſucceſſion of new ideas may be kept up, till the diſagreeable ones entirely diſappear. Thus travelling, the ſtudy of any art or ſcience, reading or writing on ſuch ſubjects as deeply engage the attention, will ſooner expel grief than the moſt ſprightly amuſements.

It has already been obſerved, that the body cannot be healthy unleſs it be exerciſed; neither can the mind. Indolence nouriſhes grief. When the mind has nothing elſe to think of but calamities, no wonder that it dwells there. Few people who purſue buſineſs with attention are hurt by grief. Inſtead therefore of abſtracting ourſelves from the world or buſineſs, when misfortunes happen, we ought to engage in it with more than uſual attention, to diſcharge with double diligence the functions of our ſtation, and to mix with friends of a ſocial and cheerful temper. Innocent amuſements are by no means to be neglected. Theſe, by leading the mind inſenſibly to the contemplation of agreeable objects, help to diſpel the gloom which misfortunes caſt over it. They make time ſeem leſs tedious, and have many other happy effects. Some perſons, when overwhelmed with grief, betake themſelves to drinking. This is making the cure worſe than the diſeaſe. It ſeldom fails to end in the ruin of fortune, character, and conſtitution.

OF LOVE.

LOVE is perhaps the ſtrongeſt of all the paſſions; at leaſt, when it becomes violent, it is leſs ſubject to the controul either of the underſtanding or will than any of the reſt. Fear, anger, and ſeveral other paſſions, are neceſſary for the preſervation of the individual; but love is neceſſary for the continuation of the ſpecies itſelf: it was therefore proper that this paſſion ſhould be deeply rooted in the human breaſt.—Though love be a ſtrong paſſion, it is ſeldom ſo rapid in its progreſs as ſeveral of the others. Few perſons fall deſperately in love all at once. We would therefore adviſe every one, before he tampers with this paſſion, to conſider

ſider well the probability of his being able to obtain the object of his love. When that is not likely, he ſhould avoid every occaſion of increaſing it. He ought immediately to fly the company of the beloved object; to apply his mind attentively to buſineſs or ſtudy; to take every kind of amuſement; and, above all, to endeavour, if poſſible, to find another object which may engage his affections, and which it may be in his power to obtain. There is no paſſion with which people are ſo ready to tamper as love, although none is more dangerous. Some men make love for amuſement, others from mere vanity, or on purpoſe to ſhow their conſequence with the fair. This is perhaps the greateſt piece of cruelty which any one can be guilty of. What we eagerly wiſh for, we eaſily credit. Hence the too-credulous fair are often betrayed into a ſituation which is truly deplorable, before they are able to diſcover that the pretended lover was only in jeſt. But there is no jeſting with this paſſion. When love is got to a certain height, it admits of no other cure but the poſſeſſion of its object, which, in this caſe, ought always, if poſſible, to be obtained. The conduct of parents with regard to the diſpoſal of their children in marriage is often very blameable. An advantageous match is the conſtant aim of parents; while their children often ſuffer a real martyrdom between their inclination and their duty. The firſt thing which parents ought to conſult, in diſpoſing of their children in marriage, is certainly their inclinations. Were due regard always paid to theſe, there would be fewer unhappy couples; and parents would not have ſo often cauſe to repent the ſeverity of their conduct, after a ruined conſtitution, a loſt character, or a diſtracted mind, has ſhown them their miſtake.

With regard to love in its uſual and more appropriate ſignification, it may be defined, "that affection which, being compounded of animal deſire, eſteem, and benevolence, becomes the bond of attachment and union between individuals of the different ſexes; and makes them feel in the ſociety of each other a ſpecies of happineſs which they experience no-where elſe." We call it an *affection* rather than a paſſion, becauſe it involves a deſire of the happineſs of its object: and that its conſtituent parts are thoſe which have been juſt enumerated, we ſhall firſt endeavour to prove, and then proceed to trace its riſe and progreſs from a ſelfiſh appe tite to agenerous ſentiment.

Animal deſire is the actual energy of the ſenſual appetite; and that it is an eſſential part of the complex affection, which is properly called love, is apparent from this conſideration, that, though a man may have ſentiments of eſteem and benevolence towards women who are both old and ugly, he never ſuppoſes himſelf to be in love with any woman towards whom he feels not the ſenſual appetite to have a ſtronger tendency than to other individuals of her ſex. On the other hand, that

2 animal

animal desire alone cannot be called the affection of love is evident; because he who gratifies such a desire without esteeming its object, and wishing to communicate at the same time that he receives enjoyment, loves not the woman, but himself. Mere animal desire has nothing in view but the species and the sex of its object; and, before it make a selection, it must be combined with sentiments very different from itself. The first sentiment with which it is combined, and by which a man is induced to prefer one woman to another, seems to be that by which we are delighted with gracefulness of person, regularity of features, and beauty of complexion. It is not indeed to be denied that there is something irresistible in female beauty. The most severe will not pretend that they do not feel an immediate prepossession in favour of a handsome woman; but this prepossession, even when combined with animal desire, does not constitute the whole of that affection which is called love. Savages feel the influence of the sensual appetite, and it is extremely probable that they have some ideas of beauty; but among savages the affection of love is seldom felt. Even among the lower orders in civil society it seems to be a very gross passion, and to have in it more of the selfishness of appetite than of the generosity of esteem. To these observations many exceptions will no doubt be found; but we speak of savages in general, and of the great body of the labouring poor, who in the choice of their mates do not study—who indeed are incapable of studying—that rectitude of mind and those delicacies of sentiment, without which neither man nor woman can deserve to be esteemed.

In the savage state, and even in the first stages of refinement, the bond of union between the sexes seems to consist of nothing more than mere animal desire and instinctive tenderness for their infant progeny. The former impels them to unite for the propagation of the species; and the latter preserves the union till the children, who are the fruit of it, are able to provide for their own subsistence. That in such unions, whether casual or permanent, there is no mutual esteem and benevolence, is apparent from the state of subjection in which women are held in rude and uncultivated nations, as well as from the manner in which marriages are in such nations contracted.

Sweetness of temper, a capital article with us in the female character, displays itself externally in mild looks and gentle manners, and is the first and perhaps the most powerful inducement to love in a cultivated mind. But such graces are scarcely discernible in a female savage; and even in the most polished woman would not be perceived by a male savage. Among savages, strength and boldness are the only valuable qualities. In these females are commonly deficient; for which reason they are contemned by the males as beings of an inferior order. The North-American tribes glory in idleness; the drudgery of labour degrades a man in their opinion, and is proper for woman only. To join young perſons in marriage is accordingly the buſineſs of the parents; and it would be unpardonable meanneſs in the bridegroom to ſhow any fondneſs for the bride. In Guiana a woman never eats with her husband, but after every meal attends him with water for waſhing; and in the Caribbee iſlands ſhe is not permitted to eat even in the preſence of her huſband. Dampier obſerves in general, that among all the wild nations with which he was acquainted the women carry the burdens, while the men walk before, and carry nothing but their arms; and that women even of the higheſt rank are not better treated. In Siberia, and even in Ruſſia, the capital excepted, men till very lately treated their wives in every reſpect like ſlaves. It might indeed be thought, that animal deſire, were there nothing elſe, ſhould have raiſed women to ſome degree of eſtimation among men; but male ſavages, utter ſtrangers to decency and refinement, gratify animal deſire with as little ceremony as they do hunger or thirſt.

Hence it was that in the early ages of ſociety a man purchaſed a woman to be his wife as one purchaſes an ox or a ſheep to be food; and valued her only as ſhe contributed to his ſenſual gratification. Inſtances innumerable might be collected from every nation of which we are acquainted with the early hiſtory; but we ſhall content ourſelves with mentioning a few. Abraham bought Rebekah, and gave her to his ſon Iſaac for a wife. Jacob, having nothing to give, ſerved Laban fourteen years for two wives. To David, demanding Saul's daughter in marriage, it was ſaid, "The king deſireth not any dowry, but an hundred foreſkins of the Philiſtines." In the Iliad, Agamemnon offers his daughter to Achilles for a wife; and ſays that he would not demand for her any price. By the laws of Ethelbert, king of England, a man who committed adultery with his neighbour's wife was obliged to pay the huſband a fine, and to buy him another wife. But it is needleſs to multiply inſtances; the practice has prevailed univerſally among nations emerging from the ſavage ſtate, or in the rudeſt ſtage of ſociety; and, wherever it prevailed, men could not poſſibly have for the ſex any of that tender regard and eſteem which conſtitute ſo eſſential a part of the complex affection of love.

But, if among ſavages and the vulgar love be unknown, it cannot poſſibly be an inſtinctive affection; and therefore it may be aſked, How it gets poſſeſſion of the human heart; and by what means we can judge whether in any particular inſtance it be real or imaginary? Theſe queſtions are of importance, and deſerve to be fully anſwered; though many circumſtances conſpire to render it no eaſy taſk to give to them ſuch anſwers as ſhall be perfectly ſatisfactory. Love can ſubſiſt only between individuals of the different ſexes. A man can hardly love two women at the ſame time; and we believe that a woman is ſtill leſs capable of loving at once more than one man. Love, therefore, has a natural tendency to make men and women pair; or, in other words, it is the ſource of marriage: but in poliſhed ſociety, where alone this affection has any place, ſo many things beſides mutual attachment are neceſſary to make the married life comfortable, that we rarely ſee young perſons uniting from the impulſe of love, and have therefore but few opportunities of tracing the riſe, progreſs, and conſequences, of the affection. We ſhall, however, throw together ſuch reflections as have occurred to us on the ſubject, not without indulging a hope, that they may be uſeful to the younger part of our readers when forming the moſt important connection in life.

We have ſaid, that the perfection of beauty, combined with animal deſire, is the firſt inducement which a man can have to prefer one woman to another. It may be added, that elegance of figure, a placid maſculine countenance, with a perſon which indicates ſtrength and agility, are the qualities which firſt tend to attach any woman to a particular man. Beauty is defined, "That particular form which is the moſt common of all particular forms to be met with in the ſame ſpecies of beings." Let us apply this definition to our own ſpecies, and try, by means of it, to aſcertain what conſtitutes the beauty of the human face. It is evident, that of countenances we find a number almoſt infinite of different forms, of which forms one only conſtitutes beauty, whilſt the reſt, however numerous, conſtitute what is not beauty, but deformity or uglineſs. To an attentive obſerver, however, it is evident, that of the numerous particular forms of uglineſs, there is not one which includes ſo many faces as are formed after that particular caſt which conſtitutes beauty. Every particular ſpecies of the animal as well as of the vegetable creation may be ſaid to have a fixed or determinate form, to which, as to a centre, nature is continually inclining. Or it may be compared to pendulums vibrating in different directions over one central point; and, as they all croſs the centre, though only one paſſes through any other point, ſo it will be found that perfect beauty is oftener produced by nature than deformity: we do not mean than deformity in general, but than any one kind and degree of deformity. To inſtance in a particular part of a human feature: the line which forms the ridge of the noſe is deemed beautiful when it is ſtraight; but this is likewiſe the central form, which is oftener found than any one particular degree of concave, convex, or any other irregular form that ſhall be propoſed. As we are then more accuſtomed to beauty than deformity, we may conclude that to be the reaſon why we approve and admire it, juſt as we approve and admire faſhions of dreſs for no other reaſon than that we are uſed to them. The ſame thing may be ſaid of colour as of form: it is cuſtom alone which determines our preference of the colour of the Europeans to that of the Ethiopians, and which makes them prefer their own colour to ours; ſo that, though habit and cuſtom cannot be the cauſe of beauty, they are certainly the cauſe of our liking it. That we do like it cannot be denied. Every one is conſcious of a pleaſing emotion when contemplating beauty either in man or woman; and, when that pleaſure is combined with the gratification of the ſenſual appetite, it is obvious that the ſum of enjoyment muſt be greatly increaſed. The perception of beauty, therefore, neceſſarily directs the energy of the ſenſual appetite to a particular object; but ſtill this combination is a mere ſelfiſh feeling, which regards its object only as the beſt of many ſimilar inſtruments of pleaſure. Before it can deſerve the name of love, it muſt be combined with eſteem, which is never beſtowed but upon moral character and internal worth; for, let a woman be ever ſo beautiful, and of courſe ever ſo deſirable as an inſtrument of ſenſual gratification, if ſhe be not poſſeſſed of the virtues and diſpoſitions which are peculiar to her ſex, ſhe will inſpire no man with a generous affection. With regard to the outlines, indeed, whether of internal diſpoſition or of external form, men and women are the ſame; but nature, intending them for mates, has given them diſpoſitions, which, though concordant, are however different, ſo as to produce together delicious harmony. The man, more robuſt, is fitted for ſevere labour, and for field-exerciſes; the woman, more delicate, is fitted for ſedentary occupations, and particularly for nurſing children. The man, bold and vigorous, is qualified for being a protector; the woman, delicate and timid, requires protection. Hence it is, that a man never admires a woman for poſſeſſing bodily ſtrength or perſonal courage; and women always deſpiſe men who are deſtitute of them. The man, as a protector, is directed by nature to govern; the woman, conſcious of inferiority, is diſpoſed to obey. Their intellectual powers correſpond to the deſtination of nature. Men have penetration and ſolid judgment to fit them for governing; women have underſtanding to make an engaging figure under good government: a greater proportion would excite dangerous rivalſhip between the ſexes, which nature has avoided by giving them different talents. Women have more imagination and ſenſibility than men, which make all their enjoyments more exquiſite; at the ſame time that they are better qualified to communicate enjoyment. Add another capital difference of diſpoſition: the gentle and inſinuating manners of the female ſex tend to ſoften the roughneſs of the other ſex; and, wherever women are indulged with any freedom, they poliſh ſooner than men.

Theſe are not the only particulars that diſtinguiſh the ſexes. With reſpect to the ultimate end of love, it is the privilege of the male, as ſuperior and protector, to make a choice: the female, preferred, has no privilege but barely to conſent or to refuſe. Whether this diſtinction be the immediate reſult of the originally-different diſpoſitions of the ſexes, or only the effect of aſſociations inevitably formed, may be questioned; but among all nations it is the practice for men to court, and for women to be courted: and, were the most beautiful woman on earth to invert this practice, she would forfeit the esteem, however by her external grace she might excite the desire, of the man whom she addressed. The great moral virtues, which may be comprehended under the general term *integrity* are all absolutely necessary to make either men or women estimable; but, to procure esteem to the female character, the *modesty* peculiar to their sex is a very essential circumstance. Nature hath provided them with it as a defence against the artful solicitations of the other sex before marriage, and also as a support of conjugal fidelity.

A woman, therefore, whose dispositions are gentle, delicate, and rather timid than bold, who is possessed of a large share of sensibility and modesty, and whose manners are soft and insinuating, must, upon moral principles, command the esteem and benevolence of every individual of the other sex who is possessed of sound understanding; but, if her person be deformed, or not such as to excite some degree of animal desire, she will attract no man's love. In like manner, a man whose moral character is good, whose understanding is acute, and whose conversation is instructive, must command the esteem of every sensible and virtuous woman; but, if his figure be disagreeable, his manners unpolished, his habits slovenly, and, above all, if he be deficient in PERSONAL COURAGE, he will hardly excite desire in the female breast. It is only when the qualities which command esteem are, in the same person, united with those which excite desire, that the individual so accomplished can be an object of love to one of the other sex; but, when these qualities are thus united, each of them increases the other in the imagination of the lover. The beauty of his mistress gives her, in his apprehension, a greater share of gentleness, modesty, and every thing which adorns the female character, than perhaps she really possesses; whilst this persuasion of her internal worth makes him, on the other hand, apprehend her beauty to be absolutely unrivalled.

The affection thus generated is more or less pure, and will be more or less permanent, according as the one or the other part of which it is compounded predominates. Where desire of possession prevails over our esteem of the person and merits of the desirable object, love loses its benevolent character; the appetite for gratification becomes ungovernable, and tends violently to its end, regardless of the misery that must follow. In that state, love is no longer a sweet agreeable affection; it becomes a selfish, painful, passion, which, like hunger and thirst, produces no happiness but in the instant of fruition; and, when fruition is over, disgust and aversion generally succeed to desire. On the other hand, where esteem, founded on a virtuous character and gentle manners, prevails over animal desire, the

the lover would not for the world gratify his appetite at the expenſe of a lady's honour or peace of mind. He wiſhes, indeed, for enjoyment; and to him enjoyment is more exquiſite than to the mere ſenſual lover, becauſe it unites ſentiment with the gratification of ſenſe; at the ſame time that, ſo far from being ſucceeded by diſguſt or averſion, it increaſes his benevolence to the woman, whoſe character and manners he eſteems, and who has contributed ſo much to his pleaſure. Benevolence to an individual, having a general end, admits of acts without number, and is ſeldom fully accompliſhed. Hence mutual love, which is compoſed chiefly of eſteem and benevolence, can hardly be of a ſhorter duration than its objects. Frequent enjoyment endears ſuch lovers to each other, and makes conſtancy a pleaſure; and, when the days of ſenſual enjoyment are over, eſteem and benevolence will remain in the mind, making ſweet, even in old age, the ſociety of that pair, in whom are collected the affections of huſband, wife, lover, friend—the tendereſt affections of human nature.

From the whole of this inveſtigation, we think it appears, that the affection between the ſexes which deſerves the name of love, is inſeparably connected with virtue and delicacy: that a man of gallantry cannot be a faithful or a generous lover; that in the breaſt of him, who has ranged from woman to woman for the mere gratification of his ſenſual appetite, deſire muſt have effaced all eſteem for the female character; and that, therefore, the maxim too generally received, "that a reformed rake makes the beſt huſband," has very ſeldom a chance to be true. We think it may likewiſe be inferred, that thouſands fancy themſelves in love who know not what love is, or how it is generated in the human breaſt; and therefore we beg leave to adviſe ſuch of our readers as may imagine themſelves to be in that ſtate, to examine their own minds, with a view to diſcover, whether, if the object of their love were old and ugly, they would ſtill eſteem them for the virtues of their character, and the propriety of their manners. This is a queſtion which deſerves to be well weighed by the young and amorous, who, in forming the matrimonial connection, are too often blindly impelled by mere animal deſire, inflamed by beauty. It may indeed happen, after the pleaſure of gratifying that deſire is gone, (and if not refined by eſteem and benevolence, go it muſt with a ſwift pace,) that a new bond of attachment may be formed upon more dignified and more laſting principles; but this is a dangerous experiment. Even ſuppoſing good ſenſe, good temper, and internal worth of every ſort, yet a new attachment, even upon ſuch qualifications, is rarely formed; becauſe it commonly, or rather always, happens, that ſuch qualifications, the only ſolid foundation of an indiſſoluble connection, if they did not originally make eſteem predominate over animal deſire, are afterwards rendered altogether inviſible by ſatiety of enjoyment creating diſguſt, which is generally the case with violent love founded on the desire of enjoyment only. As the delicate nature of female honour and decorum, and the inexpressible grace of a chaste and modest behaviour, are the surest and indeed the only means of kindling at first, and ever after of keeping alive, this tender and elegant flame, and of accomplishing the excellent ends designed by it; to attempt by fraud to violate the one, or under pretence of passion to sully and corrupt the other, and, by so doing, to expose the too-often credulous and unguarded object, with a wanton cruelty, to the hatred of her own sex and the scorn of ours, and to the lowest infamy of both,—is a conduct not only base and criminal, but inconsistent with that truly rational and refined enjoyment, the spirit and quintessence of which is derived from the bashful and sacred charms of virtue kept untainted, and therefore ever alluring to the lover's heart.

The symptoms produced by love as a disease are as follow: The eye-lids often twinkle; the eyes are hollow, and yet appear as if full with pleasure; the pulse is not peculiar to the passion, but the same with that which attends solicitude and care. When the object of this affection is thought of, particularly if the idea is sudden, the spirits are confused, the pulse changes, and its force and time are very variable: in some instances, the person is sad and watchful; in others, not being conscious of his state, he pines away, is slothful, and regardless of food. As the passion prevails, sighs grow deeper; a tremor affects the heart and pulse; the countenance is alternately pale and red; the voice is suppressed; the eyes grow dim: cold sweats break out; sleep absents itself; the secretions become disturbed, and a loss of appetite, a hectic fever, melancholy, or perhaps madness, or death, constitutes the sad catastrophe. On this subject the curious may consult Ægineta, lib. iii. cap. 7. Oribat. Synop. lib. viii. cap. 9. or a treatise professedly written on Love, as it is a distemper, by James Ferrard, Oxford, 1640.

The ancients were much addicted to amulets and potions to excite love in the object of their desire, the operation of which was violent and dangerous, and frequently deprived such as drank them of their reason. Some of the most remarkable ingredients of which they were composed were these: the hippomanes, the jynx, insects bred from putrefaction, the fish remora, the lizard, brains of a calf, the hairs on the tip of a wolf's tail, his secret parts, the bones of the left side of a toad eaten with ants, the blood of doves, bones of snakes, feathers of screech-owls, twisted cords of wool in which a person had hanged himself, rags, torches, reliques, a nest of swallows buried and famished in the earth, bones snatched from hungry bitches, the marrow of a boy famished in the midst of plenty, dried human liver; to these may be added several herbs growing out of putrid substances. Such were the ingredients that entered into the composition of that infernal draught a love-potion.

potion. The antidotes againſt love were generally *agnus caſtus*, which has the power of weakening the generative faculty; ſprinkling the duſt in which a mule has rolled herſelf; tying toads in the hide of a beaſt newly ſlain; applying amulets of minerals or herbs, which were ſuppoſed of great efficacy.

Of MELANCHOLY.

THE pathology of melancholy and mania is very obſcure; as coming on without any fever, or diſturbance in the blood's motion. Often alſo they are hereditary, depending on the original ſtructure of the body, eſpecially of the brain; the fault of which, however, cannot be detected by the niceſt anatomiſt. But it is well known, that various diſeaſes of the brain, obſtructions, tumors, either of the brain itſelf or of the cranium preſſing upon it, any injury done to the head, and, as ſome phyſicians relate, the hardneſs and dryneſs of the brain, and ſome peculiar irritations affecting the nervous ſyſtem, are capable of bringing on this malady. And indeed ſo great are the irritations affecting the nervous ſyſtem in mad people, that they often ſleep little or none for a long time. Yet even this ſo defective and imperfect knowledge of the diſeaſes of the brain and nerves, is by no means free from difficulties. For, though we know that the brain, or a certain part of it, is hurt, or that it is irritated, by a ſwelling, or a pointed bone growing into it, nobody can foretel how great, or what, may be the nature of the malady from ſuch a hurt: for examples are not wanting of people who, after loſing a large part of the brain, have recovered and lived a long time; or of thoſe who have perceived no inconvenience from a large portion of that viſcus being corrupted, until at length they have fallen ſuddenly down and died in convulſions.

Many perſons of a religious turn of mind behave as if they thought it a crime to be cheerful. They imagine the whole of religion conſiſts in certain mortifications, or denying themſelves the ſmalleſt indulgence, even of the moſt innocent amuſements. A perpetual gloom hangs upon their countenances, while the deepeſt melancholy preys upon their minds. At length the faireſt proſpects vaniſh, every thing puts on a diſmal appearance, and thoſe very objects which ought to give delight afford nothing but diſguſt. Life itſelf becomes a burden, and the unhappy wretch, perſuaded that no evil can equal what he feels, often puts an end to his own miſerable exiſtence. It is great pity that ever religion ſhould be ſo far perverted as to become the cauſe of thoſe very evils which it was deſigned to prevent. Nothing can be better calculated than true religion to raiſe and ſupport the mind of its votaries under every affliction that can befal them. It teaches them that even the ſufferings of this life are preparatory to the happineſs

of

of the next; and that all who persist in a course of virtue shall at length arrive at complete felicity.

Persons whose business it is to recommend religion to others, should beware of dwelling too much on gloomy subjects. That peace and tranquillity of mind, which true religion is calculated to inspire, is a more powerful argument in its favour, than all the terrors that can be uttered. Terror may indeed deter men from outward acts of wickedness; but can never inspire them with that love of God, and real goodness of heart, in which alone true religion consists. In short, the best way to counteract the violence of any passion, is to keep the mind closely engaged in some useful pursuit.

Of the PROGNOSTICS of DISEASES; with RULES for PRESERVING HEALTH.

PROGNOSTIC is a judgment of the event either of a state of health, or of a disease; as, Whether it shall end in life or death, be long or short, mild or malignant, &c. taken from certain symptoms thereof. When, by the following remarks, the person shall judge what disorder is coming upon him, or already present, a safe and effectual remedy will in general be found in the Medical Part of the Herbal; but, whenever that work is not sufficiently full to the purpose, I shall add such occasional observations for prevention and cure as have occurred to me in my late practice.

Hippocrates was the first who treated of medicine in a regular and rational manner, and he is therefore justly considered as the father of physic. Hippocrates remarked four stages in distempers; viz. The beginning of the disease, its augmentation, its state or height, and its declination. In such diseases as terminate fatally, death comes in place of declination. In the third stage, therefore, the change is most considerable, as it determines the fate of the sick person; and this is most commonly done by means of a *crisis*. By this word he understood any sudden change in sickness, whether for the better or for the worse, whether health or death succeed immediately. Such a change, he says, is made at that time by nature either absolving or condemning the patient. Hence we may conclude, that Hippocrates imagined diseases to be only a disturbance of the animal economy, with which Nature was perpetually at variance, and using her utmost endeavours to expel the offending cause. Her manner of acting on these occasions is to reduce to their natural state those humours whose discord occasions the disturbance of the whole body, whether in relation to their quantity, quality, mixture, motion, or any other way in which they become offensive. The principal means employed by nature for this end is what Hippocrates calls *concoction.* By this he underſtood the bringing the morbific matter lodged in the humours to ſuch a ſtate, as to be eaſily fitted for expulſion by whatever means nature might think moſt proper. When matters are brought to this paſs, whatever is ſuperfluous or hurtful immediately empties itſelf, or nature points out to phyſicians the way by which ſuch an evacuation is to be accompliſhed. The criſis takes place either by bleeding, ſtool, vomit, ſweat, urine, tumors or abſceſſes, ſcabs, pimples, ſpots, &c. But theſe evacuations are not to be looked upon as the effects of a true criſis, unleſs in conſiderable quantity; ſmall diſcharges not being ſufficient to make a criſis; which, on the contrary, are a ſign that nature is depreſſed by the load of humours, and that ſhe lets them go through weakneſs and continual irritation. What comes forth in this manner is crude, becauſe the diſtemper is yet too ſtrong; and, while matters remain in this ſtate, only a bad or imperfect criſis is to be expected. This ſhows that the diſtemper triumphs, or at leaſt is equal in ſtrength to nature, which prognoſticates death, or a prolongation of the diſeaſe. In this laſt caſe, however, nature often has an opportunity of attempting a new criſis more happy than the former, after having made freſh efforts to advance the concoction of the humours.—It muſt here be obſerved however, that, according to Hippocrates, concoction cannot be made but in a certain time, as every fruit has a limited time to ripen; for he compares the humours which nature has digeſted to fruits come to maturity. The time required for concoction depends on the differences among diſtempers mentioned above. In thoſe which Hippocrates calls very acute, the digeſtion or criſis happens by the fourth day; in thoſe which are only acute, it happens on the ſeventh, eleventh, or fourteenth, day; which laſt is the longeſt period generally allowed by Hippocrates in diſtempers that are truly acute, though in ſome places he ſtretches it to the twentieth or twenty-firſt, nay, ſometimes to the fortieth or ſixtieth, days. All diſeaſes that exceed this laſt term are called *chronical.* And while, in thoſe diſeaſes that exceed fourteen days, he conſiders every fourth day as critical, or at leaſt remarkable, by which we may judge whether the criſis on the fourth day will be favourable or not; ſo in thoſe which run from twenty to forty he reckons only the ſevenths, and in thoſe that exceed forty he begins to reckon by twenty. Beyond the hundred and twentieth he thinks that the number of days has no power over the criſis. They are then referred to the general changes of the ſeaſon; ſome terminating about the equinoxes; others about the ſolſtices; others about the riſing or ſetting of the ſtars of certain conſtellations; or, if numbers have yet any place, he reckons by months, or even whole years. Thus (he ſays) certain diſeaſes in children have their criſes in the ſeventh month after their birth, and others in their ſeventh or even their fourteenth year.

Though

Though Hippocrates mentions the twenty-firſt as one of the critical days in acute diſtempers, as already noticed ; yet, in other places of his works, he mentions alſo the twentieth. The reaſon he gives for this in one of thoſe places of his works is, that the days of ſickneſs were not quite entire. In general, however, he is much attached to the odd days : inſomuch that in one of his aphoriſms he tells us, "The ſweats that come out upon the third, fifth, ſeventh, ninth, eleventh, fourteenth, ſeventeenth, twenty-firſt, twenty-ſeventh, thirty-firſt, or thirty-fourth, days, are beneficial ; but thoſe that come out upon other days ſignify that the ſick ſhall be brought low, that his diſeaſe ſhall be very tedious, and that he ſhall be ſubject to relapſes." He farther ſays, that "the fever which leaves the ſick upon any but an odd day is uſually apt to relapſe." Sometimes, however, he confeſſes that it is otherwiſe ; and he gives an inſtance of a ſalutary criſis happening on the ſixth day. But theſe are very rare inſtances, and therefore cannot, in his opinion, overthrow the general rule. Beſides the criſis, however, or the change which determines the fate of the patient, Hippocrates often ſpeaks of another, which only changes the ſpecies of the diſtemper, without reſtoring the patient to health ; as when a vertigo is turned to an epilepſy, a tertian fever to a quartan, or to a continual, &c.

But what chiefly contributed to procure the vaſt reſpect generally paid to Hippocrates, was his accuracy in prognoſtics : thus he not only diſtinguiſhed one diſeaſe from another by the ſigns which properly belonged to each ; but by comparing the ſame ſort of diſtemper which happened to ſeveral perſons, and the accidents which uſually appeared before and after, he could often foretel a diſeaſe before it came, and afterwards give a right judgment of the event of it. By this way of prognoſticating, he came to be exceedingly admired ; and this he carried to ſuch a height, that it may juſtly be ſaid to be his maſter-piece ; and Celſus, who lived after him, remarks that ſucceeding phyſicians, though they found out ſeveral new things relating to the management of diſeaſes, yet were obliged to the writings of Hippocrates for all that they knew of ſigns ; and let us add, that we are obliged to Celſus for a true underſtanding of the works of Hippocrates. The firſt thing Hippocrates conſidered, when called to a patient, was his looks.—It was a good ſign with him to have a viſage reſembling that of a perſon in health, and the ſame with what the ſick man had before he was attacked by the diſeaſe. As it varied from this, ſo much the greater danger was apprehended. The following is the deſcription which he gives of the looks of a dying man :—"When a patient (ſays he) has his noſe ſharp, his eyes ſunk, his temples hollow, his ears cold and contracted, the ſkin of his forehead tenſe and dry, and the colour of his face tending to a pale-green or lead-colour, one may give out for certain that death is very near at hand ; unleſs the ſtrength of the patient has been exhauſted all at once by long watchings, or by a looſeneſs, loofenefs, or being a long time without eating." This obfervation has been confirmed by thofe of fucceeding phyficians, who have, from him, denominated it the *Hippocratic face.* The lips hanging relaxed and cold, are likewife looked upon by this author as a confirmation of the foregoing prognoftic. He took alfo his figns from the difpofition of the eyes in particular. When a patient cannot bear the light; when he fheds tears involuntarily; when, in fleeping, fome part of the white of the eye is feen, unlefs he ufually fleeps after that manner, or has a loofenefs upon him: thefe figns, as well as the foregoing ones, prognofticate danger. The eyes deadened, with a mift before them, or their brightnefs loft, prefages death, or great weaknefs. Eyes fparkling, fierce, and fixed, denote the patient to be delirious, or that he foon will be feized with a frenzy. When the patient fees any thing red, and like fparks of fire and lightning pafs before his eyes, you may expect an hæmorrage; and this often happens before thofe crifes which are to be attended by a lofs of blood. The condition of the patient is alfo fhown by his pofture in bed. If you find him lying on one fide, his body, neck, legs, and arms, a little contracted, which is the pofture of a man in health, it is a good fign: on the contrary, if he lies on his back, his arms ftretched out, and his legs hanging down, it is a fign of great weaknefs; and particularly when the patient flides or lets himfelf fall down towards the feet, it denotes the approach of death. When a patient in a burning fever is continually feeling about with his hands and fingers, and moves them up before his face and eyes as if to take away fomething that paffed before them; or on his bed-covering, as if picking or fearching for little ftraws, or taking away fome filth, or drawing out little flocks of wool; all this is a fign that he is delirious, and that he will die. Amongft the other figns of a prefent or approaching delirium, he alfo adds this: When a patient who naturally fpeaks little begins to talk more than he ufed to do, or when one that talks much becomes filent, this change is to be reckoned a fort of delirium, or is a fign that the patient will foon fall into one. The frequent trembling or ftarting of the tendons of the wrifts prefages likewife a delirium. As to the different forts of delirium, Hippocrates is much more afraid of thofe that run upon mournful fubjects than fuch as are accompanied with mirth.

When a patient breathes faft, and is oppreffed, it is a fign that he is in pain, and that the parts above the diaphragm are inflamed. Breathing long, or when the patient is a great while in taking his breath, fhows him to be delirious; but eafy and natural refpiration is always a good fign in acute difeafes. Hippocrates depended much on the refpiration in making his prognoftics; and therefore has taken care in feveral places to defcribe the different manner of a patient's breathing. Continual watching in acute difeafes, are figns of prefent pain, or a delirium near at hand. Hippocrates alfo drew figns from all excrements, whatever they are, that are feparated from the body of man. His moſt remarkable prognoſtics, however, were from the urine. The patient's urine, in his opinion, is beſt when the ſediment is white, ſoft to the touch, and of equal conſiſtence. If it continue ſo during the courſe of the diſtemper, and till the time of the criſis, the patient is in no danger, and will ſoon be well. This is what Hippocrates called *concocted urine,* or what denotes the concoction of the humours; and he obſerved, that this concoction of the urine ſeldom appeared thoroughly but on the days of the criſis which happily put an end to the diſtemper. "We ought (ſaid Hippocrates) to compare the urine with the purulent matter which runs from ulcers. As the pus, which is white, and of the ſame quality with the ſediment of the urine we are now ſpeaking of, is a ſign that the ulcer is on the point of cloſing, ſo that which is clear, and of another colour than white, and of an ill ſmell, is a ſign that the ulcer is virulent, and therefore very difficult to be cured: the urines that are like this we have deſcribed are only thoſe which may be named good; all the reſt are ill, and differ from one another only in the degrees of more and leſs. The firſt never appear but when nature has overcome the diſeaſe; and are a ſign of the concoction of humours, without which you cannot hope for a certain cure. On the contrary, the laſt are made as long as the crudity remains, and the humours continue unconcocted. Among the urines of this laſt ſort, the beſt are reddiſh, with a ſediment that is ſoft, and of an equal conſiſtence; which denotes, that the diſeaſe will be ſomewhat tedious, but without danger. The worſt are thoſe which are very red, and at the ſame time clear and without ſediment; or that are muddy and troubled in the making. In urine there is often a ſort of cloud hanging in the veſſel in which it is received; the higher this riſes, or the farther diſtant it is from the bottom, or the more different from the colour of the laudable ſediment above-mentioned, the more there is of crudity. That which is yellow, or of a ſandy colour, denotes abundance of bile; that which is black is the worſt, eſpecially if it has an ill ſmell, and is either altogether muddy or altogether clear. That whoſe ſediment is like large ground wheat, or little flakes or ſcales ſpread one upon another, or bran, preſages ill, eſpecially the laſt. The fat or oil that ſometimes ſwims upon the top of the urine, and appears in a form ſomething like a ſpider's web, is a ſign of a conſumption of the fleſh and ſolid parts. The making of a great quantity of urine is the ſign of a criſis, and ſometimes the quality of it ſhows how the bladder is affected. We muſt alſo obſerve, that Hippocrates compared the ſtate of the tongue with the urine; that is to ſay, when the tongue was yellow, and charged with bile, the urine he knew muſt of courſe be of the ſame colour; and, when the tongue was red and moiſt, the urine was of its natural colour. His prognoſtics from the excretions by ſtool are as follow: Thoſe that are ſoft, yellowiſh, of ſome conſiſtence, and not of an extraordinary ill ſmell, that anſwer to the quantity of what is taken inwardly, and that are voided at the uſual hours, are the beſt of all. They ought alſo to be of a thicker conſiſtence when the diſtemper is near the criſis ; and it ought to be taken for a good prognoſtic, when ſome worms, round and long, are evacuated at the ſame time with them. The prognoſis, however, may ſtill be favourable, though the matter excreted be thin and liquid, provided the evacuation be not in a very large quantity and too often, ſo as to make the patient faint. All matter that is watery, white, of a pale green or red, or frothy and viſcous, is bad. That which is blackiſh, or of a livid hue, is the moſt pernicious. That which is pure black, and nothing elſe but a diſcharge of black bile, always prognoſticates very ill; this humour, from what part ſoever it comes, ſhowing the ill diſpoſition of the inteſtines. The matter that is of ſeveral different colours denotes the length of the diſtemper, and, at the ſame time, that it may be of dangerous conſequence. Hippocrates places in the ſame claſs the matter that is bilious or yellow, mixed with blood, or green and black, or like the dregs or ſcrapings of the guts. The ſtools that conſiſt of pure bile, or entirely phlegm, he alſo looks upon to be very bad. Matter caſt up by vomiting ought to be mixed with bile and phlegm; where one of theſe humours only is obſerved, it is worſe. That which is black, livid, green, or the colour of a leek, indicates alarming conſequences. The ſame is to be ſaid of that which ſmells very ill; and, if at the ſame time it be livid, death is not far off. The vomiting of blood is very often mortal. The ſpittings which give eaſe in diſeaſes of the lungs and in pleuriſies, are thoſe that come up readily and without difficulty: and it is good if they be mixed at the beginning with much yellow: but if they appear of the ſame colour, or are red, a great while after the beginning of the diſtemper, are ſalt and acrimonious, and cauſe violent coughings, they are not good. Spittings purely yellow are bad ; and thoſe that are white, viſcous, and frothy, give no eaſe. Whiteneſs is a good ſign of concoction in regard to ſpittings; but they ought not all to be viſcous, nor too thick, nor too clear. We may make the ſame judgment of the excrements of the noſe according to their concoction and crudity. Spittings that are black, green, and red, are of bad conſequence. In inflammations of the lungs, thoſe that are mixed with bile and blood preſage well if they appear at the beginning, but are bad if they ariſe not about the ſeventh day. But the worſt ſign in theſe diſtempers is, when there is no expectoration at all, and the too great quantity of matter that is ready to be diſcharged this way makes a rattling in the breaſt. After ſpitting of blood, the diſcharge of purulent matter often follows, which brings on a conſumption, and at laſt death. A kind good ſweat is that which ariſes on the day of the criſis, and is diſcharged in abundance all over the body, and at the ſame time from all parts of the body, and thus carries off the fever.

ver. A cold ſweat is alarming, eſpecially in acute fevers, for in others it is only a ſign of long continuance. When the patient ſweats no-where but on the head and neck, it is a ſign that the diſeaſe will be long and dangerous. A gentle ſweat in ſome particular part, of the head and breaſt for inſtance, gives no relief, but denotes the ſeat of the diſtemper, or the weakneſs of the part. This kind of ſweat was called by Hippocrates *ephidroſis*. The hypochondria, or the abdomen in general, ought always to be ſoft and even, as well on the right ſide as on the left. When there is any hardneſs or unevenneſs in thoſe parts, or heat and ſwellings, or when the patient cannot bear to have it touched, it is a ſign the inteſtines are indiſpoſed.

Hippocrates alſo inquired into the ſtate of the pulſe, or the beating of the arteries. The moſt ancient phyſicians, however, and even Hippocrates himſelf, for a long time, by this word underſtood the violent pulſation that is felt in an inflamed part, without putting the fingers to it. It is obſerved by Galen, and other phyſicians, that Hippocrates touches on the ſubject of the pulſe more ſlightly than any other on which he treats. But that our celebrated phyſician underſtood ſomething even on this ſubject, is eaſily gathered from ſeveral paſſages in his writings; as when he obſerves, that in acute fevers the pulſe is very quick and very great; and when he makes mention, in the ſame place, of trembling pulſes, and thoſe that beat ſlowly; when he obſerves, that in ſome diſeaſes incident to women, when the pulſe ſtrikes the finger faintly, and in a languiſhing manner, it is a ſign of approaching death. He remarks alſo, in the Coacæ Prænotiones, that he whoſe vein, that is to ſay whoſe artery of the elbow, beats, is juſt going to run mad, or elſe that the perſon is at that time very much under the influence of anger. Many other variations of the pulſe are enumerated by phyſicians, but moſt of them uncertain, and not confirmed by experience. See the Article Pulse, in the Medical Part of the Herbal, p. 106. where the ſubject is more fully treated.

We ſhall now proceed to ſome farther remarks on the PROGNOSTICS OF PARTICULAR DISEASES.

The *tertian ague* hath one prognoſtic peculiar to itſelf, namely, dry ſcabby ulcers breaking out upon the lips; theſe ſometimes appear about the third or fourth paroxyſm; and then we may venture to foretel that the diſeaſe will go off ſpontaneouſly after the ſeventh.

The following are the prognoſtics of a *nervous fever;* and therefore, when they appear, people ſhould take precautions accordingly, by conſulting the Medical Part of the Herbal, p. 128, for a ſafe and certain preventative and cure. The patient at firſt grows ſomewhat liſtleſs, and feels ſlight chills, and ſhudders, with uncertain fluſhes of heat, and a kind of wearineſs all over, like what is felt after great fatigue. This is always attended with a ſort of heavineſs and dejection of ſpirits, and more or

leſs

lefs of a load, pain, or giddinefs of the head; a naufea, or difrelifh of every thing, foon follows, without any confiderable thirft, but frequently with urging to vomit, though little but infipid phlegm is brought up. Though a kind of lucid interval of feveral hours fometimes intervenes, yet the fymptoms return with aggravation, efpecially towards night; the head grows more giddy or heavy; the heat greater; the pulfe quicker, but weak; with an oppreffive kind of breathing. A great torpor, or obtufe pain and coldnefs, affects the hinder-part of the head frequently, and oftentimes a heavy pain is felt on the top all along the coronary future; this, and that of the back part of the head, generally attend nervous fevers, and are commonly fucceeded by fome degree of delirium. In this condition the patient often continues for five or fix days, with a heavy, pale, funk, countenance, feemingly not very fick, and yet far from being well; reftlefs, anxious, and commonly quite void of fleep, though fometimes very drowfy and heavy: but, although he appears to thofe about him actually to fleep, he is utterly infenfible of it, and denies that he doth fo. The pulfe during all this time is quick, weak, and unequal; fometimes fluttering, and fometimes for a few moments flow; nay, even intermitting, and then, with a fudden flufh in the face, immediately very quick, and perhaps foon after furprifingly calm and equal; and thus alternately.

Prognoftics of a *fcarlet fever*. With various general fymptoms of fever, the patient at firft complains of a dejection of fpirits, a flight forenefs or rather ftiffnefs in the neck, with a fenfe of ftraitnefs in the mufcles of the neck and fhoulders, as if they were bound with cords. The fecond day of the fever this forenefs in the throat increafes, and the patient finds a difficulty in fwallowing; but the difficulty feems lefs occafioned by the pain excited in the attempt, or by the ftraitnefs of the paffage, than by an inability to throw the neceffary mufcles into action. The fkin feels hot and dry, but not hard; and the patients experience frequent fmall pungent pains, as if touched with the point of a needle. The breath is hot and burning to the lips, and thirft makes them wifh to drink; but the tendency to ficknefs, and the exertions neceffary in deglution, are fo unpleafant, that they feldom care to drink much at a time. They have much uneafinefs alfo from want of reft during the night. In the morning of the third day, the face, neck, and breaft, appear redder than ufual: in a few hours this rednefs becomes univerfal; and increafes to fuch a degree of intenfity, that the face, body, and limbs, refemble a boiled lobfter in colour, and are evidently fwollen.

The figns of an impending *phrenitis,* or *inflammation of the brain,* are fully explained in the Medical Part of this work, p. 139. In this difeafe, the following are the moft fatal fymptoms: A continual and furious delirium, with watching; thin watery urine, white fæces, the urine and ſtools running off involuntarily, or a total ſuppreſſion of theſe excretions; a diſpoſition to become ſtupid, or to faint, trembling, rigour, chattering of the teeth, convulſions, hiccough, coldneſs of the extremities, trembling of the tongue, ſhrill voice, a ſudden ceſſation of pain, with apparent tranquillity. The following are favourable: Sweats, apparently critical, breaking out; a ſeeming effort of nature to terminate the diſeaſe by a diarrhœa: a large hemorrhage from the noſe; ſwellings of the glands behind the ears; hæmorrhoids.

A *vertigo* is obſerved to be both the ſymptom and forerunner of ſome dangerous diſeaſe; ſuch as apoplexy, epilepſy, or hyſteria; hæmorrhages from the noſe and other parts; ſuppreſſion of the menſes; plethora; fevers, as well as ſuch as are accompanied with debility as thoſe in which there is an increaſed impetus of the blood towards the head.—Though a vertigo be for the moſt part a ſymptom and concomitant of other diſeaſes, yet it is ſometimes a primary diſeaſe, returning at intervals, increaſing gradually, and equally impeding and deſtroying the functions of the body and mind.

A *delirium* accompanies fevers of many different kinds. Sometimes it is ſlight, eaſily removed, and ſcarcely to be accounted a bad ſign. Often, however, it is very violent, and one of the very worſt of ſigns, requiring the utmoſt care and attention. A delirium is either fierce or mild. The fierce delirium is preceded and accompanied by a redneſs of the countenance, a pain of the head, a great beating of the arteries, and noiſe in the ears; the eyes in the mean time looking red, inflamed, fierce, ſhining, and unable to bear the light; there is either no ſleep at all, or ſleep troubled with horrid dreams; the wonted manners are changed, an unuſual peeviſhneſs and ill-nature prevail. The depravation of judgment is firſt obſerved between ſleep and waking, and by the perſon's crediting his imagination, while the perceptions of ſenſe are neglected, and the ideas of memory occur in an irregular manner. Fury at laſt takes place, and ſometimes an unuſual and incredible degree of bodily ſtrength, ſo that ſeveral people can ſcarcely keep a ſingle patient in his bed. The mild delirium, on the contrary, is often accompanied with a weak pulſe, a pale collapſed countenance, and a vertigo when the patient ſits in an erect poſture; he is ſeldom angry, but often ſtupid, and ſometimes remarkably grieved and fearful. The loſs of judgment, as in the former kinds, is firſt perceived when the patient is half awake; but a temporary recovery enſues upon the admiſſion of the light and the converſation of friends. The patient mutters much to himſelf, and attends little to the things around him; at laſt, becoming quite ſtupid, he neither feels the ſenſation of hunger or thirſt, nor any of the other propenſities of nature, by which means the urine and excrements are voided involuntarily. As the diſorder increaſes, it terminates in ſubſultus tendinum, tremors, convulſions, fainting, and death. The other ſpecies of delirium alſo frequently terminates in this, when the ſpirits and ſtrength of the patient begin to fail. The ſymptoms accompanying either of theſe kinds of delirium ſhow an unuſual, inordinate, and unequal, motion of the blood through the brain, and a great change in that ſtate of it which is neceſſary to the exerciſe of the mental powers. It is ſufficiently probable, that an inflammation of the brain, more or leſs violent and general, ſometimes takes place, although the ſigns of univerſal inflammation are frequently ſlight. This we learn from the diſſection of dead bodies, which often ſhow an univerſal redneſs of the brain or of ſome of its parts, or ſometimes an effuſion or ſuppuration.

The prognoſtics of the *malignant, putrid, or ulcerous, ſore throat*, are very different in different perſons. Sometimes a rigour, with fulneſs, and ſoreneſs of the throat and painful ſtiffneſs of the neck, are the firſt ſymptoms complained of. Sometimes alternate chills and heats, with ſome degree of giddineſs, drowſineſs, or head-ach, uſher in the diſtemper. It ſeizes others with much more feveriſh ſymptoms; great pain of the head, back, and limbs; a vaſt oppreſſion of the præcordia, and continual ſighing. Some grown perſons go about for whole days in a drooping ſtate, with much uneaſineſs and anxiety, till at laſt they are obliged to take to their beds.—Thus various is the diſeaſe at the onſet. But it commonly begins with chills and heats, load and pain of the head, ſoreneſs of throat, and hoarſeneſs; ſome cough, ſickneſs at ſtomach, frequent vomiting and purging, in children eſpecially, and ſometimes very ſevere; though a contrary ſtate is more common to the adult. There is commonly a very great dejection of ſpirits, very ſudden weakneſs, great heavineſs on the breaſt, and faintneſs, from the very beginning. The pulſe in general is quick, ſmall, and fluttering; though ſometimes heavy and undulating. The eyes heavy, reddiſh, and as it were weeping; the countenance often full, fluſhed, and bloated, though ſometimes pale and ſunk. The following are the prognoſtics in the diſeaſe:—If a gentle eaſy ſweat comes on the third or fourth day; if the pulſe becomes more ſlow, firm, and equal; if the ſloughs of the fauces caſt off in a kindly manner, and appear at the bottom tolerably clean, and florid; if the breathing is more ſoft and free, and ſome degree of vigour and quickneſs return in the eyes; all is well, and a ſalutary criſis follows ſoon, by a continuance of the ſweat, and a turbid, ſubſiding, farinaceous, urine, a plentiful expectortion, and a very large deſquamation of the cuticle. But if a rigour comes on, and the exanthemata ſuddenly diſappear or turn livid; if the pulſe grows very ſmall and quick, and the ſkin remains hot and parched as it were; the breathing more difficult, the eyes dead and glaſſy, the urine pale and limpid; a phrenzy or coma may be expected to ſucceed, with

with a coldiſh clammy ſweat on the face or extremities; life will now be deſpaired of, eſpecially if a ſingultus and choaking or gulping in the throat ſhould attend, with ſudden, liquid, involuntary, livid, ſtools, intolerably fetid.

Symptoms of the *croup*, or *inflammation of the glottis.* A hoarſeneſs, with ſome ſhrillneſs and ringing ſound, both in ſpeaking and coughing, as if the noiſe came from a brazen tube. At the ſame time there is a ſenſe of pain about the larynx, ſome difficulty of reſpiration, with a whizzing ſound in inſpiration, as if the paſſage of the air were ſtraitened. The cough which attends it is commonly dry; and, if any thing be ſpit up, it is a matter of a purulent appearance, and ſometimes films reſembling portions of a membrane. With all theſe ſymptoms, there is a frequency of pulſe, a reſtleſſneſs, and an uneaſy ſenſe of heat. When the internal fauces are viewed, they are ſometimes without any appearance of inflammation; but frequently a redneſs, and even ſwelling, appears; and ſometimes there is an appearance of matter like to that rejected by coughing, together with the ſymptoms now deſcribed, and particularly a great difficulty of breathing, and a ſenſe of ſtrangling in the fauces, by which the patient is ſometimes ſuddenly taken off.

In a *pleuriſy*, the pathognomonic ſigns are a cough, a difficulty of breathing, a pain of the ſide, and a continued fever; the adjunct ſigns are the various ſorts of matter expectorated, which are ſometimes bloody, ſometimes bilious, &c. When the pains, which at firſt affected one ſide only, ſhall afterwards ſpread into the other; or when, leaving the ſide firſt affected, they paſs entirely into the other; theſe are always marks of a dangerous diſeaſe. A delirium coming on during a pneumonic inflammation is always a ſymptom denoting much danger. Bleeding is the remedy chiefly to be depended on; and may be done in either arm, as the ſurgeon finds moſt convenient; and the quantity taken away ought in general to be as large as the patient's ſtrength will allow. Yet the patient muſt keep out of bed as much as he can bear; ſhould have plenty of warm diluting drinks, impregnated with vegetable acids, accompanied with nitre or ſome other cooling neutral ſalt; and the belly alſo ought to be kept open by emollient clyſters or cooling laxative medicines. Vomiting in the beginning is dangerous; but in a ſomewhat advanced ſtate of the diſeaſe emetics have been found the beſt means of promoting expectoration. Fomentations and poultices to the pained part have been found uſeful; but bliſtering is found to be much more effectual. A bliſter, however, ought not to be applied till at leaſt one bleeding hath been premiſed, as veneſection is leſs effectual when the irritation of a bliſter is preſent. If the diſeaſe be moderate, a bliſter may be applied immediately after the firſt bleeding; but in violent caſes, where it may be preſumed that a ſecond bleeding may ſoon be neceſſary after the firſt, it will

will be proper to delay the blifter till after the fecond bleeding, when it may be fuppofed that the irritation occafioned by the blifter will be over before a third bleeding becomes neceffary. It may frequently be of ufe in this difeafe to repeat the bliftering; and in that cafe the plafters fhould always be applied fomewhere on the thorax, for when applied to more diftant parts they have little effect. The keeping the bliftered parts open, and making what is called a perpetual blifter, has much lefs effect than a repeated bliftering. When this difeafe terminates unfavourably, it often ends in an empyema, which is occafioned by the effufion of a quantity of purulent matter into the cavity of the thorax, producing a lingering and painful diforder, very often incurable. The firft fign of an empyema is a ceffation of the pain in the breaft, which before was continual: this is followed by a fenfation of weight on the diaphragm; and a fluctuation of matter, fometimes making a noife that may be heard by the by-ftanders: the acute fever is changed into a hectic, with an exacerbation at night: a continual and troublefome dry cough remains. The refpiration is exceedingly difficult, becaufe the lungs are prevented by the matter from fully expanding themfelves. The patient can lie eafily on that fide where the matter is effufed, but not on the other, becaufe then the weight of the matter on the mediaftinum produces uneafinefs. The more the hectic heat is augmented, the more is the body emaciated, and its ftrength decayed. In fome there is danger of fuffocation when they ftoop down, which goes off when they alter that pofture of the body; and in fome there is a purulent fpitting.—Thefe fymptoms are accompanied with great anxiety, palpitation of the heart, and faintings. Very few recover after an empyema has been once formed, efpecially if the operation paracentefis be neglected. After this operation is performed, if a great quantity of bloody fetid pus be difcharged, if the fever continue, and if the patient fpit up a purulent, pale, frothy, livid, or green, matter, with a decay of ftrength, there is no hope; but, when a fmall quantity of pus, of a white colour, not very fetid, is difcharged; when the fever and thirft prefently ceafe, the appetite returns, and fæces of a good confiftence are difcharged, the ftrength alfo returning in fome degree; there is then hope of a perfect recovery. If the matter be not dried up in feven weeks time, the difeafe readily changes to a fiftulous ulcer, which is very difficult to cure. An empyema affecting both fides of the thorax is more dangerous than that which affects only one.

The *inflammation of the heart* is attended with all the fymptoms before mentioned, but in a higher degree; it is befides fometimes accompanied with hydrophobic fymptoms, fainting, palpitation of the heart, a feeming madnefs, a funk and irregular pulfe, watery eyes, and a dejected countenance, with a dry black tongue.

The ſigns of an *inflammation of the ſtomach* are, great heat and pain in the epigaſtric region, extreme anxiety, an almoſt continual and painful hiccough, with a moſt painful vomiting of every thing taken into the ſtomach. This diſeaſe is always very dangerous, and the prognoſis doubtful, which alſo muſt always be in proportion to the ſeverity of the ſymptoms. A ceſſation of pain, coldneſs about the præcordia, great debility, with a languid and intermitting pulſe, and an abatement of the hiccough, denote a gangrene and ſpeedy death. From the ſenſibility of the ſtomach alſo, and its great connection with the reſt of the ſyſtem, it muſt be obvious that an inflammation of it, by whatever cauſes produced, may be attended with fatal conſequences; particularly by the great debility it produces, it may prove ſuddenly fatal, without running through the uſual courſe of inflammations.—Its tendency to admit of reſolution may be known by its having ariſen from no violent cauſe, by the moderate ſtate of the ſymptoms, and by a gradual remiſſion of theſe in the courſe of the firſt or at moſt of the ſecond week of the diſeaſe. The tendency to gangrene may be ſuſpected from the ſymptoms continuing with unremitting violence, notwithſtanding the uſe of proper remedies; and a gangrene already begun may be known by the ſymptoms above-mentioned, particularly great debility and ſudden ceſſation of pain. The tendency to ſuppuration may be known by the ſymptoms continuing but in a moderate degree for more than one or two weeks, and by a conſiderable remiſſion of the pain, while a ſenſe of weight and anxiety ſtill remains. When an abſceſs has been formed, the frequency of the pulſe is firſt abated, but ſoon after it increaſes, with frequent cold ſhivering, and an exacerbation in the afternoon and evening; followed by night-ſweats, and other ſymptoms of hectic fever. Theſe at length prove fatal, unleſs the abſceſs open into the cavity of the ſtomach, the pus be evacuated by vomiting, and the ulcer ſoon healed.

An *inflammation of the inteſtines* ſhows itſelf by a fixed pain in the abdomen, attended with fever, vomiting, and coſtiveneſs. The pain is often felt in different parts of the abdomen, but more frequently ſpreads over the whole, and is particularly violent about the navel. Inflammations of the inteſtines may ariſe from the ſame cauſe as thoſe from the ſtomach; though commonly the former will more readily occur from cold applied to the lower extremities, or to the belly itſelf. It is alſo found ſupervening on the ſpaſmodic cholic, incarcerated hernia, and volvulus. The inflammations of the inteſtines have the ſame termination with thoſe of the ſtomach; and the prognoſis in both caſes is much the ſame.

Inflammation of the liver is attended with conſiderable fever; a frequent, ſtrong, and hard, pulſe; high-coloured urine; an acute pain in the right hypochondrium, increaſed by preſſing upon the part. The pain is very often in ſuch a part of the ſide as to make it appear like a pleuriſy: and frequently, like that, is increaſed on inſpiration.

ration. The diſeaſe is alſo commonly attended with a cough, which is generally dry, but ſometimès moiſt; and, when the pain thus reſembles a pleuriſy, the patient cannot lie eaſily except upon the ſide affected. The pain is frequently extended to the clavicle, and to the top of the ſhoulder; and is attended ſometimes with hiccough, and ſometimes with vomiting. The inflammation of the liver, like others, may end by reſolution, ſuppuration, or gangrene; and the tendency to the one or to the other of thoſe events may be known from what has been already mentioned.

Inflammation of the ſpleen, comes on with a remarkable ſhivering, ſucceeded by a moſt intenſe heat and very great thirſt; a pain and tumour are perceived in the left hypochondrium, and the paroxyſms for the moſt part aſſume a quartan form. When the patients expoſe themſelves for a little to the free air, their extremities immediately grow very cold. If an hæmorrage happens, the blood flows out of the left noſtril.

It is often a very difficult matter to diſtinguiſh rheumatiſm from gout: but in *rheumatiſm* there in general occurs much leſs affection of the ſtomach; it affects chiefly the large joints, and often ſeveral of them at the ſame time: it occurs at an earlier period of life than gout; it is not obſerved to be hereditary; and it can in general be traced to ſome obvious exciting cauſe, particularly to the action of cold. To diſtinguiſh the chronic rheumatiſm from the acute: When the pains are ſtill ready to ſhift their place; when they are eſpecially ſevere in the night-time; when, at the ſame time, they are attended with ſome degree of pyrexia, and with ſome ſwelling, and eſpecially ſome redneſs, of the joints; the diſeaſe is to be conſidered as partaking of the nature of the acute rheumatiſm. But when there is no longer any degree of pyrexia remaining; when the pained joints are without redneſs; when they are cold and ſtiff; when they cannot eaſily be made to ſweat; or when, while a free and warm ſweat is brought out on the reſt of the body, it is only clammy and cold on the pained joints; and when, farther, the pains of theſe are increaſed by cold, and relieved by heat, applied to them, the caſe is to be conſidered as that of a purely chronic rheumatiſm: or perhaps more properly the firſt of the conditions now deſcribed may be termed the ſtate of irritability, and the ſecond the ſtate of atony. The chronic rheumatiſm, or rather the atonic, may affect different joints; but is eſpecially apt to affect thoſe which are ſurrounded with many muſcles, and thoſe of which the muſcles are employed in the moſt conſtant and vigorous exertions. Such is the caſe of the vertebræ of the loins, the affection of which is namd *lumbago;* or of the hip-joint, when the diſeaſe is named *iſchias* or *ſciatica.* Violent ſtrains and ſpaſms, occurring on ſudden and ſomewhat violent exertions, bring on rheumatic affections, which at firſt partake of the chronic rheumatiſm. Such are frequently the lumbago, and other affections which ſeem to be more ſeated in the muſcles than in the joints. The diſtinction of the rheumatic pains from thoſe reſembling them which occur in the ſiphilis and ſcurvy muſt be obvious, either from the ſeat of the pains or from the concomitant ſymptoms peculiar to thoſe diſeaſes.

What we call a paroxyſm of the *gout* is principally conſtituted by an inflammatory affection of ſome of the joints. This ſometimes comes on ſuddenly, without any warning, but is generally preceded by ſeveral ſymptoms; ſuch as the ceaſing of a ſweating which the feet had been commonly affected with before; an unuſual coldneſs of the feet and legs; a frequent numbneſs, alternating with a ſenſe of pricking along the whole of the lower extremities; frequent cramps of the muſcles of the legs; and an unuſual turgeſcence of the veins. While theſe ſymptoms take place in the lower extremities, the body is affected with ſome degree of torpor and languor, and the functions of the ſtomach in particular are more or leſs diſturbed. The appetite is diminiſhed; and flatulency, or other ſymptoms of indigeſtion, are felt. Theſe ſymptoms take place for ſeveral days, ſometimes for a week or two, before a paroxyſm comes on; but commonly, upon the day immediately preceding it, the appetite becomes keener than uſual. It is generally ſuppoſed, that there are ſome caſes of rheumatiſm which are ſcarcely to be diſtinguiſhed from the gout: theſe, however, Dr. Cullen thinks, are but few; and that the two diſeaſes may be for the moſt part diſtinguiſhed with great certainty, by obſerving the pre-diſpoſition, the antecedent circumſtances, the part affected, the recurrences of the diſeaſe, and its connection with the ſyſtem; which circumſtances, for the moſt part, appear very differently in the two diſeaſes.

Prognoſtics that a perſon is affected with the *plague*: 1. Great loſs of ſtrength. 2. Stupor, giddineſs, and conſequent ſtaggering, which reſembles drunkenneſs; or the head-ach and various delirium. 3. Anxiety, palpitation, ſyncope, and eſpecially the weakneſs and irregularity of the pulſe, denoting a conſiderable diſturbance in the action of the heart. 4. Nauſea and vomiting, particularly the vomiting of bile, which ſhow an accumulation of vitiated bile in the gall-bladder and biliary ducts, and from thence derived into the inteſtines and ſtomach; which alſo denote a conſiderable ſpaſm, and loſs of tone in the extreme veſſels on the ſurface of the body.

The *ſmall-pox* begins with a ſynocha, or inflammatory fever. It generally comes on about mid-day, with ſome ſymptoms of a cold ſtage, and commonly with a conſiderable languor and drowſineſs. A hot ſtage is ſoon formed, and becomes more conſiderable on the ſecond and third day. During this courſe children are liable to frequent ſtartings from their ſlumbers; and adults, if they are kept in bed, are diſpoſed to much ſweating. On the third day, children are ſometimes affected with one or two epileptic fits; and towards the end of the third day the eruption commonly appears. The principal marks by which the *chicken-pox* may be diſtinguiſhed from the ſmall-pox are, 1. The appearance, on the ſecond or third day of the eruption, of that veſicle full of ſerum upon the top of the pock. 2. The cruſt, which covers the pocks on the fifth day; at which time thofe of the fmall-pox are not at the height of their fuppuration. Foreign medical writers hardly ever mention the name of the chicken-pox; and the writers of our own country fcarcely mention any thing more of it than its name. Morton fpeaks of it as if he fuppofed it to be a very mild genuine fmall-pox. But thefe two diftempers are furely totally different from one another, not only on account of their different appearances above-mentioned, but becaufe thofe who have had the fmall-pox are capable of being infected with the chicken-pox; but thofe who have once had the chicken-pox are not capable of having it again, though to fuch as have never had this diftemper it feems as infectious as the fmall-pox. Dr. Heberden wetted a thread in the moft concocted pus-like liquor of the chicken-pox which he could find; and, after making a flight incifion, it was confined upon the arm of one who had formerly had it; the little wound healed up immediately, and fhowed no figns of any infection. From the great fimilitude between the two diftempers, it is probable, that, inftead of the fmall-pox, fome perfons have been inoculated from the chicken-pox; and that the diftemper which has fucceeded has been miftaken for the fmall-pox by hafty or inexperienced obfervers.

It is a promifing fign, in the *palfy*, when the patient feels a flight degree of painful itching in the affected parts; and, if a fever fhould arife, it bids fair to cure the palfy. When the fenfe of feeling remains, there is much more room to hope for a cure than when it is gone, as well as the power of motion. But, when we obferve the flefh to wafte, and the fkin to appear withered and dry, we may look upon the difeafe to be incurable. Convulfions fupervening on a palfy are a fatal fign.

When *fainting* happens in the beginning of any acute diftemper, it is not a good omen; but, when it takes place in the increafe or at the height of the difeafe, the danger is fomewhat lefs; but in general, when fainting comes on without any evident caufe, it is to be dreaded. In violent hæmorrhages it is favourable; as the bleeding veffels gain time to contract and recover themfelves, and thus the patient may efcape. When perfons of a full habit faint through excefs of paffion, they ought to be bled without delay, and fhould drink vinegar or lemon-juice diluted with water; and, after the bowels are emptied by a clyfter, take a paregoric draught, and go to bed.

Prognoftics from *convulfions*. Except in fome few cafes, convulfive diforders are always to be dreaded; but lefs in young people than in fuch as are advanced in life. Thofe which attack girls under the age of puberty will generally ceafe on the appearance of the menfes; and boys have likewife a chance of being relieved as they advance in life: but in grown-up people, unlefs the caufe be very evident, a cure is hardly to be expected, efpecially after the difeafe has been of long continuance. The treatment is much the fame with that of epilepfy.

The

The *epilepſy* frequently is preceded by a pain in the head, laſſitude, ſome diſturbance of the ſenſes, unquiet ſleep, unuſual dread, dimneſs of ſight, a noiſe in the ears, palpitation of the heart, coldneſs of the joints; and in ſome there is a ſenſation of formication, or a cold-air, &c. aſcending from the lower extremities toward the head. If the epilepſy comes on before the time of puberty, there are ſome hopes of its going off at that time. But it is a bad ſign when it attacks about the twenty-firſt year, and ſtill worſe if the fits grow more frequent; for then the animal functions are often deſtroyed, as well as thoſe of the mind, and the patient becomes ſtupid and fooliſh. Sometimes it will terminate in melancholy or madneſs, and ſometimes in a mortal apoplexy or palſy. It has ſometimes, however, been obſerved that epilepſies have been removed by the appearance of cutaneous diſeaſes, as the itch, ſmall-pox, meaſles, &c. therefore, if any of theſe appear, it may be reckoned a favourable prognoſtic.

Signs of a *diabetes.*—The diabetes firſt ſhows itſelf by a dryneſs of the mouth and thirſt, white frothy ſpittle, and the urine in ſomewhat larger quantity than uſual. A heat begins to be perceived in the bowels, which at firſt is a little pungent, and gradually increaſes. The thirſt continues to augment by degrees, and the patient by degrees loſes the power of retaining his urine for any length of time. The moſt ſingular phenomenon in this diſeaſe is, that the urine ſeems to be entirely or very much diveſted of an animal nature, and to be largely impregnated with a ſaccharine ſalt, ſcarcely diſtinguiſhable from that obtained from the ſugar-cane. For this diſcovery we are indebted to Dr. Dobſon of Liverpool, who made ſome experiments on the urine of a perſon labouring under a diabetes, who diſcharged 28 pints of urine every day, taking during the ſame time from 12 to 14 pounds only of ſolid and liquid food. When a perſon perceives any of the before-mentioned ſymptoms upon him, (particularly the quantity and inſipidity of the urine,) he ſhould loſe no time in taking the proper precautions; for the diabetes is rarely cured unleſs when taken at the very beginning, which is ſeldom done. Briſtol water is reckoned a ſpecific in this diſorder.

Hydrophobia.—This diſeaſe commonly does not make its attacks till a conſiderable time after the bite. In ſome few inſtances it has commenced in ſeven or eight days from the accident; but generally the patient continues in health for twenty, thirty, or forty, days, or even much longer. The wound, if not prevented, will in general be healed long before that time, frequently with the greateſt eaſe; though ſometimes it reſiſts all kinds of healing applications, and forms a running ulcer, which diſcharges a quantity of matter for many days. It has been ſaid, that the nearer the wounded place is to the ſalivary glands, the ſooner the ſymptoms of hydrophobia appear. The approach of the diſeaſe is known by the cicatrix of the wound becoming hard and elevated, and by a peculiar ſenſe of pricking at the part; pains ſhoot from it towards the throat; ſometimes it is ſurrounded with livid or red ſtreaks, and ſeems to be in a ſtate of inflammation; though frequently there is nothing remarkable to be obſerved about it. The patient becomes melancholy, loves ſolitude, and has ſickneſs at ſtomach. Sometimes the peculiar ſymptom of the diſeaſe, the dread of water, comes on all at once. We have an inſtance of one who, having taken a vomit of ipecacuanha for the ſickneſs he felt at his ſtomach, was ſeized with the hydrophobia at the time he was drinking the warm water. Sometimes the diſeaſe begins like a common ſore throat; and, the ſoreneſs daily increaſing, the hydrophobic ſymptoms ſhow themſelves like a convulſive ſpaſm of the muſcles of the fauces. In others, the mind ſeems to be primarily affected, and they have a real dread of water or any liquid before they try whether they can ſwallow it or not. Dr. James, in his Treatiſe on Canine Madneſs, mentions a boy ſent out to fill two bottles with water, who was ſo terrified by the noiſe of the liquid running into them, that he fled into the houſe crying out that he was bewitched. He mentions alſo the caſe of a farmer, who, going to draw ſome ale from a caſk, was terrified to ſuch a degree at its running into the veſſel, that he ran out in great haſte with the ſpigot in his hand. But, in whatever manner this ſymptom comes on, it is certain that the moſt painful ſenſations accompany every attempt to ſwallow liquids. Nay, the bare ſight of water, of a looking-glaſs, or of any thing clear or pellucid, will give the utmoſt uneaſineſs, and even throw the patient into convulſions. With regard to the affection of the mind itſelf in this diſeaſe, it does not appear that the patients are deprived of reaſon. Some have, merely by the dint of reſolution, conquered the dread of water, though they never could conquer the convulſive motions which the contact of liquids occaſioned: yet even this reſolution has been of no avail; for the convulſions and other ſymptoms, increaſing, have almoſt always deſtroyed the unhappy patients. However, in this diſtemper the ſymptoms are ſo various, that they cannot be enumerated; for we ſeldom read two caſes of hydrophobia which do not differ very remarkably in this reſpect. When a perſon is bitten, the prognoſis with regard to the enſuing hydrophobia is very uncertain. All thoſe who are bit do not fall into the diſeaſe; nay, Dr. Vaughan relates that, out of thirty bitten by a mad dog, only one was ſeized with the hydrophobia. During the interval betwixt the bite and the time of the diſeaſe comes on, there are no ſymptoms by which we can certainly judge whether it will appear or not.

Prognoſtics of a *dropſy of the breaſt*.—This affection, particularly with reſpect to its cauſes, is in many circumſtances ſimilar to other kinds of dropſy, particularly to aſcites. But from the ſituation of the water, which is here depoſited in the cavity of of the thorax, it may naturally be ſuppoſed that ſome peculiar ſymptoms will occur. Beſides the common ſymptoms of dropſy, (paleneſs of the countenance, ſcarcity of urine, and the like,) this diſeaſe is, in ſome inſtances, attended with a fluctuation of water within the breaſt; which, when it does occur, may be conſidered as a certain diſtinguiſhing mark of this affection. But, beſides this, it is alſo diſtinguiſhed by the remarkable affections of circulation and reſpiration with which it is attended. The breathing is peculiarly difficult, eſpecially in a recumbent poſture; and in many inſtances patients cannot breathe with tolerable eaſe unleſs when ſitting erect, or even ſtooping ſomewhat forward. The pulſe is very irregular, and has often remarkable intermiſſions. But the diſeaſe has been thought to be principally characterized by a ſudden ſtarting from ſleep, in conſequence of an almoſt inexpreſſible uneaſy ſenſation referred to the breaſt, and attended with ſtrong palpitation, which may probably ariſe from an affection either of circulation or of reſpiration. That theſe ſymptoms are common attendants of this diſeaſe is undeniable; and they are certainly the beſt characteriſtics of it with which we are yet acquainted: but it muſt be allowed that they are preſent in ſome caſes where there is no water in the breaſt; and that in other inſtances, where the diſeaſe exiſts, they are either altogether wanting, or occur only in a very ſlight degree. Certain diagnoſtics, therefore, of this diſeaſe ſtill remain to be diſcovered. When hydrothorax is preſent, from the affection of the vital functions with which it is attended, it may readily be concluded that it is a dangerous diſeaſe; and in many inſtances it proves fatal. The cure, as far as it can be accompliſhed, is obtained very much on the ſame principles as in other dropſies. Benefit is often obtained from an artificial diſcharge of water by the application of bliſters to the breaſt; but in this, as well as other dropſies, a diſcharge is chiefly effected by the natural outlets, particularly from the uſe of cathartics and diuretics. In this ſpecies of dropſy, more perhaps than in any other, recourſe has been had to the uſe of the digitalis purpurea, or fox-glove, ſo ſtrongly recommended as a diuretic by Dr. Withering in his Treatiſe reſpecting the uſe of it. There can be no doubt that this, though ſometimes productive of inconvenience, from the diſtreſſing ſickneſs and ſevere vomiting which it not unfrequently excites, even in ſmall doſes, often operates as a powerful diuretic, and produces a complete evacuation of water, after other remedies have failed. From the effects mentioned above, however, as well as from its influence on the pulſe, which it renders much ſlower, it is neceſſary that it ſhould be employed with great caution and in ſmall doſes. A dram of the dried leaves of the digitalis, macerated for four hours in half a pint of warm water, forms an infuſion which may be given in doſes of an ounce, and the dried powder of the leaves in doſes of one or two grains: theſe doſes may be gradually increaſed, and repeated twice or oftener in the day; but

but this requires to be done with great caution, left fevere vomiting or other diftreffing fymptoms fhould take place.

Scurvy.—The firft indication of the fcorbutic diathefis is generally a change of colour in the face, from the natural and healthy look to a pale and bloated complexion, with a liftleffnefs, and averfion from every fort of exercife; the gums foon after become itchy, fwell, and are apt to bleed on the flighteft touch; the breath grows offenfive; and the gums, fwelling daily more and more, turn livid, and at length become extremely fungous and putrid, as being continually in contact with the external air; which in every cafe favours the putrefaction of fubftances difpofed to run into that ftate, and is indeed abfolutely requifite for the production of actual rottennefs. The fymptoms of the fcurvy, like thofe of every other difeafe, are fomewhat different in different fubjects, according to the various circumftances of conftitution; and they do not always proceed in the fame regular courfe in every patient. But what is very remarkable in this difeafe, notwithftanding the various and immenfe load of diftrefs under which the patients labour, there is no ficknefs at the ftomach, the appetite keeps up, and the fenfes remain entire almoft to the very laft: when lying at reft, they make no complaints, and feel little diftrefs or pain: but, the moment they attempt to rife or ftir themfelves, then the breathing becomes difficult, with a kind of ftraitnefs or catching, and great oppreffion, and fometimes they have been known to fall into a fyncope. This catching of the breath upon motion, with the lofs of ftrength, dejection of fpirits, and rotten gums, are held as the effential or diftinguifhing fymptoms of the difeafe.

The *jaundice* firft fhows itfelf by a liftleffnefs and want of appetite: the patient becomes dull, oppreffed, and generally coftive. Thefe fymptoms have continued but a very fhort time, when a yellow colour begins to diffufe itfelf over the *tunica albuginea,* or white part of the eye, and the nails of the fingers; the urine becomes high coloured, with a yellowifh fediment capable of giving a yellow tint to linen; the ftools are whitifh or grey. In fome there is a moft violent pain in the epigaftric region, which is confiderably increafed after meals. In fome the difeafe degenerates into an incurable dropfy; and there have been many inftances of people who have died of the dropfy after the jaundice itfelf had been totally removed. The coming on of a gentle diarrhœa, attended with bilious ftools, together with the ceffation of pain, are figns of the difeafe being cured. We are not, however, always to conclude, becaufe the difeafe is not attended with acute pain, that it is therefore incurable; for frequently the paffage of a concretion through the biliary ducts is accompanied only with a fenfation of flight uneafinefs. If the difeafe goes off, its return muft be prevented by a courfe of tonic medicines, particularly the Peruvian bark and antifeptics: but we can by no means be certain that the jaundice will not return,

 and

and that at any interval; for there may be a number of concretions in the gall-bladder, and, though one hath paſſed, another may very quickly follow, and produce a new fit of jaundice; and thus ſome people have continued to be affected with the diſtemper, at ſhort intervals, during life.

Stone in the bladder.—The ſigns of a ſtone in the bladder are, pain, eſpecially about the ſphincter; and bloody urine, in conſequence of riding or being jolted in a carriage: a ſenſe of weight in the perinæum; an itching of the glans penis; ſlimy ſediment in the urine; and frequent ſtoppages in making water; a teneſmus alſo comes on while the urine is diſcharged.

Imaginary viſion of objects which do not exiſt.—This often takes place when the body is diſeaſed, and then the patient is ſaid to be delirious. Sometimes however, in theſe caſes, it does not amount to delirium; but the perſon imagines he ſees gnats or other inſects flying before his eyes; or ſometimes, that every thing he looks at has black ſpots in it, which laſt is a very dangerous ſign. Sometimes alſo ſparks of fire appear before his eyes; which appearances are not to be diſregarded, as they frequently precede apoplexy or epilepſy: on the other hand, it is feared that little benefit can be derived from an attention to this prognoſtic, as the fits commonly follow ſo ſuddenly.

I ſhall now proceed to deſcribe two diſorders not noted by Culpeper, or any old writer. And firſt of the

ANGINA PECTORIS.

Dr. Heberden was the firſt who deſcribed this diſeaſe, though it is extremely dangerous, and, by his account, not very rare. It ſeizes thoſe who are ſubject to it when they are walking, and particularly when they walk ſoon after eating, with a moſt diſagreeable and painful ſenſation in the breaſt, which ſeems to threaten immediate deſtruction: but, the moment they ſtand ſtill, all the uneaſineſs vaniſhes. In all other reſpects the patients at the beginning of this diſorder are well, and have no ſhortneſs of breath. After it has continued ſome months, the fits will not ceaſe inſtantaneouſly on ſtanding ſtill; and it will come on not only when the patients are walking, but when they are lying down, and oblige them to riſe up out of their beds every night for many months together. In one or two very inveterate caſes, it has been brought on by the motion of a horſe or carriage, and even by ſwallowing, coughing, going to ſtool, ſpeaking, or by any diſturbance of mind. The perſons affected were all men, almoſt all of whom were above fifty years of age, and moſt of them with a ſhort neck and inclining to be fat. Something like it, however, was obſerved in one woman, who was paralytic; and one or two young men complained of it in a ſlight degree. Other practitioners have obſerved it in very young perſons. When a fit of this ſort comes on by walking, its duration is very ſhort, as it goes off almoſt immediately upon ſtopping. If it comes on in the night, it will laſt an hour or two. Dr. Heberden met with one in whom it once continued for ſeveral days; during all which time the patient ſeemed to be in imminent danger of death. Moſt of thoſe attacked with the diſtemper died ſuddenly: though this rule was not without exceptions; and Dr. Heberden obſerved one who ſunk under a lingering illneſs of a different nature. The os ſterni is uſually pointed to as the ſeat of this malady; but it ſeems as if it was under the lower part of that bone, and at other times under the middle or upper part, but always inclining more to the left ſide; and in many caſes there is joined with it a pain about the middle of the left arm, which appears to be ſeated in the biceps muſcle.

The appearance of Dr. Heberden's paper in the Medical Tranſactions very ſoon raiſed the attention of the faculty, and produced other obſervations from phyſicians of eminence; namely, Dr. Fothergill, Dr. Wall of Worceſter, Dr. Haygarth of Cheſter, and Dr. Percival of Mancheſter. It alſo induced an unknown ſufferer to write Dr. Heberden a very ſenſible letter, deſcribing his feelings in the moſt natural manner; which unfortunately, in three weeks after the date of this anonymous epiſtle, terminated in a ſudden death, as the writer himſelf had apprehended. The youngeſt ſubject that Dr. Fothergill ever ſaw afflicted with this diſorder was about thirty years of age; and this perſon was cured. The method that ſucceeded him was a courſe of pills, compoſed of the maſs of gum-pill, ſoap, and native cinnabar, with a light chalybeate bitter: this was continued for ſome months, after which he went to Bath ſeveral ſucceſſive ſeaſons, and acquired his uſual health: he was ordered to be very ſparing in his diet; to keep the bowels open; and to uſe moderate exerciſe on horſeback, but not to take long or fatiguing walks. The only ſymptom in this patient that is mentioned, was a ſtricture about the cheſt, which came on if he was walking up hill or a little faſter than ordinary, or if he was riding a very briſk trot; for moderate exerciſe of any kind did not affect him; and this uneaſy ſenſation always obliged him to ſtop, as he felt himſelf threatened with immediate death if he had continued to go forwards. It is the ſharp conſtrictive pain acroſs the cheſt, that (according to Dr. Fothergill's obſervation) particularly marks this ſingular diſeaſe; and which is apt to ſupervene upon a certain degree of muſcular motion, or whatever agitates the nervous ſyſtem. In ſuch caſes as fell under the inſpection of Dr. Fothergill, he very ſeldom met with one that was not attended with an irregular and intermitting pulſe; and this, not only during the exacerbations, but

but often when the patient was free from pain and at reſt; but Dr. Heberden obſerves, that the pulſe is, at leaſt ſometimes, not diſturbed; and mentions his having once had an opportunity of being convinced of this circumſtance, by feeling the pulſe during the paroxyſm. But no doubt theſe varieties, as well as many other little circumſtances, will occur in this diſeaſe, as they do in every other, on account of the diverſity of the human frame; and, if thoſe which in general are found to predominate and give the diſtinguiſhing character be preſent, they will always authoriſe us in giving the name to the diſeaſe: thus, when we find the conſtrictory pain acroſs the cheſt, accompanied with a ſenſe of ſtrangling or ſuffocation; and ſtill more, if this pain ſhould ſtrike acroſs the breaſt into one or both arms; we ſhould not heſitate to pronounce the caſe an angina pectoris.

As to the nature of this diſeaſe, it appears to be purely ſpaſmodic: and this opinion will readily preſent itſelf to any one who conſiders the ſudden manner of its coming on and going off; the long intervals of perfect eaſe; the relief afforded by wine and ſpirituous cordials; the influence which paſſionate affections of the mind have over it; the eaſe which comes from varying the poſture of the head and ſhoulders, or from remaining quite motionleſs; the number of years for which it will continue, without otherwiſe diſordering health; its bearing ſo well the motion of a horſe or carriage, which circumſtance often diſtinguiſhes ſpaſmodic pains from thoſe which ariſe from ulcers; and laſtly, its coming on for the moſt part after a full meal, and in certain patients at night, juſt after the firſt ſleep, at which time the incubus, convulſive aſthma, and other ills juſtly attributed to the diſordered functions of the nerves, are peculiarly apt to return or to be aggravated. From all theſe circumſtances taken together, there can be little doubt that this affection is of a ſpaſmodic nature: but, though this ſhould be admitted, it may not be ſo eaſy to aſcertain the particular muſcles which are thus affected. The violent ſenſe of ſtrangling or choaking which ſhows the circulation through the lungs to be interrupted during the height of the paroxyſm; and the peculiar conſtrictive pain under the ſternum, always inclining (according to Dr. Heberden's obſervation) to the left ſide; together with that moſt diſtreſſing and alarming ſenſation, which, if it were to increaſe or continue, threatens an immediate extinction of life; might authoriſe us to conclude that the heart itſelf is the muſcle affected: the only objection to this idea (and, if it had been conſtantly obſerved, it would be inſurmountable) is, that the pulſe is not always interrupted during the paroxyſm. The appearances in two of the diſſections favour the opinion that the ſpaſm affects the heart; as in one ſubject the left ventricle (and, though it be not mentioned, we may preſume the right one alſo) was found as empty of blood as if it had been waſhed; and in another, the ſubſtance of the heart appeared whitiſh, not unlike a ligament; as it ſhould ſeem, in both caſes, from the force

of

of the ſpaſm ſqueezing the blood out from the veſſels and cavities. If this hypotheſis be allowed, we muſt conclude that the ſpaſm can only take place in an inferior degree, as long as the patient continues to ſurvive the paroxyſm; ſince an affection of this ſort, and in this part, of any conſiderable duration or violence, muſt inevitably prove fatal: and accordingly, as far as could be traced, the perſons who have been known to labour under this diſeaſe have in general died ſuddenly. The diſſections alſo ſhow, that, whatever may be the true ſeat of the ſpaſm, it is not neceſſary for the bringing of it on, that the heart, or its immediate appendages, ſhould be in a morbid ſtate; for in three out of the ſix that have as yet been made public, theſe parts were found in a ſound ſtate. On opening the body of the gentleman who wrote the letter to Dr. Heberden, "upon the moſt careful examination, no manifeſt cauſe of his death could be diſcovered; the heart, in particular, with its veſſels and valves, were all found in a natural condition." In the caſe communicated by Dr. Percival to the publiſhers of the Edinburgh Medical Commentaries, "the heart and aorta deſcendens were found in a ſound ſtate." And in Dr. Haygarth's patient, on opening the thorax, the lungs, pericardium, and heart, appeared perfectly ſound. Not to mention Dr. Fothergill's patient (R. M.), in whoſe body the only morbid appearance about the heart was a ſmall white ſpot near the apex. So that the cauſe, whatever its nature might have been, was at too great a diſtance, or of too ſubtle a nature, to come under the inſpection of the anatomiſt. But there was a circumſtance in two of the ſubjects that is worthy of remembrance; and which ſhows that the craſis of the blood, while they were living, muſt have been greatly injured; namely, its not coagulating, but remaining of a cream-like conſiſtence, without any ſeparation into ſerum and craſſamentum.

From all that we have ſeen hitherto publiſhed, it does not appear that any conſiderable advances have been made towards the actual cure of this anomalous ſpaſm. The very judicious and attentive Dr. Heberden (to whom the public are highly indebted for firſt making the diſorder known) confeſſes, that bleeding, vomits, and other evacuations, have not appeared to do any good: wine and cordials, taken at bed-time, will ſometimes prevent or weaken the fits; but nothing does this ſo effectually as opiates: in ſhort, the medicine uſually called nervous or cordial, ſuch as relieve and quiet convulſive motions, and invigorate the languiſhing principle of life, are what he recommends. Dr. Wall mentions one patient, out of the twelve or thirteen that he had ſeen, who applied to him early in the diſeaſe, and was relieved conſiderably by the uſe of antimonial medicines joined with the fetid gums: he was ſtill living at the time the doctor wrote his paper, (Nov. 1772.) and going about with tolerable eaſe. Two were carried off by other diſorders; all the reſt died ſuddenly. Dr. Fothergill's directions are chiefly calculated with the view to prevent the diſorder from gaining ground, and to alleviate preſent diſtreſs. Accordingly he enjoins ſuch a kind of diet as may be moſt likely to prevent irritability: in particular not to eat voraciouſly: to be ſtrictly abſtemious in reſpect to every thing heating; ſpices, ſpirits, wines, and all fermented liquors: to guard moſt ſcrupulouſly againſt paſſion, or any vehement emotions; and to make uſe of all the uſual means of eſtabliſhing and preſerving general health; to mitigate exceſſes of irritability by anodynes; or pains, if they quicken the circulation: to diſperſe flatulencies, when they diſtend the ſtomach, by moderate doſes of carminatives; amongſt which, perhaps, ſimple peppermint-water may be reckoned one of the ſafeſt. But, ſince obeſity is juſtly conſidered as a principal prediſpoſing cauſe, he inſiſts ſtrongly on the neceſſity of preventing an increaſe of fat by a vegetable diet, and uſing every other practicable method of augmenting the thinner ſecretions. Theſe were the only means which occurred to the Engliſh phyſicians of oppoſing this formidable diſeaſe.—In my own practice I have never known the Solar Tincture to fail in removing by degrees this dangerous diſorder. The caſes indeed that have occurred to me have been very few; and my uniform practice has been to order a wine-glaſs of the Tincture, diluted with water, to be taken at going to bed; and, in thoſe where the diſorder had gained great aſcendency, I preſcribed a like quantity to be taken at getting up in the morning, at leaſt an hour before breakfaſt. This has always rendered the attacks leſs violent, and at laſt totally removed them. Two table-ſpoonfuls of the Tincture undiluted ſhould be adminiſtered, if poſſible, during the height of the paroxyſm, which will generally give immediate eaſe.

Dr. Smyth of Ireland has, we are told, diſcovered that the angina pectoris may be certainly cured by iſſues, of which Dr. Macbride gives the following inſtance:

"A. B. a tall well-made man; rather large than otherwiſe: of healthy parents, except that there had been a little gout in the family; temperate; being very attentive to the buſineſs of his trade, (that of a watch-maker,) led a life uncommonly ſedentary; had, from his boyhood upwards, been remarkably ſubject to alarming inflammations of his throat, which ſeized him at leaſt once in the courſe of the year; in all other reſpects well. In 1767, (then forty-eight years of age,) he was taken, without any evident cauſe, with a ſudden and very diſpiriting throbbing under the ſternum. It ſoon afterwards increaſed, and returned upon him every third or fourth week, accompanied with great anxiety, very laborious breathing, choaking, a ſenſation of fulneſs and diſtenſion in the head, bloated and fluſhed countenance, turgid and watery eyes, and a very irregular and unequal pulſe. The paroxyſm invaded him almoſt conſtantly while he was ſitting after dinner; now and then he was ſeized with it in the morning, when walking a little faſter than uſual; and was then obliged to reſt on any object at hand. Once or twice it came on in bed; but did not oblige him to ſit up, as it was then attended with no great difficulty in breathing. In the afternoon-fits, his greateſt eaſe was from the ſupine poſture; in which he uſed to continue motionleſs for ſome hours, until, quite ſpent and worn out with anguiſh, he dropt into a ſlumber. In the intervals between theſe attacks, which at length grew ſo frequent as to return every fourth or fifth day, he was, to appearance, in perfect health. Thus matters continued for more than two years; and various antiſpaſmodic were ineffectually tried for his relief. In 1769, there ſupervened a very ſharp conſtrictory pain at the upper part of the ſternum, ſtretching equally on each ſide, attended with the former ſymptoms of anxiety, dyſpnœa, choaking, &c. and with an excruciating cramp, as he called it, that could be covered with a crown-piece, in each of his arms, between the elbow and the wriſt, exactly at the inſertion of the pronator teres; the reſt of the limb was quite free. The fits were ſometimes brought on, and always exaſperated, by any agitation of mind or body. He once attempted to ride on horſeback during the paroxyſm: but the experiment was near proving fatal to him. The difference of ſeaſon or weather made no impreſſion upon him. Still, in the intervals, his health was perfectly good; except that his eyes, which before his illneſs were remarkably ſtrong and clear, were now grown extremely tender; and that his ſight was much impaired. He had no flatulency of ſtomach, and his bowels were regular. In this ſituation, Feb. 22, 1770, he applied to me for aſſiſtance. I had ſeen, I believe, eight or ten of theſe frightful caſes before. Two of the patients dropped dead ſuddenly. They were men between forty and fifty years of age, and of a make ſomewhat fleſhy. The fate of the others I was not informed of; or, at leaſt, cannot now recollect. Having found the total inefficacy of bliſters and the whole claſs of nervous medicines in the treatment of this anomalous ſpaſm, I thought it right to attempt correcting or draining off the irritating fluid in the caſe now before us. To this purpoſe, I ordered a mixture of lime-water with a little of the compound juniper-water, and an alterative proportion of Huxham's antimonial wine: I put the patient on a plain, light, perſpirable, diet; and reſtrained him from all viſcid, flatulent, and acrimonious, articles. By purſuing this courſe, he was ſoon apparently mended; but, after he had perſiſted regularly in it for at leaſt two months, he kept for ſome time at a ſtand. I then ordered a large iſſue to be opened on each of his thighs. Only one was made. However, as ſoon as it began to diſcharge, he amended. The frequency and ſeverity of the fits abated conſiderably; and he continued improving gradually, until, at the end of eighteen months, he was reſtored to perfect health; which he has enjoyed, without the leaſt interruption, till now, except when he has been tempted (perhaps once in a twelvemonth) to tranſgreſs rules, by making a large meal on ſalted meat, or indulging himſelf in ale or rum-punch, each of which never failed to diſorder him from

 the

the beginning of his illnefs: but on even thefe occafions, he has felt no more than the flighteft motion of his former fufferings; infomuch that he would defpife the attack, if it did not appear to be of the fame ftock with his old complaint. No other caufe has had the leaft ill effect on him. Though rum was conftantly hurtful, yet punch, made with a maceration of black currants in our vulgar corn-fpirit, is a liquor that agrees remarkably well with him. He never took any medicine after the iffue began to difcharge; and I have directed that it fhall be kept open as long as he lives. The inflammations of his throat have difappeared for five years paft; he has recovered the ftrength and clearnefs of his fight; and his health feems now to be entirely re-eftablifhed."

Dr. Macbride, in a letter to Dr. Duncan, publifhed in the Edinburgh Medical Commentaries, gives the following additional obfervations on this difeafe:

"Within thefe few weeks I have, at the defire of Dr. Smyth, vifited, three or four times, a very ingenious man who keeps an academy in this city, of about thirty-four years of age, who applied to the doctor for his advice in January laft. I fhall give you his fymptoms as I had them from his own mouth, which appear to me to mark his cafe to be an angina pectoris, and as deplorable as any that I have read of. It was ftrongly diftinguifhed by the exquifite conftrictory pain of the fternum, extending to each of his arms as far as the infertion of the deltoid mufcle, extreme anxiety, laborious breathing, ftrangling, and violent palpitation of the heart, with a moft irregular pulfe. The paroxyfms were fo frequent, that he fcarcely ever efcaped a day, for fix or feven years, without one. They were ufually excited by any agitation of mind or body, though flight. He had clear intervals of health between the fits. The diftemper feems hereditary in him, as he fays his father was affected in the fame manner fome years previous to his death. He has a ftrong gouty taint, which never fhowed itfelf in his limbs; and he has led a life of uncommon fedentarinefs, from intenfe application to mathematical ftudies, attention of mind, and paffion, even from his boyifh years. Thefe circumftances may, perhaps, account for his having been taken with this difeafe at fo early an age as feventeen. A large iffue was immediately opened on each of his thighs. In a month afterwards he began to mend, and has gone on improving gradually. He can now run up ftairs brifkly, as I faw him do no later than yefterday, without hurt; can bear agitation of mind; and has no complaint, excepting a flight oppreffion of the breaft, under the fternum, which he feels fometimes in a morning, immediately after dreffing himfelf, and which he thinks is brought on by the motion ufed in putting on his clothes; though, for a complete week preceding the day on which I faw him laft, he told me that he had been entirely free from all uneafinefs, and was exulting that he had not had fuch an interval of eafe for thefe laft feven years. Dr. Smyth alfo fhowed me, in his *adverfaria*, or note-book, the cafe of a gentleman who had been under his care

in

in 1760, which he had forgotten when my book went to the prefs, but which he was reminded of the other day by a vifit from his patient. It was a genuine angina pectoris, brought on by a fedentary life, and great vexation of mind, clearly marked by the exquifite pain under the fternum, that extended acutely to the upper extremities, particularly along the left arm, together with the other fymptoms of dyfpnœa, anxiety, palpitation of the heart, &c. recited in the cafe above. The diforder went off in 1762, by large fpontaneous difcharges from the piles, but returned upon him feverely in 1765. Iffues in his thighs were then recommended to him, but not made. But, whether it was by the perfuafion of fome friend, or of his own accord, he went into a courfe of James's powder, in fmall alterative dofes, combined with a little caftor and afafœtida. This he perfifted in for about fix weeks; in the mean while, he had large acrimonious gleetings from the fcrotum, and a plentiful difcharge of ichor from the anus. From this time he began to find his complaints grow lefs and lefs diftreffing, and he has now been totally free from them for fix years paft."

Of this fhocking diforder died that eminent furgeon, Mr. John Hunter. See Encyclopædia Londinenfis, vol. x. p. 482.

STRAITNESS OF THE ŒSOPHAGUS.

This diftemper has been treated of only by Dr. Munckley, who reckons it one of the moft deplorable difeafes of the human body. Its beginning is in general fo flight as to be fcarcely worth notice, the patients perceiving only a fmall impediment to the fwallowing of folid food: they ufually continue in this ftate for many months; during which, all liquid foods, and even folids themfelves when cut fmall and fwallowed leifurely, are got down without much difficulty; by degrees the evil increafes, and the paffage through the œfophagus becomes fo narrow, that not the fmalleft folid whatever can pafs through it; but, after having been detained for fome time at the part where the obftacle is formed, is returned again with a hollow noife of a very peculiar kind, and with the appearance of convulfion. The feat of this malady is fometimes near the top of the œfophagus, and at other times farther down, nearer the fuperior orifice of the ftomach. In this laft cafe, the part of the alimentary tube which is above the obftruction is frequently fo dilated by the food which is detained in it as to be capable of containing a large quantity; and the kind of vomiting, by which it is again returned through the mouth, comes on fooner or later after the attempt to fwallow, in proportion to the nearnefs or remotenefs of the part affected. In the laft ftage of this difeafe, not even liquids themfelves can be fwallowed fo as to pafs into the ftomach, and the patient dies literally ftarved to death. On the diffection of fuch as have died in this manner, the œfophagus is found to be confiderably thickened; and in fome fo contracted within at the difeafed part, as fcarcely to admit the paffing of a common probe; in others, to adhere together in fuch a manner

manner as entirely to close up the passage, and not to be separated without great difficulty. He comes next to show what he has found to be the most efficacious method of treating this disease, which, though not uncommon, yet in general has been considered as incurable. He claims not the merit of having discovered the method of cure, but hopes that some service may arise from publishing what his experience has confirmed to him; having first received the hint from another eminent physician. The only medicine, then, from the use of which he has ever found any service, is mercury; and in cases which are recent, and where the symptoms have not risen to any great height, small doses of mercury given every night, and prevented, by purgative medicines, from affecting the mouth, have accomplished the cure. But where the complaint has been of long standing, and the symptom has come on of the food's being returned through the mouth, a more powerful method of treatment becomes necessary. In this case he has never found any thing of the least avail in removing the symptoms but mercury, used in such a manner as to raise a gentle but constant spitting, and this method he has pursued with the happiest success. If this method be commenced before the complaint has gained too much ground upon the constitution, the case is not to be despaired of; and, of those who have come under his care in this state, by much the greater part have received considerable benefit from it, and many have been entirely cured. The complaint itself, he observes, is not very uncommon; but there is no instance, to his knowledge, recorded, of success from any other manner of treating it than that he has recommended.

OBSERVATIONS ON THE MEANS OF PRESERVING HEALTH.

I. Rules for the Management of Valetudinarians.

That part of the medical system which lays down rules for the preservation of health and prevention of diseases, termed Hygeine, is not to be strictly understood as if it respected only those people who enjoy perfect health, and who are under no apprehensions of disease, for such seldom either desire or attend to medical advice; but should rather be considered as relating to valetudinarians, or to such as, though not actually sick, may yet have sufficient reason to fear that they will soon become so: hence it is that the rules must be applied to correct morbific dispositions, and to obviate the various things that are known to be the remote or possible causes of diseases. From the way in which the several temperaments are usually mentioned by systematic writers, it should seem as if they meant that every particular constitution must be referred to one or other of the four; but this is far from being reducible to practice, since by much the greater number of people have constitutions so indistinctly marked, that it is hard to ſay to which of the temperaments they belong. When we actually meet with particular perſons who have evidently either, 1. Too much ſtrength and rigidity of fibre, and too much ſenſibility; 2. Too little ſtrength, and yet too much ſenſibility; 3. Too much ſtrength, and but little ſenſibility; or, 4. But little ſenſibility joined to weakneſs;—we ſhould look on ſuch perſons as more or leſs in the valetudinary ſtate, who require that theſe morbific diſpoſitions be particularly watched, leſt they fall into thoſe diſeaſes which are allied to the different temperaments.

People of the firſt-mentioned temperament being liable to ſuffer from continued fevers, eſpecially of the inflammatory ſpecies, their ſcheme of preſerving health ſhould conſiſt in temperate living, with reſpect both to diet and exerciſe; they ſhould ſtudiouſly avoid immoderate drinking, and be remarkably cautious leſt any of the natural diſcharges be checked. People of this habit bear evacuations well, eſpecially bleeding: they ought not, however, to loſe blood but when they really require to have the quantity leſſened; becauſe too much of this evacuation would be apt to reduce the conſtitution to the ſecond-mentioned temperament, wherein ſtrength is deficient, but ſenſibility redundant.

Perſons of the ſecond temperament are remarkably prone to ſuffer from painful and ſpaſmodic diſeaſes, and are eaſily ruffled; and thoſe of the ſofter ſex who have this delicacy of habit are very much diſpoſed to hyſterical complaints. The ſcheme here ſhould be, to ſtrengthen the ſolids by moderate exerciſe, cold bathing, the Peruvian bark, and chalybeate waters; particular attention ſhould conſtantly be had to the ſtate of the digeſtive organs, to prevent them from being overloaded with any ſpecies of ſaburra which might engender flatus, or irritate the ſenſible membranes of the ſtomach and inteſtines, from whence the diſorder would ſoon be communicated to the whole nervous ſyſtem. Perſons of this conſtitution ſhould never take any of the draſtic purges, nor ſtrong emetics; neither ſhould they loſe blood but in caſes of urgent neceſſity. But a principal ſhare of management, in theſe extremely-irritable conſtitutions, conſiſts in avoiding all ſudden changes of every ſort, eſpecially thoſe with reſpect to diet and clothing, and in keeping the mind as much as poſſible in a ſtate of tranquillity. Hence the great advantages which people of this frame derive from the uſe of medicinal waters drunk on the ſpot, becauſe of that freedom from care and ſerious buſineſs of every kind which generally obtains in all the places laid out for the reception of valetudinarians.

The third-mentioned temperament, where there is an exceſs of ſtrength and but little ſenſibility, does not ſeem remarkably prone to any diſtreſſing or dangerous ſpecies of diſeaſe; and therefore it can hardly be ſuppoſed that perſons ſo circumſtanced will either of themſelves think of any particular ſcheme of management, or

have

have recourſe to the faculty for their inſtructions: ſuch conſtitutions, however, we may obſerve, bear all kinds of evacuations well, and ſometimes require them to prevent an over-fulneſs, which might end in an oppreſſion of the brain or ſome other organ of importance.

But the fourth temperament, where we have weakneſs joined to want of ſenſibility, is exceedingly apt to fall into tedious and dangerous diſeaſe, ariſing from a defect of abſorbent power in the proper ſets of veſſels, and from remiſſneſs of the circulation in general; whence corpulency, dropſy, jaundice, and different degrees of ſcorbutic affection. In order to prevent theſe, or any other ſpecies of accumulation and depravation of the animal fluids, the people of this conſtitution ſhould uſe a generous courſe of diet with briſk exerciſe, and be careful that none of the ſecretions be interrupted, nor any of the natural diſcharges ſuppreſſed. Theſe conſtitutions bear purging well, and often require it; as alſo the uſe of emetics, which are frequently found neceſſary to ſupply the place of exerciſe, by agitating the abdominal viſcera; and are of ſervice to prevent the ſtagnation of bile, or the accumulation of mucous humours, which hinder digeſtion, and clog the firſt paſſages. The free uſe of muſtard, horſe-radiſh, and the like ſort of ſtimulating dietetics, is ſerviceable in theſe torpid habits.

When the general maſs of fluids is accumulated beyond what is conducive to the perfection of health, there ariſes what the writers term a *plethora*, which may prove the ſource of different diſeaſes; and therefore, when this over-fulneſs begins to produce languor and oppreſſion, care ſhould be taken in time to reduce the body to a proper ſtandard, by abridging the food and increaſing the natural diſcharges, uſing more exerciſe, and indulging leſs in ſleep. But in oppoſite circumſtances, where the fluids have been exhauſted, we are to attempt the prevention of further waſte by the uſe of ſtrengthening ſtomachics, a nouriſhing diet, and indulgence from fatigue of body or mind. Vitiated fluids are to be conſidered as affected either with the different kinds of general acrimony, or as betraying ſigns of ſome of the ſpecies of morbific matter which give riſe to particular diſeaſes, ſuch as gout, rheumatiſm, calculus, ſcurvy, &c.

During the ſtate of infancy, we may ſometimes obſerve a remarkable acidity, which not only ſhows itſelf in the firſt paſſages, but alſo ſeems to contaminate the general maſs of fluids. As it takes its riſe, however, from weak bowels, our views, when we mean to prevent the ill conſequences, muſt be chiefly directed to ſtrengthen the digeſtive organs, as on their ſoundneſs the preparation of good chyle depends; and hence ſmall doſes of rhubarb and chalybeates (either the natural chalybeate waters mixed with milk, or the flores martiales in doſes of a few grains, according to the

the age of the child) are to be adminiſtered; and the diet is to be ſo regulated as not to add to this acid tendency: briſk exerciſe is likewiſe to be enjoined, with frictions on the ſtomach, belly, and lower extremities.

Where the fluids tend to the putreſcent ſtate, which ſhows itſelf by fetid breath, ſpongineſs and bleeding of the gums, a bloated look and livid countenance, the diet then ſhould be chiefly of freſh vegetables and ripe fruits, with wine in moderation, briſk exerciſe, and ſtrengthening bitters.

Where acrimony ſhows itſelf by itching eruptions, uncommon thirſt, and fluſhing heats, nothing will anſwer better than ſuch ſulphureous waters as the Harrowgate and Moffat in Britain, or the Lucan Swadlimbar in Ireland; at the ſame time uſing a courſe of diet that ſhall be neither acrid nor heating.

So far with reſpect to thoſe kinds of morbific matter which do not invariably produce a particular ſpecies of diſeaſe: but there are others of a ſpecific nature, ſome of which are generated in the body ſpontaneouſly, and ſeem to ariſe from error in diet, or other circumſtances of ill management with reſpect to the animal economy: and hence it is ſometimes poſſible, in ſome degree if not altogether, to prevent the ill conſequences. Thus, there are inſtances where returns of the gout have been prevented by adhering ſtrictly to a milk diet.

The rheumatiſm has alſo been ſometimes warded off by wearing a flannel ſhirt, or by uſing the cold bath without interruption.

Calculus may be retarded in its progreſs, and prevented from creating much diſtreſs, by the internal uſe of ſoap and lime-water, by ſoap-lees taken in milk or in veal-broth, or by the uſe of aerated alkaline water, which may perhaps be conſidered as being both more ſafe and more efficacious, and at the ſame time more pleaſant, than any of the other practices.

The ſcurvy may be prevented by warm clothing and perſeverance in briſk exerciſe, by drinking wine or cider; and eating freely of ſuch vegetable ſubſtances as can be had in thoſe ſituations where this diſeaſe is moſt apt to ſhow itſelf.

In conſtitutions where there is an hereditary diſpoſition to the ſcrofula, if early precautions be taken to ſtrengthen the ſolids by cold bathing, a nouriſhing courſe of diet, and moderate uſe of wine, the acrimony which riſes to the diſeaſe will probably be prevented from producing any very bad effects.

The other kinds of morbific matter, which are of a ſpecific nature, are received into the body by infection or contagion.

The infection of a putrid fever or dyſentery is beſt prevented by immediately taking an emetic on the firſt attacks of the ſickneſs or ſhivering; and, if that do not completely anſwer, let a large bliſter be applied between the ſhoulders: by this

2 method

method the nurſes and other attendants on the ſick in the naval hoſpitals have often been preſerved. As to other infectious morbific matter, we muſt refer to what has already been ſaid when treating of hydrophobia, poiſons, &c.

The ill effects which may ariſe from the different ſpecies of ſaburra, are to be obviated, in general, by the prudent adminiſtration of emetics, and carefully abſtaining from ſuch kinds of food as are known to cauſe the accumulation of noxious matters in the firſt paſſages.

Crude vegetables, milk, butter, and other oily ſubſtances, are to be avoided by perſons troubled with a ſourneſs in the ſtomach; briſk exerciſe, eſpecially riding, is to be uſed, and they are to refrain from fermented liquors: the common drink ſhould be pure water; or water with a very little of ſome ardent ſpirit, ſuch as rum or brandy. Seltzer and Vahls water are to be drunk medicinally; and aromatic bitters, infuſions, or tinctures, with the acid elixir of vitriol, from 10 to 20 drops, will be found ſerviceable, in order to ſtrengthen the fibres of the ſtomach, and promote the expulſion of its contents, thereby preventing the too-haſty fermentation of the alimentary mixture. In order to procure immediate relief, magneſia alba, or creta præparata, will ſeldom fail; the magneſia, as well as the chalk, may be made into lozenges, with a little ſugar and mucilage; and in that form may be carried about and taken occaſionally by people afflicted with the acid ſaburra.

In conſtitutions where there is an exuberance of ſtagnation of bile, and a troubleſome bitterneſs in the mouth, it is neceſſary to keep the bowels always free by taking occaſionally ſmall doſes of pure aloes, oleum ricini, cream of tartar, ſome of the common purging ſalts, or the natural purging waters.

When there is a tendency to the empyreumatic and rancid ſaburra, people ſhould carefully avoid all the various kinds of thoſe oily and high-ſeaſoned things generally termed *made diſhes*, and eat plain meat, without rich ſauces or much gravy; and in theſe caſes the moſt proper drink is pure water.

II. Rules for thoſe who enjoy perfect Health.

There can be no doubt that, in general, temperance is the true foundation of health; and yet the ancient phyſicians, as we may ſee in the rules laid down by Celſus, did not ſcruple to recommend indulgence now and then, and allowed people to exceed both in eating and drinking: but it is ſafer to proceed to exceſs in drink than in meat; and, if the debauch ſhould create any extraordinary or diſtreſſing degree of pain or ſickneſs, and a temporary fever ſhould enſue, there are two ways of ſhaking it off, either to lie in bed and encourage perſpiration, or to get on horſeback, and by briſk exerciſe reſtore the body to its natural ſtate. The choice of theſe two methods muſt always be determined by the peculiar circumſtances of the parties concerned, and from the experience which they may before have had, which agrees beſt with them.

If a perſon ſhould commit exceſs in eating, eſpecially of high-ſeaſoned things, with rich ſauces, a draught of cold water, acidulated with vitriolic acid, will take off the ſenſe of weight at the ſtomach, and aſſiſt digeſtion, by moderating and keeping within bounds the alimentary fermentation, and thus preventing the generation of too much flatus. The luxury of ices may be here of real ſervice at the tables of the great, as producing ſimilar effects with the cold water acidulated. Perſons in theſe circumſtances ought not to lay themſelves down to ſleep, but ſhould keep up and exerciſe until they are ſenſible that the ſtomach is unloaded, and that they no longer feel any oppreſſive weight about the præcordia.

If a man be obliged to faſt, he ought, if poſſible, during that time, to avoid laborious work: after ſuffering ſevere hunger, people ought not at once to gorge and fill themſelves; nor is it proper, after being over-filled, to enjoin an abſolute faſt: neither is it ſafe to reſt totally immediately after exceſſive labour, nor ſuddenly fall hard to work after having been long without motion: in a word, *all changes ſhould be made by gentle degrees*; for, though the conſtitution of the human body be ſuch that it can bear many alterations and irregularities without much danger, yet, when the tranſitions are extremely ſudden, they cannot fail of producing ſome kind or degree of diſorder.

It is alſo the advice of Celſus to vary the ſcenes of life, and not confine ourſelves to any ſettled rules: but as inaction renders the body weak and liſtleſs, and exerciſe gives vigour and ſtrength, people ſhould never long omit riding, walking, or going abroad in a carriage; fencing, playing at tennis, dancing, or other ſimilar engagements, which afford both exerciſe and amuſement, as each ſhall be found moſt agreeable or convenient, are to be uſed in their turns, according to the circumſtances and tendency to any particular ſpecies of diſeaſe. But, when the weakneſs of old age ſhall have rendered the body incapable of all theſe, then dry frictions with the fleſh-bruſh will be extremely requiſite to preſerve health, by accelerating the flow of humours through the ſmalleſt orders of veſſels, and preventing the fluids from ſtagnating too long in the cellular interſtices of the fleſhy parts.

Sleep is the great reſtorer of ſtrength; for, during this time, the nutritious particles appear to be chiefly applied to repair the waſte, and replace thoſe that have been abraded and waſhed off by the labour and exerciſe of the day. But too much indulgence in ſleep has many inconveniences, both with reſpect to body and mind, as it blunts the ſenſes, and encourages the fluids to ſtagnate in the cellular membrane;

brane; whence corpulency, and its neceſſary conſequences, languor and weakneſs. The proper time for ſleep is the night ſeaſon, when darkneſs and ſilence naturally bring it on: therefore day-ſleep, in general, is not ſo refreſhing; and to ſome people is really diſtreſsful, as creating an unuſual giddineſs and languor, eſpecially in perſons addicted to literary purſuits. Cuſtom, however, frequently renders ſleep in the day neceſſary; and in thoſe conſtitutions where it is found to give real refreſhment it ought to be indulged.

With regard to the general regimen of diet, it has always been held as a rule, that the ſofter and milder kinds of aliment are moſt proper for children and younger ſubjects; that grown perſons ſhould eat what is more ſubſtantial; and old people leſſen their quantity of ſolid food, and increaſe that of their drink.

Of FIXED AIR as a MEDICINE.

THE antiſeptic qualities of fixed air, or as it is now more generally called of the aerial or carbonic acid gas, have introduced it as a medicine in caſes of putrid diſorders, and various other complaints.—Dr. Percival obſerves, that, though fatal if inſpired in a very large quantity, it may in ſmaller quantities be breathed without danger or uneaſineſs. And it is a confirmation of this concluſion, that at Bath, where the waters copiouſly exhale this mineral ſpirit, the bathers inſpire it with impunity. At Buxton alſo, where the bath is in a cloſe vault, the effects of ſuch effluvia, if noxious, muſt certainly be perceived.

Encouraged by theſe and ſome other conſiderations, he adminiſtered fixed air in more than 30 caſes of the phthiſis pulmonalis, by directing his patients to inſpire the ſteams of an effervescing mixture of chalk and vinegar through the ſpout of a coffee-pot. The hectic fever has in ſeveral inſtances been conſiderably abated, and the matter expectorated has become leſs offenſive and better digeſted. He was not, however, ſo fortunate in any one caſe as to effect a cure; although the uſe of this air was accompanied with proper internal medicines. But Dr. Withering was more ſuccesſful. One phthiſical patient under his care, by ſuch a courſe, entirely recovered; another was rendered much better; and a third, whoſe caſe was truly deplorable, ſeemed to be kept alive by it more than two months. It may be proper to obſerve, that fixed air can only be employed with any proſpect of ſucceſs in the latter ſtages of the phthiſis pulmonalis, when a purulent expectoration takes place. After the rupture and diſcharge of vomica alſo, ſuch a remedy promiſes to be a powerful palliative. Antiſeptic fumigations and vapours have been long employed, and much extolled, in caſes of this kind. The following experiment

ment was made to determine whether their efficacy in any degree depends on the separation of fixed air from their fubftance.

One end of the bent tube was fixed in a phial full of lime-water; the other end in a bottle of the tincture of myrrh. The junctures were carefully luted; and the phial containing the tincture of myrrh was placed in water heated almoft to the boiling point by the lamp of a tea-kettle. A number of air-bubbles were feparated, but probably not of the mephitic kind; for no precipitation enfued in the lime-water. This experiment was repeated with the *tinct. tolutana. Ph. Ed.* and with *fp. vinof. camph.* and the refult was entirely the fame. The medicinal action therefore of the vapours raifed from fuch tinctures cannot be afcribed to the extrication of fixed air, of which it is probable bodies are deprived by chemical folution as well as by mixture.

If mephitic air be thus capable of correcting purulent matter in the lungs, we may reafonably infer it will be equally ufeful when applied externally to foul ulcers; and experience confirms the conclufion. Even the fanies of a cancer, when the carrot-poultice failed, has been fweetened by it, the pain mitigated, and a better digeftion produced. But, though the progrefs of the cancer feems to be checked by the fixed air, it is to be feared a cure will not be effected. A palliative remedy, however, in a difeafe fo defperate and loathfome, may be confidered a very valuable acquifition. Perhaps nitrous air might be ftill more efficacious. This fpecies of factitious air is obtained from all the metals, except zinc, by means of the nitrous acid; as a fweetener and antifeptic, it far furpaffes fixed air.

In the ulcerous fore throat, much advantage has been experienced from the vapours of effervefcing mixtures drawn into the fauces. But this remedy fhould not fuperfede the ufe of other antifeptic applications.

In malignant fevers, wines abounding with fixed air may be adminiftered to check the feptic ferment, and fweeten the putrid colluvies in the primæ viæ. If the laxative quality of fuch liquors be thought an objection to the ufe of them, wines of a greater age may be given, impregnated with aerial acid. The patient's common drink might alfo be medicated in the fame way. A putrid diarrhœa frequently occurs in the latter ftage of fuch diforders, and it is a moft alarming and dangerous fymptom. If the difcharge be ftopped by aftringents, a putrid fomes is retained in the body, which aggravates the delirium, and increafes the fever. On the contrary, if it be fuffered to take its courfe, the ftrength of the patient muft foon be exhaufted, and death unavoidably enfue. The injection of mephitic air into the inteftines, under thefe circumftances, bids fair to be highly ferviceable. And in fome cafes of this kind, the gas emitted from a mixture of chalk and oil of vitriol, conveyed into the body by the machine employed for tobacco-clyfters, quickly reftrained the

 diarrhœa,

diarrhœa, corrected the heat and fetor of the ſtools, and in a ſhort time removed every ſymptom of danger.

As a ſolvent of the calculus, its virtues have been already mentioned; but the experiments made on that ſubject do not determine the matter with ſufficient accuracy.

Of MEDICAL ELECTRICITY.

THE application of this ſubtile fluid to medicinal purpoſes was thought of ſoon after the diſcovery of the electric ſhock; and, after various turns of reputation, its medical virtues ſeem now to be pretty well eſtabliſhed. Mr. Cavallo, who has publiſhed the lateſt and the beſt treatiſe on Medical Electricity, entirely diſapproves of giving violent ſhocks, and finds it moſt efficacious to expoſe the patient to the *electrical aura*, a gentle air, diſcharged from an iron or wooden point; or, if ſhocks are given, they ſhould be very ſlight, and not exceed twelve or fourteen at a time. In this way he recommends it as effectual in a great number of diſorders. The patient may be electrified from three to ten minutes: but, if ſparks are drawn, they ſhould not exceed the number of ſhocks above mentioned.

Rheumatic diſorders, even of long ſtanding, are relieved, and ſometimes quite cured, by only drawing the electric fluid with a wooden point from the part, or by drawing ſparks through flannel. The operation ſhould be continued for about four or five minutes, repeating it once or twice every day.

The gout, extraordinary as it may appear, has certainly been cured by means of electricity, in various inſtances. The pain has been generally mitigated, and ſometimes the diſeaſe has been removed ſo well as not to return again. In thoſe caſes, the electric fluid has been thrown by means of a wooden point; although ſometimes, when the pain was too great, a metal point only has been uſed.

Deafneſs, except when it is occaſioned by obliteration or other improper configuration of the parts, is either entirely or partly cured by drawing the ſparks from the ear with the glaſs-tube director, or by drawing the fluid with a wooden point. Sometimes it is not improper to ſend exceedingly-ſmall ſhocks (for inſtance, of one thirteenth of an inch) from one ear to the other.—It has been conſtantly obſerved, that, whenever the ear is electrified, the diſcharge of the wax is conſiderably promoted.

The tooth-ach, occaſioned by cold, rheumatiſm, or inflammation, is generally relieved by drawing the electric fluid with a point, immediately from the part, and alſo externally from the face. But, when the body of the tooth is affected, electrization is of no uſe; for it ſeldom or never relieves the diſorder, and ſometimes increaſes the pain to a prodigious degree.

Inflammations of every ſort are generally relieved by a very gentle electrization. In inflammations of the eyes, the throwing of the electric fluid by means of a wooden point is often attended with great benefit; the pain being quickly abated, and the inflammation being generally diſſipated in a few days. In theſe caſes, the eye of the patient muſt be kept open; and care ſhould be taken not to bring the wooden point very near it, for fear of cauſing any ſpark. Sometimes it is ſufficient to throw the fluid with a metal point; for in theſe caſes, too great irritation ſhould be always avoided. It is not neceſſary to continue this operation for three or four minutes without intermiſſion; but, after throwing the fluid for about half a minute, a ſhort time may be allowed to the patient to reſt and to wipe his tears, which generally flow very copiouſly: then the operation may be continued again for another half minute, and ſo on for four or five times every day. The gutta ſerena has been ſometimes cured by electrization; but at the ſame time it muſt be confeſſed, it has proved ineffectual in many ſuch caſes, in which it was adminiſtered for a long time and with all poſſible attention. However, it has never been known that any body was made worſe by it. The beſt method of adminiſtering electricity in ſuch caſes, is firſt to draw the electric fluid with a wooden point for a ſhort time, and then to ſend about half a dozen ſhocks of one twentieth of an inch, from the back and lower part of the head to the forehead, very little above the eye. A remarkable diſeaſe of the eye was ſome time ago perfectly cured by electrization; it was an opacity of the vitreous humour. All the caſes of fiſtula lacrymalis, which Mr. Cavallo had known to have been electrified by perſons of ability for a ſufficient time, had been entirely cured. The method generally practiſed has been that of drawing the fluid with a wooden point, and to take very ſmall ſparks from the part. The operation may be continued for about three or four minutes every day. It is remarkable, that in thoſe caſes, after curing the fiſtula lacrymalis, no other diſeaſe was occaſioned by it, as blindneſs, inflammation, &c. by ſuppreſſing that diſcharge.

Palſies are ſeldom perfectly cured by means of electricity, eſpecially when they are of long ſtanding; but they are generally relieved to a certain degree. The method of electrifying in thoſe caſes is to draw the fluid with the wooden point, and to bring ſparks through flannel, or through the uſual coverings of the part if they are not too thick. The operation may be continued for about five minutes per day.

Ulcers, or open ſores of every kind, even of a long ſtanding, are generally diſpoſed to heal by electrization. The general effects are a diminution of the inflammation, and at firſt a promotion of the diſcharge of properly-formed matter, which diſcharge gradually leſſens, according as the limits of the ſore contract, till it be quite cured. In theſe caſes the gentleſt electrization muſt be uſed, in order to avoid too great an irritation, which is generally hurtful. To draw or throw the fluid with a wooden a wooden or even with a metal point, for three or four minutes per-day, is fully ſufficient.

Cutaneous eruptions have been ſucceſsfully treated with electrization: but in theſe caſes it muſt be obſerved, that, if the wooden point be kept too near the ſkin, ſo as to cauſe any conſiderable irritation, the eruption will be cauſed to ſpread more; but, if the point be kept at about ſix inches diſtance, or farther if the electrical machine be very powerful, the eruptions will be gradually diminiſhed, till they are quite cured. In this kind of diſeaſe, the immediate and general effect of the wooden point is to occaſion a warmth about the electrified part, which is always a ſign that the electrization is rightly adminiſtered.

The application of electricity has perfectly cured various caſes of St. Vitus's dance, or of that diſeaſe which is commonly called ſo; for it is the opinion of ſome very learned phyſicians, that the real diſeaſe called St. Vitus's dance, which formerly was more frequent than it is at preſent, is different from that which now goes under that name. In this diſeaſe, ſhocks of about one-tenth of an inch may be ſent through the body in various directions, and alſo ſparks may be taken. But, if this treatment prove very diſagreeable to the patient, then the ſhocks muſt be leſſened, and even omitted; inſtead of which, ſome other more gentle applications muſt be ſubſtituted.

Scrophulous tumours, when they are juſt beginning, are generally cured by drawing the electric fluid with a wooden or metal point from the part. This is one of thoſe kinds of diſeaſes in which the action of electricity requires particularly the aid of other medicines in order to effect a cure more eaſily; the ſcrophulous affections commonly accompany a great laxity of the habit, and a general cachexy, which muſt be obviated by proper remedies.

In cancers, the pains only are commonly alleviated by drawing the electric fluid with a wooden or metal point. Mr. Cavallo, however, mentions one caſe in which a moſt confirmed cancer of very long ſtanding, on the breaſt of a woman, had been much reduced in ſize. It is remarkable, that this patient was ſo far relieved by drawing the fluid with a metal point from the part, that the excruciating pains ſhe had ſuffered for many years did almoſt entirely diſappear; but, when the electric fluid was drawn by means of a wooden point, the pains rather increaſed.

Abſceſſes, when they are in their beginning, and in general whenever there is any tendency to form matter, are diſperſed by electrization. Lately, in a caſe in which matter was formed upon the hip, called the lumbar abſceſs, the diſeaſe was perfectly cured by means of electricity. The ſciatica has alſo been often cured by it. In all ſuch caſes, the electric fluid muſt be ſent through the part by means of two directors applied to oppoſite parts, and in immediate contact either with the ſkin, or with the coverings when theſe are very thin. It is very remarkable, that the mere paſſage

paſſage of electric fluid in this manner is generally felt by the patients afflicted with thoſe diſorders nearly as much as a ſmall ſhock is felt by a perſon in good health. Sometimes a few ſhocks have been alſo given, but it ſeems more proper to omit them; becauſe it has been found, that, inſtead of diſperſing, they rather accelerate the formation of matter.

In caſes of pulmonary inflammations, when they are in the beginning, electrization has been ſometimes beneficial; but in confirmed diſeaſes of the lungs it does not ſeem to have ever afforded any unqueſtionable benefit; however it ſeems that in ſuch caſes the power of electricity has been but ſeldom tried.

Nervous head-achs, even of a long ſtanding, are generally cured by electrization. For this diſeaſe, the electric fluid muſt be thrown with a wooden and ſometimes even with a metal point, all round the head ſucceſſively. Sometimes exceedingly-ſmall ſhocks have been adminiſtered; but theſe can ſeldom be uſed, becauſe the nerves of perſons ſubject to this diſeaſe are ſo very irritable, that the ſhocks the ſparks, and ſometimes even the throwing the electric fluid with a wooden point kept very near the head, will bring on convulſions.

The application of electricity has often been found beneficial in the dropſy when juſt beginning, or rather in the tendency to a dropſy; but it has never been of any uſe in advanced dropſies. In ſuch caſes, the electric fluid is ſent through the part, in various directions, by means of two directors; and ſparks are alſo drawn acroſs the flannel or the clothes; keeping the metal rod in contact with them, and ſhifting it continually from place to place. This operation ſhould be continued at leaſt ten minutes, and ſhould be repeated once or twice a day.—Perhaps in thoſe caſes, a ſimple electrization, (viz. to inſulate the patient, and to connect him with the prime conductor whilſt the machine is in action,) continued for a conſiderable time, as an hour or two, would be more beneficial.

Swellings in general, which do not contain any matter, are frequently cured by drawing the electric fluid with a wooden point. The operation ſhould be continued for three or four minutes every day.—It is very remarkable, that, in ſome caſes of white ſwellings quite cured by means of electricity, the bones and cartilages became disfigured.

Agues have not unfrequently been cured by electricity, ſo that ſometimes one electrization or two have been ſufficient. The moſt effectual and ſure method has been that of drawing ſparks through flannel, or the clothes, for about ten minutes or a quarter of an hour. The patients may be electrified either at the time of the fit, or a ſhort time before it is expected.

The ſuppreſſion of the menſes, which is a diſeaſe of the female ſex that often occaſions the moſt diſagreeable and alarming ſymptoms, is often ſucceſsfully and ſpeedily cured by means of electricity, even when the diſeaſe is of long ſtanding, and after the moſt powerful medicines have proved ineffectual. The caſes of this ſort in which electrization has proved uſeleſs are ſo few, and the ſucceſsful ones ſo numerous, that the application of electricity for this diſeaſe may be juſtly conſidered as an efficacious and certain remedy. Great attention and knowledge is required, in order to diſtinguiſh the arreſt of the menſes from a ſtate of pregnancy. In the former, the application of electricity, as we obſerved above, is very beneficial; whereas, in the latter, it may be attended with very diſagreeable effects; it is therefore a matter of great importance to aſcertain the real cauſe of the diſeaſe, before the electricity be applied in thoſe caſes. Pregnant women may be electrified for other diſeaſes, but always uſing very gentle means, and directing the electric fluid through other parts of the body diſtant from thoſe ſubſervient to generation. In the real ſuppreſſion of the menſes, ſmall ſhocks, i. e. about one twentieth of an inch, may be ſent through the pelvis; ſparks may be taken through the clothes from the parts adjacent to the ſeat of the diſeaſe; and alſo the electric fluid may be tranſmitted by applying the metallic or wooden extremities of two directors to the hips, in contact with the clothes; part of which may be removed in caſe they are too thick. Theſe various applications of electricity ſhould be regulated according to the conſtitution of the patient. The number of ſhocks may be about twelve or fourteen. The other applications may be continued for two or three minutes; repeating the operation every day. But either ſtrong ſhocks, or a ſtronger application of electricity than the patient can conveniently bear, ſhould be avoided; for by thoſe means ſometimes more than a ſufficient diſcharge is occaſioned, which is not eaſily cured. In caſes of uterine hæmorrhage, it is not known that the application of electricity was ever bneneficial. Perhaps a very gentle electrization, ſo as to keep the patient inſulated and connected with the prime conductor whilſt the electrical machine is in action, might be of ſome benefit.

In reſpect to unnatural diſcharges and fluxes in general, it may be obſerved, that ſome diſcharges are quite unnatural or adventitious, as the fiſtula lacrymalis, and ſome ſpecies of the venereal diſeaſe; but others are only increaſed natural diſcharges, ſuch as the menſes, perſpiration, &c. Now the power of electricity in general has been found more beneficial for the firſt than for the ſecond ſort of diſcharges, which are moſtly increaſed by it.

In the venereal diſeaſe, electrization has been generally forbidden; having commonly increaſed the pains, and other ſymptoms, rather than diminiſhed them. Indeed, conſidering that any ſort of ſtimulus has been found hurtful to perſons afflicted with that diſorder, it is no wonder that electricity has produced ſome bad effects, eſpecially in the manner it was adminiſtered ſome time ago, viz. by giving ſtrong ſhocks. However, it has been lately obſerved, that a very gentle application of electricity, as drawing the fluid by means of a wooden or metal point, is peculiarly beneficial in various caſes of this kind, even when the diſeaſe has been of long ſtanding. Having remarked above, that tumours, when juſt beginning, are diſperſed, and that unnatural diſcharges are gradually ſuppreſſed, by a judicious electrization, it is ſuperfluous to deſcribe particularly thoſe ſtates of the venereal diſeaſe in which electricity may be applied; it is only neceſſary to remind the operator to av oidany conſiderable ſtimulus in caſes of this ſort.

The application of electricity has been found alſo beneficial in other diſeaſes beſides thoſe mentioned above; but, as the facts are not ſufficiently numerous to afford the deduction of any general rules, we have not thought proper to take any particular notice of them.

We may laſtly obſerve, that, in many caſes, the help of other remedies to be preſcribed by the medical practitioner will be required to aſſiſt the action of electricity, which by itſelf would perhaps be uſeleſs; and, on the other hand, electrization may be often applied to aſſiſt the action of other remedies, as of ſudorifics, ſtrengthening medicines, &c.

Mr. Lichtenberg with a large electrophorus made ſome very curious experiments; in which, the knob of an electrified phial being drawn over the ſurface of the electric plate, finely-powdered roſin, afterwards ſifted upon the place, aſſumed the figure of ſtars and other beautiful ramifications, indicating not only an inclination to arrange itſelf in the ſame regular order with the cryſtals of ſalts, but to run out into branches like thoſe of vegetables. Theſe experiments have been repeated to great advantage by the Rev. Mr. Bennet, according to whoſe method the figures repreſented in the annexed Plate were made. The apparatus uſed for making them conſiſted only of a common Leyden phial, and a plate of glaſs 15 inches ſquare covered on one ſide with a varniſh of gum-lac diſſolved in ſpirit of wine, and ſeveral times laid over. Two ounces of ſhell-lac powdered and mixed with ſix ounces of ſpirit of wine anſwers very well for this purpoſe. The glaſs muſt be warm, and the varniſh ſpread upon it with a camel's-hair pencil. Care muſt be taken, however, not to lay it on too thick, otherwiſe the effect will not follow.—The other ſide is covered with tin-foil laid on with common paſte. When it is to be uſed, the glaſs plate is put upon a metallic ſtand with the tin-foil ſide laid undermoſt; the phial is to be charged, and the knob drawn over the varniſhed ſide. Thus any kind of figure may be drawn, or letters made, as repreſented in the plate; and from every figure beautiful ramifications will proceed, longer or ſhorter according to the ſtrength of the charge. On ſome occaſions, however, the charge may be too ſtrong, particularly where we wiſh to

9

Dodd delin. Pass Sculp.

Electrical Stars.

to reprefent letters, fo that the whole will be blended into one confufed mafs. The round figures are formed by placing metallic rings or plates upon the electrical plate, and then giving them a fpark from the electrified bottle, or fending a fhock through them. The figures may be rendered permanent by blowing off the loofe chalk, and clapping a piece of black fized paper upon them; or, if they are wanted of another colour, that may eafily be obtained by means of lake, vermilion, rofe-pink, or any of the ordinary colours ground very fine.

Electricity feems alfo to be the caufe of cryftallization; which probably is only an incipient or imperfect vegetation. Different falts affume different figures in cryftallization, and are thus moft eafily diftinguifhed from one another. Each falt is capable of affuming a very different appearance of the cryftalline kind, when only a fingle drop of the faline folution is made ufe of, and the cryftallization viewed through a microfcope. For our knowledge of this fpecies of cryftallization we are indebted to Mr. Henry Baker, who was prefented with a gold medal for the difcovery, in the year 1744. Thefe microfcopical cryftals he diftinguifhes from the large ones by the name of *configurations;* but this term feems inaccurate, and the diftinction may well enough be preferved by calling the large ones the common, and the fmall ones the microfcopical, cryftals of the falt. His method of making thefe obfervations he gives in the following words:

"I diffolve the fubject to be examined in no larger a quantity of rain or river water than I am certain it is fufficient to faturate. If it is a body eafily diffolvable, I make ufe of cold water; otherwife I make the water warm, hot, or even boiling, according as I find it neceffary. After it is perfectly diffolved, I let it reft for fome hours, till, if overcharged, the redundant faline particles may be precipitated and fettle to the bottom, or fhoot into cryftals; by which means I am moft likely to have a folution of the fame ftrength at one time as at another; that is, a folution fully charged with as much as it can hold up, and no more; and by thefe precautions the configurations appear alike, how often foever tried: whereas, if the water be lefs faturated, the proportions at different times will be fubject to more uncertainty; and, if examined before fuch feparation and precipitation of the redundant falts, little more will be feen than a confufed mafs of cryftals. The folution being thus prepared, I take up a drop of it with a goofe-quill cut in fafhion of a fcoop, and place it on a flat flip of glafs of about three quarters of an inch in width, and between three and four inches long, fpreading it on the glafs with the quill, in either a round or an oval figure, till it appears a quarter of an inch, or more, in diameter, and fo fhallow as to rife very little above the furface of the glafs. When it is fo difpofed, I hold it as level as I can over the clear part of a fire that is not too fierce, or over the flame of a candle, at a diftance proportionable to the heat it requires (which experience only

only can direct), and watch it very carefully till I difcover the faline particles beginning to gather and look white, or of fome other colour, at the extremities of the edges. Then (having adjusted the microfcope before-hand for its reception, armed with the fourth glafs, which is the fittest for moft of thofe experiments) I place it under my eye, and bring it exactly to the focus of the magnifier; and, after running over the whole drop, I fix my attention on that fide where I obferve any increafe or pufhing-forwards of cryftalline matter from the circumference towards the centre. This motion is extremely flow at the beginning, unlefs the drop has been overheated, but quickens as the water evaporates; and in many kinds, towards the conclufion, produces configurations with a fwiftnefs inconceivable, compofed of an infinity of parts, which are adjusted to each other with an elegance, regularity, and order, beyond what the exacteft pencil in the world, guided by the ruler and compaffes, can ever equal, or moft luxuriant imagination fancy. When this action once begins, the eye cannot be taken off, even for a moment, without lofing fomething worth obfervation: for the figures alter every inftant till the whole procefs is over; and, in many forts, after all feems at an end, new forms arife, differing entirely from any that appeared before, and which probably are owing to fome fmall quantity of falt of another kind, which the other feparates from, and leaves to act after itfelf has done: and, in fome fubjects, three or four different forts are obfervable, few or none of them being fimple and homogeneous. When the configurations are fully formed, and all the water evaporated, moft kinds of them are foon deftroyed again by the moifture or action of the air upon them; their points and angles lofe their fharpnefs, become uneven and defaced, and moulder, as it were, away. But fome few are permanent, and, being inclofed between glaffes, may be preferved many months, or even years, entertaining objects for the microfcope. It happens oftentimes that a drop of faline folution can hardly be fpread on the flip of glafs, by reafon of the glafs's fmoothnefs, but breaks into little globules, as it would do if the furface were greafy: this was very troublefome, till I found a way of preventing it, by rubbing the broken drop with my finger over the glafs, fo as to leave the furface fmeared with it; on which fmeared place, when dry, another drop of the folution may be fpread very eafily in what form one pleafes. It likewife fometimes happens, that, when a heated drop is placed properly enough for examination, the obferver finds he can diftinguifh nothing: which is owing to faline fteams that, rifing from the drop, cover and obfcure the object-glafs, and therefore muft immediately be wiped away with a foft cloth or leather. In all examinations by the microfcope of faline folutions, even though made in the day-time, I always employ the light of a candle, and advife every obferver to do fo likewife; for the configurations, being exceedingly tranfparent, are rendered much more diftinguifha-

Dodd del. Pass Sculp.

Crystals formed from Salts.

ble by the brown light a candle affords than by the more white and tranſparent day-light; and beſides, either by moving the candle or turning microſcope, ſuch light may be varied or directed juſt as the object requires."

In this manner were produced the beautiful cryſtallizations repreſented in the annexed Plate. They are vaſtly different from ſuch cryſtals of the ſame ſalts as are obtained by the common proceſſes; but Mr. Baker aſſures us they are no leſs conſtant and invariable than they, and that he has repeated the experiments a great number of times with the ſame ſucceſs.

Fig. 1. ſhows the microſcopical cryſtals of nitre or ſaltpetre. Theſe ſhoot from the edges, with very little heat, into flattiſh figures of various lengths, exceedingly tranſparent, and with ſtraight and parallel ſides. They are ſhown in their different degrees of progreſſion at the letters *a, b, c, d, e;* where *a* repreſents how they firſt begin. After numbers of theſe are formed, they will often diſſolve under the eye, and diſappear entirely; but, if you wait a little, new ſhoots will puſh out, and the proceſs go on afreſh. Theſe firſt figures ſometimes enlarge only with altering their ſhapes, and ſometimes form in ſuch ſort as the drop repreſents; but, if the heat has been too great, they ſhoot haſtily into ramifications very numerous and beautiful, but very difficult to be drawn; and which Mr. Baker therefore did not attempt. There ſeems all the while a violent agitation in the fluid; and moſt commonly, towards the concluſion, a few octahedra (compoſed of eight triangular planes, or two quadrangular pyramids, joined baſe to baſe) make their appearance.

2. Blue vitriol, the ſulphat of copper of the new chemiſtry, produces cryſtals round the edges, very ſhort at the beginning, but increaſing gradually, as repreſented at the figures 1, 2, 3, which denote their difference of form, and the progreſs of their growth. Theſe cryſtalline ſhoots are ſolid, regular, tranſparent, and reflect the light very beautifully from their poliſhed ſides and angles. As the watery part evaporates, numbers of long ſlender bodies like hairs are ſeen here and there, ſome lying ſide by ſide, or croſſing each other as at 4; others forming ſtar-like figures with many radiations (5, 5). This ſalt ſhoots but ſlowly, and therefore requires patience. At laſt, the true cryſtals begin to appear commonly in the middle of the drop, and very prettily branched.

3. Diſtilled verdigreaſe, or green oxyd of copper, diſſolved as above directed, and immediately applied to the microſcope, ſhows abundance of the regular figures, 1, 2, 3, 4, 5, 6, 7. If the ſolution is ſuffered to ſtand for a few hours, and a drop of it is then heated over the fire on a ſlip of a glaſs, till it begins to concrete about the ſides, and then examined, ſharp pointed ſolid figures, biſected by a line cut through the middle, from which they are cut away towards the edges, begin to appear, and ſhooting forwards (1, 1, 1). Theſe figures are often ſtriated very

prettily from the middle line to the edges obliquely (2, 2); and frequently they arise in clusters, and shooting from a centre (3, 3). These figures are a long time in growing; and, whilst they are doing so, regular crystals appear forming in several parts of the drop, of the most lively emerald-colour, and reflecting the light from their sides and angles, which are most exactly disposed, and finely polished. No crystals are formed in the middle till the water is nearly evaporated; and then they begin to form hastily, for which reason they must be carefully attended. Their common figure resembles two long *ſ ſ* crossing each other in an angle of about 60°, and shooting branches every way; each of which again protrudes other branches from one, and sometimes from both, its sides; making together an appearance like four leaves of fern conjoined by their stalks (5, 5). Separate clusters of the same sharp-pointed figures as those at the edges of the drop are also formed in the middle of it. Sometimes also they put on another form, like the leaves of dandelion (7). Very beautiful figures are likewise produced by a kind of combination of sharp-points and branches (8, 8). All the crystals are of a most beautiful green colour, but deeper or lighter according to the time of their production. The deepest are constantly produced first, and the paler ones afterwards. Towards the end of the process some circular figures are formed, extremely thin, and so slightly tinged, with green lines radiating from a centre, as to be almost colourless (9). When all seems in a manner over, bundles of hair-like bodies appear frequently scattered here and there throughout the drop, like those of blue vitriol already described.

4. Alum. The microscopical crystals of this salt prove more or less perfect according to the strength of the solution and the degree of heat employed in making the experiment. The solution of alum, however saturated with the salt, will not be found over-strong after standing some days: for in that time many crystals will have formed in it. This separation will often leave the remainder too weak for the purpose; but, by holding the vial over or near the fire, the crystals will again dissolve. After it has stood about half an hour, it may then be used. The drop put on the glass, and properly heated, exhibits commonly at first a dark cloud which appears in motion somewhere near the edge, and runs pretty swiftly both to the right and left until it is either stopped by the intervention of regular crystals, or else it proceeds both ways at once, till, having surrounded the whole drop, the two ends rush together, and join into one (*a, a*). This cloudy part, which seems to be violently agitated while it is running round, appears on a strict examination to consist of salts, shot into long and very slender lines, much finer than the smallest hair, crossing each other at right angles. As they go along, rows of solid crystals are produced from their internal edges. These are composed of many oblique plain sides (*b, b*), and which have all a tendency towards the figures of the regular crystals to be described

ſcribed preſently. But it frequently happens, that, in ſome parts of the drop, many minute and circular figures are ſeen riſing at ſome little diſtances from the edge, which, enlarging themſelves continually, appear at laſt of a ſtar-like form. The cryſtals in the middle ſeldom appear till the fluid ſeems almoſt wholly evaporated; when, on a ſudden, many ſtraight lines appear puſhing forwards, whoſe ſides or edges are jagged, and from which other ſimilar ſtraight and jagged lines ſhoot out at right angles with the firſt. Theſe again have other ſmall ones of the ſame kind ſhooting out likewiſe from themſelves; and compoſe all together a moſt beautiful and elegant configuration (D). Each of theſe lines, increaſing in breadth towards its end, appears as if it were ſomewhat club-headed (*e*, *e*, *e*). Sometimes, inſtead of ſending branches from their ſides, many of theſe lines riſe parallel to each other, reſembling a kind of paliſado, and having numberleſs minute tranſverſe lines running between them (F). But the moſt wonderful part of all, though not producible without an exact degree of heat and right management, is the dark ground-work (G). It conſiſts of an infinity of parallel lines having others croſſing them at right angles, and producing a variety ſcarcely conceivable from lines diſpoſed in no other manner; the direction of the line, (which are exquiſitely ſtraight and delicate) being ſo frequently and differently changed, that one would think it the reſult of long ſtudy and contrivance. During the time this ground-work is framing, certain lucid points preſent themſelves to view, moſt commonly on one ſide. Theſe grow continually larger, with radiations from a centre, and become ſtar-like figures as before mentioned. Some of them ſend out long tails, which give them the appearance of comets: and, at the end of all, a dark lineation in various directions darts frequently through, and occupies all or moſt of the ſpaces between them, making thereby no ill repreſentation, when viewed by candle-light, of a dark ſky illuminated with ſtars and comets. The regular cryſtals are often formed in the ſame drop with the others.

5. Borax. If a drop of ſolution of borax is held too long over the fire, it hardens on the ſlip of glaſs in ſuch a manner that no cryſtals can appear. The beſt method is to give it a briſk heat for about a ſecond, and then, applying it to the microſcope, the cryſtals will quickly form themſelves as repreſented in the figure.

6. Sal ammoniac, or muriat of ammoniac, begins with ſhooting from the edges great numbers of ſharp, but at the ſame time thick and broad, ſpiculæ; from whoſe ſides are protruded, as they riſe, many others of the ſame ſhape, but very ſhort, parallel to each other, but perpendicular to their main ſtem (1). Theſe ſpiculæ arrange themſelves in all directions, but for the moſt part obliquely to the plane from whence they riſe, and many are frequently ſeen parallel to one another (1, 1). As they continue to puſh forwards, which they do without increaſing much in breadth, ſome ſhoot

from them the ſmall ſpiculæ only (2); others divide in a ſingular manner by the ſplitting of the ſtem; others branch into ſmaller ramifications (4). Before the middle of the drop begins to ſhoot, ſeveral exceedingly-minute bodies may be diſcernable at the bottom of the fluid. Theſe in a little while riſe to the top, and ſoon diſtinguiſh their ſhape as at (5). Their growth is very quick, and for ſome time pretty equal; but at laſt ſome branch gets the better of the reſt, and forms the figure (6). The other branches enlarge but little after this, all the attraction ſeeming to be lodged in that one that firſt began to lengthen; and from this more branches being protruded, and they again protruding others, the whole appears as at (8). It is not uncommon to ſee in the middle of the drop ſome cryſtals, where, inſtead of the ſtraight ſtems above deſcribed, there is formed a kind of zig-zag, with ſpiculæ like thoſe in the other figures (7).

7. Salt of lead, or ſaccharum ſaturni. A little of this ſalt diſſolved in hot water, which it immediately renders milky, after ſtanding a quarter of an hour to ſubſide, is in a fit condition for an examination by the microſcope. A drop of it then applied on a ſlip of glaſs, and held over the fire to put the particles in action, will be ſeen forming round the edge a pretty even and regular border of a clear and tranſparent film or gluy ſubſtance (*aaaa*); which, if too ſudden and violent a heat be given, runs over the whole area of the drop, and hardens ſo on the glaſs as not to be got off without great difficulty. But, if a moderate warmth be made uſe of, which likewiſe muſt not be too long continued, this border proceeds only a little way into the drop, with a kind of radiated figure compoſed of fine lines, or rather bundles of lines, beginning from the centres to the interior edge of the border, and ſpreading out at nearly equal diſtances from each other every way, towards the exterior (*bbbb*). From the ſame centres are produced afterwards a radiation inwards, compoſed of parallelograms of different lengths and breadths; from one and ſometimes both the angles of theſe, are frequently ſeen ſhootings ſo exceedingly ſlender, that they are perhaps the beſt poſſible repreſentations of a mathematical line. The extremities of the parallelograms are generally caſt off at right angles, but they are ſometimes alſo ſeen oblique (*cccc*). Centres with the like radii iſſuing from them, and ſome of the glutinous matter for their root, are ſometimes formed in the drop, entirely detached from the edges; and in theſe it is very frequent to find a kind of ſecondary radii proceeding from ſome of the primary ones; and others from them to a great number of gradations, forming thereby a very pretty figure (D).

8. Salt of tin produces at the edges of the drop a number of octahedra, partly tranſparent, ſtanding on long necks, at ſmall diſtances from each other, with angular ſhoots between them (*aa*). At the ſame time ſolid and regular opaque cubes will be ſeen forming themſelves in other parts of the drop (*bb*). In the middle of the ſame

ſame drop, and in ſeveral other parts of it, very different figures will alſo be formed; particularly great numbers of flat, thin, tranſparent, hexangular, bodies (*ccc*): ſome among which are thicker (*e*); and a few appear more ſolid, and with ſix ſloping ſides ariſing to a point, as if cut and poliſhed (*d*). The figure (*f*) is compoſed of two high pyramids united at their baſe. Some in this kind of form are found truncated at one of their ends, and others at both. Several of the hexagonal bodies may be obſerved with ſloping ſides, forming a ſmooth triangular riſing plane, whoſe angles point to three intermediate ſides of the hexagon (*g*).

9. Epſom ſalt (ſulphat of magneſia) begins to ſhoot from the edge in jagged figures (*a*). From other parts differently-figured cryſtals extend themſelves towards the middle, ſome of which have fine lines proceeding from both ſides of a main ſtem, in an oblique direction; thoſe on one ſide ſhooting upwards in an angle of about 60°, and thoſe on the other downwards in the ſame obliquity (*f*). Others produce jags from either ſide nearly perpendicular to the main ſtem, thereby forming figures that reſemble ſome ſpecies of the polypody (*c*); but in others the jags are ſhorter (*d*). Now and then one of the main ſtems continues ſhooting to a conſiderable length, without any branchings from the ſides: but at laſt ſends out two branches from its extremity (*g*). Sometimes a figure is produced having many fine and minute lines radiating from a centre (*h*). The laſt ſhootings in the middle of the drop (*h*) are not unlike the frame-work for the flooring or roofing of a houſe, but with the angles oblique: and ſometimes a form of another kind preſents itſelf (*i*).

10. Scarborough ſalt begins to ſhoot from the edges: firſt of all in portions of quadrilateral figures, much reſembling thoſe of common ſalt; but two of their angles, inſtead of 90, are about 100°. They ſhoot in great numbers round the borders of the drop, having their ſides as nearly parallel to one another as the figure of the drop will allow: ſome proceed but a little way, others farther, before they renew the ſhoot (*aa*). In ſome places they appear more pointed and longer (*b*); and ſometimes, inſtead of the diagonal, one of the ſides is ſeen towards the edge, and the other ſhooting into the middle (*c*). The middle cryſtals (*d e f*) ſeem to be of the vitriolic kind.

11. Glauber's ſalt, or ſulphat of ſoda, produces ramifications from the ſide of the drop, like the growth of minute plants, but extremely tranſparent and elegant (*c*). Some of them, however, begin to ſhoot from a centre at ſome diſtance from the edge, and protrude branches from that centre in a contrary direction (*b*). Sometimes they ſhoot from one and ſometimes from more ſides of the central point in different varieties (*d*). Other figures are produced from different parts of the edge of the drop (*a*, *f*, *e*); but the moſt remarkable and beautiful cryſtallization forms laſt of all near the middle of the drop. It is compoſed of a number of lines proceeding from

one another at right angles, with tranſparent ſpaces and diviſions running between them, appearing altogether like ſtreets, alleys, and ſquares *(g g)*. When this cryſtallization begins, it forms with great rapidity, affording the obſerver a very agreeable entertainment: but its beauty is of very ſhort duration; in a few moments it diſſolves and vaniſhes like melted ice, which renders drawing it very difficult.

12. Salt of Jeſuits bark. The few ſhootings which this ſalt produces at the edge of the drop are of no regular figure *(a)*. The whole area becomes quickly filled with great numbers of rhombi, of different ſizes, extremely thin and tranſparent *(b)*. Some of theſe enlarge greatly, and acquire a conſiderable thickneſs, forming themſelves into ſolids of many ſides *(c c)*. Near the concluſion ſome cryſtals of ſea-ſalt are formed *(d d)*, and likewiſe a few odd triangular figures *(e)*.

13. Salt of liquorice begins ſhooting from the edge with a ſort of rhombic ſpiculæ *(a)*. Some four-branched figures like thoſe of vitriol commonly appear, but moulder away before their ramifications are completed, leaving only their ſtamina behind *(b b)*. The middle of the drop is uſually overſpread with great numbers of parallelograms, ſome exceedingly tranſparent, being mere planes; having ſometimes one, ſometimes more, of the angles canted in ſuch a manner as to produce pentagonal, hexagonal, and other, figures. Others have much thickneſs, and form parallelopipeds or priſms *(c)*. Some of the plane figures now and then protrude an irregular kind of ſhooting which appears very pretty *(d)*.

14. Salt of wormwood, or carbonat of potaſh. The firſt ſhootings of this ſalt from the edges of the drop appear of a conſiderable thickneſs in proportion to their length: their ſides are deeply and ſharply jagged or indented, being made up of many ſomewhat-obtuſe angles, and their ends point with angles of the ſame kind *(a)*. But other ſhoots frequently branch out from theſe original ones, and they again ſend forth others, making all together a very pretty appearance *(bb)*. The cryſtals of this ſalt are very different from each other, conſiſting of ſquares, rhombi, parallelograms, &c. *(c)*.

15. Salt of tobacco. If a moderate degree of heat be given to a ſolution of this ſalt, its firſt ſhootings will be from the edges of the drop, in ſlender tapering figures, ending with very ſharp points, but at conſiderable diſtances from one another. Along with theſe are formed other cryſtals, nearly of the ſame kind, but entirely detached, and farther within the drop, having the thicker ends towards the centre of the drop, and the ſharp points turned towards its edge *(a)*. When a little more heat has been given, other ſpiculæ are produced from the edge, whoſe ends ſpread on either ſide, and then terminate in a point, and which have all along their ſides triangular pointed cryſtals, placed alternately, ſo as to repreſent a zig-zag with a line drawn through its middle *(b)*. The regular cryſtals are produced in the middle of the drop, and are either hexagons or rhombi *(c)*. When the moiſture is nearly exha-

led, there are ſometimes ſeen to ſhoot from, or rather under, the ſpiculæ, upon the plane of the glaſs, a repreſentation of leaves, very ſmall at their firſt appearance, but gradually increaſing (*d*). A violent agitation may be diſcovered in the fluid by the firſt magnifier during the whole proceſs, but eſpecially at the beginning, and extremely minute cryſtals ariſing from the bottom.

16. Salt of hartſhorn. On the application of a very ſmall degree of heat, ſalt of hartſhorn ſhoots near the edges of the drop into ſolid figures ſomewhat reſembling razors or lancets, where the blade turns into the handle by a claſp (*d*). The cryſtals of this ſalt are produced with great velocity, and are ſomewhat opaque, ſhooting from the edges of the drop, on both ſides of a main ſtem, and with a kind of regularity, rugged branches like thoſe of ſome ſorts of coral (*a a*). But ſometimes, inſtead of theſe branches, ſharp ſpiculæ, ſome plain, and others jagged, are protruded to a conſiderable depth on one ſide only (*b*). As the fluid exhales, ſome one of the branching figures generally extends to a great length, producing on one ſide ſhoots that are rugged and irregular, and on the other curious regular branches reſembling thoſe of ſome plant (*c*).

17. Salt of urine ſhoots from the edges of the drop in long parallelograms like nitre (*aa*). But in other places, along the ſides of the drop ſolid angles are formed, that ſeem to be the rudiments of common ſalt (*b*). Some of the parallelograms increaſe much in ſize, and ſpread themſelves in the middle, ſo as to change their firſt figure, and become three or four times bigger than the reſt; and theſe have a dividing line that runs through their whole length from end to end, whence iſſue other ſhort lines at ſmall diſtances, oppoſite to one another, all pointing with the ſame degree of obliquity towards the baſe (*cc*). Among theſe enlarged figures ſome few ſhoot ſtill forward and tapering towards a point, but, before they form one, ſwell again, and begin as it were anew; and thus they proceed ſeveral times before their figure is quite finiſhed (*aa*). The figures 1, 2, 3, 4, 5, 6, are the regular cryſtals of this ſalt when it is allowed to diſſolve in the air, and no heat at all is given.

18. Rheum, or the clear liquor which diſtils from the noſtrils when people catch cold, is ſtrongly ſaturated with ſalt. A drop of it on a ſlip of glaſs will ſoon cryſtallize in a beautiful manner, either with or without heat; but if heated to about the warmth of the blood, and then viewed through the microſcope, many lucid points will be ſeen riſing and increaſing gradually, till their form is ſhown to be quadrangular, with two tranſparent diagonals croſſing each other (*d d*). Theſe diagonals ſhoot ſoon after far beyond the ſquare, protruding other lines at right angles from their ſides; and thus they go on to form the moſt elegant and beautiful cryſtals (*bb*, *cc*). When a drop of rheum is ſet to cryſtallize without any heat, inſtead of branched cryſtals over the whole area, ſuch are formed only in the middle; but,

about

about the edges, plant-like figures are produced, ſhooting ſeveral ſtems from one point, and reſembling a kind of moſs (E).

19. Camphor, though inſoluble in water, diſſolves very readily in ſpirit of wine. A drop of this ſolution ſpread upon a ſlip of glaſs cryſtallizes inſtantly in the beautiful manner repreſented in the figure.

20. Manna eaſily diſſolves in water, and a drop of the ſolution is a very pretty object. Its firſt ſhootings are radiations from points at the very edge of the drop: the radiating lines ſeem opaque, but are very ſlender (*a a a*). Amongſt theſe ariſe many minute tranſparent columns, whoſe ends grow wider gradually as they extend in length, and terminate at laſt with ſome degree of obliquity (*b*). Some few figures, radiating from a centre every way, and circumſcribed by an outline, are produced within the drop (*d d*). But the moſt ſurpriſing and elegant configuration is compoſed of many cluſters of radiations ſhooting one from another over great part of the drop, and making all together a figure not unlike a beautiful ſea-plant.

Of ANIMAL MAGNETISM

ANIMAL MAGNETISM is a ſympathy which exiſts between the magnet and the inſenſible perſpiration of the human body, whereby an æther, or univerſal effluvia, is made to paſs and repaſs through the pores of the cuticle, in the ſame manner as the electrical fluid paſſes through bodies; and by which many cures are performed.

The ſyſtem originated, in 1774, from a German philoſopher named Father Hehl, who greatly recommended the uſe of the magnet in medicine. M. Meſmer, a phyſician of the ſame country, by adopting the principles of Hehl, became the direct founder of the ſyſtem. He had already diſtinguiſhed himſelf by a Diſſertation on the Influence of the Stars upon the human Body, which he publicly defended in a theſis before the univerſity of Vienna. He afterwards made a tour through Germany, publiſhing every-where the great cures he performed by means of animal magnetiſm; and arrived at Paris in the beginning of the year 1778. Here he was firſt patroniſed by the author of the Dictionnaire des Merveilles de la Nature; in which work a great number of his cures were publiſhed, Meſmer himſelf receiving likewiſe an ample teſtimony of his candour and ſolid reaſoning. Our phyſician ſoon collected ſome patients; and in the month of April 1778 retired to Creteil, from whence he in a ſhort time returned with them perfectly cured. His ſucceſs was now great; and patients increaſed ſo rapidly, that the doctor was ſoon obliged to take pupils to aſſiſt him in his operatons. Theſe pupils ſucceeded equally well as Meſmer himſelf; and ſo great was their emolument, that one of them, named Deſlon, realized upwards of 100,000l. ſterling. In 1779 Meſmer publiſhed a memoir

moir on the ſubject of Animal Magnetiſm, promiſing aferwards a complete work upon the ſame, which ſhould make as great a revolution in philoſophy as it had already done in medicine.

The new ſyſtem gained ground daily; and ſoon became ſo faſhionable, that the jealouſy of the faculty was thoroughly awakened, and an application concerning it was made to the French government. In conſequence of this, a committee was appointed to inquire into the matter, conſiſting partly of phyſicians and partly of members of the Royal Academy of Sciences at Paris, with Dr. Franklin at their head. Meſmer himſelf refuſed to have any communication with the committee: but his moſt celebrated pupil Deſlon was leſs ſcrupulous, and explained the principles of his art in the following manner.

1. Animal magnetiſm is an univerſal fluid, conſtituting an abſolute plenum in nature, and the medium of all mutual influence between the celeſtial bodies, and between the earth and animal bodies. 2. It is the moſt ſubtile fluid in nature; capable of a flux and reflux, and of receiving, propagating, and continuing, all kinds of motion. 3. The animal body is ſubjected to the influences of this fluid by means of the nerves, which are immediately affected by it. 4. The human body has poles and other properties analogous to the magnet. 5. The action and virtue of animal magnetiſm may be communicated from one body to another, whether animate or inanimate. 6. It operates at a great diſtance, without the intervention of any body. 7. It is increaſed and reflected by mirrors; communicated, and propagated, and increaſed, by ſound; and may be accumulated, concentrated, and tranſported. 8. Notwithſtanding the univerſality of this fluid, all animal bodies are not equally affected by it; on the other hand, there are ſome, though but few in number, the preſence of which deſtroys all the effects of animal magnetiſm. 9. By means of this fluid, nervous diſorders are cured immediately, and others mediately; and its virtues, in ſhort, extend to the univerſal preſervation of mankind.

From this theory, M. Deſlon engaged, 1. To prove to the commiſſioners, that ſuch a thing as animal magnetiſm exiſted; 2. To prove the utility of it in the cure of diſeaſes; and to communicate to them all that he knew upon the ſubject. The commiſſioners accordingly attended in the room where the patients underwent the magnetical operations. The apparatus conſiſted of a circular platform made of oak, and raiſed about a foot and a half from the ground. At the top of it were a number of holes, in which were iron rods with moveable joints for the purpoſe of applying them to any part of the body. The patients were placed in a circle round it, each touching an iron rod, which he could apply to any part of the body at pleaſure; they were joined to one another by a cord paſſing round their bodies, the deſign being to increaſe the effect by communication. In one corner of the room was a piano-forte, on which ſome airs are played, occaſionally accompanied with a ſong. Each of the patients held in his hand an iron rod ten or twelve feet long; the intention of which was to concentrate the magnetiſm in its point, and thus to render its effects more ſenſible. Sound is another conductor of this magnetiſm; and, in order to communicate the magnetiſm to the piano-forte, nothing more is neceſſary than to bring the iron rod near it. Some magnetiſm is alſo furniſhed by the perſon who plays it; and this magnetiſm is tranſmitted to the patients by the ſounds. The internal part of the platform was ſo contrived as to concentrate the magnetiſm, and was the reſervoir whence the virtue diffuſed itſelf among the patients.

Beſides the different ways of receiving the magnetiſm already mentioned, viz. by the iron, cord, and piano-forte, the patients alſo had it directly from the doctor's finger, and a rod which he held in his hand, and which he carried about the face, head, or ſuch parts of the patient as were diſeaſed; obſerving always the direction of the poles. His principal application of magnetiſm, however, was by preſſure of the hands or fingers on the hypochondria, or lower region of the ſtomach.

The effects of theſe operations upon Deſlon's patients were very amazing. Some ſpit, coughed, and ſweat, and felt extraordinary heats in different parts of the body. Many had convulſions, which is what is called their criſis, &c.—The commiſſioners after this determined to try the experiments themſelves; for the fluid was totally imperceptible by any of the ſenſes, and they could only aſcertain its exiſtence by its ultimately curing diſeaſes, or by its obſervable effects upon the human body.

The practice having been ſince purſued in England with great ſucceſs by the ingenious Dr. Bell, I ſhall give the proceſs and effects of this diſcovery in his own words.

"There is an univerſal fluid which fills all ſpace. Every body is endowed with a certain quantity of electric fluid. There exiſts an attraction and repulſion, or ſympathy and antipathy, between animated bodies. The univerſal currents of the univerſal fluid, are the cauſe and exiſtence of bodies. One may accelerate thoſe currents in a body, and produce criſis and ſomnambuliſm, which is done by acting reciprocally upon one another, by increaſing the currents going acroſs their interſtices or pores, in conſequence of the abſolute will of the operator. As there exiſts a general and reciprocal gravitation of all celeſtial bodies towards each other, ſo there exiſts a particular and reciprocal gravitation of the conſtitutive parts of the earth towards the whole, and of that whole towards each of its parts.

"The reciprocal action of all theſe bodies is operated upon by the inſenſible perſpiration, or vapour, flowing in and out, as you ſee in a real loadſtone or in an artificial magnet, forming an outſide atmoſphere; it alſo produces currents in a more or leſs direct manner, according to the analogy of bodies. That which can act most effectually on a sickly man, is one who is in a good state of health, and is of a similar constitution—the power of man in a good state of health will be then more powerful in consequence of the latter's weakness, who receives more than he gives; it will increase the circulation, and produce beneficial effects.

"The respective position of two beings acting on one another is not indifferent. To judge what that position should be, we ought to consider each being as a whole compounded of different parts, of which each possesses a form, or particular tonical movement. It is of course by that means easily understood, that two beings have over each other the greatest influence possible, when they are so placed that their analogous parts act on one another in the most harmonical manner. It is necessary that the person who submits to be treated is willing, as well as that the operator's mind must be absolute, and think of nothing but of the different sensations he then feels. *Credite & volete.* (This is the secret.)

"Therefore, in order that two persons may act on each other in the strongest manner possible, they must be placed opposite each other; from North to South is the best; you turn your patient's face towards the South; you may treat in other directions, according to your ideas and circumstances. In that opposite position your atmospheres are joining; and you may be considered as forming but one whole, acting in an harmonic manner. When a man suffers, all the action of life is directed towards him in order to destroy the cause of suffering; likewise, when two persons are acting on each other, the whole action of that union acts on the disordered parts with a force proportioned to the increase of the mass. It may therefore be in general asserted, that the action of animal electricity and Magnetism, &c. increases in proportion to the masses.

"It is possible to direct the action of Animal Electricity and Magnetism more particularly on any individual part, by fixing your idea and directing the fluid upon the part affected. Our arms may be considered as conductors to the animal fluid, and serve to attract or repel according to our will, and establish a kind of continuity between bodies. It follows, from what has been said on the most advantageous position of two beings acting on each other in order to maintain the harmony of the whole, one ought to touch the right part with the left arm, and the right foot in contact with the left. In that position you are in affinity with your patient, your two atmospheres are joined; it shows the opposition of poles in the human body, and is nearly the same as those which may be observed in the loadstone, or artificial magnet.

"Paracelsus, as well as many other anatomists, have admitted poles in man. Mr. George Adams, in his Treatise on Magnetism, justly says, 'In some future period it may be discovered that most bodies are possessed of a polarity, as well as one direction

direction relative to the various affinity of the elements of which they are compounded. The better to conceive the poles of the human body, we ought to consider man divided into two parts, by a line drawn from the top to the pubis; all the joints of the left part may be considered as poles opposite to those corresponding herewith; the fluid passes out more sensibly, and in a greater abundance, from the extremities, as those extremities are considered as poles opposite to the right, and are the best conductors of the animal fluid.

"You may give polarity to animate and inanimate bodies; that is to say, to increase an action to a degree which they had not before, only by a friction very nearly resembling that which you give to a piece of steel before it becomes a magnet, except that it will not be so palpable. You may also change the poles in the human body pretty nearly the same as you change those of a magnet. You may also strengthen or increase the action of Animal Electricity and Magnetism by animate and inanimate bodies, as you may increase the action of an artificial magnet by adding more magnets, provided the poles are contrary: therefore every thing is filled in the universe by means of an universal fluid in which all bodies are immersed, and therefore all beings touch one another in consequence of the continual circulation by which the currents of the magnetic fluid flow out and pass in; and in consequence of this you may affect a person at a distance, provided he is of a weak habit of body, and has been in a crisis before you put the column of air into vibration which exists between the person you treat and yourself; that will affect him, as is seen or felt by the force of sounds at a concert.

"In order to be in affinity or harmony with your patient, you must touch him by the hand; as there is a circulation which forms itself between you and him, and tends to an equilibrium, it is generally by that means easier to take your patients out of their crisis. You next hold up both your hands parallel to the head, and bring them gently down as far as the pubis; you may follow the direction of the nerves; then fix your hands upon the diaphragm or stomach, where lies the greatest abundance of nerves; you may put your thumbs upon the plexus, and put the nerves in motion; you may also fix one hand upon the stomach, and draw the other towards you; by that mean yoattract or repel at pleasure.—There are various ways of manipulation, which the operator makes use of, according to circumstances. If you wish to procure sleep soon, change your position; get either to the right side of your patient or left; in that position you fix one of your hands before the head, and the other behind; keep them there with all your might, till you feel some heat in the palm of the hand. If the person is not inclined to sleep, you must charge the head in different directions, by shutting your hands as if you were boxing—then you open them quick, and this you repeat often; the person feels then a drowsiness.—You

must

Dodd del.[t]

ANIMAL MAGNETISM_ The Operator putting his Patient into a Crisis. _

must keep your hands in opposition as before; by these means the animal fluid gets into the absorbent vessels—acts also upon the nerves, which stimulates the body and produces a crisis. If you see the patient too much agitated, get opposite to him, and bring both your hands downwards from head to foot, or as if you were to fan a person, and, getting backwards, it will compose him.—Then you seek for the cause and place of the illness; or you hold the person's hand, and you ask him where he feels pain, as it is increased by treating: if he does not answer your questions properly, it is a sign he is not in a perfect state of somnambulism; you must keep him asleep longer without speaking to him.—You then seek for the seat of the disease, by extending your hand at a little distance from his body, beginning from head to foot; if your sensations are good, you may feel, with a little attention, within yourself, pains in the same part as where the person is affected—or you may feel at the end of your fingers a heat, if it is an inflammation or obstruction; if you feel a coldness, it is in the lymphatic vessels; if bilious, you feel a numbness; and many other ways which different constitutions feel;—either of these circumstances will inform you where the disease lies.—But by touching, which is the surest way, you soon become certain of the seat and cause of the disease, which sometimes lies in the opposite side to the pain, particularly in nervous affections, &c. You may touch, if you like, the cause of the disease, or charge it as you do the head; by that means you keep up the symptomatical pain, till you have rendered it critical—you second the effort of nature against the cause of the disease, and act like a stimulus, which will produce a salutary crisis, by putting the whole frame in action, which will remove any disease proceeding from obstructions, &c. after the patient finds himself composed, and the cause of the disorder diminished. When the patient is asleep, you ask him if it is time to take him out of it; if he answers Yes, draw your hands towards his head down to the feet, and rub your eyes with your thumbs several times, and wave your hand as if you were to fan a person who is too hot—you get by degrees backwards till he is recovered.

"The cause of most part of diseases is an irritability or fever, debility or obstruction; by the slowness or abolition of motion, it is an obstruction or debility, and by its acceleration produces an irritability, inflammation, and fever.

"The seat of those diseases is generally in the viscera, as the intestines, the spleen, the liver, the epiploon, the mesentery, the loins, &c. in women, the stomach, the womb, &c. These aberrations or obstructions are an impediment in the circulation of one part, which presses on the blood or lymphatic vessels, and on the nerves, which produce those spasms, on account that the fluid circulates slowly. For that reason, those persons are the soonest affected, and put into a crisis, when they are labouring under those maladies; if those vessels press upon the root of a nerve, the motion and ſenſibility of the correſponding parts are quite ſuppreſſed, as in an apoplexy, palſy, &c. There is not a better conductor for the animal fluid than the nerves, as they are ſpread all over the body; they abound more particularly in the diaphragm, ſtomachical and umbilical plexus, where lies the root of the nerves, which extend their branches (as a tree does its branches and roots in the earth) all over the body.

"Many philoſophers have thought it is in them that the ſoul lies: it is through them that the ſomnambules ſee in the dark when their eyes are ſhut.—When you treat a perſon, you muſt follow as much as poſſible the direction of the nerves; you may treat at a ſmall diſtance, and fix your hand upon the part affected, and by motion you put the column of air (which exiſts between you and your patient) into vibration, which will cauſe an irritation and produce a criſis.

"Many profeſſors make uſe of conductors, either glaſs, ſteel, ſilver, or gold; about eight inches long; they have a good effect in ſome caſes: all this proceeds from the idea of the operator. Mr. Meſmer tells us, 'When you make uſe of conductors, you muſt magnetiſe from right to right:' that is, the poles are changed.—I have repeatedly produced the ſame effect by treating from left to left, except when I have put a perſon into a ſleep without a conductor; if while aſleep I magnetiſed them from right to right, ſome have gone into a criſis, others have awaked. If you touch the forehead with your right hand, you muſt put your left in oppoſition behind; and in the ſame manner to any other part of the body, becauſe there is a re-action of fluid from one pole to the other, like a magnet; for Dr. Meſmer repreſents the human body as a magnet. If you eſtabliſh the North to the right, the left becomes the South, and the middle like the equator, which is without predominant action. I repeat it; it is moſt advantageous to be oppoſite the perſon you want to treat, in order to cure him effectually. Curing conſiſts in re-eſtabliſhing the diſturbed harmony—the general remedy is the application of animal fluid, which ſerves to re-eſtabliſh the equilibrium which is loſt in ſome part of the body. As there is but one diſeaſe, there is but one remedy; if motion is diminiſhed, it ought to be increaſed: if there is too great irritability, it ought to be decreaſed; as it is on ſolid bodies that this fluid operates, particularly on our viſcera, in order to rectify them, as they are deſtined by Nature to prepare, to diſſolve, and aſſimilate our humours, they ſhould be brought to their equilibrium by any means whatever, by employing either internal or external remedies; but we ought to be very cautious how we adminiſter them, except ſuch as the patients will order for themſelves or preſcribe for others, which are generally very ſimple. There are few remedies taken internally which are good, becauſe, when received into the ſtomach and the firſt paſſages, they experience the ſame elaboration as our aliments, the parts of which analogous to our humours are aſſimilated there by chylification, and the heterogeneous particles are expelled by the means of excretions. Thoſe remedies which may be given will often prove to have effects contrary to the intentions of the preſcriber, becauſe moſt of them are very aqueous, ſtimulate too much, and will increaſe irritation, ſpaſms, &c. and produce effects diſcordant to the harmony of the parts, which ought to be eſtabliſhed and reſtored to their proper equilibrium. If treating is not ſufficient to produce vomiting in the caſe of a perſon who has too much putridity, or abundance of bile which has been too long ſtanding, then a gentle emetic is to be given, or magneſia if there is too much acid; if alkali is predominant, order a ſolution of tartar, or of any other acid which you think will agree with your patient. In caſe of a violent cholic or coſtiveneſs, or ſore throat, injections are the beſt. Theſe are the general remedies which ought to be adminiſtered to the patients, as I am ſure that all thoſe preparations of minerals, &c. which we ſee in an apothecary's ſhop, were never intended by Nature for the human body. Modern phyſicians have, from an intereſted view, neglected the knowledge of the vegetable kingdom, more adapted by Providence for the human body. The diet of the patient is whatever Nature points out to him: it is ſhe who dictates what every man ought to follow, becauſe ſhe ſeldom deceives us in our manner of living. It is not what we eat, nor the quantity, which does good—It is what we digeſt. Animals by inſtinct will never touch any thing but what Nature has dictated to them. In this their inſtinct is far ſuperior to our reaſon. Spirituous liquors are forbidden, ſtrong green tea without milk, coffee, hot aliments, and the uſe of ſnuff, becauſe it irritates the pituitary membranes in the throat, the ſtomach, and the head, and will produce criſpation and irritation. The uſual drink may be water with a bit of toaſt in it; wine and water, or good rich wine, old ſmall beer, good porter, lemonade, or different ſyrups, all theſe may be ordered according to the caſe of the patient. The food may be good broth, either of beef, mutton, or veal, chicken boiled, and roaſted meat. Avoid ſalt or fat meat; make uſe of ſallads, good ripe fruits, &c. Gentle exerciſe in the open air, either riding or walking. Cold or warm baths are moſt excellent; the drinking of ſome mineral waters is good:—in fact, a good obſerver (though not a phyſician) may cure more people than a man of the faculty—becauſe a doctor never goes without an apothecary—they all go together hand in hand, and do more harm than good.

"If you have but one patient, and cannot move him out of bed, gather round him as many healthy perſons as you can; make them rub their hands well—then make them hold one another, and communicate to the patient: this is what I call forming a chain,—by that you communicate to him the animal fluid, which will vivify him if he is not too much debilitated. You may ſet him upon an inſulated ſtool, as when you electrify a perſon; you may ſet him upon a chair, and make a healthy perſon ſit upon the ſame chair back to back. You may magnetiſe a tree in a garden; you may have one in your room, or a ſmall reſervoir.—There are various ways, which depend upon the idea of the magnetiſer.

"There are ſeveral ways of treating and curing; in which, however, much attention and prudenee are required. But a prudent man, willing to do good to his fellow-creatures who labour under any infirmities, will never treat his patient in public, and make them walk in their ſleep, or do many other things: it is very well to convince incredulous people of the effects, but cannot do good to the patient. I will ſay alſo, that a perſon cannot treat more than two or three patients in a day to do them juſtice, and thoſe who do treat more ſeldom cure by magnetiſing alone; the patients may fancy they have been cured;—but, if they had not been ſo treated, they might alſo have been well; as their treatments are long, Nature operates, and is a better doctor. There are ſome who will firmly aſſure you they have cured people at the diſtance of two or three hundred miles off, without ever having ſeen the patients, and have put them into criſes.—I will anſwer them, they are either fools or madmen; their imagination being heated with this idea, they are like viſionaries. I knew an ingenious phyſician who ſaw every body with the yellow jaundice, and another who thought that every body had a virus in their blood, and all the patients who applied to him he treated as having a *gallicus morbus;* and another who pretended to cure every body only by looking at them: all theſe are ſome degrees of inſanity. I knew ſeveral perſons who ſuppoſed I had been treating them after I had left their houſes; they fall aſleep ſome twenty miles off, and they have related this as a fact to ſeveral of their friends, while I never thought of them; and nevertheleſs, ſuppoſe I had been treating them, and they might by chance fall aſleep, I could not with propriety relate the ſtory as a fact, becauſe it muſt be repeated often to hold good. I never reſt my judgment upon a ſingle experiment; in experimental philoſophy facts are ſtubborn, and no one can contradict them when repeated. Now I ſhall explain the manner of treating and curing effectually, on reaſonable principles, each complaint particularly.

"Suppoſe your patient has a head-ach. You ſeat him in a chair, the back towards the North, or otherwiſe; you ſit oppoſite to him; you put yourſelf in affinity with him, as I have obſerved before; you draw the general current, following the direction of the nerves; you hold your hands the ſame as if you were to hold a pen; you ſeek for the cauſe of the pain, which may lie in different parts—perhaps the patient will tell you, if you cannot find it out. If it is a hemicrania, which is owing to the foulneſs of the ſtomach, you fix your eyes upon your left hand, which you direct towards the ſtomach with your right hand—you do the ſame as if you were to turn a pancake; this you repeat ſeveral times; by that means you ſtir the atmoſphere and relax the ſtomach, and may make him vomit; you may give a little warm water to promote your operation; you may alſo treat the head by drawing the fluid downwards, if the pain has been of long ſtanding; you may order a vomit or a gentle purge, and treat them every day, and then order bitters to ſtrengthen the ſtomach. There are different head-achs: as the *cephalalgia*, when the head is affected ſlightly in one particular part; *cephalæa* is when the whole head is affected; one ſide only is called *hemicrania*, and a ſmall ſpot affected is called *clavis hyſtericus*. Theſe various head-achs ariſe from different cauſes: if it proceeds from obſtructions, criſes are very ſalutary, as they put the whole body into motion, and will remove the cauſe. Treating the part which you think affected is very neceſſary; you do the ſame with one hand or both, by drawing your hands towards you ſeveral times as before. To treat the head, you may apply your hands upon the temples, and put your thumb upon the frontal ſinus, which will often remove it.

"Deafness.—If the want of hearing proceeds from a fault in the ſtructure of the ear, there is no cure. If it proceeds from cold, fever, hard wax, or dryneſs, you may magnetiſe according to my principles. You keep yourſelf within a yard or two, according as you feel a re-action; you then fix your left hand towards the ear, and you move your right open, and bring it towards the left hand, and do the ſame as if you were to clap your hands; by that means you put the air into vibration, and, guided by your left hand as a conductor, you apply the palm of the hand upon the ears; you may put your thumb in the ear, and with your finger, as you hold a pinch of ſnuff, preſs the thumb towards the ear—you accelerate the fluid into it. You may make uſe of a conductor, either glaſs or artificial magnet, and put it into the ear, and preſs with two fingers from the baſis towards the ear; you may magnetiſe the head, by drawing the fluid towards you: all theſe means you are to make uſe of according to your ſenſations and judgment; ſometimes an injection made of Caſtile ſoap—warm brandy and water will aſſiſt you in your operation.

"Of the Tooth-Ach—This violent, though not dangerous, diſeaſe, proceeds from rheumatiſm, obſtructed perſpiration, inflammation, &c. This being the caſe, you treat according to my rules. If there is inflammation, you draw the fluid from the head; you touch the temples, the frontal ſinus, the top of the head, the articulation of the jaws, and under the chin; you may touch the tooth with your index and thumb; but a ſure way is to get an artificial magnet, and, as your patient's face is towards the South, apply the South pole upon the tooth, and touch the next teeth, and afterwards draw the fluid downwards, and you will perform a cure.

"DISEASES of the EYES.—There is nothing so difficult to cure as these diseases; and none of our organs is more subject to be affected than the sight, or from so many causes. When they proceed from obstructions in the customary evacuations, you must magnetise according to my rules. You treat the cause, also the eyes, by fixing your thumbs opposite; you press with the index the fluid into the eyes, you move your thumb opposite you; and may rub the eyes gently; you drop magnetised water into them with a quill: this you do three or four times a-day, and you order a little lemonade or syrup to your patient.

"The gutta serena, ophthalmia, cataract, specks on the eyes, and fistula lacrymalis, are very difficult and almost incurable. I have heard many magnetisers boasting of the cures they had made of these diseases. I have had more practice in that way than many of them; I confess candidly I have made but few. I shall explain the best manner of treating.—You must know first the cause, which you treat; afterwards you apply your thumbs gently on the eyes; you rub them often—you fix your thumb with the next finger at a distance from the eye; but I have had success in some cases of this kind by making use of an artificial magnet, by fixing it at the distance of half an inch from them; it has by that means removed spots and gutta serena, proceeding from the compression of the nerves by superfluous humours. I have dropped magnetised water three or four times a-day with success; a proper regimen is necessary, and some internal and external application.

"Of the EPILEPSY and HYSTERIC AFFECTIONS.—These diseases are the opprobrium of the faculty, with many others, as they cannot be cured by internal medicines, except when proceeding from obstructions, worms, or affection of the mind, &c. In those cases you treat according to the rules. First, touch the head on the top; by applying your thumb on the root of the nose, you endeavour to dissolve the obstruction which may be the cause; apply your hand upon the diaphragm, and endeavour to put the nerves in motion. You may treat at a distance also; but try to produce a crisis.—Dr. Andry, and Tourit, at Paris, have cured several epileptics by applying artificial magnets round the head, or like a horse-shoe applied upon the top of the head; when they are in a fit, apply a magnet in each hand, it will soon recover them. I have brought some to, by applying a key in their hands; they are very good in spasms, fainting, and cramps—by applying the magnet under the foot, it ceases instantly.

"Of the SCROPHULA, called the KING'S EVIL.—This disease is a disgrace to physic. Some persons have had the gift of curing by touching. I have seen in London two persons who had been touched by a man after he was hanged; they were relieved; but I really believe it was the force of imagination; being frightened by the dead man made such a revolution in the blood, that it removed the obstruction

tion in the glands. You may touch your patient in thofe parts, and draw the effluvia in order to foften the glands. If there is an ulcer, order the perfon to bathe the part with magnetifed water, and keep a bit of rag always upon the part. Sea-bathing, decoction of celery, and hemlock-juice, may be tried, befides treating.

"SORE THROAT.—Sore throat, or any inflammation in the head, is to be treated by drawing the fluid out of the part, either by putting yourfelf in oppofition, or by ftanding on one fide, and putting one hand behind the neck and the other before.

"Of the PALSY.—The palfy, when it happens to an old perfon, or has been of long ftanding, is feldom cured; but if it happens to a middling age, and one fide only is ftruck, called an *hemiplegia*, a cure will be effected by being treated foon after. You may magnetife your patient oppofite, as ufual. After you turn the fide affected towards the North, you treat the oppofite fide, which is fuppofed to be where lies the caufe: you may touch with one hand along the back-bone, or within an inch from it, along the great intercoftal, by applying your right hand upon the ftomach; you treat him about two hours; if you can put him into a crifis, which is very eafy, you may expect to cure him. You make him lift up his bad arm, or have fomebody to fupport it; you put a conductor to his hand to attract the univerfal fluid; you may infulate him, and turn the part affected towards the North; tie a filk ftring to the ceiling, at the end of it have ftrong compounded magnet, the North pole parallel to the hand; to the other hand tie likewife another ftring, at the end of which there is a large piece of iron whofe furface is larger than the magnet. Have an electrical machine, and connect the chain to the patient, then make him ftretch his arm; then touch the magnet and the piece of iron together, or one after another, to the extremities of the hands, it will cure him; I have cured feveral that way: but this does not belong to Animal Magnetifm, fay many. But has not a magnetifer a right to cure his patients as foon as poffible, and employ every means his mind fuggefts to him? It is not fo among the faculty; they muft cure or kill them *fecundem artem*, according to art. A general vomit or purge is often neceffary; the diet muft be good. If the tongue is affected, put a conductor upon it, or an artificial magnet, fuch as you make ufe of for the teeth, by preffing the fluid from the bafis towards the point on the tongue: fometimes a little gargarifm is ufeful. Electricity and the cold bath are very good.

"RHEUMATISM.—Nothing is more common in this country than this difeafe, on account of the dampnefs and change of the weather, which will abforb the electric and magnetic fluid from flying off certain parts, particularly from the feet, whence there flies out a greater abundance of fluid than from any other part of the body. It is for that reafon dogs will follow our tracks. There are very obftinate rheumatifms which proceed from different caufes ,and are difficult to cure. The method

of

of curing this disease is to magnetise the patient in opposition; try to promote perspiration, by putting him into a crisis. If the rheumatism is in a particular part of the body, you must treat the part affected either by touching or rubbing, which is the best. You may make use of an artificial magnet in the form of a horse-shoe. If the rheumatism is in the head, you apply it upon the top of it; if it is on the face and teeth, apply it on the temples; if it is in the hip, you apply it above the knee, with the poles up; if in the knees, apply it on the tarsus, with the poles up; if it is on the shoulders, you place it on the humerus, clavicula, &c. Electricity, hot and cold bath, earth-bathing, according to Dr. Graham's principles, &c. &c. Some internal and external applications will assist the operation.

"Consumption, or Decay.—This disease, so common in England, is difficult to cure; it proceeds from want of the animal fluid in the body, which wastes it to nothing; therefore it is necessary that the person who treats be very strong and healthy. His patient is like a child at the breast, pumping his animal juice; and he may be much hurt by it, like a child who sleeps with an old or unhealthy person; therefore I would advise you to treat as few as possible. Riding a young horse without a saddle, a cow, a bullock, or to be among cattle, is very good; or to sleep in a stable, and communicate a rope from the bed to the cattle, which serves as a conductor to the animal fluid.

"Diseases in the stomach are common in this country among women, owing to that pernicious custom of wearing stays; not only so, but they must have a piece of iron or steel two or three inches broad, and proportionally thick, in it, called a busk, which occasions so many diseases. They should be loose round the body. You treat the stomach by throwing fluid into it. Crises are not good for it.

"Flatulency, or wind in the stomach and bowels, arise from want of tone in those parts. It is to be treated upwards, which will make the patient break wind and produce a crisis, which is the best. After the crisis, you must treat the stomach downwards, in order to settle it; you may order carminatives. Bile on the stomach is treated upwards; also to make the patient vomit, and crises, are good for it; a glass of magnetised water afterwards will settle the stomach. In all sorts of inflammation of the lungs, liver, &c. you treat towards you, and avoid the crisis as well as when those parts are ulcerated. In the stone and gravel you treat; throw a quantity of fluid, and produce crisis; it will promote evacuation, which may do service to the patient. In external swellings, or ulcers, draw towards you, and bathe with magnetised water.

"Pregnant women, and in labour, may be treated without a crisis. I have magnetised women in labour, and put them asleep while the accoucheur was performing his duty; the woman did not recollect it, and was surprised afterwards. I have put a man aſleep who had an hydrocele; the ſurgeon performed the operation, but the patient never recollected any thing of it. Relaxation, and the blood flowing from a cut, may be ſtopped by fixing your thumb and preſſing the fore-finger over the part.

"Fevers of every kind may be cured by criſes; it is during that time that nature endeavours to get rid of what diſturbs her, either by perſpiration, vomiting, &c. Thoſe people are the beſt ſomnambuliſts, as I ſhall explain hereafter. It is very eaſily underſtood, by the method I have taken to explain the treating of the foregoing diſorders, that an ingenious magnetiſer may treat all others, as it would require a whole volume to explain them.

"Of Nervous Diseases. It is in thoſe diſeaſes that magnetiſm acts moſt forcibly, by putting the whole nervous ſyſtem in motion; it operates criſes as well as ſomnambuliſm, and offers to the attentive eye a vaſt field of obſervation. There is as great a variety in thoſe diſeaſes as there are combinations between all poſſible numbers. Different organs may be affected, and diſtinctly from others. In ſome perſons the extenſion of ſight is ſo great, that it ſeems as if they made uſe of a microſcope. Some of them can ſee, in the dark, the animal fluid flying in all directions, and appearing luminous; others will ſee the ſkin appear to them like a ſieve, and ſee the groſs humours or perſpiration as big as ſmall ſhot; and by rubbing the hands they ſee ſparks of fire coming out. Mr. Boyle mentions a perſon who, after getting half-fuddled with claret, (which I ſuppoſe relaxed the ſtomach and his nervous ſyſtem,) when he waked in the night, could ſee to read moderate print. Another who could in the night diſtinguiſh colours. Grimaldi tells us, that ſome women can, by their eyes alone, diſtinguiſh between eggs laid by black hens and thoſe by white ones. This ſingle effect will lead to many things which I ſhall relate about ſomnambules. We muſt not attribute to whim all the ſingularities which we obſerve among people affected in the nerves; it is a real cauſe, as that which determines the moſt reaſonable man. I knew a gentleman in London, who ſhook his head and arm every inſtant like a perpetual motion; a lady, I treated when in his company, had the ſame involuntary affection. There are different methods of treating thoſe diſeaſes, either by treating without criſes, or with them; ſuch people are the beſt ſomnambules. If a perſon is irritable, you treat gently, in oppoſition, by drawing a certain quantity of fluid from him; if, on the contrary, you throw the fluid towards him, you may put him into a gentle criſis; if the patient has a trembling of the limbs like the head, you treat that part: if you cannot ſucceed by treating, apply a magnetic bandeau round the head, it will ſtop it inſtantly. For trembling of the hands, you apply magnetic bracelets.

"To magnetife, or treat a perfon at a diftance, is not impoffible. The manner which feveral profeffors make ufe of, is different. There are quacks in that art who pretend to have found it out before Dr. Mefmer; but that none of them underftood it is well known. The faculty of our foul, thought, or idea, can perceive, contemplate, and unite itfelf to, any object, prefent or diftant, vifible or invifible. That it has action upon matter is well demonftrated; it acts directly upon the vivifying electric and magnetic fluid, and by its will determines it to be directed upon fuch part. We know that our foul acts upon our body, and forces any part of it to move in any direction, according to its will. This being the cafe, we may reafonably believe that it may act as well upon merely organical matter as upon animal bodies. The thought, or foul, goes to any diftance. No obftacles can refift it. It arrives and unites itfelf, by a fympathetic power, to any object it wifhes, without a mafter of ceremonies; neither the fize of the body, its ftrength, or figure, impede; all give way; the union is made in an inftant; the will, and the will only, is the caufe of it, becaufe it directs the fluid towards the difeafed and affected vifcera, by fixing them in your imagination, as much as it is poffible; and by that means it will force the magnetic fluid to touch and to penetrate to a great diftance any bodies to which the foul is willing to unite herfelf, and to reeftablifh the animal œconomy, of which fhe is the indeftructible principle. Thefe reflections fhow the poffibility, and the mean made ufe of, to treat a perfon at a diftance; of which experience will fhow the reality, and an ingenious mind may make many curious experiments: repeated trials will convince us.

"To treat a perfon at a diftance, from one houfe to another, is poffible, provided you have feen the perfon before, and put him in a crifis. The manner you do this, is to know where the perfon is, and fix the hour by your watch, and have fome friends with the patient to divert him: you muft be alone in a room, to avoid any noife, or any thing to diftract the attention of your mind. In that pofition you paint the perfon in your imagination; you reprefent in your idea the part which you fuppofe affected, and you treat in the fame manner as if the perfon were before you. That fympathy of body and mind which exifts between you and him will produce a crifis and fomnambulifm; that phænomenon is very interefting. You may alfo from the fame principles treat a perfon in the fame room, without his or her knowledge, by fixing your mind and your eyes upon the part affected, or upon the heart, ftomach, &c. and produce crifes and fomnambulifm.

"Dropsy. There are different forts of this difeafe, according to the parts which are affected. I fhall treat of the manner of curing the afcite, which is, when there is a collection of water in the belly, proceeding from obftructions, living too low, and

and ſometimes from drinking ſpirits or cold water when the body is hot. You treat the patient in oppoſition; you fix your hands upon the part, either at a diſtance, or by applying the hands on the belly; you try to produce a criſis, which is the quickeſt way. You may apply a magnetiſed bell-glaſs on the belly when the patient is in bed, the ſame on the legs if they are ſwelled, and various acceſſaries, according to the operator's fancy. Dropſy of the brain, of the breaſt, and of the legs, are treated by extracting the fluid, and promoting circulation and perſpiration.

"Of the Asthma. This diſeaſe of the lungs is very ſeldom cured when it proceeds from a bad formation of the breaſt, or is hereditary. If it comes from obſtructions, treat the lungs, and put the patient into criſes to promote circulation; but if the aſthma proceeds from another cauſe, as violent paſſions of mind, humoreal or nervous, and the patient ſpits a great deal, treat the ſtomach upwards to promote expectoration. If the patient coughs much at night, give him a glaſs of magnetiſed water going to bed, and another in the morning. Moderate exerciſe in a gentle air is very uſeful.

"Apoplexy. This ſudden loſs of the ſenſes may be cured by applying immediately, and with proper care. The cauſe is an effuſion of the blood, or a collection of watery humours. There are two ſorts, a ſanguine and a ſerous apoplexy; it is generally towards the brain that the cauſe lies, becauſe the blood does not return from the head. That being the caſe, you magnetiſe the patient either in bed or up: if he is in bed, you ſtay at his feet; you magnetiſe the head downward; you may get at his right ſide, and magnetiſe as before; you touch his head, one hand behind and the other before, and bring your hands downwards; you muſt raiſe the head of the patient high. If it is a ſanguine apoplexy, and you ſee there is no change, you may order a bleeding, or put the feet in warm flannel. Let the patient have free air. You muſt treat him four hours a-day.

"Night-Mare. This diſagreeable diſeaſe puts the patient into the greateſt torture during his ſleep; he feels often a weight upon his ſtomach, as a fiend, a cat, a dog, &c. He endeavours to cry out, and fancies himſelf going to be drowned, or to be killed. It proceeds from a weak ſtomach, nervous affections, &c. I attended a patient who uſed to be blooded every year in May. During March and April he was always ſo; but, as ſoon as he was bled, the pain was over. They are a kind of ſomnambules. You may treat the ſtomach, by throwing a quantity of fluid, in order to ſtrengthen it; alſo treat the head downwards. A glaſs of magnetiſed water, going to bed, is very good.

"Of Sensations, looked upon as a ſixth ſenſe. There are as many ſenſations as there are poſſible differences between proportions. In all ſenſations we muſt conſider

conſider three things; the cauſe producing the impreſſion, the nature and diſpoſition, of the organs receiving it, and the ſenſations which have preceded it. It is by the combination of thoſe affinities that the organs of our ſenſes may be magnified or increaſed to ſuch a degree, as to become, for every object which they preſent to us, what teleſcopes and microſcopes are to the ſight; conſequently our ſenſations are the reſult of all the effects which objects make on our organs. Our ſenſes can only draw us more or leſs near to the knowledge of objects and their nature, by a conſtant uſe and a ſerious application, in order to attain to their reality. We have a great number of ſmall organs proper to receive ſenſations; but the habit we are in of making uſe of ſome particular organs only, abſorbs the reſt. Blind people have different ſenſations from us; they will perceive a wall, or other body, before they touch it. There is no doubt but we are endowed with an internal ſenſe, which is in affinity with the univerſe, and is conſidered as an extenſion of ſight; it is by theſe means one may comprehend the poſſibility of finding the diſeaſe of another: of foreſights, predictions, and the phænomena of ſomnambules and ſybils, &c.

"It is poſſible to be affected in ſuch a manner, as to have the idea of a body at an immenſe diſtance, in the ſame manner as we ſee the ſtars, the impreſſion of which is tranſmitted to us in a right line, the ſucceſſion and continuity of a co-exiſting matter between them and our organs, bounded by the nature of their form: why ſhould it not be poſſible, by the means of an inward organ, by which we are in contact with the whole univerſe, for us to be affected by beings, the ſucceſſive motion of which is propagated to us in curve or oblique lines, in any direction? and why ſhould we not be affected by the connection of beings which ſucceed one another?

"I was acquainted with Monſieur de Botinau, who had a place under government in the iſland of St. Helena. During twenty years he made a particular ſtudy of a ſenſe unknown to us: he could perceive a fleet or a ſingle ſhip two or three hundred miles off; laſt war he deſcribed M. de Suffrein's fleet, the number of ſhips, and thoſe which had paſſed by and did not touch at the iſland. He could do more: at ſea he could tell the diſtance he was off land, as has been proved by repeated experiments in the Channel. In confirmation of this, I have ſeen the certificates granted him from the governor and principal people of the iſland, and the petition and recommendation to the miniſter, who granted him 1800 livres per annum.

"The famous Bleton, called the *ſourcier*, or ſpring-finder, whenever he walked upon ground where there was a vein of water, felt within himſelf a certain ſenſation which gave him notice there was water. Another countryman ſhook wherever there was water; the elementary, electric, or magnetic, fluid, paſſing through the pores of the earth, gave him that ſenſation.

"Of Crises. The crifes are an effort of nature againft the diforder, endeavouring to diffipate the obftacles that are in the circulation, and to reftore harmony or equilibrium in all the parts of the body. Few difeafes can be cured without a crifis, particularly when it proceeds from obftructions, &c. There are two forts of crifes. The natural one, which is attributed to nature alone, gets rid of what offends her by an increafe of movement, producing vomiting, motion, perfpiration, &c. Thefe are the moft falutary, as nature acts filently, without violence, and expels the obftacles that impede circulation, by moving gently the molecules which form thofe impediments, and go off by perfpiration, &c. The forced one is fometimes falutary in obftructions, or windy and bilious complaints. Thefe are produced when nature is infufficient to expel what offends her. The ufe of Animal Electricity and Magnetifm puts in action the whole body, and, in conjunction with her, acts efficacioufly on the patient, and he difcovers benefit and eafe, particularly if it has produced evacuations, &c. There are various means of producing them, according to the fubject, and the caufe of his difeafe. Some fay there are fix degrees of crifes; I fay there are as many as there are different conftitutions to treat. Some will alfo call it the luminous crifis, from that new fect called the illuminati. All thefe are imaginary. Suppofe you have a patient on whom you would wifh to produce a gentle crifis; you muft put yourfelf firft in affinity; then put your hand behind the head, and the other before, till the perfon is afleep. If the perfon is agitated, calm him, by drawing the fluid downwards from the head; if you treat the caufe by touching, it will increafe the pain; if you put your thumb upon the frontal finus, they will fall into a crifis. You may magnetife your watch, and to fhow what o'clock it is; they will go in. You may magnetife a flower, and give them a fmell: they will fall in. Magnetife a harpfichord: as foon as you play on it, they will go in. Put a perfon between you and the patient, and magnetife him; you will put him in. To magnetife a pond, make the patient ftay on the other fide of it; you muft ftand oppofite; make the patient hold a ftick in his hand to touch the water; you muft touch alfo the water with your magnetifed conductor; the perfon will go into a crifis immediately. Have fomebody behind him, to prevent his falling into the water: it is the beft conductor of animal fluid. To make a perfon read, be behind him; you magnetife the lines as he reads: he will go in. To make a perfon ftay behind you oppofite the looking-glafs, magnetife with a conductor the perfon in a looking-glafs, that you may fee him; the re-action of the fluid will produce a crifis. Magnetife a tree in a walk; make the perfon walk; as foon as he comes near the magnetifed tree, he will fall into a crifis. One may put a perfon in a crifis from one room to another; and, in fact, an ingenious obferver may, by what I have related, make a great many curious experiments, provided he has proper fubjects.

"Of Somnambulism. Somnambulifm is a ftate between fleeping and waking, partaking of both; the patient is a fomnambule when he can do the fame as if he were awake. Thefe natural fomnambules, who get up at night, and do many wonderful things, are well demonftrated. They are difeafed, and may be cured by treating. The magnetic fomnambules are thofe whom art has found out a mean of abforbing or fufpending fome of their external fenfes for a while, and the patient eats and drinks, goes up and down, plays upon the harpfichord, and does many things which you defire him, provided he be willing. The firft I faw was at the Marquis de Puyfegur's, in the year 1784; and all thofe who had pretended to it in this country before were impoftors; for all this we are indebted to Dr. Mefmer. Whenever any perfon has a real fomnambule, which is very eafy, by care they have a treafure. They are called by us *malades medecins*, or fick phyficians. Thefe beings fee in the dark, and go through an external atmofphere, the fame as a glow-worm; they have befides an internal atmofphere, which they make ufe of to perceive objects prefent or diftant, vifible or invifible. I have had feveral who related to me what they could perceive. They differ in many refpects according to their conftitution. One muft not depend always upon what they fay, on account of their differing fometimes. You may make them move in any direction, by your will alone; or, by moving your conductor any way upon the floor, they will follow its directions. You may make them play on any inftrument they can play upon; they will read, write, and work; all this they will do better than if awake. Being deprived of their other faculties, thefe become ftronger. No phyfician can tell the difeafe of a perfon better than a real fomnambule. They feldom fail to tell unknown perfons their difeafes, and prefcribe for them. At a future time, when the fcience is better eftablifhed, I fhall publifh a full account of the theory of fomnambulifm.

"Some will accufe me of having faid too much; but thofe who know me perfonally will never accufe me of relating any thing which I cannot demonftrate; and thofe who repeat thefe marvellous narrations hurt themfelves and the fcience in the eyes of really learned men. Thofe ftories, like tradition, which are handed down from generation to generation, and become improbable, like antiquity, lofe their former luftre. I would advife my pupils to try thofe experiments I have fhown them firft, and try others afterward.

"To make an Electric or Magnetical Apparatus. I fhall not give you a full account of the apparatus of our fociety in Paris. It is almoft like a grove. Mine, which I had in London and Dublin, is a large oak tub, eight feet in diameter, well pitched in the infide, about an inch thick, infulated upon four glafs-feet bottles of water well corked; you magnetife the bottles, and lay them down, the neck of one in the bottom of the other all around, ſo that the laſt comes to the centre. You may fill up the ſpace with broken bottles, or any vitrifiable matter, brimſtone, or reſinous matter, minerals, &c. fill it up all but ſix inches; put ſome loadſtones and artificial magnets in different directions: then cover the whole to the edge with fine dry river-ſand, put the lid over; place in the middle a poliſhed iron bar about eight feet high, with ſprigs to it, to attract the univerſal fluid which concentres itſelf in the reſervoir. At the far corner place an arbor vitæ in a box, and place under it a ſtrong magnet, the north pole upwards; the ſouth pole is fixed in a hole upon the cover, by that means you increaſe the motion of the tree, and, becoming vegetaliſed, it will grow without water. You make holes all round, about eighteen inches diſtant; put iron or braſs conductors behind, ſo as to touch the patient who comes next to it. Connect a chain of an electric machine; inſulate your patients, and make them hold hands; it will increaſe the action in them. You may treat them in that manner; you will the ſooner put them into criſes. I have had all my patients round my reſervoir in a criſis at a time. I could not attend them. You may have a tree in a box, upon inſulated feet; have a ſmall box with vitrifiable matter, and fill it with water: you may make uſe of a large bottle filled with water only, and connect a chain to it. All this apparatus may be made differently, according to the idea. Some take every morning brimſtone or lozenges; and have brimſtone in their ſleeves, and rub themſelves with different ingredients; but I never made uſe of any, and produced a great many effects.

"To magnetiſe a tree, you muſt ſtand facing the north; you muſt have a conductor which you have magnetiſed; you muſt then point it from the top of the higheſt branches to the roots; do the ſame from the other branches: if the tree is ſo large that you cannot ſee the branches on the other ſide, change your poſition from ſouth to north, and do the ſame; then approach the tree; clap your hands round it, and ſtay in that poſition five minutes, your tree becomes magnetiſed. Any patient who has been in a criſis, or ſomnambuliſm, will diſtinguiſh it. Some will go in a criſis as ſoon as they come near it; others, if they are in ſomnambuliſm, will diſcover it among the reſt.

"You may magnetiſe a myrtle, or any other ſhrub; it will appear luminous in the dark. You may magnetiſe a flower, by putting your thumbs in the middle, and eſtabliſh an equator; then drawing your thumbs to the extremities, you preſs your thumb with the next finger, and you throw the fluid upon the flower; it will appear luminous in the dark; by giving it to a perſon to ſmell, who has been in a criſis before, he will go into one again.

"To magnetiſe a conductor or a cane, put your hands in the middle of it; ſlide your hands to the extremities, your thumbs at the top, and rub the extremities with them

them; by these means you will impregnate it with an electrical fluid, that seems luminous in the dark, and as sulphurous as the electric rubbing.

"To magnetise a shilling, or a guinea, put your thumbs in the middle, and draw them to the extremities, it will appear as a ball of fire. A watch is magnetised in the same manner, by drawing your two thumbs at the top, and your index under it; establish an equator, and draw your fingers to the two poles: by showing it to a person who has been already in a crisis, he will fall in one again. They can tell you what o'clock it is in the dark; if asleep they can tell you the same, by showing a watch.

"To magnetise a harpsichord, fix your hands spread in the middle, and draw them towards the extremities; then rub the end you touch the strings with one after another, in the same manner, and thus you will impregnate it with an electric fluid. As soon as a person plays upon the harpsichord, make your patient touch it with his hand or finger; he will fall in a crisis immediately.

"To magnetise a room, or a bed, is the same. Set it to the north, facing the south; point your conductor up to the ceiling; bring it down towards you; point it to the west and east, and bring it also to your feet; the room will appear all luminous, and the bed also.

"A pond may be magnetised in the same manner, by pointing your conductor over the surface of the water, from the cardinal points; touch the water with it, and make your patient do the same; he will have a shock in falling in, and it may be of service to him. From these few experiments it is easy to conceive, that any inanimate body may be electrified or magnetised by another animal body, just as easily as by an electrifying machine, or by the force of magnets."

ARGUMENTS to PROVE, that ANIMAL MAGNETISM is the CAUSE of SYMPATHY in MAN and other ANIMALS, and in PLANTS, &c.

THAT constant flux and reflux of the vital principles and corporeal humours in man (without which both motion and life are stopped) produce those effects of sympathy and antipathy which become more natural and less miraculous; the atmospherical particle to each individual receives from the general fluid the proper attraction and repulsion. In the divers crossings of those individual atmospheres, some emanations are more attractive between two beings, and others more repulsive; so again, when one body possesses more fluid than another, it will repel; and that body which is less will make an effort to restore itself into equilibrium or sympathy with the other body. Robin Abraham Benhannes says, iron or ferruginous particles are every-where, not only in the mineral world, but in our blood and bones; now,

now, as the magnet attracts ferruginous particles, every thing of courfe is fubfervient to magnetifm by the power of attraction or fympathy.

I could relate a variety of examples to prove that fympathetic affection which prevails with people of the fame family, views, fect, or any other caufe that binds them harmonioufly together; but, as it is a fubject which every one muft have experienced, I fhall not touch further on it.

The magnetic fluid often occafions fome contractions in other parts of the body, when a mufcle has been wounded, which produces different motions in the organs of the fame body. Whether they have a fecret affinity or not is a queftion not yet determined; however, I am inclined to think they have. Thefe motions have aftonifhed many phyficians who have reflected upon this art, particularly Barthe, who has well explained them by a fubtile motion which he calls *vital fluid*, and which he might as well have called animal electricity and magnetifm.

Of ANTIPATHY.

WE do not all refemble the Trojan fhepherd, who awarded the apple to the faireft; it is not always the handfomeft woman that wins our affections; our interior emotions are involuntary feizures independent of the influence of beauty, and are the forerunners of love. So again, when two atmofpheres are in equilibrium, that is to fay, when thofe corpufcular emanations are in affinity with each other, it produceth fympathy, or attraction; but, when thofe atmofpheres are croffing each other, it produceth antipathy, or repulfion.

The difcordance of tempers, religious difputations, politics, &c. have frequently been the caufe of inveterate hatred; how can we otherwife account for that fudden averfion we feel for certain objects or perfons, if it be not in the difagreeable impreffions communicated to the nerves, and then to the brain, from the emiffion of thofe perfons or objects? This can be called by no other name than antipathy.

By antipathy many people find out the difeafes of others; they feel within themfelves, in the oppofite fide, the fame pain the other perfons have. If I put a difeafed perfon in contact with another perfon in fomnambulifm, they inftantly feel the fame pain; only, however, during the time they are in contact. It may be called fympathy; but, as they fuffer in fome proportion during that time, it is properly antipathy. It is well known there are many people who entertain an antipathy to different animals, &c.

EFFECTS of ANTIPATHY and SYMPATHY in BRUTE ANIMALS.

ANIMALS in general, like ourfelves, move at the afpect of pleafure, and fly from that of diftrefs; in fome refpects they are fenfible beings that feem to enjoy a will adequate to determine their different motions, nay, ſometimes to be poſſeſſed of the ſentiments, vices, and paſſions, of mankind, and experience likewiſe inclination and hatred, which ſeldom vary in their objects; whence proceeds that conſtant love that ſome animals ſhow for certain ſpecies, whilſt they bear the ſtrongeſt antipathy and averſion to others. They are differently affected according to their different ſpecies by corpuſcular emiſſions, but are nearly the ſame when they flow from the ſame ſpecies. Hence the one conſtantly becomes the object of the other's averſions. Thus one animal only lives to deſtroy and devour; and in his turn contributes, by his own deſtruction, to the preſervation of a ſtronger animal. Thus nature is ſupported by theſe ſucceſſive deſtructions; new combinations ariſe from the compoſitions operated in her boſom; like the phœnix, ſhe only dies to revive, and return brighter out of her own aſhes. Without thinking (as the ancients did) that a ſtring made out of the bowels of a wolf and another from a ſheep cannot agree, or, if two drums made out of their ſkins, the ſound proceeding from that of the wolf-ſkin would deprive the other of all ſound, antipathy between certain ſpecies is evidently a means allotted them by inſtinct to diſcover their prey or avoid their enemy. Thus the wolf purſues the lamb, the dove dreads the falcon, the wren the eagle, the goldfinch the toad, the hen the fox, the water-fowl the ſtork, the graſshopper the ſwallow, the blackbird the hawk, the nightingale the butcher-bird, the frog the eel, the ſnail the partridge, the oyſter the crab, the tench the pike, the fly the ſpider, and the ſpider the ſcorpion. The lion diſlikes the cock, the ape the tortoiſe, the horſe the camel, the lizard the ſerpent, the boar the ſea-calf, the martin the vulture, the owl the crow, the tunny the dolphin, the conger the lamprey, with an infinite number of others too tedious to be mentioned. The ſmell of lobſters drives bees away; the owl deſtroys the eggs of the crow, the ſtork thoſe of the bat; the weaſel thoſe of the hen; the heron and the lark are continually at war, by deſtroying each other's young. If the eagle devours the ſerpent, the latter climbs up the rocks and revenges itſelf by ſucking its enemy's eggs; the toad and the rattleſnake, under the graſs, by darting through their pores the magnetic fluid, faſcinate their prey; the weaſel in vain endeavours to avoid them; ſhe leaps from one place to another, and her ſtrength is at laſt exhauſted to no purpoſe; obliged to draw near the enemy, ſhe iſſues a diſmal cry, and, being violently attracted towards the reptile's mouth, precipitates herſelf into it, and there finds her grave. To revenge this victim, the field-ſpider ſpins her web ſuſpended over the toad: her influence troubles and at laſt lulls him to ſleep. In like manner the ſtag's breath attracts the ſerpent, and occaſions in him a giddineſs. The viper, with fiery eyes and contracted muſcles, darts venomous corpuſcles on the branch of the tree where the nightingale finds an aſylum; ſoon after, the wood-ſinger loſes his voice, is thrown into convulſions, falls down, and is devoured by the viper. It is owing to the effects of emiſſion that the hound finds out the game, and purſues it to its den, where it ſeeks for a refuge. It is by this ſame ſenſation that the partridge ſtops in the middle of a fallow ground, and forgets ſhe has the power of flying.

Animals are as ſuſceptible of ſympathetical as of an antipathetical attachments: according to ſome naturaliſts, the fox is fond of the ſerpent's company, and the duck of that of the toad; the bear avoids treading on the ant, the nightingale loves the peacock, the kite protects the cuckoo, partridges and pheaſants doat on the ſtag, and doves on teal. We are told that a lizard, elephant, and dolphin, are fond of a man; but this is nothing to the attachment of a dog to its maſter: he follows him to all places; and, ſhould he happen to loſe ſight of him, he ſtill finds out where he paſſed only by the emanation he has left in his way, (which eſcapes more abundantly through the toes, as being more porous;) and, if he meets him, by a thouſand tranſports teſtifies his joy.

Of ATTRACTION and REPULSION, otherwiſe called SYMPATHY and ANTIPATHY, in PLANTS.

PLANTS, like men, have their tranſpiration and emiſſion produced by a preſſure of a magnetic fluid which penetrates them; and they carry in all their fibres that vivifying fluid, and have alſo their particular ſpheres of attraction and repulſion. Hence that inclination that ſome vegetables ſeem to have to come nearer to each other, to grow and die together; hence that hatred that has been obſerved amongſt others, and the efforts ſeemingly made uſe of to repel each other.

The vine ſeems to improve under the elm, the olive-tree with the aloe-tree, the plantain with the fig-tree, the agaric with the cedrus, aſparagus with penny-royal, and the cocoa grows powerfully under the ſhade of ebony; the reſinous-tree is favourable to the femla, and the cotyledon and the fir-tree to the different ſpecies of aconitum and ſolanum. By a like ſympathy the poppy adorns the harveſt, the water-lily likes the ranunculus, and rue likes the water-lily; the lily ſprings delightfully by the roſe, near garlic, where it appears more ſhining, and ſmells more perfumed, notwithſtanding the ſmell of the latter is ſo offenſive; the roſe is unfavourable to onions, baſilicum dries up near rice, and cabbages die away near the cyclamen and origanum; the oak does not like the olive, the vines diſlike laurel and hemlock, and hemlock dies away near the vines. The latter brings to our recollection the doctrine of old Robin Abraham Benhannes, who in the 14th century attributed the colour of wine and its fermentation to the ferruginous particles of the grape, and to their union by magnetiſm. The effluvia from the hands or any part of man's body is the cauſe why flowers or herbs droop when touched; the ſenſitive plant

plant is a ſtriking inſtance of the force of this obſervation. The muſcicapa, or catch-fly, mimoſa, and oxalis, the flower martina, annona, dandelion, pimpernel, flower of ciſtus, helianthimum, epine venette, and caſtus opuntia, acquire a very remarkable motion by irritability.

We could take notice of numberleſs others; in fact there are none inſenſible to the emanation of ſurrounding bodies; all move in a reciprocal ſphere of attraction and repulſion. The ſun, whoſe heat attracts the magnetic fluid, dilates or contracts plants in general according to the ordinary courſe of nature, the granadille, (which in fine weather ſhows the time of the day,) the tragopogon, or goat's beard, heliotropus, the cameliorus, and chryſanthemum (or daiſy of the field), the tulip, the lily of Perſia, and a few more, ſhow by their motion the courſe of the ſun, whoſe influence attracts in their different ramifications the principles that vivify them.

When the ſun darts his ray, the enamelled flowers regtiſe and acaſſia open their leaves to receive the influence; but, if he withdraws from the horizon, you ſee their leaves claſp and their flowers decay, till the all-enlivening beam again vivifies them. There is a kind of clover put in action, by the ſolar heat, according to the different degrees of the efficient fluid. This clover will appear whitiſh in the morning, of a purple colour in the middle of the day, and towards the evening it looks yellow and pale. It is the abundance of that fluid in ſome plants which renders them ſo apt to inſpire men and other animals with deſire of love; and it is the want of it in others that appeaſes the heat of blood, and ſtops the progreſs of riſing paſſion.

CONSIDERATIONS on the INDISPOSITIONS and DISEASES of MAN.

MAN, with regard to his preſervation, ought to be conſidered,

1. In a State of Sleeping.
2. In a State of Waking.
3. In a State of Health.
4. In a State of Indiſpoſition.

If we ſurvey all nature, we find in men, animals, plants, &c. but two principles, matter and motion. The whole of the matter which conſtitutes him may be either increaſed or diminiſhed. The diminution ought to be repaired from the general maſs by the means of aliment, as food, drink, and other ſtimuli.

Motion in like manner be either increaſed or diminiſhed. The diminution occaſioned by motion, as walking or any other bodily exerciſe, is repaired by ſleep. Man ſuſtaining two kinds of loſſes, it neceſſarily follows that there muſt be two kinds of reparation in the ſtate of ſleep. Man acts like a machine whoſe principles of motion are applied inwardly, and independently of the organs of ſenſe. The

sleeping

ſleeping ſtate of man is when the uſe and functions of a conſiderable part of his individual faculties are ſuſpended for a while, during which the quantity of motion loſt while awake is repaired by the general currents in which he is placed. There are two ſorts of currents with regard to man—gravity, and the *magnetic current* from one pole to the other; that is to ſay, from head to foot, man receives and collects a certain quantity of the univerſal current as if in a reſervoir; the overplus of motion (or the overflowing of this reſervoir) determines the ſtate of waking. This exiſtence of man begins in a ſtate of ſleep; the degree of motion he receives in that ſtate, proportionate to the maſs, is employed in the formation and unfolding of his organs. As ſoon as his formation is completed, he awakes, and makes efforts on his mother powerful enough to bring him into the world. If his conſtitution is debilitated, his motion being too ſlow owing to his weakneſs, he will preſent a wrong poſition, and will not come into the world without aſſiſtance, from not having ſufficient ſtrength to turn himſelf the proper way.

Man is in a ſtate of health, when all the parts of which he is compoſed have the power of exerciſing the functions they were deſigned for with pleaſure and eaſe. If there is perfect order in all the functions, it conſtitutes a ſtate of harmony or equilibrium. Illneſs is the oppoſite ſtate, wherein harmony is diſturbed, and is either extended over the whole ſyſtem or confined to one part.

Health may be repreſented by a right line. Illneſs is a deviation from that right line: that deviation is more or leſs conſiderable, according to the ſtrength of the diſeaſe; the remedy adjuſts the order or harmony which was diſturbed; the quantity of the univerſal motion that man receives in his origin becomes tonical by being modified in the womb, and helps the unfolding of the viſcera, and all the other organical parts of his conſtitution.

This power of motion is the principle of life; this principle maintains and rectifies the functions of the viſcera. Viſcera are the conſtituent and organical parts, which prepare, rectify, and aſſimulate, all humours, determine their motion, ſecretions, and excretions. The vital principle, being a part of the univerſal motion, and obeying the common laws of the univerſal fluid, is conſequently ſubjected to the impreſſions of the influences of celeſtial, earthly, and particular, bodies with which it is ſurrounded. That faculty or property of man, which renders him ſuſceptible of all theſe impreſſions, is animal magnetiſm or animal electricity.

Man, being conſtantly in the univerſal and particular currents, is penetrated by them; the motion of the modified fluid by the different organizations of its conſtituent parts becomes tonical; it follows in that ſtate the continuity of the body to the extreme parts. From theſe extremities of the body either flow out or paſs in currents the univerſal fluid, when another body capable of receiving or returning them is placed in an oppoſite point.

1. There is a circulation formed between the currents paſſing in and out. 2. Theſe currents are ſtraitened and almoſt re-united in the ſame point; and theſe two cauſes concur together to increaſe ſucceſſively the celerity of motion.

Theſe points of emanation or introduction to or from the tonical current are poles, bearing analogy to thoſe we ſee in loadſtones or artificial magnets; conſequently there are ſome currents coming or iſſuing out of the poles which deſtroy or ſtrengthen each other; their communications being the ſame, it ſuffices to determine one for the oppoſite to be formed at the ſame time. Upon a ſuppoſed line between two poles there is a centre or point of equilibrium, the acting of which is ſuch that no direction is predominant. Theſe currents may be propagated and communicated at any diſtance whatever, either by continuities, connection of bodies and minds, as ſympathy, or that of a fluid, ſuch as air, water, ſound, &c. It is a conſtant law, that in each variety of an intermediate body, the poles are either overturned or changed.

All bodies whoſe form ends in a point or angle ſerve to receive the currents, and become their conductors. We may conſider the currents as openings or channels to convey other currents. Currents can penetrate all ſolid and liquid bodies, preſerving always the direction they have received. Theſe currents may be communicated and propagated by any means, whenever there exiſts a continuity, either ſolid or fluid, in the rays of light, and by a ſucceſſion of the vibrations of ſound. Theſe currents may be reinforced, 1. by cauſes of common motion, ſuch as the inteſtines, and local motion, ſound, noiſe, wind, &c. the electrical friction, and every other body which is a loadſtone, is already endowed with a determinate motion, by animate bodies, by trees, and all vegetables: 2. by their communication with hard bodies in which they may happen to be concentered and aſſembled, as in a reſervoir, to be afterwards at pleaſure diſtributed in every direction: 3. by the multiplication of bodies to which they are communicated, that principle being not a ſubſtance; by a modification its effect increaſes like that of fire, in proportion to its communication. If the current of animal electricity and magnetiſm concurs in its direction with the general magnetic current of the world, the increaſing of all theſe currents is the general effect which reſults from it. Theſe currents may again be reflected by looking-glaſſes, after the laws of light.

Of

Of INDISPOSITION and DISEASE.

IT has been obſerved, that man's life is a quantity of univerſal motion, which in its original becomes tonical, applied to matter, deſtined to form the organs and viſcera, and afterwards to maintain and rectify their functions. Man's life begins in motion, and ends in reſt. The entire abolition of tonical motion is death. As in all nature motion is the ſource of every combination, as well as reſt is of matter, ſo, in man, the principle of life becomes the cauſe of death.

Every unfolding and formation of an organical body depends on the various and ſucceſſive relations between motion and reſt; their equality being determined, the number of poſſible relations between the one and the other ought alſo to be determined. The diſtance between two terms or given points may be conſidered as repreſenting the duration of life; one of theſe terms or points is motion, the other reſt. The ſucceſſive progreſſion of the various proportions of the one and the other conſtitutes the progreſs and revolution of life. Proceeding thus from motion to reſt, we arrive at the point of their equilibrium; after that point we begin by degrees to die.

That progreſſion of divers modifications between motion and reſt, may have an exact proportion, or that proportion may be diſturbed. If man runs through that progreſſion without the proportion being diſturbed, he lives in a good ſtate of health, and arrives at his term without illneſs: on the contrary, as ſoon as the proportions are troubled, diſeaſe begins. Illneſs is nothing elſe but a perturbation in the progreſſion of motion and life, which may be conſidered as exiſting either in ſolid or fluid bodies. If it exiſts in ſolids, it diſturbs the harmony of the properties of organical bodies by diminiſhing the one and increaſing the other. If it exiſts in fluids, it diſturbs their local and internal motion.

The aberration from motion in ſolids, by altering their properties, diſturbs the functions of the viſcera and the various elaborations which ought to take place. The aberration from the inteſtine motions of humours produces their degeneration. The aberration from local motions produces obſtruction or debility, fever or irritation.

The ſlowneſs or abolition of motion produces obſtruction, or debility; the acceleration of motion produces fever, or irritability. The perfection of ſolids or viſcera conſiſts in the harmony of all their properties and functions; and the reſult of the functions of the viſcera is the quality of fluids with their inteſtine and local motion. To be able to rectify the general harmony of the body, we muſt rectify the functions of the viſcera; becauſe, their functions being once re-eſtabliſhed, they rectify every thing that can be ſo, and divide every thing that cannot be rectified.

That

That effort of nature or viſcera upon the humour is called *criſis,* or paroxyſm; and no diſeaſe can be cured without a criſis. In all criſes, we diſtinguiſh three ſtates, the perturbation, digeſtion, and evacuation. Diſeaſe being an aberration from harmony, that aberration or prediſpoſition may be more or leſs conſiderable, and produce more or leſs ſenſible effects, which are called *ſymptoms.* If thoſe effects are produced by the courſe of the diſeaſe, they are called *ſymptomatic ſenſations;* if on the contrary they are the efforts of nature againſt the cauſe of illneſs, they are called *critical ſymptoms.* It is of the greateſt moment to diſtinguiſh them well in practice, to prevent and ſtop the one, and favour the other.

It follows from what has been ſaid, that all cauſes of diſeaſe diſturb and alter more or leſs the proportion between matter and motion, the proportion of the viſcera, the proportion between fluids and ſolids; and conſequently they produce by their different applications a remiſſion or perturbation more or leſs conſiderable in the properties of matter. To remedy the effects of remiſſion and their perturbation, and to deſtroy or ſtop them, the remiſſion of properties muſt be provoked; that is to ſay, in animal bodies, the irritability or animal electricity muſt be increaſed by different ſtimuli. There are two methods of doing this: 1. to leſſen the obſtacles; 2. to increaſe the action of nature, by a continual, ſhaded, ſoft, and harmonic, application of magnetic currents.

A body being in harmony is hardly ſenſible to the effect of animal electricity and magnetiſm, becauſe that the application of an uniform and general action cannot alter any thing in proportions which are both exact and already conſervant with that harmony. If on the contrary a body is not in harmony, that is to ſay, if it is in that ſtate wherein proportions are diſturbed, the habit it is in to experience that diſſonance hinders it from being more ſenſible, and it becomes ſo by the application of animal electricity and magnetiſm; becauſe that diſpoſition and diſſonance are increaſed. On theſe principles it is eaſy to conceive that ſick perſons drawing near their recovery become gradually inſenſible to animal electricity and magnetiſm; that abſolute inſenſibility to its power conſtitutes the perfect cure.

It follows, from the ſame principles, that the application of animal electricity and magnetiſm muſt often increaſe the pain, as its action occaſions the ſymptomatic ſenſations to diminiſh or ceaſe; and the efforts of nature againſt the cauſes of diſeaſe being increaſed, it is abſolutely neceſſary for the critical ſymptoms to increaſe in the ſame proportion.

It is by the exact obſervation of their ſeveral effects that we are enabled perfectly to diſcern the ſymptoms. The unfolding of the ſymptoms is made in the contrary order by which the diſeaſe was formed, and may be compared to a ball of twine which winds off in the contrary order to which it was wound on.

Of HUMAN IMPREGNATION—FORMATION of the FŒTUS—ORIGIN of DISEASES—and PRINCIPLES of LIFE and DEATH.

IT was not my intention to go into this Treatiſe ſo much at large, in my preſent work; but conſidering that the ſubject is of the higheſt importance, and that the Medical Part could not be made complete without it, I have reſolved to introduce it here, though I ſhall be under the neceſſity of extending my Plan to a few more numbers. Theſe, I truſt, will not be unacceptable; ſince they will be accompanied with a ſet of very curious and valuable plates, deſigned on purpoſe to illuſtrate this intereſting ſpeculation.

In contemplating the works of creation and the word of God, unfolded to us by the light of Revelation and Scripture; by analogy, reaſon, medical experiments, and anatomy, we are enabled to trace the human œconomy farther in her retirement, and deeper in her occult retreat, than ſome medical men are willing to ſuppoſe. Impoveriſhed by a faſhionable ſtyle of living, and driven to the neceſſity of multiplying potions and fees, their object is not to heal, but nouriſh the ſeeds of human infirmity. The truth of this remark has been but too often experienced; and indeed *confeſſed* by ſome, in thoſe awful moments, when diſſimulation would be vain. Far be it from me to arraign the profeſſional character in its general capacity; it is only the medical locuſts that I wiſh to eradicate; and I am perſuaded every good man in the faculty would with heart and hand aſſiſt me in ſo laudable a purſuit. It was principally with this view, and to aſſiſt private families in the moments of extremity, that I was induced to offer thoſe ſimple modes of cure and ſelf-preſervation, ſo amply diſpenſed in my edition of Culpeper's Engliſh Phyſician. And my preſent purpoſe being to make that invaluable family-book ſtill more complete, I ſhall here explain the nature of human generation, and the principles of animal life, that I may from thence deduce the origin of hereditary diſeaſes, and point out with more facility thoſe which are accidental. And in this treatiſe I ſhall endeavour to furniſh my readers with ſuch obvious directions for *eſchewing the evil, and chooſing the good*, as, if reſolutely followed, will not fail to preſerve health and long life, and prove of no ſmall benefit to future generations.

When God created Adam, he planted in him the ſeeds of that Divine Eſſence, requiſite to propagate the human *life* and *ſoul*. Theologiſts may contradict me; yet I will not ſo much derogate from the wiſdom and omnipotence of the Creator, as to ſuppoſe he ſhould watch the impregnation of every human female, and, by ſo many ſeparate and diſtinct acts of his power, give life, ſpirit, and *ſoul*, to the fœtus. The Creator of Man, viewing with unbounded foreſight the purpoſes before him, by one act of his omnipotence blended in Adam all the faculties of the human and celestial nature; and without any doubt, when he was formed *one, in God's express image*, he possessed the means of propagating from his own essence, beings like himself. It is here difficult to associate the imperfect ideas of human reason with the mechanism of Divine Wisdom; and yet our conception may in some degree unravel the mysteries of nature by causes and speculations, which, in proportion as they captivate our senses, and raise our admiration, excite in us a reverential awe of futurity, and a grateful sensibility of the goodness and mercy of him who gave us being.

From the evidence of Scripture it is indisputably clear, that in the person of Adam the male and female properties were originally combined; as indeed we now find them in many species of the lower class of animals. In Genesis i. 27, we read, that *God created man in his own image*, i. e. of perfection; including or containing the prolific or generating powers, which are distinguished by the expression of *male and female;* and *God blessed them*, i. e. these male and female properties, *and said unto them, Increase and multiply, and replenish the earth*, i. e. with beings like Adam: for this benediction, and this command, were *antecedent* to the formation of Eve, as every one must know who reads the Scriptures.

In this *plural* capacity, therefore, Adam received the blessing of God, when he said unto him, *Be fruitful and multiply, and replenish the earth, and subdue it; and have dominion over the fish of the sea, and over the fowls of the air, &c.* The six days creation were now completed; and on the seventh day God rested from all his work; and, having formed Adam, and breathed into his nostrils the breath of life, *he became a living soul.* God also planted the garden of Eden, and put the *man* into it, to till it and to dress it; and God commanded the *man*, saying, Of every tree of the garden thou mayest freely eat; but of the tree of the knowledge of good and evil, thou shalt not eat of it; for in the day thou eatest thereof, *thou shalt surely die.* Gen. ii. 27.

Let it here be noted, that all these transactions, injunctions, and commands, had passed *before Eve was formed*, or, in other words, before the male and female essences were separated, and made the essential parts of two distinct persons. Adam likewise, before the event took place, was appointed God's viceroy over all earthly things, both animate and inanimate; the very elements being made subject to him; for *he was formed more noble than the angels, and crowned with glory and honour;* i. e. having the peculiar advantage of *multiplying his own race.* He was, as to his external form, moulded of the celestial æther; and therefore, previous to his fall, his body emanated rays of brightness and splendour, similar to those which our ideas furnish of Moses and Elias when they conversed with God. His reasoning faculty, and liv-

ing

ing ſoul, were formed of the eternal eſſence or tincture of the Divinity; being nothing leſs than what is termed the *breath of God*, that ſpark of immortality which generates *ſoul* and *body*, and is the diſtinguiſhing characteriſtic between man and beaſt. For, although brute animals inherit the five ſenſes, and poſſeſs an inſtinct to direct them in the choice of food, and to impel the propagation of their ſpecies; yet theſe are only ſenſes formed from the *out-birth*, or four elements of nature; and not from the eſſence or tincture of the Divinity, out of which the ſoul, the mental intellect, reaſon, ſenſe, and underſtanding, are all formed, and transferred to poſterity. *For with the powers God has endued man, with the ſame powers ſhall he multiply his race.*

From the foregoing paſſages we are warranted to infer, that the original man was poſſeſſed of his ſpiritual ſoul, and rational intellect, for the purpoſe of propagating the ſame to all future generations. By the force of this rational intellect, or eternal ſpirit, unclouded by the deformity of ſin, he knew and perceived the nature and property of every animated being; and to exerciſe this intellect, God brought before him every created thing, to ſee what he would call them; "*and whatſoever Adam called them, that was the name thereof.*" He knew and perceived the nature and quality of all animals; and, according to their deſignation and ſubjection to the external elements, ſo he aſſigned them thoſe names which they have ever ſince borne. Adam, however, in his primeval ſtate, was not himſelf under the influence of celeſtial or terreſtrial elements; but, on the contrary, they were ſubject to his controul. He was immortal; they corruptible. They ſprung out of time, and were elementated; he ſprung from the limbus of eternity; and into eternity the divine eſſence, or ſouls, propagated from him, muſt indiſputably return.

But man, thus created in honour and immortality, abideth not. The purpoſe of his creation was to fill the place of the rebel angels; and hence Lucifer became his mortal foe. This fallen Spirit had entered the gate of Eden, and was preparing to ſeduce Adam, when the Almighty conſtituted the teſt of his obedience; for having endowed him with a *free-will*, an innate power of *chooſing good* or *evil*, and of multiplying the ſame, it was but reaſonable to expect from him an implicit obedience, and an angelic race. He that is alone eternal and omnipotent, could not but foreſee the ſubſequent event; and it is his ſupreme goodneſs to counteract evil, by preventing its worſt conſequences. Foreſeeing that the prolific tincture, or eternal eſſence of fecundation, might be contaminated by the malignant ſpirit of Lucifer infuſing itſelf into the mind of Adam; who then, inſtead of multiplying an angelic race, would generate devils; and that, were man to fall in his individual capacity, there was no counterpart, no feminine principle, through the medium of which the

the *ſerpent's head could be bruiſed,* or a Saviour become incarnate ;—Therefore on a further ſurvey, after the works of creation had been completed, animals named, and man formed and compounded of the male and female tinctures, God ſaid, Gen. ii. 18. *It is not good that the man ſhould be alone; I will make him an help meet for him;* wherefore the rib, i. e. the feminine or conceptive eſſence, was taken out of Adam, and concentrated or moulded into a new being called *woman.* The emiſſion of this feminine eſſence or tincture threw Adam into a deep ſleep; yet, when he awoke, he knew that an eſſential principle had departed from him, and that the woman was *bone of his bone, and fleſh of his fleſh,* not having been created, *but formed out of himſelf,* whereby he only retained the animating principle, or active power of generation; whilſt the rudiments or ſeeds of future beings were conſigned to the matrix of the woman. Here then individual generation ceaſed; and Adam, without the *counterpart of himſelf,* had no longer the power to increaſe and multiply. Thus the two tinctures, or divine eſſences, animating and compounding ſoul and body, were divided; and by means only of a re-union or contact of thoſe tinctures, could generation then, or now, be performed. It is on this ground that the male and female affections are continually turned towards each other; and that the deſire of love and union ſo ſtrongly pervades every individual of the human race. Hence alſo the Tempter's reaſon for beguiling Eve; and hence the ſeducing power of love, which determined Adam to ſhare in all the horrors of her crime, ſo pathetically and affectingly deſcribed by Milton.

The fatal conſequences of the fall, we moſt ſenſibly feel, and univerſally deplore. The earth ſhook from her foundations. The order of nature was quite inverted. The ætherial and terreſtrial elements, which before were faſhioned in harmony, and acted in uniſon, were now diſcordant, intemperate, and furious. Brute preyed upon brute, and bird invaded bird. The delicious fruits and flowers of Paradiſe were exchanged for thorns and thiſtles. The ſerenity of a pellucid and ſmiling firmament, was convulſed by the thunders of an incenſed Deity, by forked lightnings, by contending ſeaſons, by devouring winds, and impetuous ſtorms. While man, ungrateful man, from the privilege of holding theſe elements in ſubjection, became ſubjected to them; and hence ſubject to all the perils and misfortunes of his fallen nature.

Here, then, began the conflict of the human paſſions, as violent and ungovernable as the elements themſelves. Here the toil and labour of the man, who ſhould *earn his bread by the ſweat of his brow,* and the tears and travail of the woman, who ſhould *conceive in pain and ſorrow,* had each their ſource. Here likewiſe, the dark catalogue of human infirmities, of diſeaſe and death, had its too early date; yet

to

to this æra, which gave birth to our manifold misfortunes, muſt we look for that benign ſource of alleviation and cure, which the relenting hand of Providence has graciouſly afforded to thoſe who will ſeek for them; *for out of the ground hath the Lord cauſed medicine to grow; and he that is wiſe will not deſpiſe them; for with ſuch doth he heal men, and taketh away their pains.* Eccl. xxxviii. 4, 7.

Since, by his fall, man became ſubject to the elements, from them he receives the conſtitution of his body; but his reaſoning intellect, and ſpiritual ſoul, are derived from the pure eſſence or tincture of the Deity, originally infuſed into the ſeed of man. To the violence and impurity of the elements, we owe the diſorders of the body; to the temptations and allurements of the devil, we juſtly impute the diſeaſes of the ſoul. Yet by due attention to our reaſoning faculty, it is no hard taſk to preſerve health, or prolong life, to the term of its *natural* diſſolution; while, by the powers of the mind, and the light of the goſpel, we may ſtill avoid the poiſon of ſin, and become members of that eternal kingdom, which is the ſure reward of the good and virtuous.

The imperfections and diſeaſes of the body, therefore, beginning with Adam, are in conſequence tranſmitted to his poſterity; and may be divided into *heriditary* and *accidental.* Hereditary complaints proceed from a certain defect of the animal powers, or imperfect ſtate of the ſanguiferous ſyſtem, at the time of copulation. The accidental conſiſt of all ſuch maladies as are communicated by the diſcordant or putrid ſtate of the elements, not only during the time the child is encompaſſed in the womb, but from its birth to the lateſt hour of its exiſtence And it might here be obſerved, that the increaſe or decreaſe of both hereditary and accidental diſeaſes, depends almoſt entirely on the purity or impurity of the blood. For, if pure, in both male and female, at the time of impregnation, the fœtus will be naturally ſtrong and healthful. So likewiſe, if after parturition, and during life, care be taken to keep the blood in an uncontaminated and elaſtic ſtate, we ſhall not only avoid the common effects of exceſſive cold, heat, and moiſture; but likewiſe that direful train of acute diſeaſes communicated by putridity and infection; or, ſhould they by chance attack us, the effect becomes ſlight and temporary. A circumſtance this, which ſurely ought to weigh perpetually on the minds of thoſe, who know how to value the bleſſing of health, or who would wiſh to live a long, an active, and a pleaſant, life. This is therefore a ſpeculation of that high importance, that I ſhall now ſhow how hereditary complaints are communicated in the act of copulation—how increaſed and foſtered in the womb—how accidental diſeaſes grow up and follow,—and how both theſe enemies to the health and happineſs of mankind may be prevented or overcome.

In regard to that union of the ſexes to which we are inſtinctively impelled, or rather in the union of thoſe eſſences or tinctures peculiar to the generative organs of male and female,

female, in the contact of which the firft moments of human exiftence commence, the moft whimfical and abfurd theories have been fet up. No branch of phyfiology has been more expofed to cenfure and miftake. While the phænomena of the heavens, of the earth, and even of the human mind itfelf, are traced with a fteady hand, and with all the dignity of philofophy, the functions of the human body, in health as well as under difeafe, though expounded with a profufion of fantaftical erudition, appear almoft in as much doubt and darknefs as in the days of Paracelfus.

Let us then proceed to review the mode by which generation is accomplifhed. I have, in the former part of this work, already explained the fyftems of Buffon and of Leeuwenhoek, in their fpeculations on the animalcules found in the feed of man and in that of brute animals; I have alfo in the medical part of Culpeper's Englifh Phyfician, fhown the mode by which generation is performed, fo far as relates to the action itfelf, and to its grofs effect. I fhall now confider it in a new light, as it concerns the propagation of foul and body, and of family-temper, likenefs, and difeafe; but, as the female is fo materially concerned in the myfterious act of impregnation, and in all its confequences, I fhall here take up the reafoning of a late ingenious anonymous author, whofe opinion exactly coincides with my own.

The extremity of the uterine fyftem, without the nymphæ, feems not, except from its aperture, and the lafcivious fufceptibility of its texture, materially requifite to generation. Immediately within the nymphæ, the *vagina,* or great canal of the uterus, begins. Before coition has difturbed its proportions, it is generally about five or fix inches long; and when thrown into a circular form, without violent diftention, its diameter is about a fixth part of its length. But as, in coition, the vagina is the immediate receptacle of the penis, it is capable of great diftention, and may be rendered of very confiderable capacity. In general, however, after frequent contact, this canal becomes much fhorter, but more proportionably increafed in its diameter; yet, being contrived by its organization for the purpofe of exciting pleafure, it can and does accommodate itfelf to whatever fize is neceffary clofely to embrace the penis in the act of copulation.

At the upper extremity of this canal, the uterus or womb is feated. It is of a pyramidal form, with its apex towards the vagina. Its greateft length, in virgins, is not more than two or three inches; and its width is fcarcely one; its internal cavity muft therefore be very fmall. It is connected to the vagina or great canal by a paffage fo fmall, that a bodkin or ftilet cannot be introduced without fome difficulty. In the broad or upper extremity of the womb, the ovaria are feated. Their fubftance is fpungy, and they contain an indefinite number of veficles of a dufkifh femi-tranfparent quality, the involucra of which are diftinct, and fimilar to the general

neral ſubſtance of the ovaria. Theſe veſicles are the *ova*, or eggs, which contain the rudiments of the fœtus, and which muſt abſolutely be impregnated with the male ſeed, before it can be poſſible for generation to take place.

Now it has been, and is, the common opinion, that, when venereal embraces take place, the whole genital ſyſtem of the male being thrown into action by libidinous fire and violent friction, by this exertion the ſemen is thrown with conſiderable vehemence from the penis, and is either forced through the mouth of the womb, and attracted by the ovaria; or, that it is received by the Fallopian tubes, and conveyed by them through a variety of convolutions, till by their fimbriæ they are conducted to the ovaria, in the manner I have already fully deſcribed in the medical part of Culpeper's Engliſh Phyſician; all which tedious and complicated proceſs is alleged to take place in the inſtant of coition.

Others again ſuppoſe, that the internal orifice of the womb becomes open and pervious during the exertion and enjoyment of copulation, and that the glands of the penis abſolutely paſs into the cavity of the womb, and eject the ſeed immediately upon the ovaria. To each of theſe theories there appear inſuperable objections. In refutation of the firſt, we need only obſerve, that the vagina, from its ſtructure, and from its organization in the act of venery, is diſpoſed ſtrongly, and in every part, to embrace the penis; and, as the glands muſt thereby be cloſely ſurrounded, although it reaches not in every perſon to the furtheſt limits of the vagina; the ſlight and momentary impetus of the ſemen will thus be very effectually reſiſted, if not totally ſubdued. If the penis be not of magnitude ſufficient to occupy the vagina to its full extent, the unoccupied ſpace muſt be ſomehow diſtended; and, let this vacuum be what it will, its reſiſtance muſt be effectual; and, if it is not diſtended, the power or preſſure which occaſions its collapſe, will over-balance the impetus of the ſemen. But ſuppoſing the virile member in all caſes to be ſo exactly proportioned as to occupy the whole length of the uterine canal, which however we know is not the caſe; yet from what principle ſhall we aſcertain that the ſeminal tube of the penis, and the apex of the womb, ſhall be made ſo exactly to correſpond as to become continuous? The ſemen, in the event of coition, is doubtleſs thrown out by the penis with ſome force; though this force will always depend upon the vigour of the male organs, and therefore muſt vary from the loweſt to the higheſt degree of vigour which theſe organs can be ſuſceptible of. But even allowing the glans penis and apex of the womb to fall into exact contact upon due penetration; and that the male ſeed is always ejected with conſiderable force from the penis, and the vigina to be no barrier to the progreſs of it; yet how is it to force its way into the cavity of the womb? The aperture which leads from the vagina or great canal into the

the womb, is in fact no aperture at all. During menstruation, indeed, it is pervious; but even then it is only capable of admitting a very small probe; and this is no argument that it is naturally, and at other times, pervious. How often too has this aperture been entirely blocked up by preternatural obstructions, and conception nevertheless taken place? Instances of this have often occurred; and the precision and authority with which they are recorded by different practitioners, leave no room to evade the argument. Hence this mode of impregnation appears not only highly objectionable, but utterly impossible; having no correspondence with the human structure, or with the economy of Nature.

After what has been said, it may appear idle to prosecute any farther refutation of the progress of the male seed by the Fallopian tubes, or through the mouth of the womb. But, as authors of the greatest respectability have believed in its progress through the tubes, and tell us they have even seen it there; it may not be improper to enquire how far this is ascertainable. The Fallopian tubes, through which the semen is said to pass, originate, by very minute perforations, through the fundus of the womb; and, increasing rapidly in their diameters, their capacities, when dilated, may be about the third part of an inch where they approach the ovaria. Here, again, they suddenly contract, leaving only a very small opening; while their main substance is still continued, and is expanded into that plaited or jagged fringe called the fimbriæ, which is contiguous to the ovaria.* I shall now ask by what law in Nature, by what effort of it, is the male semen to be conducted through this conical and convoluted canal? Can the semen now possess any active force, to introduce itself through the rigid perforations of this organ, and to overcome the collapse of the tubes? The stimulating power of the semen must soon be lost in a vessel which it has not power to distend; and we cannot suppose it capable of acting in a direction completely opposite to what is the acknowledged office of the tubes. It must be by irritability that the ovum is conveyed into the uterus from the ovaria; and we know no vessels in any part of the body whose action is double and contrary. This system therefore savours of great improbability. But we are told, by some, that they have actually seen the male semen in its unaltered state, lodged in the Fallopian tubes. These sagacious authors might as prudently have affirmed, that they had seen snow upon the canal in Hyde-park at Midsummer. They did not know, or did not choose to recollect, because it made against pre-conceived opinion, that the human seed, when subjected to heat, especially to such a moist and natural heat as those parts constantly afford, soon looses its spissitude and

* See Medical Part of Culpeper, page 17, 89, 97, &c. where all the parts, both male and female, are anatomically described.

tenacity, and becomes very ſubtilly fluid, and almoſt colourleſs. Beſides, it is univerſally acknowledged, that a conſiderable part of the ſemen is almoſt always, immediately after coition, rejected by the female. When we attend to the many inſtances of credulity and impoſition in the theories of generation, we need not marvel at the aptitude and facility with which pretended diſcoveries creep into notice, and the ſolemnity with which they obtrude themſelves into ſyſtems.

All the foregoing arguments againſt the poſſibility of a pervious communication between the vagina and the uterus, are alſo concluſive againſt the ſuggeſtion, that the penis, in the act of coition, penetrates into the cavity of the womb. Nor is the aſſertion of thoſe who contend that this orifice, by the turgidity of the parts during coition, naturally opens and dilates itſelf to receive the male ſeed, marked with the leaſt degree of probability. How is this dilatation of the orifice to be effected? Though the whole uterine ſyſtem, during the venereal act, be rendered ſtiff and turgid by animal deſire and influent blood, yet it is more probable that this turgidity would rather compreſs than dilate the orifice; and the ſtructure and texture of the womb ſeem exceedingly unfavourable to ſudden dilatation by any means whatever. In an unimpregnated or virgin ſtate, the womb is ſo ſmall that its ſides coaleſce or adhere together, and it has no hollow appearance whatever, though, from the texture and elaſticity of its fabric, it may be thrown into a globular form, which will conſtitute a cavity. But in coition, with all its occult and uncommon phenomena, what charm have we left to overcome this coaleſcence, and form this cavity, by opening or ſeparating the membranous ſides of the womb? Will it here be ſaid that the forcible ejection of the male ſemen will effect this purpoſe; or that the ſtiff and turgid ſtate of the penis itſelf will force its way into the fabric ſo remote and delicate? Though females may entertain ſanguine ideas of theſe things, we muſt ſuppoſe the vigilant anatomiſt, toiling through the unalarming and chilly organs of the dead, ought to furniſh a more rational hypotheſis, whence to deduce the active principle and admirable proceſs of human impregnation.

Authors have been always eager to eſtabliſh the certainty of a conſiderable afflux of blood to the female organs, and conſequent turgidity during the voluptuous communication of the ſexes; and this has been a wonderful prop to many abſurd conjectures. This afflux, and conſequent turgidity, they ſuppoſe originates, like the erection of the penis, from the ſtrength of libidinous ideas, and other locally-irritating cauſes; and is intended by nature to induce a tenſion in the female organs, that the progreſs of the ſemen may thereby be facilitated. This tenſion, again, they ſuppoſe induces ſome kind of conſtriction, which is ſaid to ſupport the action of the different parts of the genital ſyſtem, but particularly of the Fallopian tubes. Theſe tubes, it is ſaid, are remarkably diſtended, during coition, by the blood ruſhing into the numerous veſſels which creep between their coats, by which means they are erected, and their fimbriated terminations applied to the ovaria; and it is gravely added, that diſſections of gravid women, and the comparative anatomy of brutes, corroborate the opinion. Were it not for the ſerious reſpect with which this anatomical obſervation hath for a length of time been favoured, nobody ſurely would be at the pains of detecting the abſurdity. Allowing that this turgidity, with all its concomitant circumſtances, really happens in the *living* ſubject, how can it poſſibly exiſt in a carcaſe flaccid with death, and, as is always the caſe in a human anatomiſed body, where death muſt have taken place ſome conſiderable time before?

But this turgidity, though it ſometimes may happen, and yet in a degree very limited to what is alleged, does not always happen; and, when it really does take place, it ſeems rather to be the companion and promoter of libidinous gratification, than a principal and eſſential promoter of conception. To many women the embraces of the male are extremely, if not completely, indifferent; and to ſome they are abſolutely diſagreeable; yet even theſe women are prolific. There is no difficulty in ſuggeſting a very ſufficient and natural reaſon why the parts of the female directly ſubjected to the action of the penis, during the venereal congreſs, ſhould become turgid with influent blood, and ſometimes be conſtricted. Nature, though ſhe ſeems in general unfriendly to exceſſive luſt, ſometimes permits it; and theſe are the means ſhe ſeems to have appointed for heightening it. Beſides, it is proper that the animal inſtinct, which prompts the reproduction of the ſpecies, ſhould not be diſappointed in its gratification, however brutal theſe ſenſations and ideas may appear to the purified philoſopher. Theſe means then, however they may contribute to the mutual ſenſibility of the ſexes, in the voluptuous gratification of animal pleaſure, appear to have no real influence on the proceſs of generation, after the venereal congreſs has ceaſed; nay, we have reaſon to believe that their action or influence does not extend beyond the limits of the vagina, except in common with the reſt of the genital ſyſtem, even during that congreſs. If an afflux of blood to theſe parts were always to be attended with theſe effects, what violence muſt the ovaria be expoſed to by reiterated coition, and by every return of the menſtrual diſcharge? During the menſtrual afflux, a very conſiderable diſtenſion muſt ſurely take place over the greateſt part, if not the whole, of the genital ſyſtem; and, as this turgidity is the principal reaſon aſſigned for the action of the tubes, by what means are the fimbriæ diverted from exerciſing thoſe functions which turgidity, though from another cauſe, at another time ſo ſucceſsfully inſtigates? Alſo, how happens it that grateful copulation is not always productive, and the contrary; that the fimbriæ, in

every

every venereal act, do not operate upon the ovaria, and thereby produce more fœtuses, or a waste of the ova? and that the organs themselves are not incapacitated, or diminished in their energy, by such repeated exertions? We have every reason then to conclude, that the tension and constriction of the female organs, induced by the afflux of blood during coition, if of consequence, are intended solely to promote animal gratification; and, that they have no direct influence on the actual progress of the semen through the above-described communications to the ovaria.

Upon the whole, it is certainly no ways equivocal, that the semen cannot, in any manner, be applied to the ovaria by means of the fimbriæ; that it cannot ascend or advance through the convolutions of the Fallopian tubes; that it cannot divaricate and traverse the compressed uterus; and that it cannot even operate a passage through therigid bulwark of the cervix uteri. The probability of the progress of the *aura seminalis*, through the same paths, is destroyed by the same arguments; and the whimsical opinions founded on the presence of animalcules in the semen, and on the organic bodies furnished by the semen of both sexes, and, uniting in the uterus, as far as this alleged aperture is concerned, must stand or fall by the same fate. It may seem however strange, that a doctrine so ancient, and so universally believed, should be so easily overthrown; and it may furnish, to the speculative reader, unfavourable ideas of the present state of medical literature. He may indeed wonder, that, while every science has become rational and respectable by the exertions of their cultivators, medicine alone has been able to resist the diligence of a thousand years; although it has been wrested from the hands of nurses, and its profession become dignified and lucrative, it can scarcely be said, at this day, to afford one unquestionable idea. In the volumes of physiology, compiled by the most learned physicians, and drawn from the most learned sources, will the unconcerned philosopher find the dogmata of medicine consistent with Nature, or with common sense?

But since the semen, in some shape or other, contains that animating principle which is indispensably necessary to generation; and since the ovaria as indisputably produced something from whence a living creature is to be evolved, it becomes demonstrably clear, that the influence of the male seed must be powerfully incorporated with the female, and directed to the ovaria, before this effect can possibly take place. We have already seen how this cannot happen; let us now endeavour to point out a rational medium by which it may be accomplished. For this purpose we must again return to the vagina, or canal of the uterus, as being the principal organ, on the part of the female, which actually contributes to propagation; and without the full and complete use of which, impregnation cannot take place. It therefore demands a very minute and attentive investigation.

The vagina is elaſtic, and ſomewhat membranous, compoſed of muſcular fibres, blood-veſſels, nerves, and lymphatics. It commences from beneath, at the nymphæ, and, riſing obliquely about five inches, is loſt upon the uterus. Its capacity is very different in different ſubjects, and in no very diſtant periods of life in the ſame ſubject. A very reſpectable anatomiſt finiſhes his deſcription of it by ſaying, it is *membro virili ſecundum omnes dimenſiones accommodabilis.* Its inner membrane, though very uneven, is delicately ſmooth, and, from its nervous texture, exquiſitely ſenſible; the outer membrane is more ſpongy and muſcular; and, the whole body of the canal is very plentifully ſupplied with blood-veſſels, nerves, and lymphatics. We know little more of the lymphatics of theſe parts, than that they are more numerous proportionally than in any other part of the body. Thoſe which originate in the exterior parts of the female genital ſyſtem, traverſe the inguinal glands, while the deep-ſeated ones take a much more direct courſe to their place of union with the lacteals; but of theſe we ſhall be more particular, when we adduce our obſervations in favour of a very powerful *abſorption* ſubſiſting in the vagina.

The entrance into the canal of the uterus from without, is guarded by the nymphæ, which form an eminence on each ſide, ſo peculiarly conſtructed and arranged, that we muſt think lightly of the phyſiologiſt who could ſuppoſe them to be only appendages in office to the urethra. Indeed, as Nature frequently operates more than one end by a particular ſtructure, we ſhall not pretend to limit the ſecondary or inferior offices which the nymphæ may promote; but we ſee much reaſon to believe them created to aſſiſt powerfully in preventing the ſpeedy eſcape of the male ſemen, and thereby expoſing it longer to the action of the abſorbent ſyſtem. A multitude of circumſtances corroborate this belief; and it will not be impaired by the allegation, that theſe ridges by no means conſtitute a regular and complete valve. Immediately within this barrier, a ſtructure, on the ſame principles as thoſe of the nymphæ, but more elegant and powerful, commences; and it is continued over the ſurface of the vagina, gradually growing finer, till it is loſt in ſmoothneſs near the upper extremity of the canal. This ſtructure is the rugæ of the vagina, ſo accurately drawn and deſcribed by Haller and others; but degraded by ſome anatomiſts, who mark it only as uſeful in exciting venereal enjoyment, or admitting expanſion during coition and parturition. It is inſinuating a mean and diſgraceful reflection on the important order and operations of Nature to ſuppoſe, that theſe rugæ, which are not caſually arranged, but are regulated with as much preciſion and uniformity as we can trace in any other part of the genital ſyſtem;—I ſay it is nugatory and preſumptuous to aſſert, that this intricate, extenſive, and beautiful, arrangement, has been ſo minutely laboured for no other purpoſe, but merely to

excite

excite a greater titillation during the grofs and libidinous commerce of the fexes, and a greater extenfion during parturition. This ftricture may indeed promote thefe fecondary purpofes; but it is intended for much nobler ends. Had thefe rugæ been conftructed merely for fimple contraction and dilatation, they would have covered equally the whole furface of the vagina, which certainly does not happen; neither, if thefe had been their principal ufes, would they be fo foon and fo eafily obliterated. We believe, then, that the rugæ of the vagina are thus contrived principally to protract the femen in that vifcus after the penis is withdrawn, and thereby to favour *abforption*; efpecially as the qualities of the femen coincide wonderfully with thefe intentions.

The femen, as it is fecreted from the blood in the tefticles, is very different from that heterogeneous mixture which is expelled by the urethra in coition; though, by the alteration, its fecundating quality is not improved. When it is conveyed into the vefic1es, it is of a thin confiftence, of a pale yellowifh colour, and little in quantity. In thefe vefic1es it is fomewhat infpiffated, and its colour heightened; and, after it is mixed with the liquor of the proftrate glands, it becomes ftill thicker, and of a more whitifh colour. This confiftence which the femen acquires in its progrefs from the tefticles, may produce other flight properties; but the principal intention of it feems to be, to correfpond more effectually with the abforbent power of the vagina: for thus, by the increafed tenacity of the femen, the remora of its fecundating part muft be protracted in the vagina, while at the fame time the abforbents are allowed more time to attack thofe active fubtile parts intended to be carried into the circulating fyftem. We may add here, in order farther to confirm the opinion concerning the ufe of the tenacity of the femen, that, when too little of this mucilage is derived from the glands, or when it is of a depraved or thin quality, the whole mixture efcapes the machinery of the vagina too rapidly, and hence coition becomes unproductive. This is the feminal ferofity, as it is called, held to be one of the few caufes of fterility in man. And we may add farther, that, when the confent and power of procreation begins to fail on the part of the woman, the crenulations of the vagina are then always vifibly decayed, whether affected by the advances of age, or by imprudently-reiterated venery. But what are we to think of a very refpectable author, who gravely tells us, that the femen, by ftagnation, and by the addition of the cream-like liquor of the proftrate glands, is better fuited to the projecting effort of the urethra in the event of coition? Indeed, it is not to be denied, that the increafe in quantity of the feminal mixture may enable the projectile power of the urethra, with its aiding mufcles, to act with greater efficacy; but a boy would laugh in my face were I to tell him, that by adding to the weight and tenacity of water, his ſquirt would throw it much farther. To act in concert, then, with theſe unqueſtionable qualities of the ſemen, the ſurface of the vagina, by means of its rugæ, from their elevation and arrangement, muſt have a very conſiderable effect in heightening the remora we have deſcribed. No doubt, if Nature had only had in view the prevention of the regreſs of the ſemen, we might have met with a much ſimpler mechaniſm; but, as to this part very different offices, and all of them material, were allotted, it has been intricately qualified for them all.—Thus, upon the whole, we ſee an admirable diſpoſition in the ſemen, and in the ſurface of the vagina, to facilitate and promote the action of the abſorbent veſſels.

Though the *abſorbent ſyſtem* has not been traced with the ſame minuteneſs and ſucceſs which have followed the inveſtigation of the ſanguiferous ſyſtem, it is however known to be very general, and very powerful; and it is remarkably ſo in the cavity of the pelvis. How, otherwiſe, is that effuſion which is conſtantly going on, in order to lubricate the whole genital ſyſtem in the female, and to prevent the coaleſcence or concretion of its ſides, reſumed? In thoſe unfortunate females whoſe menſes have taken place, but in whom likewiſe the expulſion of them has been prevented by the unruptured hymen, or by unnatural membranes blocking up the paſſage, much of the blood has always been reſorbed; and in thoſe whoſe diſeaſe has exiſted long, and where the thick parts of the blood have begun to be broken down, the colluvies has been reſorbed, and a train of ſymptoms induced, not to be accounted for by the mere turgidity which this obſtruction occaſioned. The infection and progreſs of ſyphilis or confirmed lues, not only eſtabliſh the certainty of a very rapid and powerful abſorption in the vagina; but alſo exhibit the power and influence of the irregularities of its ſurface. It is ſurely very evident, that the chief application of the venereal virus, whether in gonorrhœa or ſyphilis, but eſpecially in gonorrhœa, muſt be near the farther extremity of the vagina, though there can be no doubt but the ulcerated glans may often affect the exterior parts by its introduction; but, in a confirmed lues, the fundus of the vagina is rarely the ſeat of ulcer, and it is never affected in gonorrhœa. Here, the ſurface of the vagina being moſtly ſmooth, the poiſon runs downwards, till falling upon the rugæ, it is there intercepted and retarded. Here then the poiſon is multiplied, and leiſurely applied to the mouths of the lymphatics, through which it is carried into the blood; where, aſſimulating together, it contaminates the whole maſs. Though the progreſs of the ſyphilitic poiſon is not always thus regular, the variations do not affect the opinion. When the lymphatics, and their glands, are vigorous and eaſy permeable; when the application of the venereal virus is within the nymphæ; and, when it is ſufficiently active, the firſt ſymptoms of diſeaſe ariſe from general contamination; and,

and, were this poiſon always very mild, and taken up by the abſorbents within the nymphæ, there is no doubt but the whole maſs would almoſt always be diſeaſed, without much chance of ulcer or preceding bubo. But there are many circumſtances which tend to retard the ſpeedy abſorption of the ſyphilitic virus, even when it is extremely active; and, among theſe, the inflammation which in general it muſt induce, is not perhaps the leaſt conſiderable; but theſe cannot affect the abſorption of the ſeminal fluid of the male. The ſyphilitic virus too, may, from the laxity and lubricity of the vagina, (a circumſtance very general in immodeſt women,) not only eſcape abſorption, but may be carried outwards, to exerciſe its energy on the external parts. And it is from theſe reaſons partly, that immodeſt women are ſo little diſpoſed to conception, and that modeſt women, when ſubjected to venereal infection, generally experience the more latent and violent ſpecies of this diſeaſe. And, as a greater ſurface of abſorbents is expoſed in the female to the contaminating influence of the diſeaſed male organs, and as the greateſt part of the female genital ſyſtem has a much readier intercourſe with the blood than through the inguinal glands, we meet with this ſpecies of ſyphilis much oftener in women than in men. The cure of ſyphilis, too, by ſpecific remedies introduced into the vagina, fully demonſtrates the ſtrength and activity of the lymphatics in this canal. Is there then a ready and eſtabliſhed communication, for diſeaſe, and for its remedies, between the vagina and the general circulating ſyſtem of the blood, while a mild fluid, yet poſſeſſed of activity infinitely beyond that of any poiſon, and created for the higheſt and beſt of purpoſes, is not permitted to traverſe the ſame channels? Many other corroborating circumſtances, both in fact and in analogy, might be adduced here, were not theſe arguments in themſelves concluſive.

In a due ſtate of health there is what may be called an inteſtine motion in the blood, occaſioning and promoting its commixture, as well as its ſeparation. In all general diſeaſes, and even in many which are called local, this inteſtine motion is heightened, diminiſhed, or deranged; and in the exanthematous or eruptive diſorders, it muſt be remarkably ſo. In ſyphilis, though this diſeaſe is not directly exanthematous, there muſt be exceſſive diſturbance, and certain depravation, prevailing throughout the whole ſyſtem, before ſuch complete deſtruction can be brought upon it. In theſe caſes of diſeaſe, where vehement infection, with all its conſequences, is overturning all before it, we have always found, that milder infections could make no impreſſion. Hence the practitioner never heſitates to ingraft the ſmall-pox, though the patient may have already received the diſeaſe, either by natural contagion, or by prior inoculation: hence a milder diſeaſe is often removed by a ſeverer one; hence ſlow conſumption is always retarded, and often overcome, by fecundation;

tion; and hence fecundation itſelf, as the feebler ſtimulus, is often prevented by the anticipating diſturbance of ſyphilis, or of ſimilar diſeaſes vehemently pre-occupying the circulating ſyſtem. It is this anticipation, this prior poſſeſſion, and change in the circulating blood, which reaſonably and emphatically accounts for the want of influence in the human ſemen upon the female after impregnation has fully taken place, or while the mother is providing milk. And we might account for the production of twins, triplets, and thoſe rare inſtances of more numerous progeny, from the ſame circumſtances. One, two, or more, ova may indeed be ſo ripe as to meet completely the fecundating impulſe of the male ſemen at one time; and it is perhaps more ſtrange that the different fœtuſes ſhould be maturated and expelled about the ſame time, than if a greater period intervened between the expulſion of each; and might not a ſecond intercourſe of the ſexes be ſucceſsful, when the female circulating maſs was not fully pre-occupied by the influence of the firſt? But the extent and influence of prior infection, or impregnation of the blood, has been better obſerved in the venereal than in any other diſeaſe, or natural occurrence. Women, whoſe general ſyſtem is vitiated by the ſyphilitic virus, are always incapable of conception; or if the vitiation is not complete, but in a ſlight degree, an imperfect fecundation may take place; but its product determines the want of energy, and the unqualified ſtate of the mother from whence it drew its principal arrangement. Theſe ideas are corroborated by the mode of cure adopted in the circumſtances we have been deſcribing, and by the general effects of it.

Thus we have endeavoured, and we hope with ſucceſs, to eſtabliſh the truth of a ſtrong power of abſorption in the genital ſyſtem of the female, originating in the vagina; and a diſpoſition in the whole maſs of blood, to be affected according to the properties of what may be mingled with it. And as, from the preſent ſtate of anatomical knowledge, we have no right to ſuſpect any other mode than this of abſorption, by which the unrejected and finer parts of the ſemen can in any ſhape, and with any effect, be determined towards the ovaria, let us ſee how this can be farther aſcertained by what we may ſuppoſe to be the effect of the abſorbed ſemen, and the future appearances of impregnation.

In human creatures the evolution of all their parts is gradual, and the work of time. From the moment in which the ovarian nucleus receives the vivifying impulſe from the ſemen, till the period of puberty; from the dawn of its exiſtence, to the completion of its figure and its powers; its alterations are ſo many, and ſo varied, that our idea of the germ is not recogniſable in that of the infant, and our idea of the infant again is loſt in that of the perfect animal. A gelatinous particle, without neceſſary form and texture, becomes a ſtupendous fabric, ſo intricate and elaborate, though at the ſame time perfect and complete, that human ingenuity and reaſon have toiled almoſt fruitleſsly for thouſands of years in inveſtigating the progreſs. It has indeed been averred by ſome, that all the different organs of the animal in its complete ſtate are original and diſtinct in the embryo, and are only unfolded and rendered more evident by its increaſe. This ſurely is not the caſe. The animal is certainly endowed with power of completing itſelf; and can, from inorganized parts, produce an organized ſtructure. The parts are only evolved and perfected as they become uſeful in the different ſtages; and the evolution of many of them can be prevented without the deſtruction of life, or exceſſive prejudice to thoſe already evolved. If the different organs, or rather principles, are at firſt perfect, why are thoſe effects which depend upon them not perfect alſo? Why is the ſtate of infancy a ſtate of idiotiſm? why is the temper of youth capricious and flexible? and why are the temper and paſſions of the adult but barely diſcernible in the preceding ſtages?

As we are of opinion then, that the different organs are matured only as they become requiſite and neceſſary, conſequently, we believe the evolution of the generative organs in both ſexes muſt be among the laſt efforts of the increaſe and completion of the body. This evolution could not have taken place earlier: if it had, the mind muſt have been affected by thoſe impulſes which announce the maturation of theſe organs, by which we know the mind, body, and ſoul, are connected. In the male, the foundation and powers of maturation, of that ſtrength, and of thoſe more rational qualities which belong to him, are laid to ripen with puberty: hence communication with the female, before theſe are finally arranged and ſecured, proves inefficient, and entails upon him debility both of body and mind. The ſame thing holds, as far as the ſame ends are concerned, with reſpect to the female; and we cannot ſuppoſe that Nature could be ſo idly eccentric, as to puniſh the female with a diſpoſition or propenſity to procreate, before the body was capable of undergoing the various diſorders and dangers of pregnancy and parturition. For the ſame reaſons, none of the ordinary organs of ſenſe are qualified to receive or communicate diſtinct impreſſions, till the brain, the ſeat of the ſoul, as the heart is of life, has acquired thoſe properties which muſt fit it for its arduous offices. It is only when the different organs of ſenſe have been completely evolved, and all their parts ſound and juſt, that the power of the mind is effectuated and eſtabliſhed. This faculty, though it ſeems eſſentially different from Reaſon, is no doubt the origin of it; for the extenſion of common ſenſe, from memory, or rather from compariſon, and what may be called the balance of the ſenſes, conſtitutes what is called Reaſon and Judgment. While the organs are incomplete, from infancy or from diſeaſe, their communication with the underſtanding is alſo incomplete. Thoſe who have been born blind, or whoſe eyes have been deſtroyed in infancy before they were become uſeful, have none of thoſe ideas which depend upon the eye; it is the ſame with the deaf, and in all caſes of ideas depending upon one ſenſe: and we may add, the early caſtrated have no comprehenſion of, or propenſity to, the gratification of love. Do not theſe things ſhow—and a thouſand other circumſtances might be adduced to ſtrengthen the proof—that the mind acquires its powers only as the parts of the body are unfolded, and confirmed; that the body is perfected only as the mind is qualified to receive its impreſſions; and that the parts of the body are perfected by one another?

During infancy and youth, ſtrictly, the ovaria are ſimple inorganic maſſes, partaking of no more life than is barely ſufficient to ſuſtain them, and connect them with that energy and progreſs of conſtitution which are afterwards to unfold all their properties. At the period of puberty, thus denominated from the change which takes place in the genital ſyſtem at this time of life, this progreſs and developement of the ovaria is finiſhed by Nature; and thoſe bodies are generated and completed within them, which will exiſt without impregnation by the male, but which this impregnation alone can finally maturate and evolve. That theſe bodies are not generated at an earlier date, Anatomy as well as Reaſon, founded on the foregoing arguments, aſſure us; and, that the ova of all the fœtuſes, which the female is afterwards to produce, are generated at that time, ſeems equally certain. Though this change in the ovaria is the moſt eſſential, the whole genital ſyſtem alſo undergoes a very material change. The ſimple alterations of ſtructure and dimenſions in the different parts of this ſyſtem, though they are neceſſary and ſubſervient to generation and parturition, yet they are not ſo material, either in themſelves, or to our purpoſe, as to require a minute deſcription. This, however, is not the caſe with reſpect to the menſes. It is chiefly with a view to the nutrition of the fœtus that the extra-ſanguification in the female is provided by Nature; which is determined to the genital ſyſtem, in the ſame manner as the other fluids are determined to other outlets; but, as the continued drilling off of this extra blood would be exceedingly inconvenient and diſguſting, Nature has prepared, as it were, a ciſtern for its reception. What may be ſufficient to bring on the hæmorrhage, however, is only accumulated; and the general redundancy, induced by the obſtruction and accumulation, ſubſides gradually as the hæmorrhage goes on. This is the manner of menſtruation in the unimpregnated female, and theſe are the reaſons why it aſſumes a periodical form. In the impregnated female again, the preparation of extra blood ſtill continues, but its conſumption becomes very different. By the extenſion of the uterus, and by the waſte occaſioned by the nouriſhment of the fœtus and its involucra, the ſurcharge or extra-preparation of blood is nearly balanced, or is taken up as it is prepared; and hence the periodical efforts are almoſt loſt. In the firſt months of of pregnancy, however, the uterine ſyſtem is not always able to conſume the ſurcharge of blood, and thereby take off the periodical effort; and hence it is that the loſs of the fœtus happens moſt generally in the early months, and at the uſual period of the menſes, unleſs ſome accident has ſupervened. And it is nearly from the ſame reaſons that miſcarriage is ſo often to be apprehended in the latter months of pregnancy, and that the fœtus is afterwards expelled from the womb. When the fœtus has acquired all that bulk and ſtrength which the capacity and powers of the uterus can confer, and when a change of circulation and mode of life becomes neceſſary to it, the uterus and fœtus become plethoric; a general accumulation ſucceeds; and the periodical efforts of the menſes return. During the middle months of pregnancy the fœtus is in a ſtate of rapid growth, and is capable of conſuming all the blood which the mother can furniſh; but there is neither room nor waſte, in the latter months, for the blood which the mother is conſtantly pouring in; and hence ariſes that plethora, both in mother and child, which is to inſtigate the effort to parturition, which occaſions the effuſion after parturition, and which is to ſupply the extended circulation of the born child.

But, beſides the utility of menſtruation to the fœtus, there is a very evident connection between it and impregnation. To ſpeak of it as a proof of the ripened qualifications of the female, is to ſay nothing; its immediate action is eſſential to conception. In the human female, it is well known that coition is almoſt only ſucceſsful immediately after this evacuation has ſubſided. Who will reconcile this—and it is no modern and groundleſs obſervation—to the conſequence which has been aſcribed to turgidity and tenſion, which we have already adverted to? Almoſt every woman who has frequently undergone pregnancy, and who has attended judiciouſly to the phænomena of that ſituation, calculates from the laſt ceſſation of the menſes. At this time, or rather very ſoon after it, the plethoric tumult of the general ſyſtem has completely ſubſided, and the abſorbed ſemen gets quiet and unanticipated poſſeſſion of the circulating blood; and at the ſame time the gradually-returning plethora promotes its action, and perhaps its determination to the ovaria. When the menſes are interrupted, or profuſe and frequent, impregnation ſeldom takes place; and it admits not of a doubt, that, when the determination of this blood is towards the mammæ, in the form of milk, coition is unſucceſsful; and, as ſoon as its determination to the urine ſyſtem is reſtored, other things being favourable, copulation ſucceeds. We may add as a known fact, that continuing to give ſuck after the uſual period will occupy the plethora, and prevent its determination, in the form of blood, to the uterine ſyſtem; and this practice has often been had recourſe to, in order to prevent conception. Sometimes there is reaſon to believe that conception has taken place

6 while

while the plethoric determination to the breafts continued. I am rather difpofed to believe, that in fuch cafes its return to the uterine fyftem had re-commenced; for about the fame time the milk generally lofes its alimentary qualities, and gradually dwindles away.

But we have faid enough to defcribe and fubftantiate thofe parts of the female which puberty has prepared for generation. We fhall now confider its effects on the male. It need not be repeated, that the feminal fluid is an exceedingly penetrating and active fluid. Its effects, after it is generated, even upon the male, demonftrate its activity and influence, far beyond the precincts wherein we believe it to be accumulated. After puberty, the fecretion of it, during even indifferent health, is continually going on; and thofe collections of it in its refervoirs, which are not thrown out by venereal exercife, or by other means lefs decent, are reforbed and mingled with the general mafs. What is actually reforbed about the period of puberty, before the fyftem has been habituated to it, or faturated with it, produces very curious and remarkable effects over the whole body. The flefh and fkin, from being tender, delicate, and irritable, become coarfe and firm; the body in general lofes its fucculency; and a new exiftence feems to take place. The voice, a proof of the tenfion and rigidity of the mufcular fibre, lofing its tendernefs and inequalities, becomes ungratefully harfh; and the mind itfelf, actuated by the progrefs of the body, and forgetting all its former inclinations and attachments, acquires diftinctly new propenfities and paffions. Thefe changes are not entirely the effect of ordinarily-progreffive age and ftrength; neither are they promoted by intercourfe with the world; for caftration will anticipate them, and premature venery, or even gradual familiarity, and early onanifm, will diminifh them. Boys who have been fubjected to caftration never acquire either that ftrength of body or capacity of mind which dignifies the complete male; and the fame cruel and unnatural operation performed on brute animals, diminifhes their bodily ftrength, their courage, and the fiercenefs of their temper.

If fuch are the effects of the feminal fluid when reforbed by the male, how powerful muft it be when fuddenly mingled, and moft probably in greater quantity, with the circulating fluids of the attracting female! Coition, or rather the abforption of the feminal fluid of the male by the female, even when not fuceeded by impregnation, induces an alteration very general over the female fyftem. The local influence of which may be inferred from the general change which it is capable of inducing during complete health; from the relief which it effectuates in many fpecies of difeafe; and from the general vivacity and cheerfulnefs diffufed over the whole animal frame. It would be prolix to go over every difeafe which will warrant thefe opinions; yet, in the eye of common obfervation, the fallow and inanimate female,

female, by coition, often becomes plump and robuſt, and beautiful and active; while the widow, or married woman deprived of commerce with her huſband, gradually returns to the imperfections and peculiarities of ſingle life; and that the ancient virgin, all her life deprived of this animating effluvium, is generally conſumed with infirmity, ill-temper, or diſeaſe. It is well known, too, that the want of coition at the time of life when Nature ſeems to require it, induces many diſorders in females; and that the uſe of it removes theſe, and even other diſeaſes. Chloroſis, or the whites, almoſt always attack females immediately after puberty; and, even when the violence of its ſymptoms have not been diſcerned till a later period, its origin can always be traced back to that time. When the human ſyſtem is completely evolved, and all its parts have acquired their full growth, a balance is produced between the circulating and ſolid ſyſtems; though, from the ideas we have ſuggeſted concerning the menſes, this balance in the female cannot ſtrictly be called complete. It is only complete in her when in perfect health, and in an impregnated ſtate; at other times, the catamenia, as preponderating againſt the powers of the ſolid ſyſtem, in proportion to the degree of their period, diſturb the equilibrium, and thereby more or leſs induce a ſtate inconſiſtent with perfect health. But, when the propelling power of growth has ceaſed before the ſolids, either from actual diſeaſe, or want of uniformity in either period, or acceſſion with reſpect to the progreſs of the circulating ſyſtem, have acquired their proper vigour and tone; and when the catamenia has aſſumed its deſtination before it is accompanied by the general as well as local energy which is requiſite to expel it, an univerſal want of balance comes on; the blood loſes its ſtimulating influence on the vitiated ſolids, and theſe, in their turn, act feebly on the diſtempered blood. Accordingly, in the cure of this diſeaſe, no matter whether adopted from particular theories or from experience, medicines are directed to reſtore vigour to the ſolids, and conſiſtence and ſtimulus to the circulating maſs. Nature proceeds in the ſame manner; and the beneficial effects of coition in the cure of this diſeaſe have been too material to eſcape obſervation. It may be alleged, that theſe effects depend entirely upon local influence; and that even voluptuous gratification, by quieting the turbulence of paſſion, is of conſequence in the cure. We ſhall not ſay that theſe things are unavailing; for it appears that the relief obtained is chiefly owing to the increaſed inteſtine motion, and conſequent ſtimulus, communicated to the blood by the abſorbed ſemen, whereby the ſolids themſelves are ultimately reſtored; and we are the more confirmed in this opinion, becauſe all theſe fortunate effects attend, whether coition be ſucceeded by impregnation or not. Hyſterics, and other diſeaſes, would furniſh us with ſimilar explanations and ſimilar cures.

Let us now advance a little nearer our object. It is beyond a doubt, that, in whatever manner the semen acts upon the female, it does not act suddenly, notwithstanding the general assertions of many authors. However productive coition may be, the fecundated product of the ovaria is not immediatelydisengaged. We dare not avouch this fact from observations made on the human subject, because such observations never have been attempted, nor ever can with the smallest probability of success: but the dissection of brutes, by the most eminent anatomists, with a direct view to the elucidation of this fact, ascertains it as far as such evidence can be admitted. In the dissection of small animals by De Graff, he found no discernible alteration in the uterus during the first forty hours after coition, but a gradual change was perceivable in the ovaria; and what he supposed the ripened origin of the future animal, at the end of that time, losing its transparency, become opaque and ruddy. After that time, the fimbriæ were found closely applied to the ovaria; the cavities from whence the ova had been expressed were discernible; and about the third day the ova were discovered in the uterus. In large animals, and in those whose time of uterine gestation was longer, it was found that the progress which we have been describing was proportionably slower. The same experiments have been made by different anatomists, and perhaps with very different views; and, though they have not always been managed with the same judgment and dexterity, yet all of them more or less confirm the idea that there is a very considerable lapse of time intervening between productive copulation and the expulsion of the ovum from the ovaria. But, if this be the case with animals which soon arrive at puberty, and which, like human creatures, copulate not perfectly before puberty,—whose lives are short, and their progress in equal periods of time more rapid than those in man,—by parity of reason, it must happen, that in women the period between impregnation and the expulsion of the fecundated product of the ovaria must be considerably greater than what has been observed to take place in these animals. If all this is true—how are we to suppose Nature to be employed during this interval? We believe it is during this period that the whole female constitution is labouring under the fecundating influence of the seminal fluid taken into the blood by the *absorbents;* while the ovaria are largely participating, and their product ripening, by means of the general stimulating process. And the same process which maturates the ovum tends to facilitate its exclusion. The ovaria, as well as their product, are at this time enlarged, and other changes, subject to the examination of our senses, induced. It is no proof against the reality of this general alteration in the circumstances of the circulating system, and consequent revolution in the ovaria, that the whole is accomplished with but little visible disturbance, either local or universal. In other

2 cases

cafes of material alteration in the mafs of blood, equal quietnefs and obfcurity prevail. In fcrophulous or fcorbutic taints; in the inoculated fmall-pox, or when they are produced by contagion; the poifon filently and flowly diffufes itfelf throughout the whole mafs, and a highly-morbid ftate is imperceptibly induced. Thus, an active and infinuating poifon intimately mixes itfelf with all the containing, perhaps, as well as contained, parts, perverts their natures, and is ready to fall upon and deftroy the very powers of life, before one fymptom of its action or of its influence has been difcerned. It is the fame in a confirmed lues, and it is even more remarkable in the hydrophobia derived from the bite of a mad dog; and the whole round of contagious difeafes have the fame unalarming, yet certain, progrefs and termination.

That the final influence of this elaborate procefs fhould be determined particularly, and at all times, to the ovaria, is no way marvellous. To qualify the ovaria for this, they are fupplied with a congeries of blood-veffels and nerves, at puberty larger and more numerous than what is allotted to any other part of fimilar magnitude. Were the ovaria merely a receptacle for the ova, which the venereal orgafm communicated by the nerves, or by the impulfion of the applied femen, was to lacerate; what ufe would there be for fo intricate and extenfive an arrangement of blood-veffels and nerves? But we may farther remark, that every diftinct procefs in the human body, either during health or difeafe, tends to one particular and diftinct purpofe. The kidneys do not fecrete bile, nor does the liver ftrain off the ufelefs or hurtful parts of the blood which are deftined to pafs off by the emulgents; neither do the falivary and bronchial glands promifcuoufly pour out mucus or faliva; the variolous virus does not produce a morbillous eruption, fyphilitic caries, or fcrophulous ulcer; why then fhould the fecundated blood unconcernedly and promifcuoufly determine its energy to the fkin, the lymphatics, or the fubftance of the bones? We know none of the operations in the human body, deftined for the ordinary purpofes of life and health, or for the removal of difeafe, but in a greater or lefs degree involve the machinery of the whole fyftem. A fingle mouthful of food, while it is prepared, purified, and applied to its ultimate purpofes, is fubjected to the action of all the known parts of the body, and without doubt to all thofe parts the properties of which we are unacquainted with; a draught of cold water fpreads its influence almoft inftantaneoufly from one extremity to the other; the flighteft wound difturbs even the remoteft parts, and is followed, not unfrequently, with the moft unhappy effects; an almoft invifible quantity of poifon fets the whole frame in torture, and all the active powers of the body inftinctively exert themfelves to folicit its expulfion.—Can we diftinguifh thefe things, and admire them, and then fuppofe ſuppoſe that the moſt material operation of the human body—the renovation of itſelf—is to be accompliſhed in a corner, and with infinitely leſs formality and ſolemnity than a ſpittle is caſt upon the wind? The evident means are ſufficiently degraded; we need not exert our ingenuity to degrade them farther.

It is during this interval, between productive coition and the excluſion of the ovum from the ovaria, that likeneſs, hereditary diſeaſes, and the like, are communicated and acquired. Inſtead of that influence which the imagination of the mother is ſuppoſed to poſſeſs over the form of the child, might we not ſuſpect, that the ſeminal fluid of the male, co-operating, *during this interval*, with the influence of the female upon the ovum, inſtigated a likeneſs, according to the influence of the male and female tinctures, in the united principles? It is during this period only that the diſeaſes of the male can be communicated to the child; and, if we admit not of this interval and general operation of the ſeminal fluid, we cannot ſee how they can be communicated, though thoſe of the mother may be communicated then or at a much later period, conſidering how the child is nouriſhed while it is in the uterus and at the breaſt. It may be urged againſt this early and effectual acquiſition of likeneſs, that the fœtus does not acquire even the diviſion of its largeſt members till long after its excluſion from the ovaria: but then we are confident, that, as the fœtus takes all its form and other properties from the active ſubtilty of theſe blended tinctures, we cannot ſee any reaſon why it ſhould not poſſeſs this hereditary faculty, in common with the reſt. If likeneſs depend upon the imagination of the female, how happens it that the children of thoſe whoſe profligate manners render the father uncertain, and whoſe affections ceaſe with the inſtant of libidinous gratification, are as frequently diſtinguiſhable by their likeneſs as thoſe children who have been born under none of theſe misfortunes? If the features are not planted during this period, and if imagination be not idle or uſeleſs, how was the ſix-fingered family, mentioned by Maupertuis, continued? When a female of that family married a man who had only the uſual number of fingers, the deformity of her family became uncertain, or ceaſed; and we muſt ſuppoſe her imagination could not have been inactive or diminiſhed, whether alarmed by the fear of continuing a deformed race, or inſtigated by the vanity of tranſmitting a remarkable peculiarity. Were imagination, in a pregnant woman, ſo powerful as many have endeavoured to repreſent it, the mother, profligate at heart, though not actually wicked, would always betray the apoſtaſy of her affections; and even a virtuous woman might divulge that ſhe had looked with as much eagerneſs at a handſome ſtranger as ſhe had looked at the aquiline noſe or other prominent feature of her huſband.

But,

But, admitting that the ſeminal fluid of every male poſſeſſes ſome kind of influence peculiar to that male, and connected with his form, as well as his conſtitution; in the ſame or in ſome ſimilar manner it contains, notwithſtanding the elaborateneſs of its preparation, the ſtamina of diſeaſes, ſome of which often lie longer dormant than even the features of individuals; that the ova are as peculiarly conſtructed, by the conſtitution of the female, as any other parts which depend upon gradual and ſolitary evolution and that theſe, operating upon each other by the intervention of the general ſyſtem of the female, may, according to the power or prevalence of either, affect the features and figure of the incipient animal, or rather the inorganized maſs from which the features and figure of the animal are afterwards to be evolved: admitting all theſe things, will national, or even more extenſive, ſimilitude corroborate the opinion?

While men continue in the ſame climate, and even in the ſame diſtrict, an uniform particularity of features and figure prevails among them, little affected by all thoſe changes which improve or degrade the mind; but, when they migrate, or when they are corrupted by the migration of others, this national diſtinction in time is loſt, though in the latter caſe it ſeems to be recoverable, unleſs the cauſe of change be continued.—The beautiful form and features of the ancient Greeks are at this day diſcernible in their deſcendants, though they are debaſed by intercourſe with ſtrangers, and by forms of government ultimately affecting their conſtitutions; the deſcendants of the few who by chance or deſign have been obliged to ſettle among the ugly tribes in the extremities of the North, have, by their intercourſe with theſe tribes, and by neceſſarily accommodating themſelves to the ſame modes of life, beſides other circumſtances, become equally ugly; and the Jew himſelf, though he abhors to mingle with a different nation, and though his mode of life is nearly the ſame in all climates, yet the ſettlement of his anceſtors in any one particular climate for ſome centuries will very ſenſibly impair the characteriſtic features of his people. As equally in point, and leſs liable to queſtion, we may mention the following ſimilar obſervations. A Scotchman, an Engliſhman, a Frenchman, or a Dutchman, may, even without their peculiarities of dreſs, be almoſt always diſtinguiſhed in their very pictures; the ſturdy and generous Briton, notwithſtanding the ſhortneſs of the period, and the uninterrupted intercourſe, is traced with uncertainty in the effeminate and cruel Virginian: and the Negroes in North America, whoſe familes have continued ſince the firſt importation of theſe unhappy creatures, and whoſe modes of living, excluſive of their ſlavery, are not materially changed, are much leſs remarkable for the flat noſe, big lips, ugly legs, and long heels, than their anceſtors were, or than thoſe who are directly imported from the ſame original nation.

From these observations it seems allowable to infer, that, though climate, manners, occupation, or imitation, cannot materially affect the form or features of the existing animal, yet these circumstances, becoming the lot of a series of animals, may, by inducing a change in the general mass both of the male and female, be the remote cause of a change in their product.

After what has been premised, it seems rational to conclude, that the prolific fluid, in coition, is neither carried through the Fallopian tubes, nor protruded through the aperture of the uterus, to the ovaria; but that it is taken up by the absorbent vessels, and conveyed into the sanguiferous system, where indeed every active principle that can possibly affect the human constitution is also conveyed. That, after circulating through the blood, it is by its natural impulse, and the additional stimulus acquired from the mother, forced through the corresponding vessels into the ovaria; where, if it finds one or more of the ova in a state fit or ripe for impregnation, conception takes place accordingly: and either one or more are impregnated, as the maturated state of the ovaria might happen to be. But if none of the ova or eggs are in a state sufficiently mature, or chance to be injured by any offending humours, by debility, or disease, in either of these cases impregnation is frustrated, just the same as happens to an addled egg, or to a damaged grain of corn thrown into the earth.

On the other hand, if the male organ be deficient in vigour, or the semen be defective in quantity, consistency, or active power, it then fails of stimulating the female fluid, and is incapable of influencing impregnation. In order therefore that the act of copulation should be productive, the male must unquestionably convey to the female an elaborate tincture, which possesses the essence of his whole system, as well mental as corporeal. In this act, the utmost energy and powers of the mind, of the body, and of the soul, are intimately connected; and all contribute their particular influence to the seed; of which every father must be sensible, when he recollects the action of the heart, the seat of life—of the brain, the seat of the soul—and of the whole powers of the body, concentrated and impelled, as it were, through the genital system.—That this liquor comprehends the active principles of *body and soul*, will not I think be doubted in those who give the foregoing arguments their proper weight; and that it conveys with it, more or less, the direct image of the parent, I take to be confirmed by the evidence of Scripture; where we are told that one absolute and unequivocal form was given to man, in the express image of the Deity. So that man, thus organized and commissioned, was doubtless to convey to future generations that divine image or signature which God had graciously stamped upon him. For this purpose the seed of man, or efficient principle

of

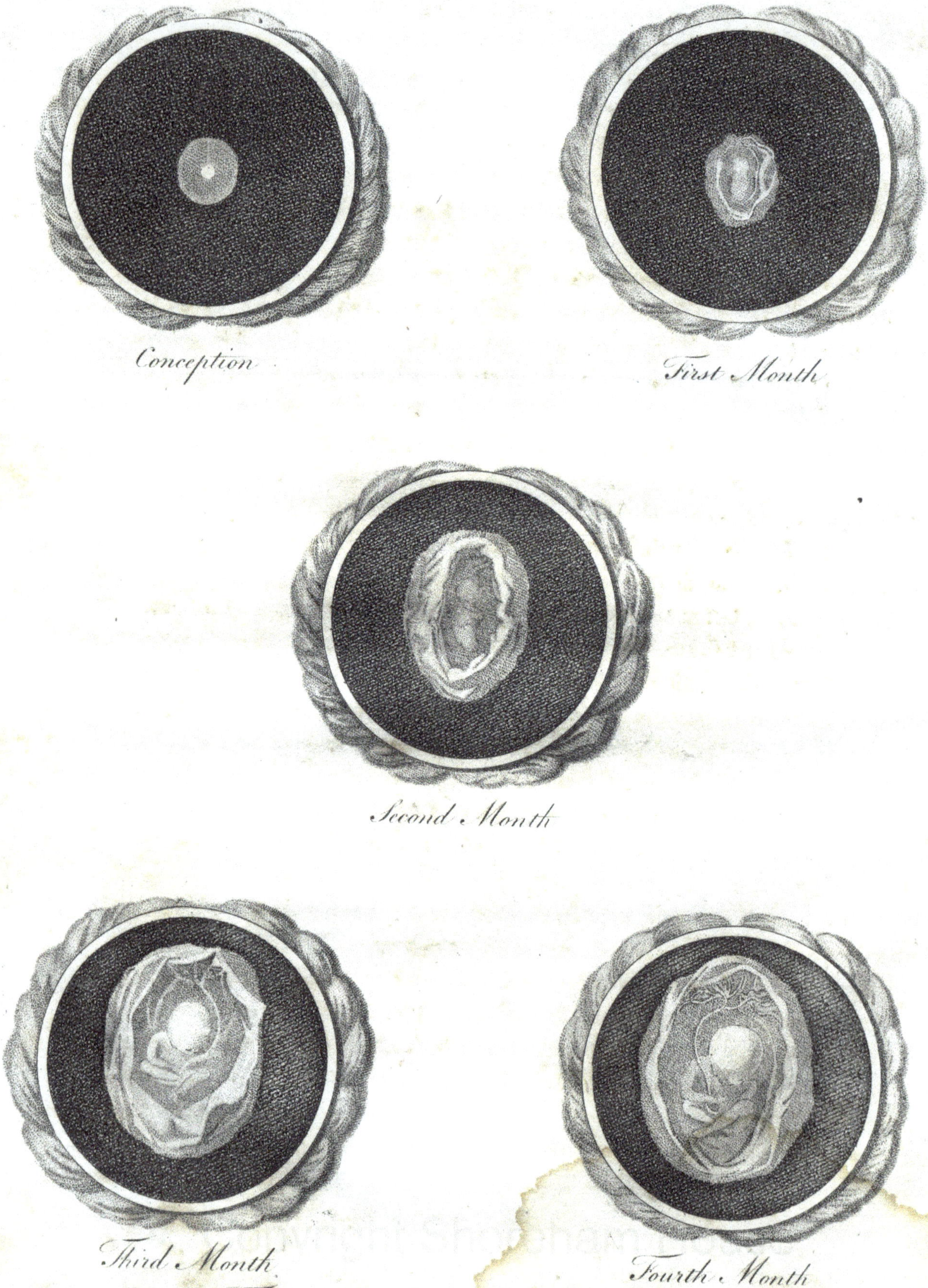

Dodd. del.

Formation of the Human Fœtus. Plate I.

J. Page, sculp.

of generation, muſt be mingled with the vegetative fluid of the female; and, being attracted or taken up by the abſorbent veſſels from the uterine canal, paſſes immediately into the circulating ſyſtem, where, aſſimilating with the peculiar temperature of the mother, and acquiring new energy from the enlivening quality of the blood, it is directed through its natural channels to the ovaria, impregnating the germ by its active quality, and conveying to it the peculiarities it had derived from the conſtitutions, forms, tempers, and diſpoſitions, of the parents, with the ſeeds of whatever diſeaſes, impurities, or taints, were lurking in their blood. For from the blood and brain is the male ſeed primarily elaborated, and into the female maſs is thi thrown and aſſimilated, before impregnation can poſſibly take place. In the courſe of ſix days, I conclude the united tinctures to have travelled through the whole circulating ſyſtem—to have participated of the hereditary forms and peculiarities of the mother, and to have propelled the ovum or egg from its ſeat in the ovaria to a ſuſpended ſituation in the womb, hanging by a minute thread, that afterwards becomes the umbilical veſſel, or aperture through which nouriſhment and life is conveyed from the mother to the child. The firſt viſible ſtate of conception, which reſembles the lucid appearance of a drop of water, tending to coagulation, is correctly ſhown in the firſt figure of the annexed Plate, preciſely in the ſtate it was extracted from the uterus of a female who died on the ſixth day after contact with the male.

At the time the ovum, or rudiments of the embryo, deſcends into the womb, it is indeed very minute; but at the end of about thirty days, we may partly diſcover the firſt lineaments of the fœtus, though ſmall and imperfect, being then only about the ſize of a houſe-fly. Two little veſicles appear in an almoſt tranſparent jelly; the ſmaller of which is intended to become the head of the fœtus, and the larger one is deſtined for the trunk; but neither the limbs nor extremities are yet to be ſeen; the umbilical cord appears only as a minute thread, and the placenta, which only reſembles a cloud above, has no ramifications, or appearances of blood-veſſels. This ſtate of the embryo is expreſſed in the ſecond figure of the annexed Plate.

Towards the end of the ſecond month, the fœtus is upwards of an inch in length, and the features of the face begin to be evolved. The noſe appears like a ſmall prominent line; and we are able to diſcover another line under it, which is deſtined for the ſeparation of the lips. Two black points appear in the place of eyes, and two minute holes mark the formation of the ears. At the ſides of the trunk, both above and below, we ſee four minute protuberances, which are the rudiments of the arms and legs. The veins of the placenta are alſo now partly viſible; as may be ſeen in No. 3 of the Plate.

In

In the third month the human form may be decidedly afcertained; all the parts of the face can be diftinguifhed; the fhape of the body is clearly marked out; the haunches and the abdomen are elevated, and the hands and feet are plainly to be diftinguifhed. The upper extremities are obferved to increafe fafter than the lower ones; and the feparation of the fingers may be perceived before that of the toes. The veins of the placenta are now diftended, and are feen to communicate with the umbilical tube. This ftate of geftation is faithfully delineated in No. 4 of the annexed Engraving.

In the fourth month the fœtus feems to be completed in all its parts, and is about four inches in magnitude. The fingers and toes, which at firft coalefced, are now feparated from each other, and the inteftines appear, in all their windings and convolutions, like little threads. The veins of the placenta begin to be filled with blood, and the umbilical cord is confiderably enlarged; as may be feen in the fifth figure of the fubjoined Plate.

In the fifth month, the bodily conformation being perfected in all its parts, and a complete circulation of the blood induced, the mother quickens. The fœtus now affumes a more upright figure, which correfponds with the fhape of the uterus. Its head is found more elevated, its lower extremities are more diftended, its knees are drawn upwards, with its arms refting upon them. It now meafures from feven to eight inches in length, and is defcribed in the firft figure of the fecond fubjoined Plate.

Towards the end of the fixth month, the fœtus begins to vary its pofition in the womb, and will frequently be found to incline either to the right or to the left fide of the mother. It will by this time be increafed to nine or ten inches; and its ufual pofture, after quickening, may be feen in the fecond figure of the fecond annexed Plate.

In the feventh month the child acquires ftrength and folidity; as may be demonftrated by thofe painful throws and twitchings which its mother feels from time to time; and it is now increafed to eleven or twelve inches.

In the eighth month it generally meafures from fourteen to fixteen inches; and in the ninth month, or towards the end of its full time, it is increafed from eighteen to twenty-two inches, or more; when the head, by becoming fpecifically heavier than the other parts, is gradually impelled downwards, and, falling into the birth, brings on what is termed the pains of parturition, or natural labour. For the exact pofition of the child in the womb during thefe three laft months, as well as the former, fee the correfponding figures in the two annexed Engravings, the whole of which were correctly drawn from real fœtuffes, extracted from the wombs of different women.

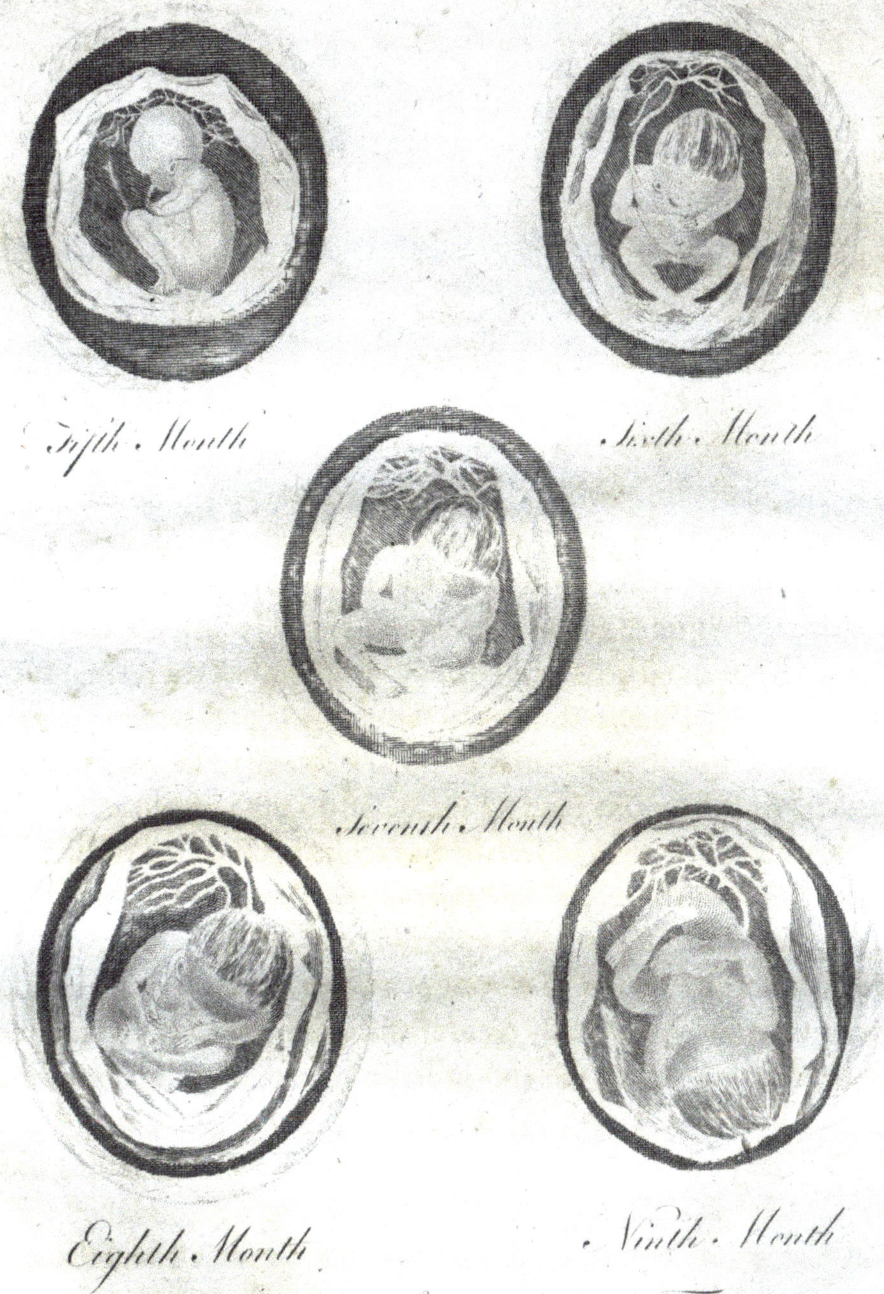

Formation of the Human Fœtus. Plate II.

Dodd del. Pass sculp.

Published as the Act directs June 20th 1794.

The nouriſhment of the fœtus during all this time is derived from the placenta, which is originally formed out of that part of the ovum which is next the fundus uteri. The remaining part of the ovum is covered by a membrane called *ſpongy chorion;* within which is another called *true chorion,* which includes a third, termed *amnios.* This contains a liquor, or watery fluid, in which the fœtus floats till the time of its birth. Before the child acquires a diſtinct and regular form, it is called *embryo;* but from the time all its parts become viſible, it takes and retains the name of *fœtus* till its birth. During the progreſs of impregnation, the uterus ſuffers conſiderable changes; but, though it enlarges as the ovum increaſes, yet, in regard to its contents, it is never full; for, in early geſtation, theſe are confined to the fundus only: and, though the capacity of the womb increaſes, yet it is not mechanically ſtretched, for the thickneſs of its ſides do not diminiſh; there is a proportional increaſe of the quantity of fluids, and therefore pretty much the ſame thickneſs remains as before impregnation. The gravid uterus, or pregnant womb, is of different ſizes in different women; and muſt vary according to the bulk of the fœtus and involucra. The ſituation will alſo vary according to the increaſe of its contents, and the poſition of the body. For the firſt two or three months the cavity of the fundus is triangular, as before impregnation; but, as the uterus ſtretches, it gradually acquires a more rounded form. In general the uterus never riſes directly upwards, but inclines a little obliquely, moſt commonly to the right ſide; its poſition is never, however, ſo oblique as to prove the ſole cauſe either of preventing or retarding delivery; its increaſe of bulk does not ſeem to ariſe merely from diſtenſion, but to depend on the ſame cauſe and increaſe as the extenſion of the ſkin in a growing child. This is proved from ſome late inſtances of extra-uterine fœtuſes, where the uterus, though there were no contents, was nearly of the ſame ſize, from the additional quantity of nouriſhment tranſmitted, as if the ovum had been contained within its cavity. The internal ſurface, which is generally pretty ſmooth, except where the placenta adheres, is lined with a tender effloreſcence of the uterus, which, after delivery, appears as if torn, and is thrown off with the cleanſings. This is the *membrana decidua* of Dr. Hunter; which he deſcribes as a lamella from the inner ſurface of the uterus; though Signor Scarpa, with more probability, conſiders it as being compoſed of an inſpiſſated coagulable lymph.

Though the uterus, from the moment of conception, is gradually diſtended, by which conſiderable changes are occaſioned, it is very difficult to judge of pregnancy from appearances in the early months. For the firſt three months the os tincæ feels ſmooth and even, and its orifice as ſmall as in the virgin ſtate. When any difference can be perceived about the fourth or fifth month, from the deſcent of the fundus through the pelvis, the tubercle or projecting part of the os tincæ will ſeem larger, longer, and more expanded; but, after this period, it ſhortens, particularly at its fore-parts and ſides, and its orifice or labia begin to ſeparate, ſo as to have its conical appearance deſtroyed. The cervix, which in the early months is nearly ſhut, now begins to ſtretch and to be diſtended to the os tincæ; but during the whole term of utero-geſtation, the mouth of the uterus is ſtrongly cemented with a ropy mucus, which lines it and the cervix, and begins to be diſcharged on the approach of labour. In the laſt week, when the cervix uteri is completely diſtended, the uterine orifice begins to form an elliptical tube, inſtead of a fiſſure, or to aſſume the appearance of a ring on a large globe; and often at this time, eſpecially in pendulous bellies, diſappears entirely, ſo as to be out of the reach of the finger in touching. Hence the os uteri is not in the direction of the axis of the womb, as has generally been ſuppoſed.

About the fourth, or between the fourth and fifth, month, the fundus uteri begins to riſe above the pubes, or brim of the pelvis, and its cervix to be diſtended nearly one-third. In the fifth month the belly ſwells like a ball, with the ſkin tenſe, the fundus about half way between the pubes and navel, and the neck one-half diſtended. After the ſixth month the greateſt part of the cervix uteri dilates, ſo as to make almoſt one cavity with the fundus. In the ſeventh month the fundus advances as far as the umbilicus. In the eighth it reaches mid-way between the navel and ſcrobiculus cordis; and in the ninth to the ſcrobiculus itſelf, the neck then being entirely diſtended, which, with the os tincæ, become the weakeſt part of the uterus. Thus at full time the uterus occupies all the umbilical and hypogaſtric regions; its ſhape is almoſt pyriform, that is, more rounded above than below, and having a ſtricture on that part which is ſurrounded by the brim of the pelvis. The appendages of the uterus ſuffer very little change during pregnancy, except the ligamenta lata, which diminiſh in breadth as the uterus enlarges, and at full time are almoſt entirely obliterated.

The various diſeaſes incident to the uterine ſyſtem, and other morbid affections of the abdominal viſcera, in weak and ſickly females, will frequently excite the ſymptoms, and aſſume the appearance, of real pregnancy. Complaints ariſing from a ſimple obſtruction are ſometimes miſtaken for thoſe of breeding; when a tumor about the region of the uterus is alſo formed, and gradually becomes more and more bulky, the ſymptoms it occaſions are ſo ſtrongly marked, and the reſemblance to pregnancy ſo very ſtriking, that the ignorant patient is often deceived, and even the experienced phyſician impoſed on.

Scirrhous, polypous, or ſarcomatous, tumours, in or about the uterus or pelvis; dropſy or ventoſity of the uterus or tubes; ſteatoma or dropſy of the ovaria, and ventral conception; are the common cauſes of ſuch fallacious appearances. In many of theſe caſes the menſes diſappear; nauſea, retchings, and other ſymptoms of breeding, enſue; flatus in the bowels will be miſtaken for the motion of the child; and, in the advanced ſtages of the diſeaſe, the preſſure of the ſwelling on the adjacent parts. Tumefaction and hardneſs of the breaſts ſupervene, and ſometimes a viſcid or ſerous fluid diſtils from the nipple; circumſtances that ſtrongly confirm the woman in her opinion, till time, or the dreadful conſequences that often enſue, at laſt convince her of her fatal miſtake.

Other kinds of ſpurious gravidity, leſs hazardous in their nature than any of the preceding, are commonly known by the names of *falſe conceptions* and *moles*: the former of theſe is nothing more than the diſſolution of the fœtus in the early months; the placenta is afterwards retained in the womb, and, from the addition of coagula, or in conſequence of diſeaſe, is excluded in an indurate or enlarged ſtate; when it remains longer, and comes off in the form of a fleſhy or ſcirrhous-like maſs, without having any cavity in the centre, it is diſtinguiſhed by thename of *mole*. Mere coagula of blood, retained in the uterus after delivery, or after immoderate floodings at any period of life, and ſqueezed, by the preſſure of the uterus, into a fibrous or compact form, conſtitute another ſpecies of mole, that more frequently occurs than any of the former. Theſe, though they may aſſume the appearances of gravidity, are generally, however, expelled ſpontaneouſly, and are ſeldom followed with dangerous conſequences. But, when two or more of the ova deſcend into the uterus, attach themſelves ſo near one another as to adhere in whole or in part, ſo as to form only one body, with membranes and water in common, this body will form a confuſed irregular maſs, which is called a *monſter*; and thus a monſter may be either defective in its organic parts, or be ſupplied with a ſupernumerary ſet of parts derived from another ovum. This proceeds from a defect or accident in nature, which it is entirely beyond the power of medicine to rectify or prevent.

It would ſeem, however, from a due contemplation of the foregoing facts, from the frame and ſtructure of females, and from the ultimate end and purpoſe of their conformation, that almoſt every malady reſulting from a ſtate of pregnancy, except the laſt-mentioned, may be in a great meaſure prevented or removed. The natural temperature of women differs in a very conſiderable degree from that of men, inaſmuch as their blood and juices are determined to an oppoſite and diſtinct purpoſe; and hence it is that obſtructions of the menſes, their exceſs, or privation of the office intended them, conſtitute thoſe peculiar maladies which we term *Diſeaſes of Women.*

men. The natural temperature of the male, is *hot* and *dry;* that of the female, *cold* and *moift.* The action of the procreative tincture of man is SOLAR, i. e. of the heating and quickening faculty; that of the woman is LUNAR, i. e. of a cool and vegetative quality. As the fun heats, and gives prolific energy to the fruits of the earth, fo man fecundates and gives life to the prolific tincture of the woman. Thus the male, as the microcofm, or epitome of the celeftial fyftem, poffeffes an inherent fimilitude with the fun, which vivifies and quickens; and thus the female, poffeffing an inherent fimilitude with the moon, vegetates and brings forth the fruit of her womb, and not only feels the influence and fympathy of that luminary in her monthly difcharges, but in all the travail and viciffitudes of pregnancy. To the fame fource likewife we trace the caufe, and decide the queftion, Whether the fruit of the womb be male or female? for, if the male feed be predominant, heat will abound, and the male fœtus will be generated; but, if the cooling moifture of the woman overcomes the mafculine heat in the male feed, a female is then produced. The old and exploded notion of this caufe depending on the child's falling to the right or left fide of the mother, is too abfurd to weigh a moment on the mind of any reafonable enquirer.

We difcover likewife that the male, being conftituted of the folar temperature, is naturally fubjected to thofe infirmities of body and mind which refult from the elements of fire and air; while thofe of the female are of lunar tendency, arifing from the elements of water and earth. Of thefe four elements our grofs or material part is formed, and by their due and proper commixture in the conftitution, or circulating mafs, are life and health eftablifhed; whilft, on the contrary, by their difcordant, defective, or predominant, power, difeafe and death are produced. Now the male abounding in heat, and the female in moifture, is the reafon why many diforders incident to man are alleviated by contact with the woman, as thofe of the woman are by contact with the man. In the grand fcale of Nature, we find the meridian heat and fcorching rays of the fun are qualified and corrected by the cooling moifture and mild influence of the midnight moon; but, when either of thefe are obftructed in their effect, by the intervention of accidental caufes, by ftorms, by tempefts, or unfeafonable blafts, we then endeavour to reprefs by art the evil confequences that are likely to enfue. Juft fo in the human economy, the grand purpofe and defign of medicine is to correct and modify the difcordant elements in the conftitution, and give that vigour and tone to the vital powers, which conftitute the genuine principles of health and life.

From what has been fuggefted we might fafely infer, that the conftitution and temperature of the female require a medicine of an oppofite action and tendency to that

that adapted to the male, and which ought to be compounded of elements congenial to the intentions of Nature, calculated to purge the uterus, to purify the seminal fluid, and give stimulus to the catamenia; which, if not put in motion by the functions of nature, becomes dull and stagnant, and vitiates the whole circulating mass; whence those disorders, peculiarly incident to the most amiable, as being the most virtuous, of women, are confessedly derived; and for the cure and prevention of which, a peculiar and distinct remedy has long been wanting.

These, and other considerations, influenced by the known power of second causes, and their faculty of acting upon the mechanism of the human frame, induced me to attempt the chemical preparation of two subtile Tinctures, constituted of a co-mixture of the purest elements of which our blood is composed, and adapted to the particular temperature and constitutions of the opposite sexes. That intended for the use of Man I call the SOLAR TINCTURE, as being congenial to the seminal functions and vital principles of his constitution. That adapted to Woman I call the LUNAR TINCTURE, as being calculated to act upon the menstrual and vegetative fluids, and as being compounded of those elements which make up the frame and temperature of her body. The invention of these Tinctures hath been the result of a long and laborious application to the study of unveiled Nature—of the properties of fire, air, earth, and water, in the propagation of animal and vegetable life, and in the composition of medicine; in which, though these elements form the PABULUM of the universe, yet the art of collecting, uniting, and assimilating them with the vital fluids, seems to be unknown among modern chemists, and has escaped the observation of medical science. The fixidity of these Tinctures at once establishes their power and efficacy beyond all others; for they can never be affected by change of weather or climate, nor by heat or cold; nor will they suffer any diminution of their strength or virtue by remaining open, or uncorked; a circumstance which cannot be affirmed of any other fluid at present known, throughout the world.

I shall now proceed to show the action of the LUNAR Tincture on Female constitutions; and as this medicine is only intended to remedy such complaints as particularly relate to pregnancy, and the menstrual discharge, I shall omit to notice any other maladies, until I come to treat of the SOLAR Tincture; which, though essentially directed to give tone and vigour to the constitution of the male, is nevertheless equally efficacious to the female in removing all disorders of the blood and lymph, that are alike common to valetudinarians of both sexes. No complaint in the female habit, therefore, comes under our present enquiry, till at or near the age of puberty. Until this important period of the sex arrives, the Rules heretofore laid down in the Medical Part of my new edition of Culpeper's Family Physician, for the management

nagement and future health of young ladies, deſerve a very cloſe and ſerious attention. The evident diſtinction between the male and female in their ſtructure and deſign, in their bodily ſtrength and vigour, and in the procreative fluids, demands the utmoſt attention from themſelves, and the tendereſt care from the phyſician. Nor can we too often or too earneſtly caution parents and guardians againſt the evils of that abſurd though faſhionable ſtyle of bringing up young ladies, by confining them almoſt entirely to their apartments, keeping them on poor low diet, and uſing artificial means to make them ſpare and delicate, which contributes more to their prejudice than all the incidental diſeaſes to which they are otherwiſe ſubject. Theſe refinements in female education, beſides deſtroying their ruddy complexion, (which is often the deſign of it,) relaxes their ſolids, impoveriſhes their blood, weakens their minds, and diſorders all the functions of their body, whereby they are often rendered incapable of conception, and denied the felicity of becoming mothers. On the contrary, it ought to be the ſtudy, as it certainly is the duty, of all that have girls under their care, to indulge them in every innocent diverſion, and in every active exerciſe, that can give freedom to the limbs, or agility to the body; all of which have a natural tendency to exhilarate their ſpirits, to promote digeſtion, to ſtimulate their blood and juices, and, at the proper age, to bring on a free and eaſy diſcharge of the menſtrual flux.

Though it is univerſally admitted, that this flux is abſolutely neceſſary to nouriſh and ſupport the fœtus, and that without it human generation cannot be carried on; and that it is conſequently and obviouſly peculiar to the female uterine ſyſtem; yet is it curious to obſerve the various abſurd and contradictory opinions ſome phyſicians have laboured to eſtabliſh, merely, one would ſuppoſe, to bewilder the underſtanding, and ſubject delicate females ſtill more to that erroneous or miſguided treatment, in which their health, their life, and every earthly bleſſing, are too frequently involved.

Dr. Bohn, and Dr. Freind, inſiſt that this flux is nothing more than a plenitude of the common maſs of blood, which nature throws off only for relief againſt the too abundant quantity. Dr. Freind ſuppoſes, that this plenitude ariſes from a coacervation in the blood-veſſels of a ſuperfluity of aliment, which, he thinks, remains over and above what is expended by the ordinary ways; and that women have this plethora, and not men, becauſe their bodies are more humid, and their veſſels, eſpecially the extremities of them, more tender, and their manner of living generally more inactive than that of men; and that theſe things, concurring, are the occaſion that women do not perſpire ſufficiently to carry off the ſuperfluous alimentary parts, till they are accumulated in ſuch quantities as to diſtend the veſſels, and force their way through the capillary arteries of the uterus. It is ſuppoſed to hap-

pen

pen to women more than to the females of other ſpecies, which have the ſame parts, becauſe of the erect poſture of the former, and the vagina and other canals being perpendicular to the horizon; ſo that the preſſure of the blood is directed towards their orifices: whereas in brutes they are parallel to the horizon, and the preſſure is wholly on the ſides of thoſe veſſels. The diſcharge, he thinks, happens in this part rather than in any other, as being more favoured by the ſtructure of the veſſels; the arteries being very numerous, and the veins ſinuous and winding, and therefore more apt to retard the impetus of the blood; and conſequently, in a plethoric caſe, to occaſion the rupture of the extremities of the veſſels, which may laſt, till, by a ſufficient diſcharge, the veſſels are eaſed of their overload. To this he adds the conſideration of the ſoft pulpous texture of the uterus, and the vaſt number of veins and arteries with which it is filled. Hence a healthy maid, being arrived at her growth, begins to prepare more nutriment than is required for the ſupport of the body; which, as there is not to be any farther accretion, muſt of neceſſity fill the veſſels, and eſpecially thoſe of the uterus and breaſts, they being the leaſt compreſſed. Theſe will be dilated more than the others; whence, the lateral vaſcules evacuating their humour into the cavity of the uterus, it will be filled and extended. Hence a pain, heat, and heavineſs, will be felt about the loins, pubes, &c. and the veſſels of the uterus, at the ſame time, will be ſo dilated as to emit blood in the cavity of the uterus, and its mouth will be lubricated and looſened, and blood iſſue out. As the quantity of blood is diminiſhed, the veſſels will be leſs preſſed, and will contract themſelves cloſer, ſo as again to retain the blood, and let paſs the groſſer part of the ſerum; till at length only the uſual ſerum paſſes. Again, there are more humours prepared, which are more eaſily lodged in veſſels once dilated; and hence the menſes go and return at various periods in various perſons.

This hypotheſis is judiciouſly oppoſed by Dr. Drake, who maintains, that there is no ſuch plenitude, or at leaſt that it is not neceſſary to menſtruation; arguing, that, if the menſes were owing to a plethora ſo accumulated, the ſymptoms would ariſe gradually, and the heavineſs, ſtiffneſs, and inactivity, neceſſary ſymptoms of a plethora, would be felt long before the periods were completed, and women would begin to be heavy and indiſpoſed ſoon after evacuation, and the ſymptoms would increaſe daily; which is contrary to all experience, many women, who have them regularly and eaſily, having no warning, nor any other rule to prevent an indecent ſurpriſe, than the meaſure of the time; in which, ſome that have ſlipped have been put to confuſion and ſhifts no ways conſiſtent with the notice a plethoric body would give. He adds, that even in thoſe who are difficultly purged this way, the ſymptoms, though very vexatious and tedious, do not make ſuch regular approaches

proaches as a gradual accumulation neceſſarily requires. If we conſider what violent ſymptoms come on in an hour, we ſhall be extremely puzzled to find the mighty acceſſion of matter, which ſhould, in an hour or a day's time, make ſuch great alterations. According to the hypotheſis, the laſt hour contributed no more than the firſt; and of conſequence, the alteration ſhould not be greater in the one than in the other, ſetting aſide the bare eruption.

There are others who give into the doctrine of fermentation, and maintain the evacuation in thoſe parts to be an effect of an effervefcence or ebullition of the blood. This opinion has been maintained by Dr. Charleton, and by Bale, De Graaf, and Drake, the two firſt of whom ſuppoſe a ferment peculiar to women, which produces this flux, and affects that part only, or at leaſt principally. Dr. Graaf, leſs particular in his notion, only ſuppoſes an effervefcence of the blood, raiſed by ſome ferment, without aſſigning how it acts, or what it is. The ſudden turgeſcence of the blood occaſioned them all to think, that it aroſe from ſomething till then extraneous to the blood; and led them to the parts principally affected to ſeek for an imaginary ferment, which no anatomical enquiry could ever ſhow, or find any receptacle for, nor any reaſoning neceſſarily infer. Again, that heat which frequently accompanies this turgeſcence, led them to think the caſe more than a plethora, and that there was ſome extraordinary inteſtine motion at that time.

Dr. Drake contends, that it is not only neceſſary there ſhould be a ferment, but a receptacle alſo for this ferment; concluding, from the ſuddenneſs and violence of the ſymptoms, that a great quantity muſt be conveyed into the blood in a ſhort time, and conſequently that it muſt have been ready gathered in ſome receptacle, where, while it was lodged, its action was reſtrained. He pretends to aſcertain the place both of the one and the other, making the gall-bladder to be the receptacle, and the bile the ferment. The liquor he thinks well adapted to raiſe a fermentation in the blood, when diſcharged into it in quantity; and, as it is contained in a receptacle that does not admit of a continual iſſue, it may be there reſerved, till in a certain period of time the bladder becoming turgid and full, through the compreſſion of the incumbent viſcera, it emits the gall; which, by the way of the lacteals inſinuating itſelf into the blood, may raiſe that effervefcence which occaſions the aperture of the uterine arteries. To confirm this, he alleges, that perſons of a bilious conſtitution have the menſes either more plentifully, or more frequently, than others; and that diſtempers manifeſtly bilious are attended with ſymptoms reſembling thoſe of women labouring under difficult menſtruation. But, if this argument be admitted, men would have the menſes as well as women. To this however he anſwers, that men do not abound in bile ſo much as women, the pores of the former being more open, and carrying off more of the ſerous part of the blood, which is the vehicle

vehicle of all the other humours, and conſequently a greater part of each is diſcharged through them than in women, wherein the ſuperfluity muſt either continue to circulate with the blood, or be gathered into proper receptacles, which is the caſe in the bile. The ſame reaſon he gives why menſtruation ſhould not be in brutes: the pores of theſe being manifeſtly more open than thoſe of women, as appears from the quantity of hair which they bear, for the vegetation whereof a large cavity, and a wider aperture of the glands, is neceſſary, than where no ſuch thing is produced: yet there is ſome difference between the males and females even among theſe, ſome of the latter having their menſes, ſuch as the orang outang,* &c. though not ſo often, nor in the ſame form and quantity, as women. But without dwelling on theſe abſtract reaſonings, the abſurdity of which will be obvious to every perſon who turns to the foregoing ſyſtem of human impregnation, we need only remark, that there are two critical periods in every woman's life, that completely deſtroy their hypotheſis. Theſe are, that at the age of fourteen or fifteen, the menſes begin to flow; but ſubſide at the age of forty or fifty. At their commencement, we generally find the difficulty, and conſequent diſeaſe, ariſe from their *deficiency;* whereas, according to the foregoing doctrine, they would then *always* flow with the greateſt freedom. At the period when they ſhould ceaſe, they are apt to come in ſuch abundance as to bring on a flooding, which not only endangers, but too frequently deſtroys, life—a fatal conſequence that could not poſſibly happen, were the above arguments true.

Of FEMININE, or LUNAR, DISEASES.

THAT the vegetative or procreative faculties of women are univerſally governed by the lunations of the moon, their own experience, as well as the demonſtrations given in my ILLUSTRATION of ASTROLOGY, indiſputably prove. The firſt ſhow of the catamenia, if it be natural, invariably comes with the new or full moon; or ſometimes, though very ſeldom, at the commencement of her firſt or laſt quarters; and this effort of nature is juſtly conſidered as the ſure ſign of a procreating ability, and of complete puberty. Whenever this ſeaſon arrives, whether early or late, the conſtitution of every female undergoes a conſiderable change, and the greateſt care and attention are then neceſſary, ſince the future health and happineſs of every woman depends, in a great meaſure, upon her conduct at this period. It is the duty of mothers, and of thoſe who are intruſted with the education of girls, to inſtruct them early in the conduct and management of themſelves at this critical moment. Falſe modeſty, inattention, and ignorance of what is beneficial

* See this curious ſubject, concerning the orang outang, and other animals reſembling the human ſpecies, treated at large, both hiſtorically and philoſophically, in my new Syſtem of Natural Hiſtory, vol. ii. 8vo.

or hurtful at this time, are the ſources of many diſeaſes and misfortunes, which a very little attention might now prevent. Nor is care leſs neceſſary in the ſubſequent returns of this diſcharge. Taking improper food, violent agitatious of the mind, or catching cold, is often ſufficient to ruin the health, or to render the female for ever after incapable of procreation.

In order to eſcape the chloroſis, and other ſimilar diſeaſes incident to young women at that period when the menſes commence, let them avoid indolence and inactivity, and accuſtom themſelves to exerciſe in the open air as much as poſſible. The diſcharge in the beginning is ſeldom ſo inſtantaneous as to ſurpriſe them unawares. The eruption is generally preceded by ſymptoms that indicate its approach; ſuch as a ſenſe of heat, weight, and dull pain, in the loins; diſtention and hardneſs of the breaſts, head-ach, loſs of appetite, laſſitude, paleneſs of the countenance, and ſometimes a ſlight degree of fever. When theſe ſymptoms occur, every thing ſhould be carefully avoided which may obſtruct the diſcharge, and all gentle means uſed to promote it; as ſitting frequently over the ſteam of warm water, drinking warm diluting liquors, &c. When the menſes have begun to flow, great care ſhould be taken to avoid every thing that tends to obſtruct them; ſuch as fiſh, and all kinds of food that are hard of digeſtion, and cold acid liquors. Damps are likewiſe hurtful at this period; as alſo anger, fear, grief, and other affections of the mind. From whatever cauſe this flux is obſtructed, except in the ſtate of pregnancy, proper means ſhould be inſtantly uſed to reſtore it; and if exerciſe in a dry, open, and rather cool, air, wholeſome diet, generous liquors in a weak and languid ſtate of the body, cheerful company, and amuſement, fail, recourſe muſt be had to medicine. In all ſuch caſes, blood-letting muſt be carefully avoided; but let the patient take from 20 to 30 drops of the Lunar Tincture, in a wine-glaſs of warm water, or penny-royal tea, every morning before breakfaſt, every day at noon, and every night before going to bed, until the intention be anſwered, which will uſually take place in three or four days, without the aſſiſtance of any other medicine whatever. But it ſometimes happens in relaxed conſtitutions, that the menſtrual diſcharge, on its firſt appearance, is vitiated, and over-abundant; the conſequence of which is, that the patient becomes weak, the colour pale, the appetite impaired, and the digeſtion languid, ſo that dropſy or conſumption is likely to enſue. Effectually to prevent theſe, let the patient be kept two or three days in bed, with her head low, and obſerve a ſlender diet, principally of white meats, and her drink red-port negus. Every night and morning, for ten or twelve days, let her take one table-ſpoonful of the Solar Tincture, diluted in double the quantity of decoction of nettle-roots, or of the greater comfrey; and after the flux has abated,

abated, and her health and ſtrength ſeem to return, let her only take a table-ſpoonful of the Solar Tincture every other day at noon, in a glaſs of cold ſpring-water; which wonderfully contributes to reſtore a due conſiſtency to the circulating maſs, promotes digeſtion, and invigorates the ſpirits. Before the cuſtomary period returns, ſhe muſt diſcontinue the Solar Tincture; and, if there be the leaſt appearance of irregularity or obſtruction, let her again take night and morning, for two or three days, from 20 to 30 drops of the LUNAR TINCTURE, in a glaſs of pennyroyal tea, and ſhe will quickly find a regular habit, and her health amazingly eſtabliſhed. In obſtinate or neglected caſes, where the menſes have ſeceded, and, after an irregular appearance, have turned wholly into the habit, both theſe Tinctures ſhould be uſed with a leſs ſparing hand, particularly under circumſtances in any reſpect ſimilar to the following remarkable

CASE.

Being called to the aſſiſtance of a young lady of fifteen years of age, I was informed her menſes had made an irregular appearance about five or ſix times, coming firſt with the full and then with the new moon, and afterwards at the diſtance of two or three months apart, until they totally diſappeared, and turned back upon the habit. No notice was taken, until the patient was ſeized with a violent bleeding at the noſe, attended with fever, and epileptic fits. After being under the care of an eminent phyſician for ſeveral months, who directed veneſection, and almoſt every cuſtomary application, to no kind of purpoſe, the diſorder fixed in her neck, forming a large tumour, the acrimony of which fell upon her lungs, and threw her into ſtrong convulſions. In this extremity I was ſent for. Perceiving the whole ſyſtem deranged by ſpaſmodic affections, and a locked jaw almoſt finally completed, my firſt object was to relieve the vital organs, by giving force and elaſticity to the circulating maſs. With this view, I with difficulty forced open the mouth, and adminiſtered one table-ſpoonful of the Solar Tincture *undiluted;* and within half an hour, to the aſtoniſhment of her friends, I had the pleaſure of ſeeing every convulſive ſymptom die away, and of hearing the patient's voice, of which ſhe had been totally deprived for upwards of a week before. Two hours after, another ſpoonful of the Solar Tincture was taken with additional ſucceſs; and the patient afterwards continued this medicine in the quantity of a table-ſpoonful, in a wine-glaſs of warm water, three times a-day, for ſix days, at the expiration of which time her appetite and ſtrength were ſurpriſingly returned; and ſhe was then put under a regular courſe of the Lunar Tincture. Twenty drops in a wine-glaſs of pennyroyal tea were taken every night and morning for thirteen ſucceſſive days, and on the morning following, it being the full moon, with which her menſes originally came, ſhe

ſhe had the conſolation to find that every obſtruction was removed, and that the due courſe of nature was completely re-eſtabliſhed. The glandular ſwellings gradually ſubſided, her natural complexion quickly returned, and ſhe now continues in blooming health, perfectly regular, free from all obſtructions, and from every conſequent complaint, thankful for the bleſſings of her recovery, and deſirous of communicating the means to any unfortunate female under ſimilar affliction; and to whom reference may at any time be had, by application to the author.

CHLOROSIS, or GREEN-SICKNESS; by ſome called the LOVE-FEVER.

THIS diſeaſe uſually attacks virgins a little after the time of puberty, and firſt ſhows itſelf by ſymptoms of *dyſpepſia*, or bad digeſtion. But a diſtinguiſhing ſymptom is, that the appetite is entirely vitiated, and the patient will eat lime, chalk, aſhes, ſalt, &c. very greedily; while at the ſame time there is not only a total inappetence to proper food, but it will even excite nauſea and vomiting. In the beginning of the diſeaſe, the urine is pale, and afterwards turbid; the face becomes pale, and then aſſumes a greeniſh colour; ſometimes it becomes livid or yellow; the eyes are ſunk, and have a livid circle round them; the lips loſe their fine red colour; the pulſe is quick, weak, and low, though the heat is little ſhort of a fever, but the veins are ſcarcely filled; the feet are frequently cold, ſwell at night, and the whole body ſeems covered with a ſoft ſwelling; the breathing is difficult: nor is the mind free from agitation as well as the body; it becomes irritated by the ſlighteſt cauſes; and ſometimes the patient loves ſolitude, and becomes ſad and melancholy. There is a retention of the menſes throughout the whole courſe of the diſorder; which eventually fixes on the vital organs, and death enſues.

The above complaint indiſputably ariſes from ſtifling or ſuppreſſing the calls of nature at this vernal ſeaſon, or juvenile ſpring of life, when the primary command of God, "*increaſe and multiply,*" is moſt ſenſibly impreſſed upon the whole human fabric. Every tube and veſſel appertaining to the genital ſyſtem, being now filled with ſpermatic or procreative liquor, excites in the female a powerful, yet perhaps involuntary, irritation of the parts, which ſtrongly ſolicits the means of diſcharging their load, that can only be done by venereal embraces. Theſe, from prudential reaſons, being often neceſſarily denied, the prolific tinctures ſeize upon the ſtomach and viſcera, pen back and vitiate the catamenia, choke and clog the perſpirative veſſels, whereby the venal, arterial, and nervous, fluids, become ſtagnant; and a leucophlegmatia, or white flabby dropſical tumour, pervades the whole body, and quickly devotes the unhappy patient to the arms of death. Thus, I am ſorry to remark, are thouſands of the moſt delicate and lovely women plunged into eternity, in the very

very bloſſom of life, when female excellence is but budding forth, big with the promiſed fruit of deliciouſneſs and joy! How much then does it become the duty of parents and guardians, who have daughters or wards in ſituations like theſe, and where no very groſs objection can ariſe, to ſuffer them to marry with the men they love, or otherwiſe to provide ſuitable matches for them; ſince this will effect the moſt rational and moſt natural cure, by removing the cauſe of the complaint altogether. If, however, matrimony be not then convenient, nor likely in a ſhort time to take place, recourſe muſt forthwith be had to proper regimen, and medical aid, otherwiſe delirium or conſumption will quickly enſue. The beſt method of regimen is laid down in the medical part of my edition of Culpeper, page 217. which, if well obſerved, in addition to the following courſe, will generally perform a cure. Take leaves of mugwort, briony, and penny-royal, of each a handful; infuſe them four days in two quarts of ſoft water, and then pour off the clear liquor for uſe. Take a gill-glaſs three parts full, with thirty drops of the Lunar Tincture added to it, three times a-day, viz. morning, noon, and night, till the decoction be all uſed. Then reduce the doſe to twenty drops of the Tincture in a wine-glaſs of cold ſpring-water morning and evening, for fifteen days; after which it may be taken only once a-day, or every other day, until the patient find herſelf free from every ſymptom of the diſeaſe. For this malady, it is the only ſpecific hitherto known; it unclogs the ſpermatic tubes; purges and cools the uterus and vagina; promotes the menſtrual diſcharge, cleanſes the urinary paſſages, diſſolves viſcid humours in the blood, ſharpens the appetite, ſtimulates the nerves, and invigorates the ſpirits, which in all ſtages of *chloroſis* are ſo apt to be depreſſed. When this diſorder is not very obſtinate, nor far advanced, let the patient take from twenty to thirty drops of the Lunar Tincture, in a wine-glaſs of cold ſpring-water, for thirty or forty days ſucceſſively, and it will perform a cure without the trouble of preparing the decoction. In this malady, I have lately had the happineſs of completing an elegant cure, which I mention here, merely for the information of ſuch unfortunate maids as may be languiſhing under the ſame deplorable circumſtances. The following is a literal ſtatement of the

C A S E.

A young lady, turned of ſeventeen, had been afflicted with *chloroſis* almoſt three years. In the early part of the malady, ſhe conceived an unconquerable appetite for wood-cinders, concreted mortar, tobacco-pipes, ſealing-wax, &c. The courſes appeared at different intervals of the diſeaſe, but always irregularly, and more or leſs in a vitiated ſtate. About half a year preceding my attendance, this flux had totally ceaſed; but, upon the approach of every new moon, with which her menſes originally came, ſhe was afflicted with pains in the back and loins, heavineſs and turgidity about the region of the womb, and other cuſtomary ſymptoms of the catamenia; yet not the ſmalleſt ſhow could be brought to appear. A little before this, the lady's affections had been placed on a young man in the neighbourhood; but whoſe ſituation in life was by no means on a ſcale adapted to the views of her father and family. The moment therefore this attachment was diſcovered, the lady was confined to her apartment, and not ſuffered to take either exerciſe or freſh air, but when it ſuited for ſome truſty attendant to accompany her. This confinement brought on a ſettled melancholy, a green ſallow complexion, dejected ſpirits, univerſal laſſitude, and waſting of the fleſh. The morbid ſtate of her body having thus undermined her conſtitution, without attracting either her own or her father's obſervation, the diſorder fell upon the vital organs, and with ſo rapid a progreſs, that within twenty-four hours ſhe was ſeized with an ardent fever, attended with loſs of appetite, delirium, and a total privation of ſpeech. In this ſhocking ſtate, ſhe had the alternate advice of three phyſicians of the firſt reſpectability; but the diſorder increaſing, and putting on the moſt dangerous ſymptoms, after having baffled their utmoſt ſkill, a conſultation was had, and the miſerable patient was conſigned to the grave.

Under theſe deplorable circumſtances, it was my lot to be called in; and, upon a cloſe examination of the patient, ſcarcely any viſible ſigns of life remained. The pulſe had nearly ſubſided. The action of the heart and lungs could ſcarcely be diſcerned. The eyes were ſunk and fixed; yet retained an uncommon look of expreſſion and ſentiment. At this time ſhe had a large bliſter round her neck, another on the pit of her ſtomach; a third, very large, between her ſhoulders; a fourth on the head; a fifth and ſixth inſide the ankles and legs. Veneſection had been ſo often repeated, that ſcarcely blood enough remained to ſupport the heat and action of the heart. In this exhauſted ſtate, I only adminiſtered three table-ſpoonfuls of the Solar Tincture, *undiluted*, at intervals of little more than an hour apart; and in the ſpace of four hours after, I had the heart-felt ſatisfaction of ſeeing the energy of the blood reſtored; pulſation gradually reſumed its action, the lungs were dilated; reſpiration became free; and a profuſe ſweat, which the Tincture induced, fortunately opened the perſpiratory veſſels; and the patient began to give evident ſigns of eaſe and ſenſibility. Warm nouriſhing food was afterwards taken in ſmall quantities; and I was enabled to remove the bliſters, and perform the dreſſings, without pain or torture to the languid patient. The Solar Tincture was now adminiſtered every day for ten days, in the quantity of a table-ſpoonful in a wine-glaſs of warm barley-water, three times in the day, and once in the night, whenever watchfulneſs came on. About the middle of the ſeventh day, ſhe began to articulate,

though

though not a word had been uttered for upwards of ſix weeks before; and on the tenth day, her voice and bodily functions were ſo far reſtored, that I deemed it ſafe to give her an interval of ſix days reſt, without any medicine whatever. I had the happineſs to find my expectations completely anſwered; for nature, aſſiſted by nouriſhing food, effected more than a profuſion of drugs; ſo that in little more than twenty days my patient was able to walk about her room, and to put herſelf under a courſe of the Lunar Tincture. This ſhe perſiſted in, with nouriſhing diet, ſeconded by occaſional but very gentle airings in the carriage, for near a month longer; when on the approach of the enſuing new moon, to the unſpeakable joy of her friends, the menſtrual flux reſumed its natural courſe: the comfort and relief of which were ſo viſible to the patient, that ſhe in ecſtacy exclaimed, "*My ſufferings are at an end.*" This lady has ever ſince continued to improve in health and ſpirits in ſo ſurpriſing a degree, that looking back on her late miſerable and reduced ſtate of body, forms a contraſt ſo great as almoſt to exceed belief. Yet the lady and her worthy parent are at all times ready to authenticate the fact to any reputable enquirer, or to the friends of any unfortunate female labouring under a ſimilar affliction.

Of the FLUOR ALBUS, or WHITES.

THE fluor albus, female weakneſs, or whites, as it is commonly called, is a diſeaſe of the womb and its contiguous parts; from which a pale-coloured greeniſh or yellow fluid is diſcharged, attended with loſs of ſtrength, pain in the loins, bad digeſtion, and a wan ſickly aſpect. The quantity, colour, and conſiſtence, of the diſcharge, chiefly depend upon the time of its duration, the patient's habit of body, and the nature of the cauſe by which it was produced. Weakly women of lax ſolids, who have had many children, and have long laboured under ill health, are of all the moſt ſubject to this diſagreeable diſeaſe; from which they unfortunately ſuffer more ſevere penance than others, as the niceſt ſenſations are often connected with ſuch a delicacy of bodily frame as ſubjects them to it. In Holland it is very frequent, and in a manner peculiar to the place, from the dampneſs of its ſituation; the ſurrounding air being ſo overcharged with moiſture, as to relax the body, ſtop perſpiration, and throw it upon the bowels or womb; producing in the firſt a diarrhœa or flux, in the laſt the fluor albus or female weakneſs. The diſcharge often proceeds from the veſſels ſubſervient to menſtruation; becauſe in delicate habits, where thoſe veſſels are weak, and conſequently remain too long uncontracted, the fluor albus ſometimes immediately follows the menſes, and goes off by degrees as they gradually cloſe. It alſo comes from the mucous glands of the womb, as is particularly evident in very young females of eight or ten years old; in whom, though very rarely, it has

has been obſerved, and where it muſt then neceſſarily have eſcaped from thoſe parts, as the uterine veſſels are not ſufficiently enlarged for its paſſage at ſo early a period. Sometimes, as in women with child, it proceeds from the paſſage to the womb, and not from the womb itſelf; which, during pregnancy, is cloſely ſealed up, ſo that nothing can paſs from thence till the time of labour. The application of thoſe inſtruments called peſſaries, from the pain and irritation they occaſion, are alſo apt to bring on this diſcharge. The fluor albus has been ſuppoſed to ſupply the want of the menſes; becauſe, where the firſt prevails, the laſt are generally either irregular, or totally wanting: but it might more properly be ſaid, that the preſence of the fluor albus, which is a preternatural evacuation, occaſions the abſence of that which is natural; as is evident from the return of the menſes after the fluor albus has been cured. Indeed, when this diſcharge appears about the age of thirteen or fourteen, and returns once a-month, with ſymptoms like thoſe of the menſes, then it may be deemed ſtrictly natural, and ought not to be ſtopped. The fluor albus may be diſtinguiſhed into two kinds. The firſt ariſes from a ſimple weakneſs, or the relaxation of the ſolids; which may either be general, where the whole bodily ſyſtem is enervated and unſtrung; or partial, where the womb only is affected, in conſequence of hard labour, frequent miſcarriages, a ſuppreſſion or immoderate quantity of the menſes, or a ſprain of the back or loins. In the firſt caſe, the diſcharge, being generally mild, may be eaſily taken away. In the ſecond, it may proceed from a vitiated or impure blood, where the body from thence is loaded with groſs humours, which nature, for her own ſecurity and relief, thus endeavours to carry off. In ſuch caſes, the diſcharge is often of a reddiſh colour, like that from old ulcerous ſores; being ſometimes ſo ſharp, as to excoriate the contiguous parts, and occaſion a ſmarting, and heat of urine. A deep-ſeated darting pain, with a forcing down, attending ſuch a diſcharge, is a very dangerous and alarming ſign, and indicates an ulceration or cancerous ſtate of the womb. This malignant ſtate of the diſeaſe, if of long continuance, is extremely difficult of cure; and diſpoſes the patient to barrenneſs, a bearing down, dropſy, or conſumption. In ſhort, as this is a malady of the moſt diſagreeable kind, which by long continuance or neglect becomes difficult of cure, and often proves fatal, it were to be wiſhed that women, on ſuch occaſions, would be more attentive to their own ſafety, by uſing all poſſible means, in due time, to prevent the diſorder.

As women are ſometimes connected with thoſe who do not conſcientiouſly regard their ſafety, it is a circumſtance of the utmoſt conſequence to diſtinguiſh a freſh venereal infection from the fluor albus, or whites: for, if the firſt be miſtaken for the

laſt,

laſt, and be either neglected or improperly treated, the worſt conſequences may ariſe. In addition therefore to what I have ſtated in page 219 of the Medical Part of my edition of Culpeper, the following ſigns will ſerve to inform the patient whether there be occaſion for her doubts or not. A freſh infection, called gonorrhœa, is malignant and inflammatory; the fluor albus moſt commonly ariſes from relaxation and bodily weakneſs; and therefore the remedies proper in the firſt diſorder, would render the laſt more violent, by locking up and confining the infectious matter. In the gonorrhœa, the diſcharge chiefly proceeds from the parts contiguous to the uninary paſſage, and continues whilſt the menſes flow; but in the fluor albus it is ſupplied from the cavity of the womb and its paſſage, and then the menſes are ſeldom regular. In the gonorrhœa, an itching, inflammation, and heat of urine, are the forerunners of the diſcharge; the orifice of the urinary paſſage is prominent, and the patient is affected with a frequent irritation to make water. In the fluor albus, pains in the loins, and loſs of ſtrength, attend the diſcharge; and, if any inflammation or heat of urine follow, they happen in a leſs degree, and only after a long continuance of the diſcharge, which, becoming ſharp and acrimonious, excoriates the ſurrounding parts. In the gonorrhœa, the diſcharge ſuddenly appears without any evident cauſe; but in the fluor albus it comes on more ſlowly, and is often produced by irregularities of the menſes, frequent abortion, ſprains, or long-continued illneſs. In the gonorrhœa, the diſcharge is greeniſh or yellow, leſs in quantity, and not attended with the ſame ſymptoms of weakneſs. In the fluor albus, it is alſo often of the ſame colour, eſpecially in bad habits of body, and after long continuance; but is uſually more offenſive, and redundant in quantity. The whites often afflict maids of a weakly conſtitution, as well as married women and widows; and indeed there are few of the ſex, eſpecially ſuch as are ſickly, who have not known it more or leſs. For whatever diſeaſe renders the blood poor, foul, or viſcous, and reduces a woman to a languid condition, is commonly ſucceeded by the whites, which, when they come in this manner, continue to weaken the body more and more, and are in great danger, without ſpeedy remedy, of wearing away the patient, and making her a miſerable victim to death. Let no woman, therefore, neglect this diſorder, when ſhe finds it on her, but endeavour to obtain an immediate cure. The regimen and general management are pointed out in the Medical Part of the work juſt referred to, p. 220; and, in lieu of all other medicines, make a decoction of tormentil-root, biſtort, comfrey, and red-roſe leaves; take a gill-glaſs three parts full, and add to it thirty or forty drops of the Lunar Tincture, which muſt be perſiſted in morning, noon, and night, for ten days; then take it morning and evening only for ten days more; after which diſcontinue

 the

the decoction, and take the Tincture every morning for a month, twenty drops in a wine-glaſs of cold ſpring-water, the diſeaſe will be found gradually to abate; and, upon any ſymptoms of a return of it, take from fifteen to twenty drops of the Tincture in a wine-glaſs of cold water every morning for a week, and it will go entirely off; as hath been verified in a great number of patients, who are ready to teſtify that they owe their cure, even in the moſt obſtinate caſes, entirely to the Lunar Tincture.

Of BARRENNESS, or INFERTILITY.

BARRENNESS is ſuch a ſtate of a woman's body, as indiſpoſes it, upon the uſe of the natural means, to conceive and propagate her ſpecies. This proceeds from many ſources, which may be reduced to theſe two general heads: Firſt, An indiſpoſition of the parts to receive the male ſemen in the act of copulation, or that vital effluvium ſtreaming from it, which alone can impregnate the ovaria. Secondly, An inaptitude in the blood to retain and nouriſh the vital principle after it is communicated, ſo as to make it grow and expand its parts, till it becomes a proper fœtus. Conception is alſo hindered by a hectic, hydropic, or feveriſh, ſickly habit; by a deficiency or obſtruction of the monthly courſes, which impoveriſhes the fluids; by the whites, which, continuing too long, relax the glands of the womb, and drown, as it were, the prolific particles; and too often by a *vice* which utterly deſtroys the tone and vigour of the parts; as is fully exemplified in the Medical Part of my Culpeper, p. 221.

Preparatory to the cure of infertility, it is proper to uſe evacuations, unleſs any particular ſymptom ſhows them to be dangerous. Bleeding, lenient purgatives, ſuch as the ſolutive electuary, and a gentle vomit of ipecacuanha, eſpecially if the perſon be plethoric or cacochymic, cannot but be of great ſervice; then proceed with the following ſtrengthening electuary: Take roots of ſatyrion and eringo candied, of each one ounce; powders of cinnamon, ſweet fennel ſeeds, and preſerved ginger, of each half an ounce; mace, roots of contrayerva, and Spaniſh angelica, of each one dram; troches of vipers, one ounce; juice of kermes, ſix drams; tincture of cantharides, half a dram; ſyrup of cloves, a ſufficient quantity to make an electuary. Let the quantity of a large nutmeg be taken every morning early, about five o'clock every afternoon, and at night going to bed; and, immediately after taking the electuary, drink a wine-glaſs full of the following infuſion, adding to it from twenty to thirty drops of the Lunar Tincture, viz. Take cinnamon powdered, one ounce; of ſweet fennel-ſeeds bruiſed, and lavender-flowers, of each half an ounce; Spaniſh angelica root, ginger, contrayerva, mace, and cochineal, of each one dram and a half; canary wine, two quarts: infuse according to art for two or three days, and strain off the infusion for use. Continue the electuary for ten days successively; then omit a week, and continue it for ten days more; after which continue the infusion and Tincture only, three times a day, for ten days more: then take it only twice a-day for a month, or as long as the case requires, adding from fifteen to thirty drops of the Tincture to each glass, as the age or constitution of the patient may require. This course will be found most excellent for barrenness and debility; particularly while thus assisted by the Lunar Tincture; which will greatly warm and rectify the blood and juices, increase the animal spirits, invigorate and revive the whole human machine, and not only raise the appetite to venereal embraces, but remove the usual impediments to fertility; prepare the womb for performing its office, and the ova for impregnation. The Tincture warms, comforts, and excites, the generative parts to admiration, and seldom fails of curing all common occasions of barrenness in a month or six weeks, if duly followed; as a proof of which I beg leave to add the pleasing circumstances of the following singular

CASE.

A YOUNG LADY of rank and fortune, but of a delicate frame, entered into the marriage state about four years ago. Instead of deriving from it that blissful gratification which gives the honoured name of Mother, she became weak, languid, pale, and melancholy. The whole nervous system was relaxed—the natural functions of the body were suspended—œdematous tumours obstructed the sanguiferous passages, whence incurable barrenness and lingering consumption were the sad prospects left in view. In this melancholy state of body and mind, by advice of her physician, when all hopes were at an end, she was put under a regular course of the LUNAR TINCTURE; which, to the astonishment of all, gradually deterged the obstructed vessels—propelled the animal juices through the system—strengthened and braced the nerves—induced a regular habit—restored the sparkling eye and rosy cheek, and gave new vigour to the animal functions—the result of which has been, that before the end of the ensuing year, after her health was thus recovered, the lady became the happy mother of a SON and HEIR, to the inexpressible joy of an affectionate husband and a sympathising family!

INDIS-

INDISPOSITIONS ATTENDANT ON PREGNANCY.

THOUGH pregnancy is not a diseafe, but rather a natural alteration of the animal œconomy, which every female is formed to undergo, yet it is attended with a variety of complaints which require great attention; but for the cure or alleviation of which, medical aid has proved very deficient. In thefe complaints, however, the Lunar Tincture exerts moft extraordinary properties, and excels whatever has been heretofore offered under a medical form. It is an univerfal purifier of thofe heterogeneous particles which produce naufea, and arife from the combining efforts of the mafculine and feminine tinctures; from whence, according to the groffnefs of the procreative effences at the time of conception, proceed vomiting, pains in the head and ftomach, fainting, &c. occafioned by the jarring elements arifing from the difproportion in the heat and active principle of the conftituent parts of the male and female feed; which is not only attended with great debility and depreffion to the mother, in her whole nervous fyftem, but often with hereditary difeafes, and dreadful confequences to the infant offspring. Indeed fo great has been the conflict of the male and female procreative tinctures for the maftery or predominant power, while paffing through the circulating mafs or habit of the mother, that the moft curious and aftonifhing phenomena have, on many occafions, been obferved to refult from it.—In a fmall village in Somerfetfhire, in the year 1759, a girl was born with the hair on her head of two remarkably diftinct colours: the right fide, from an exact parallel line which divided the fkull into two equal parts, was almoft black; but the left fide, from the fame line, was of a reddifh yellow. As fhe grew up, the dark hair became of a jet black, exactly like that of her father; whilft the other became of a ftrong carrotty red, precifely refembling that of her mother; and, after the age of puberty, the hair on the privities, and under the arm-pits, as well as on the arms and legs, was diverfified in the fame manner; that on the right fide, all the way down from head to foot, being black; whilft that on the left was entirely red. The young woman lived till the 28th year of her age, and was reforted to as a great curiofity.

Another well-known yet remarkable inftance of this conflict of the male and female procreative tinctures at the time of impregnation, was the cafe of a man who a few years fince kept a public-houfe in Tooley-ftreet, Southwark. His father was a white man, belonging to one of the Weft-India packets; and his mother was a negro-girl, whom he had taken a fancy to, and purchafed on the arrival of one of the Guinea flave-fhips at the ifland of Jamaica. He brought her with him to London, and in the courſe of the enſuing year ſhe was delivered of a ſon, the whole right ſide of which was white like the father; but the whole of the left ſide was black like the mother. As he grew up, this viſible diſtinction became more ſtrongly marked; and, during the time he kept the above public-houſe in Tooley-ſtreet, he was reſorted to by an immenſe concourſe of people, who flocked there to ſpend their mite, in order to be ſatisfied that ſo great a curioſity really exiſted. The whole of his body appeared to be interſected by an exact parallel line, by which the efforts of conception ſeem to have united the male and female tinctures in preciſe equilibrio, without ſuffering them to intermix in coagula, or in impregnating and expelling the ovum from the ovaria, to its ſuſpended ſtate in the uterus. Hence the hair on the right ſide was long and brown, like that of the father; and half the face, neck, body, and privities, with the arm, thigh, leg, and foot, on the right ſide, were white; while the correſponding parts on the left ſide were black, like the mother, with half the hair on the privities and head black and woolly, exactly like that of a true negro.

A ſtill more curious and ſtriking example of this aſtoniſhing effort in the male and female procreative fluids, is verified in the caſe of Mr. John Clark, of Preſcot-ſtreet, Goodman's-fields. His father was a native of Africa, who by dint of good fortune had amaſſed a conſiderable ſum of money, and ſettled in London. He married a remarkably healthy white woman, a native of Devonſhire, who had been ſome time his ſervant. By her he had two ſons and three daughters, who were mulattos, except the eldeſt ſon, who was the firſt born, and the perſon here alluded to. From the head to the navel, all round his body, he was remarkably fair, had a fine ſkin, handſome round features, light-brown hair, and ſanguine complexion, like his mother; but from the navel downwards he was completely black, with ſhort black woolly hair on the privities, exactly like the father. At the age of thirty he married a young lady of good family and fortune, but of a delicate diſpoſition. For near three months he had the addreſs to conceal this deformity of colour from the knowledge of his wife, by wearing fleſh-coloured ſilk drawers and ſtockings, which he pretended were lined with flannel to keep off the rheumatiſm, with which he had been ſorely afflicted, even to a degree that endangered his life, every time he attempted to leave them off. It happened, however, from ſome neglect of concealment before going to ſleep, that the curioſity of his wife was ſtrongly excited; and the opportunity proving favourable in other reſpects, it being quite day-light in the morning, and her huſband faſt aſleep, ſhe eagerly proceeded to ſatisfy her doubts. Gently turning down the bed-clothes, and removing the other impediments in the way of a complete inſpection, ſhe no ſooner diſcovered the real ſtate of things, than ſhe ſhrieked out vehemently, and fainted away! The huſband, thus ſuddenly awakened, beheld his wife in a fit, and ſaw with ſorrow and regret the conſequences of a diſcovery which reſulted from his own neglect. He immediately aroſe, called up the ſervants, and procured medical aſſiſtance with all convenient ſpeed; but in vain—the ſudden ſurpriſe, added to the mortification and terror of mind, had ſo powerful an effect, that the lady died in convulſions, nearly two months gone with child. I have often lamented that fortune did not throw me in the way at this critical juncture, for two reaſons: in the firſt place, I have the vanity to think I could have ſaved the patient's life; but, had I failed, in the ſecond, I would have perſuaded Mr. Clark, from motives of philoſophical ſpeculation, and for the improvement of medical ſcience, to have ſuffered me to open the womb of this unfortunate lady, in order to extract the fœtus; which, under the circumſtances of this uncommon conformation of the father, might have enabled me to throw a new light on this very curious ſubject of occult enquiry, perhaps ſo as to have accounted, more obviouſly, for the jarring conflicts and ſtruggling efforts of the maſculine and feminine tinctures; to which alone we are to look to for the formation of hermaphrodites, the production of monſters, &c.

Sympathy and antipathy moſt certainly operate very powerfully on females in the early ſtate of pregnancy, and might, as was then ſuggeſted, have had a principal ſhare in carrying off the above unhappy patient, while no means were uſed to counteract their influence on the maſs of blood. Sudden frights, longing and loathing, and all marks on the fœtus, are obviouſly derived from this cauſe, and can only be corrected by giving energy and ſtimulus to the circulating ſyſtem, whereby the functions both of mind and body are ſtrengthened, and the nervous fluid fortified and protected againſt the ſudden impreſſion of external objects. It ſeems to be admitted by many eminent practitioners, that the diſeaſes incident to a pregnant ſtate in the early months, ariſe from ſympathy; whilſt thoſe peculiar to the more advanced ſtages of geſtation, are produced by the ſtretching and preſſure of the uterus on the contiguous viſcera. Thu heart-burn and diarrhœa, tenſion and pains of the breaſt, nauſea and head-ach, deſire of unnatural food, tremors, and dejected ſpirits, fainting and hyſteric fits, premature menſtruation, and conſequent abortion, proceed from the firſt of theſe cauſes; while coſtiveneſs, ſtrangury, cramp, and cholic, appear to reſult from the other. And, though the celebrated Dr. Stahl, Dr. Cullen, and others, have ſo much differed as to the theory of theſe diſeaſes, yet they all agree that gentle opiates, aromatic infuſions, ſtrengthening bitters, and medicines calculated to give energy to the languid ſtate of the circulation, and to purify the groſs and viſcid elements which oppreſs the ſtomach and viſcera, are the only proper remedies

medies to be adminiftered. Now the Lunar Tincture poffeffes the aromatic and aftringent virtues in an admirable degree; and is elegantly adapted to invigorate and affift the active faculties of nature, in expelling all vifcid humours from the ftomach and bowels; and, being compounded of the moft fubtle and occult elements, which preferve the vital principle, it hence produces the moft falutary effects on all women in a ftate of pregnancy, by ftimulating the procreative faculty to the formation of the fineft children; correcting and purifying the procreative fluid from infection or difeafe; preventing moles or falfe conceptions, removing all loathings, longings, or vomiting, and effectually preventing abortion, from any caufe whatever.—For thefe reafons, when a woman enters into the ftate of matrimony, fhe would do well to take twenty drops of the Lunar Tincture every other morning, to promote conception; fhe fhould then continue it three times a-week, from conception to the end of the fourth month; then it may be omitted till a fortnight before her time, when fhe fhould take twenty drops in a wine-glafs of cold fpring water every morning till her labour, at which time it will wonderfully ftrengthen her, affift her throws, facilitate the birth, promote the lochia, and carry off the after-pains. She might take it occafionally during the month, in any fymptoms of cold, fever, or hyfterics, diluted in a wine-glafs of warm barley-water, about the middle of the day.

Women who are fubject to mifcarriages, fhould never fail to take this medicine, from the time they have reafon to believe they are pregnant, until a full month after they have quickened. It may be taken once, twice, or thrice, a day, or every other day, as the urgency of the cafe may require, from twenty to thirty drops, in a glafs of forge-water, or in foft fpring-water in which common oak-bark has been fteeped; and fhe will effectually get over all caufes of abortion. Women after fudden mifcarriages, or bad labours, will find wonderful relief by taking twenty drops of it in a wine-glafs of warm barley-water, for a week or ten days. Nurfes, alfo, whofe milk is griping, or defective, fhould take it once or twice a-day, or as often as occafion may require. The intention will quickly be experienced, the milk will be purified and augmented, and all the fluid fecretions promoted in a manner productive of found health, both to the mother and child.—In cafes where œdematous fwellings of the legs and labia are occafioned by the interruption of the refluent blood from the preffure of the diftended uterus on the vena cava—in violent floodings—in nervous fpafms—in epileptic fits, and in obftinate convulfions, where the *vis vitæ* muft be fupported by replenifhing the veffels with the utmoft fpeed —recourfe fhould be had to the Solar Tincture, which in the moft dangerous cafes has been found to give immediate relief; and, if duly perfifted in, according to the bill of directions, will fcarcely ever fail to effect a cure.

STATE

STATE OF WOMEN AT THE TURN OF LIFE.

THE moſt critical and dangerous time of a woman's life is that wherein the menſes ceaſe to flow, which uſually happens between forty and fifty years of age. The great change that this produces, by ſo copious a drain being returned into the habit, without previous preparation, is the ſole cauſe of its danger. Every woman muſt be more or leſs ſenſible when this period arrives, and ſhould conduct herſelf accordingly; for, when the menſes are about to go off, they appear for the moſt part irregularly, both in time and quantity; once in a fortnight, three, five, or ſix, weeks; ſometimes very ſparingly, and other times in immoderate quantities. For want only of neceſſary care and attention, during the time that the menſes thus give ſymptoms of their departure, many and various are the complaints that enſue; amongſt which are cold chills, ſucceeded by violent fluſhings of the face, and heats of the extremities; reſtleſs nights, troubleſome dreams, and unequal ſpirits; inflammations of the bowels; ſpaſmodic affections; ſtiffneſs in the limbs, ſwelled ankles, ſore legs, with pains and inflammation; the piles, and other ſymptoms of plenitude. But all this might eaſily be prevented, by attending to a due regimen, and taking theſe Tinctures as occaſion may require. Whenever a woman has reaſon to ſuſpect her menſes are about to leave her, let her loſe four, five, or ſix, ounces of blood, as her habit of body will admit; then let her make a decoction, by taking gentian-roots, one pound; ſenna and orange-peel, of each half a pound; pour upon them a gallon of hot water, and, after it has ſtood twenty-four hours, pour off the liquor for uſe. Let her take from twenty to forty drops of the Lunar Tincture in a gill-glaſs of the above decoction, every night and morning for ten days; then let her continue it every morning for ten days more, and afterwards once every two or three days, or oftener if the terms are of an ill colour and ſcent, until they are corrected. This courſe muſt be followed every ſpring and fall, for a month or ſix weeks ſucceſſively, by all women who find their menſes come irregularly, or too ſparing, until they entirely ceaſe; after which let the patient put herſelf under a courſe of the Solar Tincture for a month or ſix weeks, taking one ſpoonful in a wine-glaſs of warm water every night and morning for a week; then let it be taken only once a-day, in cold water, for the reſidue of the time; and, if ſhe takes occaſionally two table-ſpoonfuls of the Solar Tincture, diluted in a tumbler of warm water, as a beverage after dinner or ſupper, inſtead of wine or brandy and water, it will be productive of great benefit in eſtabliſhing a healthful ſtate of the blood, and carrying off the viſcid humours generally produced by the menſtrual flux returning into the habit.

Should

Should it at this time happen, which it often does, that the terms flow too abundantly, and produce a flooding, the patient muſt immediately loſe ſix or eight ounces of blood, and be kept as much as poſſible at reſt, with her head low, until the medicine has had time to take effect; let her diet be ſpare, but not too lax; and let her apply to the following courſe: Take conſerve of red roſes, marmalade of quinces, juice of kermes, candied nutmegs, ſyrup of quinces, and ſyrup of coral, of each half an ounce; aromaticum roſalbum, and aſtringent ſaffron of iron, of each two drams; oil of cinnamon, ſix drops: mix into an electuary, (which might be made up by any apothecary, if the receipt be ſent him;) and take the quantity of a large nutmeg every day at noon for ſix, eight, or ten, days, or longer, as the urgency of the caſe may require, drinking immediately after it twenty drops of the Lunar Tincture in a wine-glaſs of warm water; the flooding, by this means, will gradually abate, the feveriſh ſymptoms will go off, the back will be ſtrengthened, the womb-veſſels cleanſed, and the patient wonderfully reſtored. After the tenth day, in moſt caſes, the electuary might be diſcontinued; and the Lunar Tincture ſhould then be taken every morning for a month, from fifteen to twenty drops, according to the conſtitution of the patient; by which time the parts will be braced, comforted, and coiled up; ſo as to fear no danger of a relapſe. About a month after, let her undergo a courſe of the Solar Tincture, for the purpoſe of rectifying and ſtimulating the maſs of blood; this ſhould be taken for a month; a table-ſpoonful night and morning in a wine-glaſs of cold ſpring-water for the firſt ten days; and then once a-day only for the reſidue of the time; the good effects of which will be ſenſibly and quickly felt.

The intention of nature in returning this flux back into the habit, is to nouriſh and preſerve life, not to deſtroy it. Until the age of puberty, girls require this blood for the ſuſtentation and nouriſhment of their bodies; when that is ſufficiently eſtabliſhed, it is applied to the purpoſes of nouriſhing the fœtus, and of ſuckling the infant after it is born. When child-bearing ceaſes, and the eve of life comes on, the flux is returned back, to comfort and preſerve it; therefore, if women were but careful to obſerve a regular courſe before this flux returns upon them, by adopting the methods I have preſcribed, and by taking the medicine ſpring and fall for two or three years previous to the time, they might not only eſcape the perils and dangers attendant on this period, but would lay the foundation of a ſettled ſtate of health, and enjoy a ſound conſtitution of body to extreme old age.

OF MASCULINE, OR SOLAR, DISEASES.

SOLAR difeafes are all fuch as proceed from a hot and dry caufe, and have their origin in the blood and lymph. For, as the beams flowing from the fun are the fountain of life and heat to the great world, or univerfal fyftem of nature, fo the blood, flowing from the heart, is the fountain of life and heat to the little world, or univerfal fyftem of the *microcofm*, or body of man. And again, as the ftream of rays from the fun regulates the feafons, and produces the variety of climates, fo the ftream of blood in man's body, as affected by the fun, regulates and diverfifies the form and figure of the whole race of human beings. As feafons and climates are fubject to the external elements, which are ftill governed by the fuperior influence of the fun, fo they are rendered either mild, healthful, and productive, or turbulent, peftilential, and barren. Juft fo the whole circulating mafs is affected by change of climates and feafons, and by all the variations and agitations of the external elements; and hence difeafes are induced in the blood, and are either mild, ardent, or acute, in proportion as the fanguiferous fluid becomes diftempered and impaired by the action of the ambient, or contiguous atmofphere. Thus we perceive the folar influence on the human frame, and difcover that the origin of difeafe is in the blood; for, no longer than this vital ftream is kept in due circulation, pure and uncontaminated, can animal life be fuftained, or the body preferved in health and vigour.

From the exprefs words of Scripture, Levit. xvii. 11, 14. Deut. xii. 23. we are warranted to infer, that "*in the BLOOD is the LIFE;*" and there is not a doubt but the living principle of the blood conftitutes the life of the body. Of this opinion was the celebrated Hervey, as well as many of the ancient philofophers and phyficians; and the late Mr. John Hunter declared himfelf to be of the fame way of thinking. We find the blood unites living parts, in fome circumftances, as certainly as the yet-recent juices of the branch of one tree unite it with that of another. Were either of thefe fluids to be confidered as extraneous or dead matters, they would act as ftimuli, and no union would take place in the animal or vegetable kingdoms. This argument Mr. Hunter eftablifhed by the following experiment. Having taken off the tefticle from a living cock, he introduced it into the belly of a living hen. Many weeks afterwards, upon injecting the liver of the hen, he injected in the tefticle of the cock likewife, which had come in contact with the liver, and adhered to it. In the nature of things there is not a more intimate connection between life and a folid than between life and a fluid. For, although we are more accuftomed to connect it with the one than the other, yet the only real difference

difference which can be ſhown between a ſolid and a fluid is, that the particles of the one are leſs moveable among themſelves than thoſe of the other. Beſides, we often ſee the ſame body fluid in one caſe and ſolid in another. The blood will alſo become vaſcular like other living parts. Mr. Hunter affirms, that, after amputations, the coagula in the extremities of arteries form veſſels, and may be injected by injecting theſe arteries; and he had a preparation by which he could demonſtrate veſſels riſing from the centre of what had been only a coagulum of blood, and opening into a ſtream of circulating blood. If blood be taken from the arm in the moſt intenſe cold which the human body can bear, it raiſes the thermometer to the ſame height as blood taken in the moſt ſultry heat. This is a ſtrong proof of the blood's being alive; for living bodies alone have the power of reſiſting great degrees both of heat and cold, and of maintaining in almoſt every ſituation, while in health, that temperature which we diſtinguiſh by the name of animal heat. Blood is likewiſe capable of being acted upon by a ſtimulus; for it coagulates from expoſure, as certainly as the cavities of the abdomen and thorax inflame from the ſame cauſe. The more it is alive, that is, the more the animal is in health, it coagulates the ſooner on expoſure; and the more it has loſt of its living principle, as in the caſe of violent inflammations, the leſs is it ſenſible to the ſtimulus produced from its being expoſed, and it coagulates the later. We may likewiſe obſerve, that the blood preſerves life in different parts of the body. When the nerves going to a part are tied or cut, the part becomes paralytic, and loſes all power of motion; but it does not mortify. If the artery be cut, the part dies, and mortification enſues. What keeps it alive in the firſt caſe? nothing but the living principle, which alone can keep it alive; and this phenomenon is inexplicable on any other ſuppoſition than that the life is contained in the blood. Another argument is drawn by Mr. Hunter from a caſe of a fractured os humeri. A man was brought into St. George's hoſpital for a ſimple fracture of the os humeri, or arm; and died about a month after the accident. As the bones had not united, Mr. Hunter injected the arm after death. He found that the cavity between the extremities of the bones was filled up with blood which had coagulated. This blood was become vaſcular, or full of veſſels; in ſome places it was very much ſo. He does not maintain that all coagulated blood becomes vaſcular: and indeed the reaſon is obvious; for it is often thrown out and coagulated in parts where its becoming vaſcular could anſwer no end in the ſyſtem: as, for example, in the cavities of aneuriſmal facs. If it be ſuppoſed, that, in ſuch caſes as that juſt now mentioned, the veſſels are not formed in the coagulum, but come from the neighbouring arteries, it is equally an argument that the blood is alive: for the ſubſtance into which veſſels ſhoot muſt be ſo. The

very

very idea, that ſuch a quantity of dead matter as the whole maſs of blood circulates in a living body, is abſolutely abſurd.

Thoſe who have ventured to oppoſe this doctrine, and the evidence of Scripture with it, conſider the brain and nervous ſyſtem as the fountain of life; and that, ſo far from receiving its life from the blood, the nervous ſyſtem is capable of inſtantaneouſly changing the craſis of the blood, or any other animal fluid; and, though the nervous ſyſtem cannot continue its action for any length of time if the action of the blood-veſſels is ſuſpended, yet the heart and blood-veſſels cannot act for a ſingle moment without the influence of the nervous fluid. For this reaſon, ſay they, it is plain we muſt ſuppoſe the nervous ſyſtem, and not the blood, to contain properly the life of the animal, and conſequently to be the principal vital organ. The ſecretion of the vital fluid from the blood by means of the brain, is, by the ſupporters of this argument, denied. They ſay, that any fluid ſecreted from the blood muſt be aqueous, inelaſtic, and inactive; whereas the nervous fluid is full of vigour, elaſtic, and volatile in the higheſt degree. The great neceſſity for the circulation of the blood through all parts of the body, notwithſtanding the preſence of the nervous fluid in the ſame parts, they ſay is, becauſe ſome degree of tenſion is neceſſary to be given to the fibres, in order to fit them for the influx of the nervous fluid; and this tenſion they receive from the repletion of the blood-veſſels, which are every where diſperſed along with the nerves.

To follow this opinion through every argument, would prove tedious and unneceſſary, as the following ſhort obſervations will decide the matter abſolutely againſt the patrons of the nervous ſyſtem. In the firſt place, then, if we can prove the life of the human body to have been communicated from a fluid to the nervous ſyſtem, the analogical argument will be very ſtrongly in favour of the ſuppoſition that the caſe is ſo ſtill. Now that the caſe once was ſo, is moſt evident; for the human body, as well as the body of every other living creature, in its firſt ſtate, I have ſhown to be a gelatinous maſs, without muſcles, nerves, or blood-veſſels. Nevertheleſs this gelatinous matter, even at that time, contained the nervous fluid. Of this there can be no doubt, becauſe the nerves are formed out of it, and have their power originally from it; and what is remarkable, the brain is obſerved to be that part of the animal which is firſt formed. Of this gelatinous or procreative fluid we can give no further account, than it is the nutritious matter from which the whole body appears to be formed. At the original formation of man and other animals, therefore, the nutritious matter was made the ſubſtratum of the whole body, conſiſting of muſcles, nerves, blood-veſſels, &c. nay more, it was the immediate efficient cauſe of the nervous power itſelf. Again, in the formation of the embryo, we we fee a vital principle exifting as it were at large, and forming to itfelf a kind of regulator to its own motions, or a habitation in which it choofes to refide, rather than to act at random in the fluid. This habitation, or regulator, is undoubtedly the nervous fyftem; but at the fame time, it is no lefs evident that a nutritious fluid is the immediate origin of thefe fame nerves, and of that very nervous fluid. Now we know, that the fluid which in the womb nourifhes the bodies of all embryo animals, is neceffarily equivalent to the blood which nourifhes the bodies of adult ones; and confequently, as foon as the blood became the only nutritious juice of the body, at that fame time the nervous fluid took up its refidence there, and from the blood diffufed itfelf along the nerves, where it was regulated exactly according to the model originally formed in the embryo. Perhaps it may be faid, that the vital power, when once it hath taken poffeffion of the human or any other body, requires no addition or fupply, but continues there in the fame quantity from firft to laft. If we fuppofe the nervous power to be immaterial, this will indeed be the cafe, and there is an end of reafoning upon the fubject; but, if we call this power a volatile and elaftic fluid, it is plain that there will be more occafion for recruits to fuch a power than to any other fluid of the body, as its volatility and elafticity will promote its efcape in great quantities through every pore of the body. It may perhaps be objected, that it is abfurd to fuppofe the blood capable of putting matter in fuch a form as to direct its own motions in a particular way: but even of this we have a pofitive proof in the cafe of the electric fluid. For if any quantity of this matter has a tendency to go from one place to another where it meets with difficulty, through the air for inftance, it will throw fmall conducting fubftances before it, in order to facilitate its progrefs. Alfo, if a number of fmall and light conducting fubftances are laid between two metallic bodies, fo as to form a circle, for example; a fhock of electricity will diftroy that circle, and place the fmall conducting fubftances nearer to a ftraight line between the two metals, as if the fluid knew there was a fhorter paffage, and refolved to take that, if it fhould have occafion to return. Laftly, it is univerfally allowed, that the brain is a fecretory organ, made up of an infinite number of fmall glands, which have no other excretories than the medullary fibres and nerves. As a confiderable quantity of blood is carried to the brain, and the minute arteries end in thefe fmall glands, it follows, that the nervous fluid *muft come from the blood.* Now, there is no gland whatever, in the human or any other body, but will difcharge the fluid it is appointed to fecrete, in very confiderable quantity, if its excretory is cut. Upon the cutting of a nerve, therefore, the fluid fecreted by the brain ought to be difcharged; but no fuch difcharge is vifible. A fmall quantity of glairy matter is indeed difcharged from the large nerves;

but this can be no other than the nutritious juice neceſſary for their ſupport. This makes it plain, even to demonſtration, that the fluid ſecreted in the brain *is inviſible* in its nature; and, as we know the nervous fluid hath its reſidence in the brain, it is very probable, to uſe no ſtronger expreſſion, that it is the peculiar province of the brain to ſecrete this fluid from the blood, and conſequently that the blood originally contains the vital principle.

This fact being eſtabliſhed, I ſhall now endeavour to deſcribe the action of quickening, or mode by which life is communicated to the child in the womb, which uſually takes place in the fifth month of pregnancy. Opportunities, however, of diſſecting the human gravid uterus at or near this critical juncture occurring but ſeldom, it is with great difficulty that a ſubject of this delicate and abſtruſe nature can be treated with perſpicuity, and is the principal cauſe why it has not been attempted by former phyſiologiſts. I have already ſhown, that the rudiments of the embryo puts forth four membranes, viz. the placenta, the navel-ſtring, the chorion, and amnios, which contains the fluid above-mentioned, in which the fœtus floats. Until the period of quickening arrives, the embryo poſſeſſes only vegetative life, ſimilar to that of a common plant; and its growth is nouriſhed and preſerved by the fluid in which it ſwims, until the nerves, veins, arteries, and vital organs, are entirely formed, and the circulation of its mother's blood is completed through them, which is conducted in the following manner.

The placenta is the medium by which the blood from the heart of the mother is communicated to that of the child; but to check its too rapid progreſs, which would overwhelm the tender veſſels of the infant frame, the texture of the placenta is formed ſimilar to that of a ſponge, round like a cake, of conſiderable dimenſions, and capable of great abſorption, being chiefly made up of the ramifications of the umbilical arteries and vein, and partly of the extremities of the uterine veſſels. The arteries of the uterus diſcharge their contents into the ſubſtance of this cake; and the veins of the placenta, receiving the blood either by a direct communication of veſſels, or by abſorption, at length form the umbilical vein, which paſſes on to the ſinus of the vena porta, and from thence to the vena cava, and heart of the infant, by means of the canalis venoſus, a communication that is cloſed up in the adult. But the circulation of the blood through the heart is not conducted in the fœtus as in the adult: in the latter, the blood is carried from the right auricle of the heart through the pulmonary artery, and is returned to the left auricle by the pulmonary vein; but a dilatation of the lungs is eſſential to the paſſage of the blood through the pulmonary veſſels, and this dilatation cannot take place till after the child is born, and has reſpired. This deficiency is therefore ſupplied in the fœtus by an

immediate

immediate communication between the right and left auricle, through an oval opening, in the ſeptum which divides the two auricles, called *foramen ovale.* The blood in the fœtus is likewiſe tranſmitted from the pulmonary artery to the aorta, by means of a duct called *canalis arterioſus*, which, like the canalis venoſus and foramen ovale, gradually cloſes after birth. The blood is returned again from the fœtus to the mother through two arteries called umbilical arteries, which ariſe from the iliacs. Theſe two veſſels, taking a winding courſe with the vein, form with that and the membranes by which they are ſurrounded, what is called the umbilical chord. Theſe arteries, after ramifying through the ſubſtance of the placenta, diſcharge their blood into the veins of the uterus, in the ſame manner as the uterine arteries diſcharged their blood into the branches of the umbilical vein. So that, after quickening, the blood of the mother is conſtantly paſſing in at one ſide of the placenta, and out again at the other, for the nouriſhment of the child.

Now what we call the *action of quickening*, is that inſtantaneous, yet undeſcribable, motion of the vital principle, which, the inſtant the fœtus has acquired a ſufficient degree of animal heat, and is completely formed in all its parts, ruſhes like an electric ſhock, or flaſh of lightning, conducted by the ſanguiferous and nervous fluids, from the heart and brain of the mother, to the heart and brain of the child. At this moment the circulation begins; the infant fabric is completely ſet in motion, and the child becomes a living *ſoul.* As ſoon, therefore, as the circulation commences, the child ſtarts into life; and the inſtant the circulation ceaſes, life ceaſes alſo. This *act of quickening* is therefore derived from the blood; and is ſo ſenſibly felt by the mother, that ſhe often faints, or feels an internal depreſſion of her animal and vital powers, which may be ſaid, in ſome meaſure, to have departed from her. But the act of quickening does not take place in all women at the ſame period, nor always in the ſame women at the ſame diſtance of time from her conception; nor is it governed by any given number of weeks or days after conception has taken place; but depends entirely on that inſtant of time, when the joint influence of animal heat, and an entire completion of the nerves, veins, arteries, and other parts and organs, of the fœtus, are fitted and ready to receive and ſupport a due circulation of the blood and juices; for this, and this alone, is the ſource of quickening, and the beginning of animal life. Strong and healthy women will therefore quicken ſooner than the weak and delicate, by reaſon that their procreative and ſtimulating powers are more robuſt, and can ſooner contribute that portion of animal heat which is neceſſary to the entire completion of the fœtus in all its parts; and which will happen ſooner or later, according to the health and ſtrength of the pregnant woman, and her ſufficiency of menſtrual blood to ſupport the demand. For this flux will

now

now be wholly taken up by the new ſubject, until the hour of birth; after which it either renews its monthly evacuation, as being redundant in the mother; or, if ſhe ſuckles the child, it is then determined to the mammæ, and is converted into milk.

Such is this curious and moſt admirable contrivance of nature, for the re-production and propagation of mankind; and ſuch the nature and event of that myſterious action of quickening, which has hitherto been involved in ſo much darkneſs and obſcurity, as to lead the unthinking multitude to ſuppoſe, that giving life to the fœtus was in every inſtance a new and diſtinct interpoſition of the Deity, inſtead of religiouſly imputing it to that primary exertion of his omnipotence, which, in the original formation of Adam, implanted in his nature the power of re-producing his like, and of imparting life and ſoul to his ſpecies, by a fixed and immutable decree, to be continued down from father to ſon, to the final end and conſummation of this ſublunary world. If the ſeed of Adam had not been originally endued with the gift of imparting life and ſpirit to his future generations, how could the ſouls of his deſcendants be ſubjected to original ſin? Were any one child deſcended from the race of Adam to receive the gift of life and ſoul from a ſubſequent exertion of the power of God, it would become a new and diſtinct act of creation, and the offspring could not poſſibly be contaminated by the Fall, nor be ſubjected to the miſeries and misfortunes reſulting from it, as having received its being from an independent cauſe.

I have, to the beſt of my ability, endeavoured to illuſtrate this occult proceſs of Nature, by means of the annexed copper-plate engraving, taken from the viſcera and womb of an afflicted female, who fainted and died at the time of quickening, the fœtus itſelf being now preſerved in ſpirits. The ſtructure of the g avid uterus is, however, extremely difficult to be ſhown, and the more ſo under theſe peculiar circumſtances. In the wombs of women who die after this period, or at the time of labour, or ſoon after delivery, fibres running in various directions are obſervable more or leſs circular, that ſeem to ariſe from three diſtinct origins, namely, from the place where the placenta adheres, and from the aperture and orifice of each of the tubes; with all the veins and veſſels communicating to and from the placenta and the mother, ſurcharged with blood; but it is almoſt impoſſible to demonſtrate regular plans of veſſels and fibres, continued any length, without an interruption which involves us in doubt, and deſtroys that view of the admirable connexion which nature has formed between the vital organs of the mother and child in a ſtate of advanced pregnancy.

From the foregoing obſervations we may ſafely conclude, that the maſs of blood is the univerſal medium by which life is propagated, and health preſerved, to every claſs

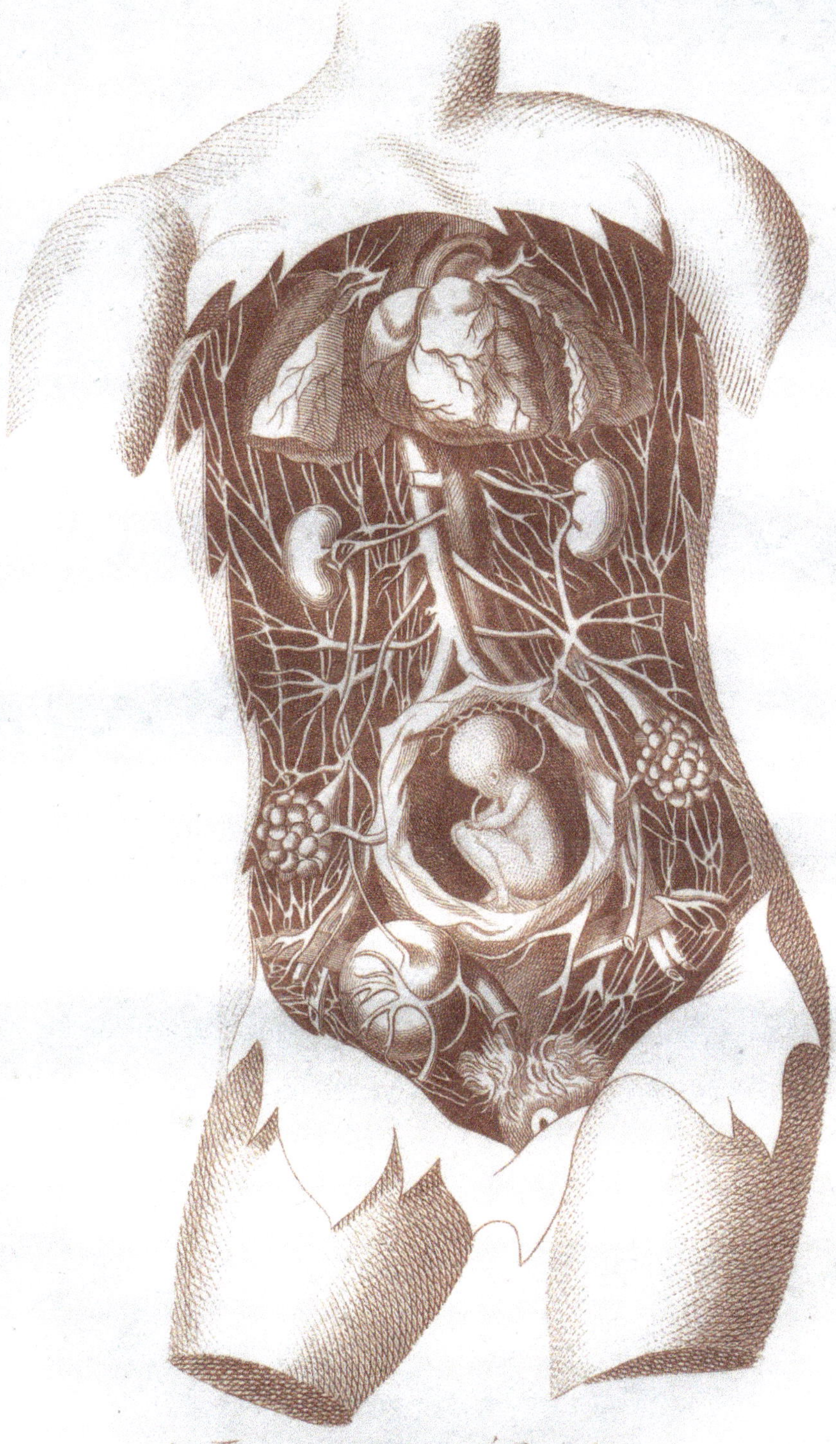

The Action of Quickening.

claſs of beings; and that, in its impure or infected ſtate, it is the ſource from whence the endleſs number of hereditary diſeaſes derive their origin. Whatever fault impairs the parent blood, fails not to taint the tender habit of its young; whence it has become an eſtabliſhed maxim, that, as healthy parents naturally produce healthy children, ſo diſeaſed parents as naturally produce a diſeaſed offſpring. Some of theſe diſeaſes appear in the earlieſt infancy; others occur equally at all ages; whilſt others lurk unſuſpected in the habit to extreme old age, or even to a new generation, ſlowly impairing the vital organs, and gradually undermining the conſtitution, before their ſource, and fatal tendency, can poſſibly be diſcovered. There are ſome diſeaſes indeed, which, though born with us, cannot be ſaid to be derived from the parent, as when a fœtus receives ſome hurt by an injury done to the mother; while others, neither born with us nor having any foundation in the conſtitution, are ſucked in with the nurſe's milk. Let it then be the care of every parent, who from ſome local misfortune is ſo far compelled to depart from the ties of nature as to abandon her tender offspring to the breaſt of another, to be ſatisfied, as far as human foreſight and medical penetration can reach, that the conſtitution and blood of the nurſe be free from ſcrophula and every other hereditary impurity.

Accidental diſeaſes, though not derived from the parents, nevertheleſs in general ſpring from the blood; which, conſtituting or propagating animal life through every part of the body, is neceſſarily expoſed to every external offending cauſe, from which impreſſion particular accidental diſeaſes enſue. The climate, itſelf, under which people live, will often produce theſe affections in the blood; and every particular climate hath more or leſs a tendency to produce a particular diſeaſe, either from its exceſs of heat or cold, or from the mutability of the weather. An immenſe number of diſeaſes are alſo produced in the blood by impure air, or ſuch as is loaded with putrid, marſhy, and other noxious, vapours. The ſame thing likewiſe happens from high-ſeaſoned or corrupted aliment, whether meat or drink; though even the beſt and moſt nutritious aliment will hurt, if taken in too great a quantity; not to mention poiſons, which are endowed with ſuch pernicious qualities, that, even when taken in the ſmalleſt quantity, they produce the moſt grievous ferment in the blood, ending perhaps with death itſelf. There are likewiſe other accidents and dangers to which mankind are expoſed, that ingraft innumerable diſeaſes in the maſs of blood; ſuch as the bite of venomous reptiles, or of a mad dog; an injudicious inoculation or mis-treatment of the ſmall-pox, or meaſles; the pſora, or itch; the venereal infection; alſo broken limbs, wounds, and contuſions;

ſions; which, though proceeding from an external cauſe at firſt, fail not to impair the blood, and often terminate in internal diſeaſes and premature death.

Man, however, is not left without defence againſt ſo many and ſuch great dangers. The human body is poſſeſſed of a moſt wonderful power, by which it preſerves itſelf from diſeaſes, keeps off many, and in a very ſhort time cures ſome already begun, while others are by the ſame means more ſlowly brought to a happy concluſion. This power, called the *autrocrateia*, or *vis medicatrix naturæ*, is well known both to phyſicians and philoſophers, by whom it is moſt juſtly celebrated; for this alone is ſufficient for curing many diſeaſes, and is of ſervice in all. Nay, even the beſt medicines operate only by exciting and properly directing this expulſive force, by which the excrementitious humours from the aliments and blood are expelled, through the proper channels of evacuation, through the excretory ducts, chiefly by means of the *inſenſible perſpiration*, by which power the offending humours from the blood and juices are perpetually flying off. But though phyſicians juſtly put confidence in this power, and though it generally cures diſeaſes of a ſlighter kind, yet it is not to be thought that thoſe of a more grievous tendency are to be left to the unaſſiſted efforts of Nature. Phyſicians have therefore a two-fold error to avoid, namely, either deſpiſing the powers of the *vis medicatrix* too much, which, if left alone, would work a radical and perfect cure; or, putting too great confidence in theſe exertions of nature, they are left unſeconded and alone, till the virulence of infection or diſeaſe undermines the conſtitution, and bears down all before it.

The grand and perpetual means by which the foul and offending humours in the blood and juices are continually carried off, is undoubtedly through the perſpirative pores and veſſels, which it is highly compatible with ſound health to keep open, and for which purpoſe medicaments are principally uſed. When this evacuation is copious and groſs enough to be diſcerned by the eye, as in ſweat, the perſpiration is ſaid to be *ſenſible;* but where it is ſo volatile as to eſcape the notice of the ſenſes, as is the caſe in the ordinary ſtate of the body, it is called *inſenſible perſpiration.*—The veſſels through which the perſpiration is performed lie obliquely open under the ſquamæ or ſcales of the cuticle or ſcarf-ſkin. They are inconceivably ſmall; from a calculation of Leeuwenhoek it appears, that the mouths of one hundred and twenty-five thouſand of them may be covered with a common grain of ſand. The moſt conſiderable of theſe pores are the orifices of the ducts ariſing from the miliary glands. Through theſe veſſels there is continually tranſuding a ſubtle humour, from every point of the body, and throughout the whole expanſe of the cuticle. The matter evacuated this way is found by certain experience to be more than equal to that evacuated all the other ways, i. e. by ſtool, urine, &c. Sanctorious found in Italy, under the circumſtances of a moderate diet, middle age, and eaſy life, that the matter inſenſibly perſpired was five-eighths of that which was taken in for food: ſo that there only remained three-eighths for nutrition, and for the excrements of the noſe, ears, inteſtines, bladder, &c.

The ſame author ſhows, that as much is evacuated by inſenſible perſpiration in one day as by ſtool in fourteen days; particularly that, in the ſpace of a night's time, about ſixteen ounces are ordinarily diſcharged by urine, four ounces by ſtool, and above forty ounces by inſenſible perſpiration. He alſo obſerves, that, if a man eat and drink eight pounds in a day, five pounds of it are ſpent in inſenſible perſpiration; and adds, as to the times, that within five hours after eating there is perſpired about one pound; from the fifth to the twelfth hour about three pounds; and from the twelfth to the ſixteenth ſcarcely half a pound. M. Dodart, from a number of experiments made thirty-three years ſucceſſively, proves that we perſpire much more in youth than in age. In ſome perſons the perſpiration is ſo copious, that they void very little of the coarſer excrements, though they eat heartily. The benefits of inſenſible perſpiration are ſo great, that without it animal life could not be preſerved. The general cauſe of perſpiraion is the circulation and heat of the blood, which enables it to throw off the offending matter. The great ſubtlety, equability, and plenty, of the matter thus perſpired, its increaſe after ſleep, &c. conſtitute the grand ſymptoms of a perfect ſtate of health; and the chief means of preſerving the ſame. On the contrary, the departing from theſe is the firſt ſure ſign of approaching diſeaſe.

Perſpiration is performed, preſerved and increaſed, by the viſcera, veſſels, and fibres; by motion or exerciſe as far as the firſt appearance of ſweat; by moderate uſe of venery; by ſleep of ſeven or eight hours, the body well covered, yet not loaded with bed-clothes: cheerfulneſs; light, fermented, yet ſolid, food, not fat; pure, not heavy, air, &c. The contraries of all theſe, as alſo the increaſe of the other excretions, diminiſh, prevent, and deprave, it. Hence we ſee the cauſe and effect of this perſpirable matter, its uſe in preſerving the parts ſoft and flexible, and in ſupplying what is loſt, but chiefly in preſerving the nervous papillæ moiſt, freſh, lively, and fit to be affected by objects, and to tranſmit their impreſſions. Hence it is, that upon a ſtoppage of the uſual perſpiration there ariſe ſo many indiſpoſitions, particularly fevers, agues, rheums, &c. Too much perſpiration occaſions weakneſs, and ſwoonings; whilſt too little, or none at all, occaſions the capillary veſſels to dry, wither, and periſh. Hence alſo the larger emunctories come to be obſtructed; hence the circulation is diſturbed, ſharp humours retained; and

hence

hence putridity, crudity, fevers, inflammations, and imposthumes. Cold prevents perspiration, by constringing the pores of the skin, and thickening the liquors circulating in the cutaneous glands; heat, on the contrary, augments it, both by opening the excretory ducts of the glands, and by increasing the fluidity and velocity of the humours. To determine the state and condition of the perspiration, so necessary for judging of those of the body, Sanctorius invented a weighing-chair, whereby he examined the quantity, degree, &c. of perspiration in several circumstances of the body, under several temperatures of the air, and in the several intervals of eating, drinking, sleeping, &c.

Some of the more extraordinary phenomena observed in this speculation, are, that, for some time after eating, the perspiration is least of all; that between the fifth and twelfth hour after meals perspiration is greatest; that riding either on horseback, in a coach, or ship, &c. swift motion on the ice, &c. but, above all, a brisk friction of the skin, promote perspiration surprisingly; and that perspiration is naturally always much less in women than in men. Perspiration is influenced by the passions of the mind. Thus anger and joy increase, and fear and sadness lessen, both perspiration and urine. Anger causes a strong motion in the membranes of the heart, and quickens its contraction and dilatation, and thereby quickens the contraction and dilatation of the blood-vessels and secerning ducts, and of consequence increases the discharges of perspiration and urine; and that more or less, in proportion to the strength and continuance of the passion. Joy affects these discharges in like manner as anger. In the passions of fear and sorrow, perspiration and urine are lessened, by the depression of the activity of the soul under those passions. The proportion of perspiration to urine is increased by all those exercises which increase the motion of the blood, and warm the skin.

We have an account of a person who, by passing many nights in astronomical speculations, had his perspiration so obstructed by the cold and damp of the air in Holland, that a shirt he had worn for five or six weeks was as clean as if it had been worn but one day. The consequence of this was, that he gathered subcutaneous waters, but was cured in time. The garments best calculated to encourage and promote insensible perspiration, to keep the mouths of the minute vessels open, and to guard the body from the too sudden and violent effects of cold are those made of flannel. Whence flannel shirts and waistcoats, or a square piece of flannel worn over the breast or pit of the stomach, particularly in the winter months, are productive of such beneficial effects to weakly and debilitated constitutions, and act as a valuable preservative to the hale and robust. In the annexed copper-plate engraving, I have endeavoured to show the manner in which the insensible perspiration

issues

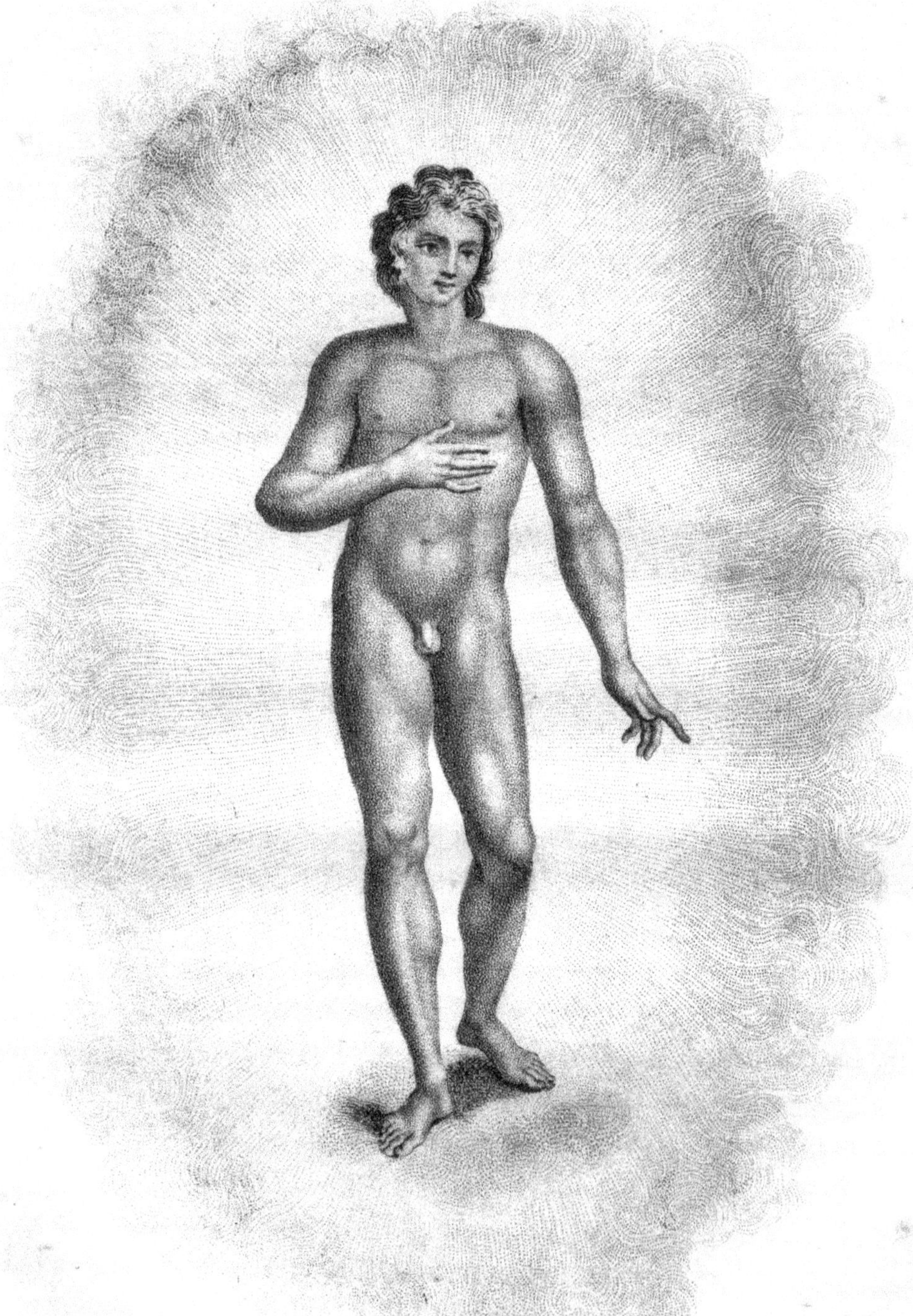

The Insensible Perspiration

Published as the Act directs June 20 1794, by E. Sibly

iſſues from the pores of the body, which can only be diſcerned by means of a lens; being of ſo volatile and ſubtle a quality, that it paſſes through our garments with the utmoſt eaſe, particularly if woollen; and it even aſcends through the bed-clothes like a miſt, in the greateſt abundance when we are aſleep, and the animal functions are at reſt.

In this manner Nature, from all caſual obſtructions, endeavours to relieve herſelf; and, ſo long as diſeaſes are recent, and of a mild tendency, they are uſually carried off by this means, without requiring any aid from medicine. When, however, diſeaſes are of long ſtanding, and the humours in the blood become too foul and viſcous to be thrown off by the *vis medicatrix naturæ*, the whole habit is quickly vitiated, and the circulating maſs becomes morbid; yet even in this infected ſtate, the vital heat and activity of the blood ſtrives to purify itſelf, by determining theſe morbid particles to the ſkin, where they form ſcabs, ulcers, pimples, and other ſpots, as in the ſcrophula, leproſy, ſmall-pox, meaſles, ſyphilis, &c. or elſe the virulent matter is directed inwards, where falling upon the lungs and other viſcera, death quickly enſues. Here then we may view the ſhocking conſequences which reſult to thoſe who enter into matrimony under a tainted or infected ſtate of the blood. Indeed perſons that are afflicted with the leproſy, ſcrophula, or king's evil, ſhould never marry until a perfect cure has been effected, and a pure and healthful ſtate of the blood induced. To enter into wedlock under a venereal taint, is a moſt unwiſe, a moſt cruel, and an ungenerous act. A man with only a ſlight infection, by contact with the woman, will himſelf perhaps experience a perfect cure, in conſequence of the foul and infectious matter being drawn from the parts of the female organs, ſeconded by the action of the rugæ and abſorbent veſſels on the ſurface of the vagina. But the unhappy female is ſure to take the diſorder; and, ſhould ſhe prove with child, ſhe not only carries the poiſonous infection into the marrow of her own bones, but brings an infant offspring into the world, devoted to miſery and diſeaſe; for whatever foul or infectious humour is implanted in the parent blood, it is immediately carried by the circulation to the vital organs of the child, juſt as the flame of one candle is by contact communicated to another. Nor can we be ſurpriſed at theſe things, if we only reflect on what has already been adduced, and contemplate the ſyſtem and œconomy of the human frame. Conſider only the powerful effects of a few grains of cantharides, which, if externally applied, act as a burning cauſtic; but, if taken into the ſtomach, inſtantly overturn the natural courſe of the circulation, by forcing the whole maſs of blood into the extremities, but more particularly, with great vehemence and turgidity, into the private parts; for which reaſon cantharides are taken with intent to cure the weakneſs and debility of the penis; but the truth is, that greater debility, and an emaciated conſtitution, are ſure to follow, and not unfrequently inſtant death.

If, then, ſo powerful an effect can be wrought on the blood by ſwallowing a few irritating particles of a ſmall inſect, may we not juſtly infer, that by infuſing into the circulating maſs particles congenial to itſelf, the utmoſt relief may be afforded to it, even in its moſt depraved and inactive ſtate? From this conſideration alone, we may venture to pronounce, that all diſorders originating in the blood might either be prevented or repelled, could ſuch a medium be diſcovered, by which we might infuſe immediately into the maſs a combination of ſuch elemental principles as the blood and juices themſelves conſiſt of in their pureſt and moſt elaſtic ſtate; for this, in fact, is the aim of all medicines; but which they miſs, by being adminiſtered in their groſs form, and being obliged to paſs the ſeveral digeſtive operations of the ſtomach, before they can reach the blood, whereby the principal part of their occult virtue is loſt among the food, or ſecreted in ſuch ſmall quantities as to produce very little effect. But a medium, poſſeſſing theſe congenial principles, ready digeſted, and ſo combined as to be taken inſtantly, and without diminution, into the habit, would not only keep the cruor and the ſerum in due proportion, which is ſo eſſential to health, but would ſtimulate, correct, purify, and augment, the blood, as its reduced or diſordered ſtate might from time to time require. Such a medium, after infinite labour, and unlimited experience, I pronounce the Solar Tincture to be; and ſuch will be found its operative effects, under whatever circumſtances it may be adminiſtered, in any climate or ſeaſon; the innocent and balſamic qualities of which are as grateful to the internal organs of the human frame as the ſolar rays are cheering to the external; and it affords me no ſmall gratification to aſſert, that, in offering it to the public, I invade no man's property, nor imitate any medicine at preſent known in public or private practice.—The experiments I have made with it upon a variety of diſeaſed wretched objects, exceed belief; and I ſhall ſtill continue to adminiſter it gratis to the poor, who are given over by others, or who have not the means of applying for medical aſſiſtance.

The infinite variety of complaints an impure or infected ſtate of the blood induces, almoſt exceeds belief; and hence the new and deceptive forms a ſcrophulous or ſcorbutic taint puts on, which often deceive the moſt eminent of the faculty, and baffle the beſt intention towards a cure. An impure or ſcrophulous taint will invade the nobleſt organs of the human frame, before the patient can be aware of his danger. In the firſt ſtage of its viſible effects, a weary pain ſeizes the joints and muſcles, attended with a waſting of the legs and loins. In the ſecond ſtage, the gums ſwell, grow painful, hot, and irritable, and bleed upon the ſlighteſt preſſure;

the

the roots of the teeth become bare and loose, and the breath nauseous. In the third stage, the gums grow putrid, the teeth black and rotten, the sublingular veins become varicose, and the breath cadaverous; fœtid blood distils from the lips, gums, mouth, nose, lungs, stomach, liver, spleen, pancreas, intestines, womb, kidneys, &c. scabs and ulcers break out in all parts of the body; and the joints, bones, and viscera, become morbid. In the fourth stage, putrid, eruptive, and spotted, fevers, ensue, which end in an atrophy; or else follow diarrhœas, dysentery, dropsy, consumption, palsy, contractions, melancholy, and all the long and direful train of nervous disorders, which to describe would fill a volume.

To counteract this most virulent of all chronic complaints, the utmost exertions of human skill have been employed. The remedies prescribed in its different stages are almost innumerable. The object is to reduce the virulence of the infection, and to eradicate its seeds from the blood and lymph; to which end the mildest and most simple medicines are recommended. Mineral and tar waters, for their warm and stimulating quality; milk or whey, from their similitude to the chyle; the cold bath, for bracing the solids and quickening the circulation; antiscorbutic vegetables, &c. for purging and sweetening the blood, such as scurvy-grass, water-cresses, wormwood, hemlock, centaury, vervain, water-trefoil, juniper-berries, the Peruvian bark, sassafras, guaiacum, aloes, assa-fœtida, camomile, diascordium, saffron, senna, rhubarb, manna, Æthiop's mineral, hartshorn, native cinnabar, antimony, &c. When these fail, mercury, or a mercurial salivation, is looked upon as the only cure; which, in fact, is but to give the human frame its last vehement shock, and to send the wretched patient in agonies to the grave!

The intention of all these remedies is to impregnate the blood with qualities opposite to those with which it is infected; and this must be done in a superior degree of force and power, before a cure can be completed. But these medicaments are often administered under such nauseous forms, and in so crude and unqualified a state, that they not only torture the patient, but miss entirely their intended aim. The nauseous taste of medicine is nothing but its grosser particles; which, instead of entering the stomach, to irritate and oppress its organs, ought to be drawn off by chemical process; for it is the occult virtue of every drug, not its grosser part, that performs the cure. Now the peculiar excellence of the Solar Tincture is, that it combines the essential and occult virtues of all the scorbutic vegetables, ready digested, concocted, purified, and resolved into an elegant balsamic essence, pleasing to the taste and grateful to the stomach. It flies immediately to the heart, whether internally or externally applied, blends and assimilates with the venal and arterial blood, which it generates, corrects, warms, purifies,

rifies, animates, and impels through the whole ſyſtem. It cleanſes all the viſcera, and glandular parts, particularly the lungs and kidneys; ſtimulates the fibres, whereby the gaſtric juice and digeſtion are promoted; diſſolves viſcid humours, and expels infection. It exerts very conſiderable effects on the whole nervous ſyſtem, ſenſibly raiſes the pulſe, ſtrengthens the ſolids, and invigorates the animal ſpirits. It penetrates into the moſt intimate parts, opens the mouths of the minuter veſſels, reſtores the natural perſpiration, and promotes all the fluid ſecretions. In every ſtage of infectious diſeaſes, and in all ſudden epidemical diſorders, which uſually follow from a wet, putrid, and unwholeſome, ſtate of the atmoſphere, it is an abſolute ſpecific; and, as a preventive, an alterative, and purifier of the blood, it has not its equal in the world. It quickly relieves every common malady originating in the blood, ſuch as relaxations, debility, laſſitude, tremors, ſinking of the ſpirits, and all thoſe nervous affections which haraſs and oppreſs the weak, ſedentary, and delicate; and are often the conſequences of high living and luxuriant indulgences, without bodily exerciſe and freſh air. In all theſe caſes, the Solar Tincture is calculated to warm and ſteady the cold tremulous nerves; to ſheath and invigorate the muſcular ſyſtem; to animate the ſpirits; and renovate the whole man; whereby the chill watery fluids become rich and balſamic, and the circulating maſs reſumes its healthful ſtate. It is an infallible cure for joint-achs, cramps, ſpaſms, rheumatic gout, nervous head-ach, agues, and all diſorders ariſing from obſtructed perſpiration. In complaints of the breaſt, ſtomach, and bowels, it gives immediate relief; and, in aſthmatic and conſumptive caſes, is an elegant and expeditious cure. It will ſtop mortification in very advanced ſtages, by expelling the poiſonous matter, and correcting the juices of the whole body. It requires no argument to convince, more than a ſingle trial; after which, I think, no family who value their health or life will chooſe to be without it; particularly under any of the following afflictions:

SCROPHULA, SCURVY, OR KING's EVIL.

IN the firſt and ſecond ſtages of this diſorder, a ſmall table-ſpoonful of the Tincture, taken in a wine-glaſs of cold ſpring water night and morning, will prevent the further progreſs of the diſeaſe, and in a very ſhort time reſtore the blood to its healthful ſtate, the effects of which will be ſo obvious to the patient, that he will be at no difficulty when to diſcontinue the medicine. In the third ſtage, it is often requiſite that the medicine be internally and externally applied. The mouth ſhould be frequently waſhed with the Tincture diluted in warm-water; and it will very ſoon expel the poiſon from the gums. If the viſcera be in a morbid ſtate, which

may

may be known by the excrements, or foulneſs of expectoration, it will be neceſſary to take the medicine, night and morning, for ſeveral days, in the quantity of a table ſpoonful *undiluted;* and, at noon, a table-ſpoonful in the ſame quantity of warm water. The ſcabs, whether dry or moiſt, ſhould be frequently waſhed with the Tincture, *undiluted*, which, being abſorbed by the minuter veſſels, and taken into the habit, will expel the humour, and clear away the ſcurf. If tumours or foul ulcers occur, waſh them frequently with a dilution of the medicine in the ſame quantity of warm-water, until the heat and virulence be abated; then apply the Tincture *undiluted,* with lint or fine rags, by which means the infectious matter will be totally eradicated, the blood and juices purified, and the ulcers healed.

In the fourth ſtage, whatever may chance to be the ſad malady to which the diſorder ultimately turns, a ſtrict attention to regimen, exerciſe, and freſh air, as far as the ſtrength and condition of the patient will admit, muſt be particularly attended to. And, in all theſe caſes, the beſt and moſt ſimple methods of treatment are laid down in the Medical Part of this work, p. 168, &c. to which I beg leave to refer every patient in this dreadful ſtage of the diſeaſe; and, in aid of the advice there given, let the Solar Tincture be regularly perſiſted in every night and morning, in the quantity of one table-ſpoonful in as much warm water; and, at twelve o'clock at noon, take a table-ſpoonful *undiluted.*—Let this be continued eight or ten days; then take a table-ſpoonful diluted in warm water three times a-day, morning, noon, and night, till the nerves and organs begin to reſume their healthful tone; then let the doſes be gradually abated to a ſpoonful in water every other morning, which ſhould be continued till health is perfectly re-eſtabliſhed; and which, by God's bleſſing, will generally happen, even in theſe deſperate caſes, in the courſe of a month or ſix weeks.—As a preventive of all foul or ſcrophulous taints in the habit, and as an alterative, and purifier of the blood, it may be occaſionally taken every other morning for a week together, particularly in the ſpring and fall, in the quantity of a table-ſpoonful in a wine-glaſs of cold ſpring water; or it may be occaſionally taken as a beverage after dinner or ſupper, mixed in a tumbler with warm water, and made palatable with ſugar. It will be found pleaſant to the taſte, and grateful to the ſtomach, superior to any ſpirits, or punch. The many inſtances of elegant and uncommon cures effected by the Solar Tincture, on perſons of the firſt eminence, may be inſpected at any time, on application at my houſe. But at the particular requeſt of the parties, I have here added the following remarkable

C A S E.

Mr. R. Pinder, of Bramſtone, near Bridlington, in Yorkſhire, had been long afflicted with a violent ſcorbutic humour in his blood, which threw out ſometimes dry, and sometimes moist, scabs and tumours on the skin. Being neglected, it at length pervaded the whole system, till, turning inwardly, it fell upon his lungs, and reduced him to the last stage of a consumption. In this deplorable state, given over by the faculty, left totally emaciated, and incapable of turning in his bed, he fortunately had recourse to the Solar Tincture. The first dose was given *undiluted*, which threw him into a fine perspiration, and composed him to sleep, which had long been a stranger to his eyes. After one large bottle had been administered, agreeable to the bill of directions, at the end of a week he was so much restored, that with very little assistance he was enabled to put on his own clothes; and, after continuing the medicine for little more than a month, he was able to walk abroad. And now, after having continued the Tincture night and morning, and occasionally using it as a beverage made similar to warm brandy and water, he has quite recovered his former health and strength; being, to the surprise of every body who beheld him in his late emaciated condition, as robust and as hearty as it is well possible for a man to be.

DEBILITATED, TAINTED, AND ENFEEBLED, CONSTITUTIONS.

MUSCULAR debility was a misfortune but little known to our forefathers. Whether immured in venereal embraces, or sacrificing at the shrine of Bacchus, moderation and seasonable hours directed the measure of their enjoyment. If revelry or voluptuousness by chance unstrung their nerves, gymnastic exercises and field-sports, or the more pleasurable delights of the chace, quickly restored them to their proper tone,—gave new vigour to the blood,—health to the cheek,—and lighted up afresh the flame of love. But now, how strange is the reverse. Habituated to effeminacy, and fed with dainties,—revelling all night with wine, and stretched on beds of down all day,—shut up in stews and brothels, scarcely breathing wholesome air,—clasped in the arms of tainted or diseased females, until enjoyment palls upon the senses, and the muscular powers absolutely refuse their office;—no wonder so many men are found old in every thing but years; whose constitutions are fairly worn down, blood stagnant, solids relaxed, secretions diverted from their proper course, muscles debilitated, eyes sunk, cheek pallid, and spirits gone. These are not half the evils resulting from this fashionable source of destructive folly. It may not be amiss, however, to describe the remarkable cases of a few, of whom the Solar Tincture has made perfect cures, by infusing a new portion of health into the mass of blood; sincerely hoping, that a more wise and manly course of life will shortly eradicate these disgraceful complaints, and restore to the ladies a genuine race of Englishmen and Britons.

CASES.

Premature Debility.—A gentleman in the army, under thirty years of age, complained to me that he had all at once *become incapable of enjoying his wife.* Suſpecting the nature of his diſorder, I deſired him to be open and candid, to relate to me his *real* ſituation, and not a *pretended* one, which was only to impoſe on his own underſtanding. He thanked me for the rebuke—ſaid he would be frank, and in a few words declared, That from exceſſive luſt, and continual debauch, he had loſt his virility; and, to add to the misfortune, he was on the eve of being married. In other reſpects he felt no diminution in his health or conſtitution; and, from external appearances, this was ſurely the laſt imperfection that could have been ſuſpected. His complexion was vigorous and lively, his fleſh firm, and conformation excellent; yet, notwithſtanding this, he was impotent to ſuch a degree, that neither the ſtrength of his own deſires, nor the excitations of the female, could affect the part. It often happens, that, though the organs remain ſound, yet, if the nervous and ſeminal fluids have degenerated from a healthful ſtate—if they are impoveriſhed by being too much drained, or turned into an unnatural courſe—they cannot then perform their office, by reaſon that their moving powers and ſtimulus on the blood are become too weak to direct their force and action in the manner nature requires in the act of copulation. I therefore enjoined him, to abſtain entirely from all attempts of the kind, for three months at leaſt; directed the ointment as in p. 240 of the Medical Part of this work, with the Solar Tincture three times a-day for two months; then twice a-day, until he found it no longer neceſſary. After taking ſix large bottles, he generouſly thanked me for a more hale and robuſt ſtate of body than he ever remembered to have enjoyed before. He has ſince ſent me ſeveral patients, in almoſt as debilitated a ſtate as himſelf, who are now ready to unite with him in giving full teſtimony to the renovating powers and peculiar efficacy of the Solar Tincture.

A RELAXED HABIT.

LITTLE more than three months ago, a gentleman, about fifty years of age, lately returned from the Eaſt-Indies, applied to me for the cure of what he termed a *broken conſtitution.* He had made very free with the ſable beauties of Bengal —had undergone a mercurial ſalivation, and appeared to be ſinking under an univerſal languor and debility of the whole muſcular ſyſtem. The ſphincter of the bladder was ſo weakened, that the urinary ſecretion came from him by drops, in ſo perpetual and involuntary a manner, as not to be perceived until the moiſture of one ſet of cloths became ſo ſenſibly afflicting as made it neceſſary to ſupply

freſh

fresh ones, which usually happened every hour. The corporeal functions were dissipated and relaxed, the tone of the stomach and viscera was nearly gone, the tremulous nerves reluctantly performed their office, and the circulation was become stagnant and morbid. I advised an immediate recourse to the most nourishing food, with strong port-wine negus for his drink, and the Solar Tincture to be taken four times a-day for the first month; three times a-day for the second month, and once or twice a-day afterwards, as occasion might seem to render necessary. Before the expiration of twenty days, the sphincter muscle acquired its proper tone, the pulse became strong and regular, and the nervous tremors were considerably abated. By the end of the second month, a renovation of the whole animal œconomy seemed to have taken place, and a visible accumulation of the blood and juices had retrieved the circulation. Before the expiration of three months, I had the gratification to see this patient completely restored to such a state of bodily health and strength, as utterly astonished himself, after taking only eight large bottles of the Solar Tincture.

HYPOCHONDRIACAL DEBILITY, OR WEAK NERVES.

A GENTLEMAN in Oxfordshire lately came to town on purpose to consult me in his complaint. He appeared to be near thirty years of age, of middling stature, but of a weakly constitution. He had for upwards of seven years past paid his addresses to a lady, whom he had long promised, and very much desired, to marry; but, whenever he proposed in his mind to fix the day, or whenever it happened that he attempted to salute or embrace her, he was seized with an unaccountable tremor of the whole body, his spirits sunk, his virility left him, and a violent palpitation of the heart ensued. In short, he was so distrustful of his own powers, that he confessed it was the fear of not being able to perform the rites of the marriage-bed that had been the real and the sole cause of thus protracting his wedding-day. This is certainly a most singular instance of the hypochondriacal affection, and of its derangement of the nervous system. The debility induced by it, seems to arise from the weaker energy of the brain, the fault of which, however, cannot be detected by the nicest anatomist. For this reason, we do not well know how such defect should be restored; but as nature, seemingly for this purpose, excites the motion of the heart and arteries, we must ascribe the continuance of such debility to the too-weak reaction of the sanguiferous system. The heart will generally palpitate from a violent excitement of the nerves, especially when the blood is endowed with too small a share of stimulus. Hence palpitation from any affection of the mind, and from hysterics in women. Under whatever circumstances this hypochondriacal affection happens, it debilitates the whole animal machine, and renders the person unable to perform the proper offices of life. The prostration of spirits, weakness, and languor, are often

often ſurpriſingly great, though the pulſe ſeems tolerably ſtrong, as being heightened by animal deſire. . The effect, however, is ſure to produce a languid circulation, the blood ſeeming to adhere, with uncommon energy, about the region of the heart. I ſuſpect it is in theſe caſes that cantharides are moſt frequently uſed. The patient acknowledged, after ſome heſitation, that he had tried them; but they only produced an involuntary though violent erection, by no means adapted to the cure, nor to the purpoſe he intended. Hence this remedy is not only inadequate, but extremely dangerous; for it too much exhauſts the vital powers, and is followed by a vaſt dejection of ſpirits, tremors, ſtartings of the tendons, &c. which bring on rigours, cold clammy ſweats, ſyncope, and often premature death.

The means, therefore, which nature points out for the cure of this ſpecies of debility, are directed to ſupport and increaſe the action of the blood through the heart and arteries; and the remedies to be employed are tonics and ſtimulants. Of all the ſtimulants, which in this conſtitutional defect may be advantageouſly employed, port-wine ſeems to be the moſt eligible. It has the advantage of being grateful to the palate and ſtomach, and of having its ſtimulant parts ſo much diluted, that it can be conveniently given at all times and ſeaſons, and may be employed with ſufficient caution; but it is of little ſervice unleſs taken pretty largely.—It may be ſuſpected that wine has an operation analagous to that of opium; and on good grounds. But we can diſtinctly mark its ſtimulant power only; which renders its effects in the phrenitic delirium manifeſtly hurtful, but in caſes of debility as remarkably uſeful.—Hence I directed the Solar Tincture to be taken morning, noon, and night, in ſtrong doſes, for the firſt month; once a-day, or oftener, at the diſcretion of the patient, until the end of the third month; but to drink every day after dinner, a pint of generous port; and to inform me at intervals the change he might find in his conſtitution. He took with him a dozen large bottles of the Solar Tincture; and before a month elapſed, I had the pleaſure of receiving an epiſtle of unfeigned thanks. He found himſelf ſo much reſtored by the courſe I laid him under, that, before the expiration of the three months, he married the lady; and I have no douht will very ſhortly have iſſue.—I have been ſomewhat more elaborate in deſcribing the particulars of this caſe, having reaſon to believe it is not an uncommon malady; and would therefore wiſh to enable every patient to become as much as poſſible a judge of his own infirmity.

NOCTURNAL EMISSIONS, OR INCONTINENCE OF THE SEMEN.

A YOUNG man, of robuſt make, and in the prime of life, being under twenty-ſix years of age, applied to me for relief in the above unfortunate complaint. It appeared, that, from the time of puberty, he had found a weakneſs in the part, and an occasional discharge of the seed, upon the slightest irritation. As he grew up to greater maturity, the malady increased upon him. Upon every attempt to have contact with a female, semen passed involuntarily from him, before even a complete erection could take place, whereby his purpose was continually defeated. This defect grew upon him, until the bare sight or thought of any thing which tended to excite venereal desires brought away the seed; yet it had no affinity whatever to a gleet, because the emission never occurred but either in the attempt or in the desire of copulation, or under the influence of lascivious dreams. In proportion as this weakness grew upon him, his desire of familiarity with the sex became the stronger; and, I am inclined to think, was the principal reason of the increase of the malady, and of the nocturnal emissions, which happened more or less every time he went to sleep. This incessant discharge had reduced him to a meagre visage, sallow complexion, hollow eyes, depression of spirits, and slow fever; and a galloping consumption would soon have followed. I directed the Solar Tincture every morning at sun-rising, at mid-day, and at six o'clock in the afternoon, in the quantity of a wine-glass full, with one-third warm water; and every night at going to bed, twenty drops of liquid laudanum, for the purpose of making his sleep too strong to be affected by the influence of dreams. This course, assisted by a strengthening regimen of calf's-foot jelly, veal-broth, and strong port-wine negus, had very quickly the desired effect. His sleep was perfectly sound and calm, and, after the first night, he could not recollect the return of any nocturnal emission. The strengthening ointment, directed in page 240 of the Medical Part of my work, was used every other morning; and within the space of only two months, the seminal vessels were completely braced up, and the disorder so totally removed, as not to leave a single symptom of his former weakness.

ONANISM.

A YOUTH, apparently under age, applied to me for the cure of a disorder, which, he said, had deprived him of the power of erection, and of all sensation in the privities. In so young a subject, I could not suppose this want of tone to arise from a general debility of the nervous system, particularly as no other symptoms warranted the conclusion. I had a strong suspicion it was the effect of Onanism, or secret venery, which usually ends in this species of absolute impotency; but this he denied. He told me he had some time ago contracted the foul distemper, and through shame, and the dread of its coming to the knowledge of his friends, he had neglected to disclose his misfortune to any person, until the present malady was brought on. Of the foul distemper, however, I could find no other symptom than a simple gleet; and, upon putting the necessary questions, not a single reply corre-

sponded

ſponded with the uſual effects of that diſorder. After half an hour's cloſe examination, I brought him to confeſs what I above ſuſpected, that he had ſo much addicted himſelf to this ſhameful and deſtructive vice, that the ſeminal veſſels were completely relaxed; the erectories, the nerves, and glans, of the penis, had entirely loſt their tone; an involuntary diſcharge of the ſemen, without irritation, or turgidity of the parts, had long taken place, and brought on a want of appetite, an impoveriſhed ſtate of the blood, and an univerſal laſſitude of the body. The lecture I gave him upon this occaſion, will never, I truſt, be effaced from his memory; and he has ſince faithfully promiſed that it ſhall not. I directed the ſtrengthening electuary and ointment, in page 239 and 240 of the Medical Part of this Work, to be uſed as therein preſcribed; then to take, four times a-day, a table-ſpoonful of the Solar Tincture in an equal quantity of warm water, for a month at leaſt; then three times a-day for the ſecond month, and twice a-day, in cold ſpring-water, for the two months following; which gradually coiled up the debilitated parts, gave elaſticity to the blood, retrieved the ſenſation of the glans, and the ſympathetic office of the erectories, braced the nerves, ligaments, and tendons, and gave that due tone and energy to the muſcular ſyſtem, which in leſs than four months reſtored the patient to perfect health and vigour.

An IMPURE or TAINTED HABIT.

THIS malady, ſo common among our diſſipated youth, generally ariſes from a venereal complaint badly cured. Indeed the ſcrophula, the king's evil, the leproſy, and other foul humours, when too long ſuffered to prey upon the blood, will naturally induce this conſequence; yet ninety-nine caſes out of every hundred, are found to reſult from the improper uſe of mercury, either taken too abundantly into the ſtomach, or too often applied externally, in the venereal diſeaſe. A gentleman in the militia very lately came to me under this misfortune, who had abſolutely worn down the organs of his ſtomach by taking medicines for its cure, without obtaining the ſmalleſt relief. He was no ſooner warm in bed, than deep-ſeated nocturnal pains attacked his arms, ſhins, and head, which many of the faculty miſtook for rheumatiſm. The membranes, muſcles, and ligaments, of the joints, were ſcarcely ever free from pain; whilſt carious ulcers occaſionally broke out upon the ulna, tibia, and bones of the cranium. Theſe ſymptoms had alſo deceived ſeveral of the faculty, who, taking his complaint to be a confirmed lues, ſtill added to the malady, by loading him with freſh doſes of mercury. The truth is, that this diſorder was by no means of a venereal nature, but was rather the conſequence of the remedy than of the diſeaſe, ſince it aroſe entirely from the long and repeated doſes of mercury his body had ſuſtained, and which was grounded in his habit by

ſalivation.

ſalivation. The mercury had inſinuated itſelf into the marrow of his bones, had vitiated every fluid ſecretion, and tainted the very air he breathed. Under ſuch circumſtances I will allow, it is very difficult, if not almoſt impoſſible, for a phyſician, upon a ſuperficial inſpection, abſolutely to decide, whether the original diſeaſe hath been altogether overcome; yet ſurely he ought attentively to diſtinguiſh and conſider the ſeveral ſymptoms apart; and then, by comparing them with each other, a clear judgment may be formed upon the general review. Finding, by this method, the real ſtate of the patient's caſe, I ordered him a nouriſhing diet, gentle exerciſe, and an abſolute denial of the ſmalleſt intercourſe with women. To this he readily ſubmitted, putting himſelf under a regular courſe of the Solar Tincture, which he took three times a-day, in the quantity of a wine-glaſs three parts full, filled up with warm water, for the firſt month. At the expiration of this time he paid me a viſit, when his company was infinitely *more agreeable*, becauſe the pleaſing odour of health had ſuperſeded the nauſeous effluvia of his diſeaſe. I now only enjoined him to follow the ſame regimen and *abstemious* mode of living for a month or two longer, taking the Tincture diluted in a glaſs of cold ſpring-water once or twice a-day, as he might find himſelf inclined. This he rigidly attended to; and I have now the pleaſure to declare, that only nine large bottles of the Solar Tincture have reſtored this gentleman from the moſt dangerous and deplorable ſtate of a tainted and corrupted habit, to ſound health, and a renovated ſtate of the blood and juices.

A TAINTED HABIT IN A STATE OF PREGNANCY.

THIS is the moſt ſhocking caſe my practice or experience ever produced. The patient was taken in labour, and in the act of parturition, the child preſented its right arm, which ſeparated from the body, while the operator was returning it into the womb. The life of the mother being deſpaired of, I was ſent for; when, on inſpection, I quickly perceived conception had taken place under an infected ſtate of one of the parents. I performed the reſidue of the operation myſelf, and brought away the fœtus without a further ſeparation of the joints, but with great difficulty, ſince it was ulcerated and half rotten with diſeaſe. By a moſt tender and judicious treatment of the woman, aſſiſted by the Lunar Tincture, her life was preſerved; and in the ſpace of five weeks ſhe appeared to have regained her health and ſtrength; when, to the aſtoniſhment of every one, ſhe fell into a violent ſalivation. Being ſent for upon this ſingular occaſion, I thought it right to interrogate the huſband; when, after a vaſt deal of heſitation and diſſembling, he confeſſed having had connection with his wife under a venereal infection; and, with a view to prevent the conſequences, he had prevailed on her to ſwallow ſtrong doſes of mercury, which I have reaſon to ſuppoſe lay dormant in the body until after her delivery; when the efforts

efforts of nature being no longer directed to the preservation of the child, suffered the mercury to attack the salival glands, and to produce the effect we have just described. I ordered her a spare but nourishing diet; worked off the mercury in the customary way, and then began a course of the Solar Tincture. A table-spoonful, in an equal quantity of warm water, was taken four times a-day for the first week; then three times a-day until the end of the month; afterwards twice a-day in cold spring-water for a month longer; and then once or twice a-day, or every other day, as the patient found convenient; by this means she happily experienced a complete cure in less than three months, and now enjoys a perfect state of health, desirous of certifying the fact to any unfortunate female, who, under similar circumstances, wishes to call upon me for that purpose. Indeed every woman, who has the misfortune to suspect even the smallest taint of a similar nature to be lurking in her blood, should put herself under a course of the Solar Tincture, and persist in it every night and morning, in the quantity of a table-spoonful diluted in a wine-glass of cold spring-water, during the whole nine months state of pregnancy.

The above case brings to my recollection a very singular instance of an accidental salivation, brought upon a young lady by a foreign substance irritating one of the parotid glands; the particulars of which I shall here insert, for the sake of those who may happen to be under similar circumstances.—In the month of April, 1751, a young lady, about the age of sixteen years, of a delicate habit, but subject to no particular complaints, perceived the beginning of a disease which afterwards proved most obstinate and loathsome, viz. an incessant spitting. The quantity of this discharge was different at different times, varying from one pint to two pints and a half in twenty-four hours. As to its quality, it seemed to be no other than the ordinary secretion of the salival glands. By so large and constant an evacuation, her strength became extremely impaired; and the most efficacious medicines had proved useless. She had taken large quantities of the Peruvian bark, both alone and combined with preparations of iron; and afterwards the fetid gums, opium, amber, alum, and the Neville-Holt water, had in succession been given her. In the mean time an exact regimen had been prescribed, she had been ordered to ride constantly; and to confine herself to a mucilaginous diet, such as veal, calves' feet, &c. Likewise a gently-opening medicine had now and then been interposed. The disease still continued unaltered; she had afterwards tried the *tinctura saturnina;* and had, at the same time, been encouraged to chew the Peruvian bark, and to swallow the saliva. But all these attempts had been vain; and after she had taken some or other of the medicines above-mentioned until the end of September, 1753, namely, above two years, it appeared to her physician, Sir George Baker, unreasonable to expect relief in such a case from any internal medicines whatever. He now conceived a ſuſpicion, that ſome extraneous body, having accidentally found its way into the *meatus auditorius*, might poſſibly be the cauſe of this extraordinary ſecretion, by keeping up a continued irritation in the parotid glands. With this view he examined her ears, and extracted from them a quantity of fetid wool. How, or when, it came thither, no account could be given. To this ſubſtance he attributed the beginning of the ſalivation, notwithſtanding that the diſeaſe did not immediately abate on the removal of the wool; as it appeared to be no improbable ſuppoſition that the diſcharge might be continued by the force of habit, though the original cauſe no longer remained. It ſeemed therefore expedient to introduce ſome other habit, in the place of the increaſed ſecretion of ſaliva; which habit might afterwards be gradually left off. With this intention, he prevailed on the patient to chew perpetually a little dry bread, and to ſwallow it with her ſpittle. In a few weeks, it became neceſſary for her to chew the bread only at certain hours in the day; and thus, after two months, ſhe became entirely free from a moſt diſguſtful and tedious diſorder. It is worthy of obſervation, that, at firſt, the ſwallowing of ſo much ſaliva frequently occaſioned a nauſea; and that then, for a few hours, ſhe was obliged to ſpit it out as uſual; and that, during the greateſt part of the time, when ſhe chewed the bread, ſhe had a ſtool or two every day more than common.

TABES DORSALIS, OR CONSUMPTION OF THE BACK.

A YOUNG gentleman, twenty-two years of age, applied to me in the above diſorder, which had worn him down to a mere ſkeleton. The tabes is ſeldom diſtinguiſhed by any remarkable fever, cough, or difficulty of breathing; but is attended with want of appetite, a weak digeſtion, and a morbid ſtate of the blood, whence the body grows languid, and waſtes by degrees. Sometimes this ſpecies of conſumption is brought on by a venereal ulcer; but it moſt commonly proceeds from exceſſive evacuations of the ſemen, which was the caſe with this patient. He had too early addicted himſelf to an intercourſe with lewd women, which eventually brought on an involuntary ſhedding of the ſeed, which came from him on the leaſt exertion, whether of walking, riding, lifting a weight, or even of pulling off his clothes.—I ordered him a ſtrong nutritious diet, with a table-ſpoonful of the Solar Tincture four times a-day, in the ſame quantity of warm water, which he purſued for a month. He found his ſtrength was ſo much recovered, that I could ſafely adviſe moderate exerciſe both on horſeback and on foot. The gleet, however, was uncommonly obſtinate; and the Tincture was continued for the ſecond month in the ſame quantity. By this time the parts were conſiderably braced; he could run or jump without perceiving the ſmalleſt emiſſion; and the healthful colour of his

cheek

cheek began to return.—He now perſiſted in the Tincture, only three times a-day, for a month longer; after which the doſe was reduced to night and morning for another month; he then took it twice a-day for two months more, at the end of which period every ſymptom of the complaint was removed, he had fully recovered his fleſh and ſtrength; and now preſerves it by taking the Solar Tincture as a beverage, made after the manner of brandy and water. This diſorder has in general been deemed incurable. It is true, that even in its early attacks, it is ſo eſſentially neceſſary to abſtain from venereal embraces, that, without it, the beſt remedies will prove altogether uſeleſs; hence the tabes dorſalis ſo often proves mortal, becauſe the patient has ſeldom reſolution enough to diſpenſe with his amours.

RHEUMATIC GOUT.

THIS diſeaſe is generally brought on by alternate heats and colds in the blood, whereby a humour is produced which attacks the joints and muſcles, ſometimes accompanied with diſcolorations and ſwellings, and at other times without either; but it is always attended with excruciating pain. Mr. John Brandham, of Bridlington Quay, was attacked in this manner; when, after ſome time, the ſevere pain of his joints, falling into his legs and thighs, deprived him of the uſe of his limbs, and confined him entirely to his bed. He was ſoon after ſeized with a violent pain in his head and ſtomach, which ſo much affected his reſpiration, that inſtant death was expected. In this extremity, half a wine-glaſs of the Solar Tincture was adminiſtered, *undiluted*, which removed the danger, and gave his ſtomach immediate eaſe. A table ſpoonful, in the ſame quantity of warm water, was then given every third hour during the ſucceeding day and night, by which the pains were conſiderably abated. He continued the medicine four times a-day for a month longer; at the expiration of which time he experienced a perfect cure, and has never ſince found the ſmalleſt return of his complaint; of which he is deſirous of ſatisfying any enquirer, who chooſes to apply for that purpoſe.

AGUES, CONVULSIONS, CHOLIC, BLOODY-FLUX, and VIOLENT SPASMS IN THE STOMACH AND BOWELS.

DURING the fit, let one or two table-ſpoonfuls of the Solar Tincture, *undiluted*, be adminiſtered ſucceſſively, as the extremity of the caſe may require; and afterwards let the patient continue the medicine, night and morning, in the quantity of a table-ſpoonful in a wine-glaſs of warm water, or oftener, as the obſtinacy of the caſe may render neceſſary, and in a very ſhort time a perfect cure will be experienced; a few inſtances of which I ſhall add, in the words of thoſe who have tranſmitted me the facts.

To

To E. SIBLY, M. D.

SIR,—A few nights ago, I was attacked in bed with a violent pain in my ſtomach and bowels, which alternately produced ſuch a ſucceſſion of convulſive ſpaſms and cold chills, that I really thought I was ſeized for death. Fortunately a bottle of your Solar Tincture was in the houſe, purchaſed the day before by my ſon, of which my ſervant gave me a table-ſpoonful and a half, unmixed with water. The inſtant effect it had on my ſtomach, I could only compare to electricity; for, to the aſtoniſhment of all about me, the ſpaſms inſtantly ceaſed, a gentle perſpiration came on, in which ſtate I fell aſleep, and did not awake till the morning, when I found myſelf entirely free from pain. On getting up, I took a ſpoonful more of the Tincture in an equal quantity of warm water; and have not ſince experienced the ſmalleſt return of the diſorder. Requeſting you will make this known, for the benefit of others, I remain, with grateful eſteem, &c.

No. 25, Philpot-lane, Fenchurch-ſtreet, Feb. 12. M. ARMSTRONG.

To E. SIBLY, M. D.

SIR,—In gratitude, I cannot but thank you for that excellent medicine, the Solar Tincture. It has ſaved my life. I was ſuddenly ſeized with a violent cholic, which brought on a mortification of the bowels. The efforts of the faculty were tried in vain, and I was given over. In theſe moments of extremity, my exiſtence was preſerved by only two ſpoonfuls of your medicine, *undiluted,* which inſtantly relieved me from the rack of torture. After two more doſes, the obſtruction was removed by natural evacuation, and a few hours reſtored me to my uſual ſtate of good health. I intreat you to publiſh this for the public good, and ſhall be ever gratefully your's,

Clifton, near Briſtol, Feb. 24. JOHN POWELL.

To E. SIBLY, M. D.

SIR,—Actuated by a principle of gratitude, I cannot omit acquainting you of an extraordinary cure performed on me by means of your Solar Tincture.—I had for ſome time been afflicted with the dyſentery or bloody-flux, and was reduced to a very weak and languid ſtate, without deriving any benefit from the preſcriptions of the faculty. This induced me to make trial of your Solar Tincture; when, after taking only two ſmall bottles, I found myſelf perfectly recovered; therefore, by publiſhing this to the world, you will confer a favour on your grateful, &c.

WILLIAM JACKSON.

No. 8, Windmill-ſtreet, Tottenham-court-road, May 15.

DISEASES

DISEASES OF THE BREAST AND LUNGS, ASTHMA, DROPSY, OR CONSUMPTION.

TAKE one ſpoonful of the Tincture, night and morning, for twenty days ſucceſſively, diluted in two ſpoonfuls of cold ſpring-water; then reduce it to the ſame doſe every other day, which will in general remove the malady in the courſe of a month; but, if the dropſy or conſumption have been far advanced, it will be neceſſary to continue the medicine for one, two, or even three, months longer, reducing the number of doſes in proportion as health and ſtrength appear to return, and as the blood ſhall have reſumed its proper conſiſtency, and a briſker circulation. In theſe complaints, it will not be amiſs to take the Tincture in a tumbler of warm water, as a beverage, for ſome time after the cure is perfected, as it will infallibly prevent the blood from returning to its watery and impoveriſhed ſtate, and will rareſy and expel the viſcid coheſions in the pulmonary veſſels. In theſe diſorders, the Solar Tincture may be ſafely admiſtered to females even during obſtructions of the catamenia, as hath lately been experienced by perfecting an admirable cure on a lady in Graſton-ſtreet.

This lady was afflicted with obſtructions of the liver and ſpleen, inſomuch that ſhe could not walk up one pair of ſtairs without much pain, and ſhortneſs of breath. Her menſes were obſtructed; and twice or thrice a-day ſhe was attacked with aſthmatic ſpaſms, accompanied with febrile ſymptoms. This affliction being of a peculiar nature, I was obliged to preſcribe both the Solar and Lunar Tinctures, in the following manner: Whenever the fever came on, ſhe took a doſe of the Solar Tincture; and every morning and evening, ſixty drops of the Lunar Tincture in a gill of mugwort tea; and in twenty-one days ſhe was perfectly recovered, and reſtored to her uſual colour and vivacity, to the great joy of her parents and friends.

MENTAL DEPRESSION, OR LOWNESS OF SPIRITS.

THIS may be conſidered the primary diſorder of the nervous train; and, if reſiſted in time, may in moſt caſes be eaſily cured. For this purpoſe take a table-ſpoonful of the Solar Tincture, diluted in a wine-glaſs of cold ſpring-water, every forenoon at eleven or twelve o'clock, for fourteen ſucceſſive days; then uſe it in every two or three days for a month, and the complaint will be entirely removed, as all patients will ſenſibly feel, by their alertneſs, activity, and unuſual flow of natural ſpirits; of which the following caſe may ſerve as an example:

To E. SIBLY, M. D.

SIR,—From a full conviction of the efficacy of your Solar Tincture, I cheerfully come forward to inform you, that, having been much afflicted with depreſſion of

 ſpirits,

goes off, fo that the patients can drink freely. But this does not always happen; for Mr. Vaughan mentions the cafe of a patient, in whom, "when he had in appearance ceafed to breathe, the fpafmus cynicus was obfervable, with an odd convulfive motion in the mufcles of the face; and the ftrange contrariety which took place in the action of thefe produced the moft horrid affemblage of features that can well be conceived. Of this patient alfo it was remarkable, that in the laft hours of his life he ceafed to call for drink, which had been his conftant requeft; but was perpetually afking for fomething to eat."

The hydrophobia feems to be a fymptom peculiar to the human race; for the mad animals which communicate the infection do not feem to have any dread of water. Notwithftanding this, dipping is the common remedy for the cure of dogs and men. With regard to the fymptoms of madnefs in dogs, they are very equivocal; and thofe particularly enumerated by fome authors, are only fuch as might be expected in dogs much heated or agitated by being violently purfued and ftruck. One fymptom indeed, if it could be depended upon, would determine the matter; namely, that all other dogs avoid and run away from one that is mad; and even large dogs will not attack one of the fmalleft fize who is infected with this difeafe. Upon this fuppofition they point out a method of difcovering whether a dog, who hath been killed, was really mad or not; namely, by rubbing a piece of meat along the infide of his mouth, and then offering it to a found dog. If the latter eats it, it is a fign the dog was not mad; but, if the other rejects it with a kind of howling noife, it is certain that he was. Dr. James tell us, that among dogs the difeafe is infectious by ftaying in the fame place; and that, after a kennel has been once infected, the dogs put into it will be for a confiderable time afterwards in danger of going mad alfo. A remedy for this, he fays, is, to keep geefe for fome time in the kennel. He rejects as falfe the opinion that dogs when going mad will not bark; though he owns that there is a very confiderable change in their bark, which becomes hoarfe and hollow.

With regard to the immediate caufe among mankind, there is not the leaft doubt that the hydrophobia is occafioned by the faliva of the mad animal being mixed with the blood. It does not appear that this can operate through the cuticula; but, when that is rubbed off, the fmalleft quantity is fufficient to communicate the difeafe, and a flight fcratch with the teeth of a mad animal has been found as pernicious as a large wound. It is certain alfo, that the infection has been communicated by the bites of dogs, cats, wolves, foxes, weafels, fwine, and even cocks and hens, when in a ftate of madnefs. But it does not appear that the diftemper is communicable from one hydrophobious perfon to another, by means of the bite, or any other way.

It

have. This therefore they perpetually blow off with violence, which in a patient of Dr. Fothergill's occaſioned a noiſe not unlike the hollow barking of a dog, and which he conjectures might have given riſe to the common notion that hydrophobious patients bark like dogs. They have an inſatiable thirſt; but are unable to get down any drink, except with the utmoſt difficulty; though ſometimes they can ſwallow bread ſoaked in liquids, ſlices of orange, or other fruits. There is a pain under the *ſcrobiculus cordis,* as in the tetanus; and the patients mournfully point to that place as the ſeat of the diſeaſe. Dr. Vaughan is of opinion that it is this pain, rather than any difficulty in ſwallowing, which diſtreſſes the patient on every attempt to drink. The voice is commonly plaintive and mournful; but Dr. Vaughan tells us there is a mixture of fierceneſs and timidity in the countenance which he cannot deſcribe, but by which he could know a hydrophobious perſon without aſking any queſtions. Some ſeem to have at times a furious delirium, and an inclination to ſpit at or bite the by-ſtanders; while others ſhow no ſuch inclination, but will even ſuffer people to wipe the inſide of their mouths with the corner of a handkerchief in order to clear away the viſcid ſaliva which is ready to ſuffocate them. In ſome male patients there is an involuntary erection of the penis, and emiſſion of the ſemen; and the urine is forced away by the frequent return of the ſpaſms.

In a letter from Dr. Wolf, of Warſaw, to Henry Baker, F. R. S. dated Warſaw, Sept. 26th, 1767, we have the following melancholy account of the caſes of five perſons who died of the hydrophobia: None of them quite loſt their ſenſes; but they were all talking without intermiſſion, praying, lamenting, deſpairing, curſing, ſighing, ſpitting a frothy ſaliva, ſcreeching, ſometimes belching, and retching, but rarely vomiting. Every member is convulſed by fits, but moſt violently from the navel up to the breaſt and œſophagus. The fit comes on every quarter of an hour; the fauces are not red, nor the tongue dry. The pulſe is not at all feveriſh; and, when the fit is over, nearly like a ſound pulſe. The face grows pale, then brown, and during the fit almoſt black; the lips livid; the head is drowſy, and the ears tingling; the urine limpid. At laſt they grow weary; the fits are leſs violent, and ceaſe towards the end; the pulſe becomes weak, intermittent, and not very quick; they ſweat, and at laſt the whole body becomes cold. They compoſe themſelves quietly as if to get ſleep, and ſo they expire. A general obſervation was, that the lint and dreſſings of the wounds, even when dry, were always black, and that when the pus was very good in colour and appearance. In one of Dr. Wolf's patients who recovered, the blood ſtank intolerably as it was drawn from a vein; and one of Mr. Vaughan's patients complained of an intolerable fœtid ſmell proceeding from the wounded part, though nobody but himſelf could perceive it. In general, the violent convulſions ceaſe a ſhort time before death; and even the hydrophobia goes

goes off, ſo that the patients can drink freely. But this does not always happen; for Mr. Vaughan mentions the caſe of a patient, in whom, " when he had in appearance ceaſed to breathe, the ſpaſmus cynicus was obſervable, with an odd convulſive motion in the muſcles of the face; and the ſtrange contrariety which took place in the action of theſe produced the moſt horrid aſſemblage of features that can well be conceived. Of this patient alſo it was remarkable, that in the laſt hours of his life he ceaſed to call for drink, which had been his conſtant requeſt; but was perpetually aſking for ſomething to eat."

The hydrophobia ſeems to be a ſymptom peculiar to the human race; for the mad animals which communicate the infection do not ſeem to have any dread of water. Notwithſtanding this, dipping is the common remedy for the cure of dogs and men. With regard to the ſymptoms of madneſs in dogs, they are very equivocal; and thoſe particularly enumerated by ſome authors, are only ſuch as might be expected in dogs much heated or agitated by being violently purſued and ſtruck. One ſymptom indeed, if it could be depended upon, would determine the matter; namely, that all other dogs avoid and run away from one that is mad; and even large dogs will not attack one of the ſmalleſt ſize who is infected with this diſeaſe. Upon this ſuppoſition they point out a method of diſcovering whether a dog, who hath been killed, was really mad or not; namely, by rubbing a piece of meat along the inſide of his mouth, and then offering it to a ſound dog. If the latter eats it, it is a ſign the dog was not mad; but, if the other rejects it with a kind of howling noiſe, it is certain that he was. Dr. James tell us, that among dogs the diſeaſe is infectious by ſtaying in the ſame place; and that, after a kennel has been once infected, the dogs put into it will be for a conſiderable time afterwards in danger of going mad alſo. A remedy for this, he ſays, is, to keep geeſe for ſome time in the kennel. He rejects as falſe the opinion that dogs when going mad will not bark; though he owns that there is a very conſiderable change in their bark, which becomes hoarſe and hollow.

With regard to the immediate cauſe among mankind, there is not the leaſt doubt that the hydrophobia is occaſioned by the ſaliva of the mad animal being mixed with the blood. It does not appear that this can operate through the cuticula; but, when that is rubbed off, the ſmalleſt quantity is ſufficient to communicate the diſeaſe, and a ſlight ſcratch with the teeth of a mad animal has been found as pernicious as a large wound. It is certain alſo, that the infection has been communicated by the bites of dogs, cats, wolves, foxes, weaſels, ſwine, and even cocks and hens, when in a ſtate of madneſs. But it does not appear that the diſtemper is communicable from one hydrophobious perſon to another, by means of the bite, or any other way.

It

It has been generally allowed by practitioners, that, though the hydrophobia may be prevented, yet it can seldom be cured after the disease has made its appearance. The most essential part of the treatment therefore depends on an immediate use of the proper means of prevention. For this purpose some advise the instant cutting out the part bitten, which must certainly be an effectual mode, provided we could be sure the poison had not reached beyond the wound. When, however, we consider the rapidity with which the blood and juices flow, it seems impossible we can ever wholly depend on such an operation. I should nevertheless advise it to be done; after which let the part be well soaked with the Solar Tincture; and, to fortify the blood, let the patient immediately swallow a table spoonful every three hours, *undiluted*, for the first day; and the same dose night and morning, for a month following. Let the part be again soaked with the Tincture four times a-day, for three or four days; and I am satisfied a safe and perfect cure may be relied on. For the bite of adders, snakes, &c. bathing the part, and taking the medicine *undiluted*, will counteract the virulence of the poison, and preserve the patient from further injury.

For GUN-SHOT WOUNDS, CUTS, STABS, &c.

GENTLEMEN in the army and navy, and all persons liable to gun-shots, stabs, wounds, &c. should never be without the Solar Tincture. Its salutary effects on the blood, in all these cases, are really surprising. It totally prevents, and will even stop, mortification, in very advanced stages. It quickly supplies the greatest loss of blood; fortifies the heart, cherishes the vital organs, and heals and unites the flesh in an uncommon degree. If taken internally, and poured at the same time into the wound, it is quickly propelled through the heart by the veins and arteries; and thus renovates the exhausted spirits, and preserves life. Its effect on a few simple wounds may be seen in the following cases.

To E. SIBLY, M. D.

SIR,—For the sake of those liable to accidents, I think it right to inform you of a most remarkable cure performed by your Solar Tincture, on a very deep and dangerous wound made on Mrs. Cook by a case-knife, of more than the depth of my fore-finger. After trying every means in vain to stop the blood, I sent for a bottle of your Solar Tincture, and well bathed the wound therewith. The blood and Tincture readily assimilated, and formed a crust on the orifice of the wound, which very soon stopped the effusion of blood. But what is most remarkable, the wound was completely healed in less than six days, and is now so perfectly closed, as to be almost imperceptible. You are welcome to publish this, and in so doing will oblige, &c.

Seymour-street, *Portland-square*, *April* 14. WILLIAM COOK.

To E. SIBLY, M. D.

SIR,—In juſtice to my own feelings, I cannot but acquaint you with a cure performed by your Solar Tincture, in a very uncommon manner. As I was travelling in the ſtage to Boxley-Abbey, near Maidſtone, in Kent, a gentleman, who ſat next me, putting his head out of the window, received a violent cut acroſs the eye with the coachman's whip, which produced an immediate ſwelling and inflammation, attended with ſo much agony, that he declared the pain was inſupportable. I had purchaſed a bottle of your Solar Tincture while in town, knowing it had performed many ſurpriſing cures in my neighbourhood. This I immediately opened, and applied to the inflamed part; and, after waſhing the eye well with it, I bound a white handkerchief tight over it, wetted with the Tincture. In leſs than ten minutes the anguiſh was greatly aſſuaged; and in the courſe of three hours it was quite well. The gentleman expreſſed the utmoſt aſtoniſhment at the celerity of the cure, as did every paſſenger in the coach. I wiſh this to be made as public as poſſible, for the benefit of thoſe who are liable to accidents; and am with reſpect, &c.

M. STABLES.

I ſhall only remark further, with reſpect to wounds, bruiſes, &c. that a ſhort time ago, as a coach was driving furiouſly out of Cavendiſh-ſquare, the horſes unfortunately beat down a girl of eight years of age, the daughter of Robert and Elizabeth Larken, of Clipſton-ſtreet; and, the wheels paſſing over her body, ſhe was taken up to all appearance dead. The ſpectators were for carrying her immediately to the hoſpital; but, the accident happening very near my houſe, I was ſent for. I avoided letting blood, but bathed the bruiſed parts thoroughly with the Solar Tincture, and introduced half a ſpoonful, *undiluted*, into her ſtomach. It was now about nine o'clock at night. She was compoſed and aſleep before ten, being overcome by the medicine. A ſpoonful more of the Tincture was given her at different periods of the night, the ſudorific power of which brought on a plentiful perſpiration. At ten o'clock the next morning ſhe awoke, and got up, and was ſo well recovered as to be able to play about with her companions, in all reſpects the ſame as if nothing had happened. The girl, and her parents, are pleaſed with every opportunity of recounting the circumſtances of this event, to any enquirers.

Let it not be ſaid, that, becauſe this medicine appears to be preſcribed for many diſorders, it can be good for none.—I affirm, that every complaint for which it is recommended, *originates in the blood*, or in *obſtructed perſpiration.* The action of the Solar Tincture is on the blood and juices; it ſtrikes at the root, not at the branches; by which peculiar advantage it effects a cure when other medicines fail. And, though there is a medicine, ſold in regular practice at a guinea an ounce, which

which poffeffes no one virtue comparable to the Solar Tincture; yet the proprietor, unwilling to adopt fuch examples, or to withhold from the afflicted in every line of of life the benefits of his difcovery, has determined to render it to the public at only 7s. 6d. the fmall, and 13s. the large, bottles, duty included, with ample directions in every complaint for which it ought to be adminiftered.—A fingle bottle will in many cafes perform a fpeedy cure, when, in the ordinary courfe of medical practice, it would occupy a month, and coft many pounds for unneceffary attendance, and excefs of drugs.

OF THE

PRINCIPLES

OF

LIFE AND DEATH.

LIFE denotes the animated ftate of nature; and, in human beings, exifts as long as an union of the foul and body lafts. With us, therefore, life continues, until fuch feparation has really taken place; which can no more be faid to have happened during the paroxifm of a fit, or of a blow which for a time deprives us of fenfation, or in the *early* period of an unnatural or fudden death, than during the time we are afleep. It is the want of proper fkill at fuch times that too often occafions death to take place, when life abfolutely exifts in the blood, and might with little care have been preferved. Death is therefore the act of feparation of the foul from the body; in which fenfe it ftands oppofed to life, which confifts in the union thereof. An animal body, by the actions infeparable from life, undergoes a continual change, and receives its diffolution by degrees. Its fmalleft fibres become rigid; its minuter veffels veſſels grow into ſolid fibres no longer pervious to the fluids; its greater veſſels grow hard and narrow; and every thing becomes contracted, cloſed, and bound up: whence the dryneſs, immobility, and extenuation, obſerved in old age. By ſuch means the offices of the minuter veſſels are deſtroyed; the humours ſtagnate, harden, and at length coaleſce with the ſolids. Thus are the ſubtileſt fluids in the body intercepted and loſt, the concoction weakened, and the reparation prevented; only the blood continues to run ſlowly through the greater veſſels, aſſiduous to preſerve life, even after the animal functions are deſtroyed. At length, in the proceſs of theſe changes, death becomes inevitable, as the neceſſary conſequences of life, But it is rare indeed that life is thus long protracted, or that death ſucceeds merely from the natural decays and impairment of old age. Accidental diſeaſes, and our neglect of preſerving health, cut the work ſhort.

The ſigns of death are often very uncertain. If we conſult what Winſlow or Bruchier have ſaid on this ſubject, we ſhall be convinced, that between life and death the ſhade is ſo very undiſtinguiſhable, that even all the powers of art can ſcarcely determine where the one ends and the other begins. The colour of the viſage, the warmth of the body, and ſuppleneſs of the joints, are but uncertain ſigns of life ſtill ſubſiſting; while, on the contrary, the paleneſs of the complexion, the coldneſs of the body, the ſtiffneſs of the extremities, the ceſſation of all motion, and the total inſenſibility of the parts, are but uncertain marks of death begun. In the ſame manner alſo, with regard to the pulſe and breathing; theſe motions are ſo often kept under, that it is impoſſible to perceive them. By bringing a looking-glaſs near to the mouth of the perſon ſuppoſed to be dead, people often expect to find whether he breathes or not. But this is a very uncertain experiment: the glaſs is frequently ſullied by the vapour of the dead man's body; and often the perſon is ſtill alive, though the glaſs is no way tarniſhed. In the ſame manner, neither noiſes in the ears, nor pungent ſpirits applied to the noſtrils, give certain ſigns of the diſcontinuance of life; and there are many inſtances of perſons who have endured them all, and afterwards recovered without any external aſſiſtance, to the aſtoniſhment of the ſpectators. This ſurely ought to be a caution againſt haſty burials, eſpecially in caſes of ſudden death; for it is ſhocking to reflect, that ſome hundreds of valuable members of ſociety are annually torn from their diſconſolate families by ſome accidental ſudden cauſe, and hurried thoughtleſsly to the grave, in whom the principles of life were capable of being revived! This lamentable truth has been eſtabliſhed by the happy ſucceſs of the Humane Society, from whoſe laudable exertions ſeveral hundred perſons have been reſtored to life, who, to all viſible appearance, *were paſt recovery*. Every age and country affords ſome inſtances of perſons having been recovered, even alter lying long for dead; and from the number of thofe preferved by mere lucky accidents, it is evident ftill greater numbers might be faved by timely pains and fkill. Thofe who have contemplated the ftructure of the human machine know, that its diffolution cannot *naturally* happen but by that gradual decay of the whole fyftem above defcribed, when the veffels are become impervious to the fluids, the circulation weakened or deftroyed, and the vital organs no longer able to perform their office. But, when their functions are merely fufpended by fome fudden fhock, it may be likened to the ftate of a watch ftopped by a fall, which refumes its motion the inftant that injury is repaired. In the animal œconomy, "*the BLOOD is the LIFE;*" Levit. xvii. 11, 14. Deut. xii. 23. therefore, if its circulation be fufpended or deftroyed, death follows. But, if the blood can be re-agitated, and its circulation refumed, life will of neceffity be reftored. For this reafon, whenever any accident has happened, by which fudden death appears to have taken place, whether by blows, fits, falls, fuffocation, ftrangulation, drowning, apoplexy, convulfion-fits, thunder and lightning, affaffination, duelling, or the like, let the unfortunate perfon be carried into a warm houfe, and laid by the fire, or put into a warm bed; let two or three table-fpoonfuls of the Solar Tincture be introduced as early as poffible into the ftomach, and rubbed profufely in, by a warm hand, upon the fpine of the back, loins, breaft, and region of the heart, and poured into the wound, if there be any; the warm ftimulating quality of the medicine, affifted by the external heat and friction, will quickly roufe the ftagnant blood and juices, particularly in the grand refervoir the heart, where, rarefying, preffing every way, and being refifted by the valves, it will fwell fo as to fill the flaccid right auricle of the heart, which by the fhock had become empty and at reft; and thus, ftimulating its fibres, will put them in motion. The right auricle being thus filled, and ftimulated into contraction, fills the ventricle; which, by this means being irritated, likewife contracts and empties itfelf into the pulmonary artery; and, the moment this is done, the circulation begins again where it left off; and the lungs, being filled by the dephlogifticated air contained in the medicine, begin to act, and life is reftored, provided the organs and juices are in a fit difpofition for it; which they undoubtedly are much oftener than is imagined. Nor is this ftimulating action of the Tincture upon the heart at all furprifing; for every medical man knows, or ought to know, that the heart, even when taken out of the body, if it be pricked with a pin, or hath warm water thrown upon it, will beat afrefh, and endeavour to exert its functions, though for fome time before it had been motionlefs. No perfon therefore ought to be confidered *dead*, until the energy of the blood is fo far gone, that it can never again be agitated fo as to fill and ftimulate into contraction the right *finus venofus* and auricle of the heart.

When the patient is thus far recovered, he ought to be treated with great care and tenderneſs; and ſome warm milk, wine and water, elder-flower-tea, or any nouriſhing ſpoon-meat, ſhould be given to him as ſoon as he appears capable of taking food. In ſome caſes it may be neceſſary to open the temporal artery and the external jugular, or to bleed in the arm; but this ſhould never be done, if it can ſafely be diſpenſed with, as it certainly weakens the animal principle, which it is the firſt object of this medicine to ſtrengthen. Under different circumſtances, and as particular occaſions may require, the rules laid down in p. 196 of the Medical Part of this work, and recommended by the Humane Society, will be found of conſiderable advantage. Above all, let me entreat an anxious perſeverance in this ſublimeſt of all virtues—the attempt to recover periſhing lives. Humanity calls for it in the moſt moving accents; and what can inſpire a good heart with more ſincere, perfect, conſcientious, and commendable, ſatisfaction, than a retroſpect of ſuch endeavours as have been generouſly exerted and ſucceſsfully contributed to recover, perhaps to reſtore, the life of a fellow-creature from the moſt deprecated calamity—ſudden death, with its alarming retinue of threatening conſequences to thoſe who die unprepared? ſince, by thus preſerving a ſinner to a future period, perhaps a ſoul may emerge in full maturity to felicity which ſhall have no end!

To demonſtrate the reanimating power of the medicine, experiments may be made upon a fowl, lamb, cat, dog, or other animal, by plunging them under water until they are apparently dead, or piercing them through the head, or any part of the body except the heart; by ſuffocation, or an electrical ſhock: for ſudden death, howſoever it happens, whether by drowning or otherwiſe, is much the ſame as to its effects on the vital organs; conſequently they are all to be treated in a ſimilar manner.

Upon the whole it is evident, that by contemplating the œconomy and harmony of our ſtructure, both external and internal, we may quickly diſcern a proper line of conduct for the conſervation of health, and the prolongation of life; and we ſhall alſo perceive a more auguſt view of the marvellous works of divine wiſdom in the ſtructure of the human frame, than we ſhall perhaps again find in the whole compaſs of nature. The gift of health was evidently the deſign of our benevolent Creator in the conſtruction of our bodies; it is therefore no leſs our duty than our intereſt to preſerve this bleſſing to our lateſt moments, as the ſeaſoning and fund which gives the reliſh to all our other enjoyments. To enumerate the various abuſes of health, which take place from our earlieſt infancy, particularly among the rich and gay, and which are continued through the ſucceeding ſtages of modiſh life, would fill a volume. Suffice it to obſerve, that they prevail more particularly among people who are the moſt highly poliſhed and refined. To compare their artificial mode of living with that of nature, would afford a very ſtriking contraſt, and ſupply an obvious reaſon why perſons in the lower orders of ſociety are generally the longeſt livers, and enjoy the beſt ſtate of health; and hence we are warranted to conclude, that a large proportion of the diſeaſes to which we are ſubjected are produced by ourſelves.

Notwithſtanding this unaccountable abuſe of our health, yet the want of it unfits us for moſt of the common avocations of life, and is more eſpecially an enemy to the ſocial and humane affections, as it generally renders the unhappy ſufferer peeviſh and ſullen, diſguſted at the allotments of Providence, and apt to induce ſuicide, by ſuggeſting gloomy and ſuſpicious ſentiments of the Almighty. It obſtructs the free exerciſe and full improvement of our reaſon, makes us a burden to our friends, and uſeleſs to ſociety. Whereas the uninterrupted enjoyment of health is a conſtant ſource of good humour, and good humour is a great friend to openneſs and benignity of heart; enables us to encounter the various ills and diſappointments of this world with more courage, or to ſuſtuin them with more patience; and, in ſhort, conduces much, if we are otherwiſe duly qualified, to our acting our part in every exigency of life with more firmneſs, conſiſtency, and dignity. Therefore it imports us much to preſerve and improve the habit of its enjoyment, without which every other external entertainment is taſteleſs, and moſt other advantages are of little avail. To this end, we ought above all things to cultivate prudence, temperance, ſobriety, fortitude, and equanimity of temper; for without a prudent care of the body, and a ſteady government of the mind, to guard the one from diſeaſe, and the other from the feuds of paſſion and prejudice, ſound health is unattainable. By *temperance* we enjoy the real gratifications of life, without ſuffering any conſequent inconvenience. Sobriety enables us to be content with ſimple and frugal fare, and protects us from the pain and diſgrace of intoxication. Fortitude enables us to bear thoſe infirmities which prudence and ſobriety cannot ſhun, and baniſhes all dread of imaginary evils from our thoughts. Equanimity of temper contributes greatly to the happineſs of life, as well as to the preſervation of health, by keeping the mind from anxiety and perturbation, and arming us againſt the calumnies and animoſities of human nature. Violent paſſions, and the exceſſes they induce, gradually impair and wear away the conſtitution; whilſt the calm and placid ſtate of a temperate mind, and the healthful exerciſe of the body, preſerve the natural functions in full vigour and harmony, and exhilarate the ſpirits, which are the chief inſtruments of action. The worſt conſequences that could poſſibly reſult from a ſtrict adherence to this regimen, would be that of exterminating a ſwarm of locuſts, and rendering the diſcovery of my medicine of leſs importance to the community.

Of the CRISIS, or CRITICAL TURN, of a DISEASE.

THE Crisis of a Disease is no other than the struggle betwixt nature and the infirmity, which of them shall prevail. If nature at the time of the crisis overcomes the malignity of the disease, it is a sure sign it will be cured; but, if the sickness prevails, it is then a pernicious crisis, and shows sudden alterations for the worse. Every sudden and vehement motion of the disease may be called a crisis; therefore days critical, decretory, and crismal, are all one and the same thing, and import no more than a certain and more sure judgment of the infirmity afflicting, either more powerful, or less vehement, at those times when the true crisis happens; therefore a crisis is to be calculated from that moment of time when the disease first invaded the patient. And on this ground I shall make some observations to prove the truth of what I have now to deliver, and of what I have before so often proved, that I cannot but admire the wonderful providence of God, who disposeth all things by number, weight, and measure, prescribeth to the whole system of nature so immutable a law, that it were as easy for the heaven and the earth to return to their original chaos, as to break and infringe that immutable law, unless the divine will and pleasure alter it miraculously.

We discriminate two sorts of diseases; acute, and chronic. Of acute diseases some are simply acute, others peracute, that is very acute; others again are peracute, or exceedingly acute. Those that are simply acute are finished in eight, ten, eleven, fourteen, twenty, or twenty-one, days. They are terminated in the time the Moon traceth the twelve celestial signs of the zodiac, viz. in twenty-seven days and eight hours.

Those acute diseases which suffer changes are very fickle; for sometimes they increase, and sometimes they are remitted, according as the Moon meets with the beams of either benefic or baneful planets; and sometimes they change out of acute diseases into chronic; and thus a continued fever may change into an hectic fever, or an intermittant fever into a continual fever; and these diseases terminate in forty days.

Very acute diseases are such as terminate in five, six, seven, or eight, days; amongst which is the disease called peripneumonia, or inflammation of the lungs. Exceedingly acute diseases are such as end in three or four days at farthest, as pestilences, apoplexies, &c.

Chronic diseases follow the motion of the Sun, and it is about ninety days before the first crisis begins to appear; for in that time the Sun comes to the proper quartile of the place he was in at the decumbiture; as appears in hectic fevers, dropsies, and the like. But, when he comes to those degrees from the decumbiture which are

are called indicative, or intercidental, which are both one, or judicial, (as may be seen in the Table,) some alteration will appear, whereby a man may judge of the crisis to come. For the patient will be well, if the Sun be well configurated with benign planets; but worse, if in aspect with evil ones; and this rule is infallible, if you consider it from the nativity throughout the whole course of a man's life; for *diseases are the particular attendants of the inequality of the elements in every human being.*

Also a crisis may be perfect, or imperfect. A perfect crisis is when the disease appears plain, and perfectly to be judged of; and this is sometimes hopeful, and sometimes desperate. Hopeful, when there is a great probability of health and recovery; desperate, when there are palpable signs of death. An imperfect crisis is when the disease is changed upon every light occasion; as if Mars be the author of the disease, and in a double-bodied sign; in this case the disease will be variable.

That crisis may be deemed safe, which comes without pernicious aspects; but that is doubtful and dangerous which comes with malignant aspects; what these aspects are, with the significators of every disease, and the mode of ascertaining them, are already explained in my ILLUSTRATION OF ASTROLOGY. I have there shown, that to judge of a disease, it is necessary to observe the motion of the Sun, Moon, and lord of the ascendant. With respect to the lord of the ascendant, observe, before you give judgment, what application he makes to any planet, either by conjunction, quartile, or opposition; or, should he apply to more than one planet, look to which of them he approaches nearest, and then count how many degrees of longitude are between them; and, if the disease be acute, then for every degree add a day; but, if chronic, a week, month, or year, according to the situation of Jupiter, Venus, Mercury, or the Moon, at a perfect crisis.

Now the time called critical is always evil, because of the contrariety of the sign the Moon is then in to the sign she was in at the decumbiture, which induceth the contrariety of her nature to the opposite place; therefore at such a time there ariseth a controversy and contest between the disease and nature. The Moon upholds nature in acute diseases; and hence is the reason that a bad crisis will always happen, if she be afflicted upon a critical day by the bodies or evil beams of Saturn or Mars, or by the lord of the eighth house, or by the lord of the fourth house, if he be a malevolent, because he signifies the grave. But, if the Moon at the time of the crisis behold the lord of the ascendant, or be configurated with the benefic planets, health ensues, and the malady will be vanquished and overcome in the conflict.

If the disease terminates not upon the first crisis, observe how the Moon will be configurated on the second crisis, and judge by the same rules. If it terminates not then, as will sometimes happen, view the third crisis, and judge by that the same way.

way. If your judgment, ſupported by reaſon and the former rules, declare that the diſeaſe will not terminate one way or other, neither in health nor death; then examine the face of the heavens at the time the Moon returns to the place ſhe was in at the decumbiture, which is at the end of twenty-ſeven days, eight hours, and ſome minutes; and judge according as the Moon ſhall be then configurated with benefic or malignant planets; for this of neceſſity terminates all acute diſeaſes; though we may obſerve that not one in a hundred holds on ſo long, nor one out of twenty continues half ſo long.

If the acute diſeaſe ends not in a month, it is then turned into a chronic diſeaſe; and muſt be judged of by the Sun. The rules for judging chronic diſeaſes by the Sun are ſimilar to thoſe by which we judge of acute diſeaſes by the Moon. Now, for the right diſtinction and calculation of time to judge of the progreſs of a diſeaſe in this way, obſerve the following method: See what degree the Moon was in at the decumbiture, by an Ephemeris, and add twenty-two degrees thirty minutes, which is called the indicative time, becauſe it informs the phyſician the nature of the diſeaſe; for upon theſe indicative days the diſeaſe is uſually remitted and mitigated. To this indicative time add twenty-two degrees thirty minutes more, and this points out the judicial day, viz. juſt forty-five degrees from the place of the Moon when the patient fell ſick, being the half of a criſis, and manifeſts, according as the Moon happens to be aſpected, whether a good or a bad criſis will enſue. To the judicial day add twenty-two degrees thirty minutes more, and it makes ſixty-ſeven degrees thirty minutes, which produces the ſecond indicative day, as falling between the criſis and judicial day. From this the phyſician may expect indications how the diſeaſe will finally ſhow itſelf. To this add twenty-two degrees thirty minutes more, and you have the perfect criſis of the diſeaſe from the decumbiture, viz. ninety degrees, or one quarter of the zodiac. At this time nature will manifeſt, according to the planets that are in aſpect to the Moon, whether the ſick perſon will have a good or bad criſis; and, adding twenty-two degrees thirty minutes more, it makes the next judicial day, when the Moon approacheth to it; and ſoon, through the whole twelve ſigns of the zodiac, and over it again, if the diſeaſe terminate not in that time, as will plainly appear by the following Table, which ſhows when the Moon comes to an indicative or to a judicial day, that is, a ſemiquartile, or half a criſis; and when to a true quartile, and when to an oppoſition, which is called a full criſis; and ſo to all the indicative and judicial days during the ſickneſs, &c.

EXAMPLE.

Suppoſe the true place of the Moon, at the time a perſon falls ſick, be ſixteen degrees of Gemini, which will be found in the fourth column of the following Table,

ble, ſo that ſixteen degrees of Gemini will be the Moon's radical place in the decumbiture. Over againſt 16 degrees, to the right hand, I find 8 30, and over the head thereof I find ♋; ſo that, when the Moon came to eight degrees thirty minutes of Cancer, it was the firſt indicative day, wherein the phyſician might expect to ſee how the diſeaſe would ſhow itſelf. Upon every criſis or indicative day, make ſpecial obſervation what planet the Moon is in configuration with; if with a benevolent planet, expect ſome remiſſneſs in the diſeaſe; but, if with a malevolent, the contrary effect will follow. Next, on the right hand to 8 30 of ♋, you will find 1 ♌, which ſhows that, when the Moon comes to the firſt degree of Leo, ſhe will be in ſemiquartile to her firſt place; and this is, as before ſtated, half a criſis, at which time the diſeaſe will more or leſs manifeſt itſelf according to ſuch configurations as the Moon is found to make with the other planets at the time ſhe comes to the firſt degree of Leo. In the next column on the right hand, you ſee 23 30, and over it ♌. This points out the indicative day, wherein the phyſician is enabled further to judge of the increaſe or decreaſe of the diſeaſe. In the next column you find 16, and over it ♍, which indicates that, when the Moon came to the ſixteenth degree of Virgo, there was a true criſis, whereby the diſeaſe might be more fully inveſtigated, and a judgment framed according to the aſpects the Moon in that degree had to the good or evil planets; for from hence will the patient or phyſician deſcry a better or worſe criſis, in progreſſive order. And thus, in the continued line or column, you may run round the face of the Heavens, obſerving the configurations of the Moon when ſhe comes to thoſe places of the zodiac wherein ſhe makes the indicative, judicial, and critical, days, and what planet or planets ſhe is then in contact with, and whether in the decumbiture they promiſe good or evil. Beſides this, you muſt obſerve on what day the Moon, or the lord of the aſcendant, tranſits the cuſp of the ſixth, ſeventh, and eighth, houſes, and how ſhe is then aſpected with the benevolent or malign planets; and obſerve whether ſhe be combuſt, or in *via combuſta*, which is from the twentieth degree of Gemini to the firſt of Cancer, in the northern part of the zodiac, and in the ſouthern from the ſixth degree of Sagittarius to the ſixteenth of the ſame conſtellation; and from the twenty-fourth degree of Sagittarius to the fifth degree of Capricorn, or in conjunction, quartile, or oppoſition, of Saturn, or Mars, or of a combuſt planet, or of ſome fixed ſtar of a malignant nature; for in all theſe caſes an indication is given of death, or of long and ſevere ſickneſs, according to the number of teſtimonies and aſtral indications, as expreſſed in the rules given in my ILLUSTRATION of ASTROLOGY; but in which the following moſt valuable Table was omitted.

A LU-

A LUNAR TABLE,

Which, by entering with the Degree of the Moon at the Time any Perſon falls ſick, will point out at one View the Indicative, Judicial, and Critical, Day, of the Diſeaſe.

♈	♈	♉	♊	♋	♋	♌	♍	♎	♎	♏	♐	♑	♑	♒	♓
0 30	23	15 30	8	0 30	23	15 30	8	0 30	23	15 30	8	0 30	23	15 30	8
1 30	24	16 30	9	1 30	24	16 30	9	1 30	24	16 30	9	1 30	24	16 30	9
2 30	25	17 30	10	2 30	25	17 30	10	2 30	25	17 30	10	2 30	25	17 30	10
3 30	26	18 30	11	3 30	26	18 30	11	3 30	26	18 30	11	3 30	26	18 30	11
4 30	27	19 30	12	4 30	27	19 30	12	4 30	27	19 30	12	4 30	27	19 30	12
5 30	28	20 30	13	5 30	28	20 30	13	5 30	28	20 30	13	5 30	28	20 30	13
6 30	29	21 30	14	6 30	29	21 30	14	6 30	29	21 30	14	6 30	29	21 30	14
7 30	30	22 30	15	7 30	30	22 30	15	7 30	30	22 30	15	7 30	30	22 30	15
	♉				♌				♏				♒		
8 30	1	23 30	16	8 30	1	23 30	16	8 30	1	23 30	16	8 30	1	23 30	16
9 30	2	24 30	17	9 30	2	24 30	17	9 30	2	24 30	17	9 30	2	24 30	17
10 30	3	25 30	18	10 30	3	25 30	18	10 30	3	25 30	18	10 30	3	25 30	18
11 30	4	26 30	19	11 30	4	26 30	19	11 30	4	26 30	19	11 30	4	26 30	19
12 30	5	27 30	20	12 30	5	27 30	20	12 30	5	27 30	20	12 30	5	27 30	20
13 30	6	28 30	21	13 30	6	28 30	21	13 30	6	28 30	21	13 30	6	28 30	21
14 30	7	29 30	22	14 30	7	29 30	22	14 30	7	29 30	22	14 30	7	29 30	22
		♊				♍				♐				♓	
15 30	8	1 30	23	15 30	8	0 30	23	15 30	8	0 30	23	15 30	8	0 30	23
16 30	9	2 30	24	16 30	9	1 30	24	16 30	9	1 30	24	16 30	9	1 30	24
17 30	10	3 30	25	17 30	10	2 30	25	17 30	10	2 30	25	17 30	10	2 30	25
18 30	11	4 30	26	18 30	11	3 30	26	18 30	11	3 30	26	18 30	11	3 30	26
19 30	12	5 30	27	19 30	12	4 30	27	19 30	12	4 30	27	19 30	12	4 30	27
20 30	13	6 30	28	20 30	13	5 30	28	20 30	13	5 30	28	20 30	13	5 30	28
21 30	14	7 30	29	21 30	14	6 30	29	21 30	14	6 30	29	21 30	14	6 30	29
22 30	15	8 30	30	22 30	15	7 30	30	22 30	15	7 30	30	22 30	15	7 30	30
			♋				♎				♑				♈
23 30	16	9 30	1	23 30	16	8 30	1	23 30	16	8 30	1	23 30	16	8 30	1
24 30	17	10 30	2	24 30	17	9 30	2	24 30	17	9 30	2	24 30	17	9 30	2
25 30	18	11 30	3	25 30	18	10 30	3	25 30	18	10 30	3	25 30	18	10 30	3
26 30	19	12 30	4	26 30	19	11 30	4	26 30	19	11 30	4	26 30	19	11 30	4
27 30	20	13 30	5	27 30	20	12 30	5	27 30	20	12 30	5	27 30	20	12 30	5
28 30	21	14 30	6	28 30	21	13 30	6	28 30	21	13 30	6	28 30	21	13 30	6
29 30	22	15 30	7	29 30	22	14 30	7	29 30	22	14 30	7	29 30	22	14 30	7

With respect to moderate or slight diseases, Hippocrates asserted, in the first place, "That contraries, or opposites, are the remedies for each other;" and this maxim he explains by an aphorism; in which he says, that evacuations cure those distempers which come from repletion, and repletion those that are caused by evacuation. So heat is destroyed by cold, and cold by heat, &c. In the second place, he asserted, that "physic is an addition of what is wanting, and a subtraction or retrenchment of what is superfluous:" an axiom which is explained by this, viz. that there are some juices or humours, which in particular cases ought to be evacuated, or driven out of the body, or dried up; and some others which ought to be restored to the body, or caused to be produced there again. As to the method to be taken for this addition or retrenchment, he gives this general caution, That you ought to be careful how you fill up, or evacuate, all at once, or too quickly, or too much; and that it is equally dangerous to heat or cool again on a sudden; or rather, you ought not to do it: every thing that runs to an excess being an enemy to nature. In the fourth place, Hippocrates allowed that we ought sometimes to dilate, and sometimes to lock up: to dilate, or open the passages by which the humours are voided naturally, when they are not sufficiently opened, or when they are closed; and, on the contrary, to lock up or straiten the passages that are relaxed, when the juices that pass there ought not to pass, or when they pass in too great quantity. He adds, that we ought sometimes to smooth, and sometimes to make rough; sometimes to harden, and sometimes to soften again; sometimes to make more fine or supple; sometimes to thicken; sometimes to rouse up, and at other times to stupify or take away the sense; all in relation to the solid parts of the body, or to the humours. He gives also this farther lesson, That we ought to have regard to the course the humours take, from whence they come, and whither they go; and in consequence of that, when they go where they ought not, that we make them take a turn about, or carry them another way, almost like the turning the course of a river; or, upon other occasions, that we endeavour if possible to recal, or make the same humours return back again; drawing upward such as have a tendency downward, and drawing downward such as tend upward. We ought also to carry off, by convenient ways, that which is necessary to be carried off; and not let the humours once evacuated enter into the vessels again. Hippocrates gives also the following instruction; That, when we do any thing according to reason, though the success be not answerable, we ought not too easily, or too hastily, to alter the manner of acting, as long as the reasons for it are yet good. But, as this maxim might sometimes prove deceitful, he gives the following as a corrector to it: "We ought (says he) to mind with a great deal of attention what gives ease, and what creates pain; what is easily supported, and what cannot be endured." We ought not to do any thing raſhly; but ought often to pauſe, or wait, without doing any thing: by this way, if you do the patient no good, you will at leaſt do him no hurt.

Theſe are the principal and moſt general maxims of the practice of Hippocrates, and which proceed upon the ſuppoſition, that nature cures all ſlight diſeaſes. When, however, they are acute or ſevere, they demand the utmoſt ingenuity and ſkill of the phyſician to moderate their violence; and it is then that their termination may be known by conſulting the foregoing Table.

Of the UTILITY of the preceding TABLE.

IN order to ſhow the great utility and convenience of this Lunar Table, in deciding the event of any particular fit of illneſs, I ſhall here ſtate ſome real predictions which were made, during the indiſpoſition of ſome of my patients, and for whom the following horoſcopes were erected.

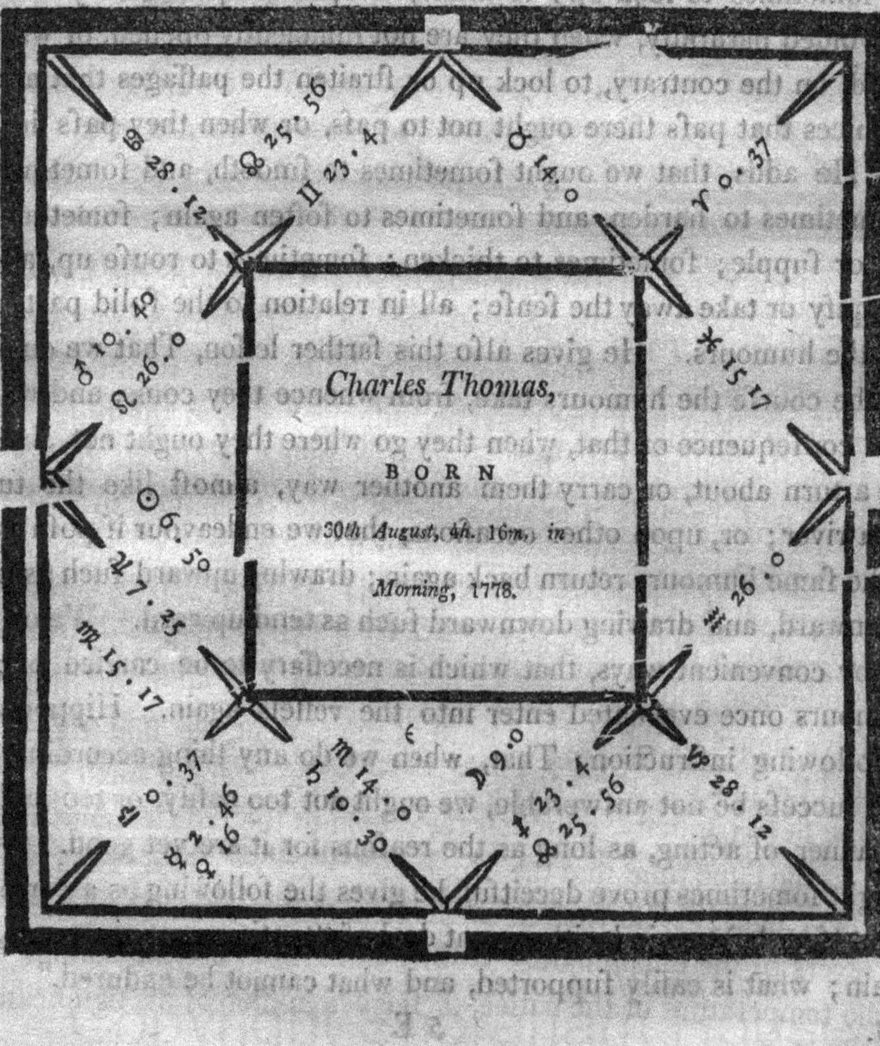

Charles

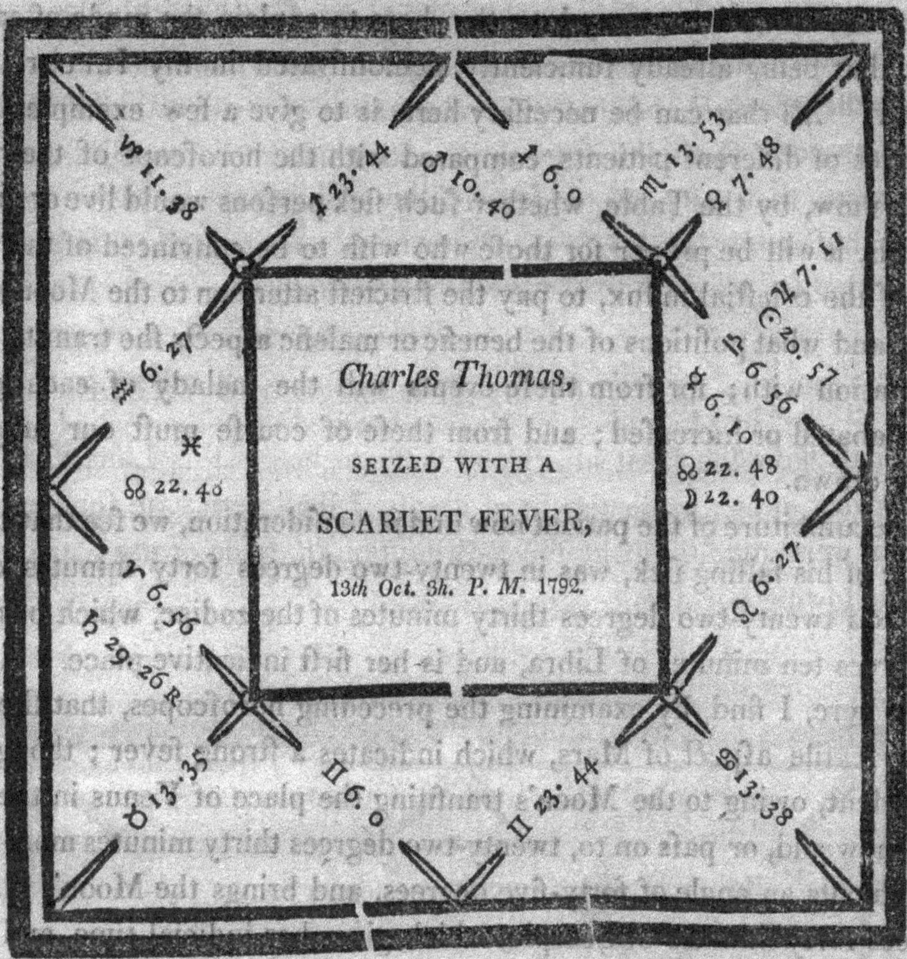

I have placed the horoſcope of this patient's nativity before the figure of his decumbiture; in order to ſhow, by way of analogy, that ſuch an indiſpoſition would certainly take place about this time, from the poſition of the ſignificators, and their particular configurations with the heavenly intelligencers, at the time of his birth. For this reaſon, the figure of the ſick perſon's nativity ſhould always be inſpected where it can be had, becauſe it enables us the better to judge, in many intricate caſes, whether the diſeaſe will terminate with life or death; for although, in moſt common maladies, it is poſſible to determine this queſtion pretty accurately, by the help of the preceding Table, without the radical figure of birth, yet, where that can be had, our judgment will in general be more certain, and often infallible.

In the above figure of the decumbiture of the patient, we find the Moon hath lately tranſited the place of the Sun and Jupiter in the figure of birth; and that this place is in the fiery triplicity, afflicting the Moon in the radical point with a quartile aſpect; at the ſame time that the Moon and Mars beholding each other with a trine, from fiery ſigns, at the time of birth, clearly ſhows that the native would be ſubject to fevers of the inflammatory kind. But I ſhall decline making any comments on the temperature of the native, or the deſignation of his ſignificators at the time

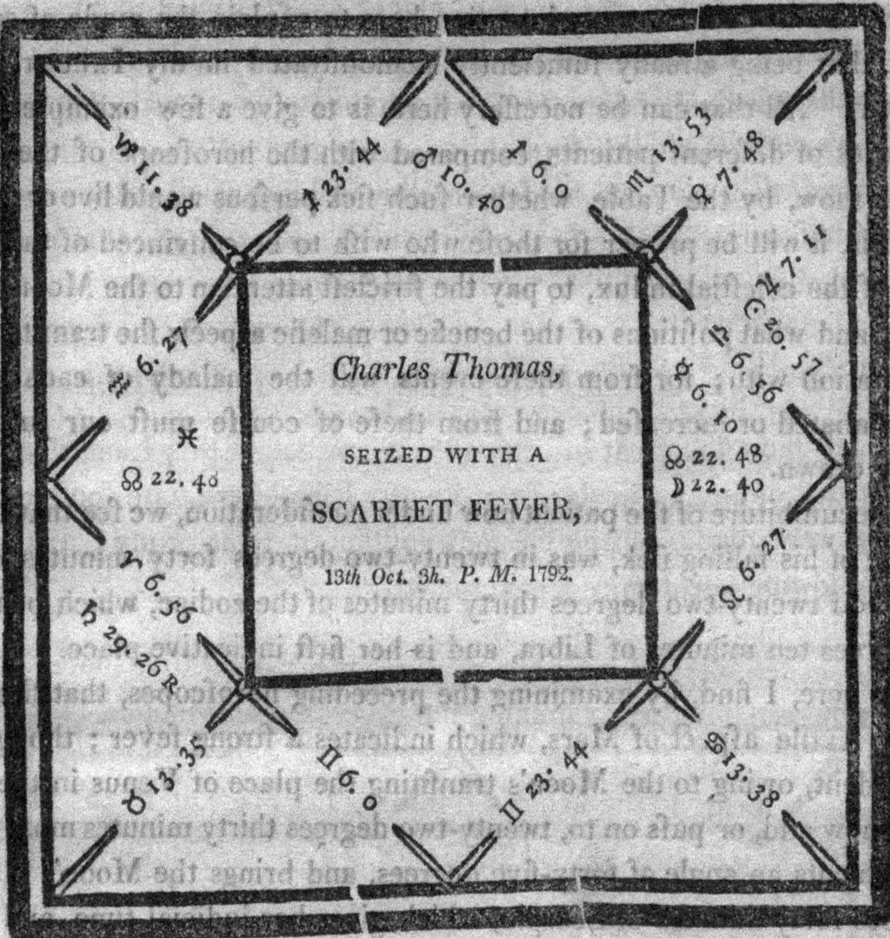

I have placed the horoſcope of this patient's nativity before the figure of his decumbiture; in order to ſhow, by way of analogy, that ſuch an indiſpoſition would certainly take place about this time, from the poſition of the ſignificators, and their particular configurations with the heavenly intelligencers, at the time of his birth. For this reaſon, the figure of the ſick perſon's nativity ſhould always be inſpected where it can be had, becauſe it enables us the better to judge, in many intricate caſes, whether the diſeaſe will terminate with life or death; for although, in moſt common maladies, it is poſſible to determine this queſtion pretty accurately, by the help of the preceding Table, without the radical figure of birth, yet, where that can be had, our judgment will in general be more certain, and often infallible.

In the above figure of the decumbiture of the patient, we find the Moon hath lately tranſited the place of the Sun and Jupiter in the figure of birth; and that this place is in the fiery triplicity, afflicting the Moon in the radical point with a quartile aſpect; at the ſame time that the Moon and Mars beholding each other with a trine, from fiery ſigns, at the time of birth, clearly ſhows that the native would be ſubject to fevers of the inflammatory kind. But I ſhall decline making any comments on the temperature of the native, or the deſignation of his ſignificators at the time

Saturn, to which ſhe was approaching, in twenty-nine degrees twenty-ſix minutes of Sagittarius. When this aſpect was formed, the patient had viſibly recovered, and the brain was never after affected. I now proceed twenty-two degrees thirty minutes more on the zodiac, which brings the Moon to her next indicative day, in fifteen degrees ten minutes of Capricorn. Here the two celeſtial luminaries form a quartile, a diſcordant aſpect, which gave the patient a relapſe. Proceeding the next twenty-two degrees thirty minutes, I come to the Moon's judicial place, in ſeven degrees forty minutes of Aquaries. Here we find a mundane trine, formed by Jupiter and Venus with the Moon, and a zodiacal trine of Mercury, a plain demonſtration that the diſorder muſt abate, and that a favourable criſis would enſue. To the ſeven degrees forty minutes of Aquaries, I add twenty-two degrees thirty minutes more, which brings the Moon to her ſecond indicative place, in ten minutes of Piſces. Viewing the decumbiture, I now find the Moon approaches to a trine aſpect under the benign influence of Jupiter and Venus, which overcomes the quartile of Mars, and indicates that the diſeaſe would be completely conquered by the next criſis. To aſcertain the truth and manner of this, I proceed onwards twenty-two degrees thirty minutes more, which brings the Moon to twenty-two degrees forty minutes of Aquaries, where the ſecond grand criſis was to be produced. Now upon maturely inſpecting the decumbiture, I find the Moon, at the time this patient was ſeized with his diſorder, was placed in her north node, and contributed to the evil effects of the other configurations; but at the time of this grand criſis, ſhe is fortunately poſited in her ſouth node, thereby helping to depreſs the vitiated humours of the body, and to overcome the diſeaſe. This poſition, contributing to the favourable influence of the other configurations, reſtored the patient from his bed of ſickneſs, and his ſtrength gradually increaſed; ſo that by the time the Moon formed her conjunction with Saturn, as expreſſed on the face of the heavens in the figure of his decumbiture, the maſs of blood was purified from all feveriſh ſymptoms, and the patient was reſtored to his accuſtomed health and ſtrength. And thus we may ſee, that by erecting the decumbiture, or figure of the poſitions of the heavenly bodies, at the time any patient is ſeized, and proceeding in this manner to aſcertain the influence of the good or evil aſpects on the indicative, judicial, and critical, days, we ſhall, without difficulty, be able to determine whether the diſeaſe will prove ſlight or dangerous, and be directed accordingly in our regimen and mode of treatment. But in order to make this ſpeculation ſtill more obvious to the young practitioner, as well as to the curious reader, I ſhall now proceed to examine the decumbiture of a patient, whoſe diſeaſe was more malignant, and proved fatal.

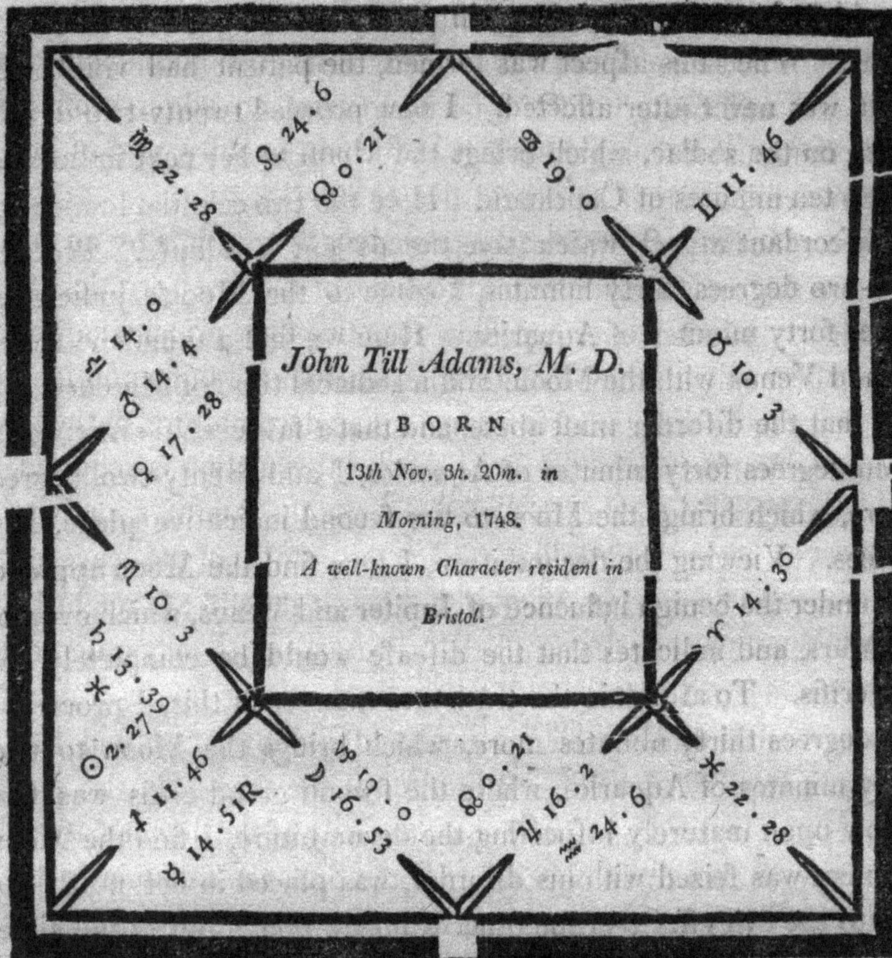
John Till Adams, M. D.
BORN
13th Nov. 3h. 20m. in
Morning, 1748.
A well-known Character resident in
Bristol.

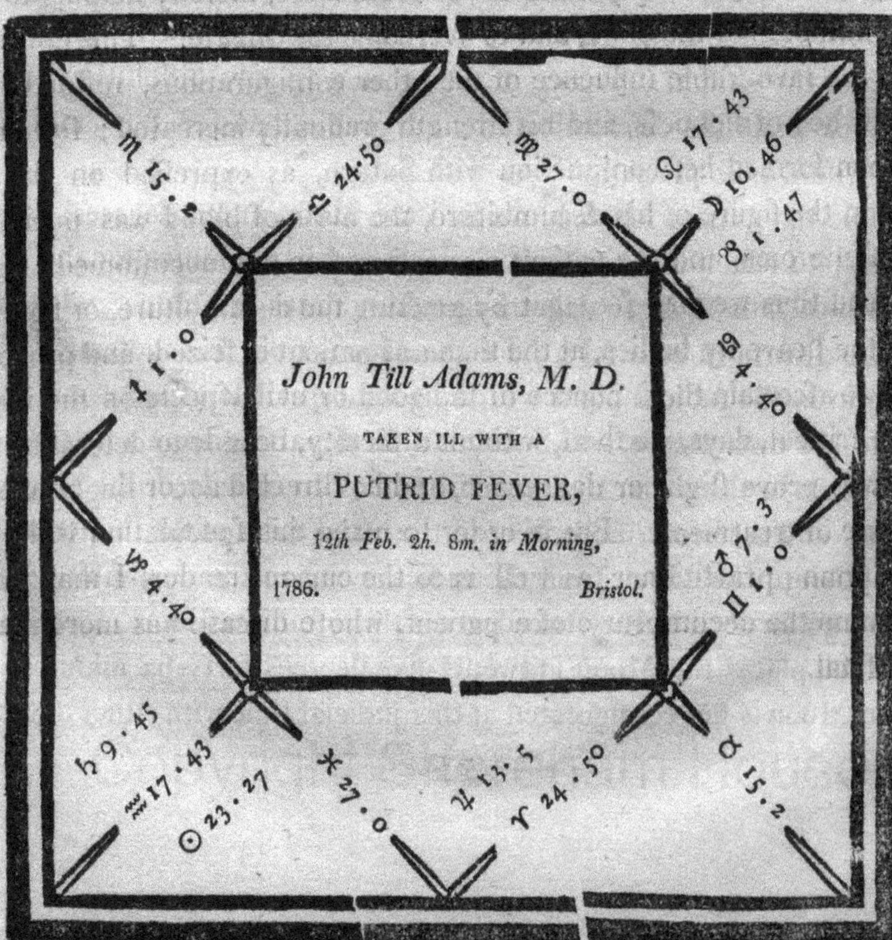
John Till Adams, M. D.
TAKEN ILL WITH A
PUTRID FEVER,
12th Feb. 2h. 8m. in Morning,
1786. Bristol.

These figures I erected while resident in Bristol, at the request of my good friend Dr. Till Adams, who being seized with a malignant fever, accompanied with dangerous symptoms, and being himself a friend to, and an admirer of, the Occult Sciences, was desirous of seeing the result of such an enquiry, and of judging himself, by these means, whether he should live or die.

In considering the sidereal effect of the preceding figures, it is by no means requisite to calculate the genethliacal prognostications of the seven erratics at the time of the native's birth. It is however necessary to notice their principal aspects and positions in the horoscope, in order to determine whether the same positions are transmitted, or similar or adverse aspects formed in the decumbiture, at the time the patient is taken ill; but no further or more minute speculation is required, since we are neither considering the effect of directions, nor the fate of a nativity; but are endeavouring to prove, that, by only observing the position of the heavens at the time the patient is taken ill, the probable termination of the disease might be foretold, and whether it would end in life or death. First, then, we may observe, the Moon is situated in the eighth house, termed by the ancients the house of death, because of its obscurity and position under the earth. Besides this, we find the Moon in opposition to Saturn, who was her dispositor at the time of birth; and from this aspect she forms an opposition with Venus, the lady of the doctor's ascendant; and immediately approaches to an opposition of the Sun, the fountain of life. These are three evil directions by position, and furnish a very unfavourable prospect of the event of the disease. For the Moon, the giver of radical moisture, afflicted by the adverse rays of the Sun, the author of vital heat, fails not to produce such a putrifaction of the animal juices, as to bring on a speedy dissolution of the body. Let us then examine the decumbiture by our Table, and see how and when this fatal event would take place.

At the time the patient fell sick, we find the Moon in ten degrees forty-six minutes of Leo; to which add twenty-two degrees thirty minutes for the first indicative time, which falls in three degrees sixteen minutes of Virgo; and indicates the nature of the disease to be a fever; for the Moon, from this indicative place, beholds the Sun in the radical point with a baneful quartile aspect; and the Moon, according to her own nature, was Saturnine, as departing from a sextile configuration with Saturn in the radix, to a quartile with Venus, lady of the ascendant, and giver of life; and therefore, according to the astral rules of the immortal Ptolomy, this first motion of the Moon from the radical point of the decumbiture indicated evil. I now go forward twenty-two degrees thirty minutes more, for the first judicial day, which places the Moon in twenty-five degrees forty-six minutes of Virgo. Now as the Moon is not configurated at this judicial time with either of the planets, neither

neither by transit in the nativity, nor aspect in the decumbiture; and her judicial place falling in the twelfth house, the house of affliction; we cannot draw any favourable judgment from these circumstances; but, on the contrary, a dangerous crisis is to be expected. To this judicial time, we add twenty-two degrees thirty minutes more, and it brings the Moon sixty-seven degrees from the place she occupied when the patient was seized with the disorder; and this is her second indicative place, which falls in eighteen degrees sixteen minutes of Libra. Now, if we inspect the foregoing horoscopes, we shall find the Moon, in approaching to this point, has just departed from a baneful opposition with Jupiter, which, having the direct opposite effect of a conjunction with that benevolent planet, which represents the heart and vital principle, shows a contaminated or morbid state of the blood and lymph. We likewise perceive the Moon is in quartile to her own radical place, transiting at the same time the body of Venus, and making this aspect the harbinger of a fatally-approaching crisis. To determine this fact, I proceed twenty-two degrees thirty minutes further in the zodiac, which brings the Moon to ten degrees forty-six minutes of the sign Scorpio, at which point of time the crisis, or critical day, of this patient's disease occurred. Now by inspecting the figures, we shall perceive this crisis is ushered in by such evil configurations of the heavenly bodies, the second causes under nature, as would not only heighten the malady, and put it out of the power of medicine to subdue, but would infallibly terminate in death. In the first place we shall notice, that the Moon transits the place of Saturn in the radical point; secondly, she is configurated in a malefic quartile aspect of Saturn in the decumbiture; thirdly, she is within orbs of a baneful quartile of Venus, lady of the patient's ascendant; and fourthly, she is rapidly approaching to a quartile configuration of the Sun, which is inimical to life and motion, without any one friendly aspect of the benefic planet Jupiter intervening, to lessen or repel the malefic influence. Such, therefore, are the testimonies, that under any kind of malady and wherever they occur, infallibly portend the death of the patient; and they accordingly put an end to the existence of this much-respected man, whose integrity in his profession had gained him universal esteem, and renewed in him the inestimable character of the immortal Culpeper, who, never, with a view to gain, gave two medicines for the cure of an afflicted fellow-creature, when one was sufficient. But death levels all distinctions; and, in strict conformity with the time and manner pointed out by the above decumbiture, it conducted the soul of this excellent man from an earthly to an heavenly habitation, on the 20th of February, 1786, at the time the Moon formed her quartile aspect with the Sun, which was in eight days from the time he was seized with the fever, and six days after it was foretold by the preceding horoscopes; from whence, having foreseen the doctor's fate, I composed an an Elegy on his death, while he was yet alive, which I got printed, and publiſhed on the very day he expired; thus manifeſting to the world, with the patient's earneſt approbation, an incontrovertible inſtance of the verity of aſtral prediction.

Having thus far endeavoured to prove the utility of the Lunar Table, by the indiſpoſition and recovery of Charles Thomas, a pupil of Mr. Hall, engraver to his majeſty, in the one caſe; and by the ſickneſs and death of Dr. Till Adams, in the other; I ſhall now, for the farther ſatisfaction of the reader, prove, that it is poſſible to judge whether a patient will live or die, from the horoſcope of the decumbiture only, without knowing or recurring to the horoſcope of the patient's nativity, or time of birth.—To this end, the following axiom muſt ever be remembered: That if we find, at the time any perſon is ſeized with illneſs, that the Moon is afflicted by more than one planet; and that on the next critical day ſhe forms a congreſs with the malefic planets Saturn and Mars, either by conjunction, quartile, or oppoſition, the ſick perſon ſhall die on the day and hour in which the afflicted Moon comes to the interficient point of the zodiac; as the great Ptolomy declareth in his 16th Aphoriſm: "We muſt behold the motion of the Moon as ſhe paſſeth through the critical, judicial, and mortal, days; for, if ſhe be in them fortunate, it fareth well with the patient; but, if unfortunate, the contrary." I ſhall exemplify this b the following example.

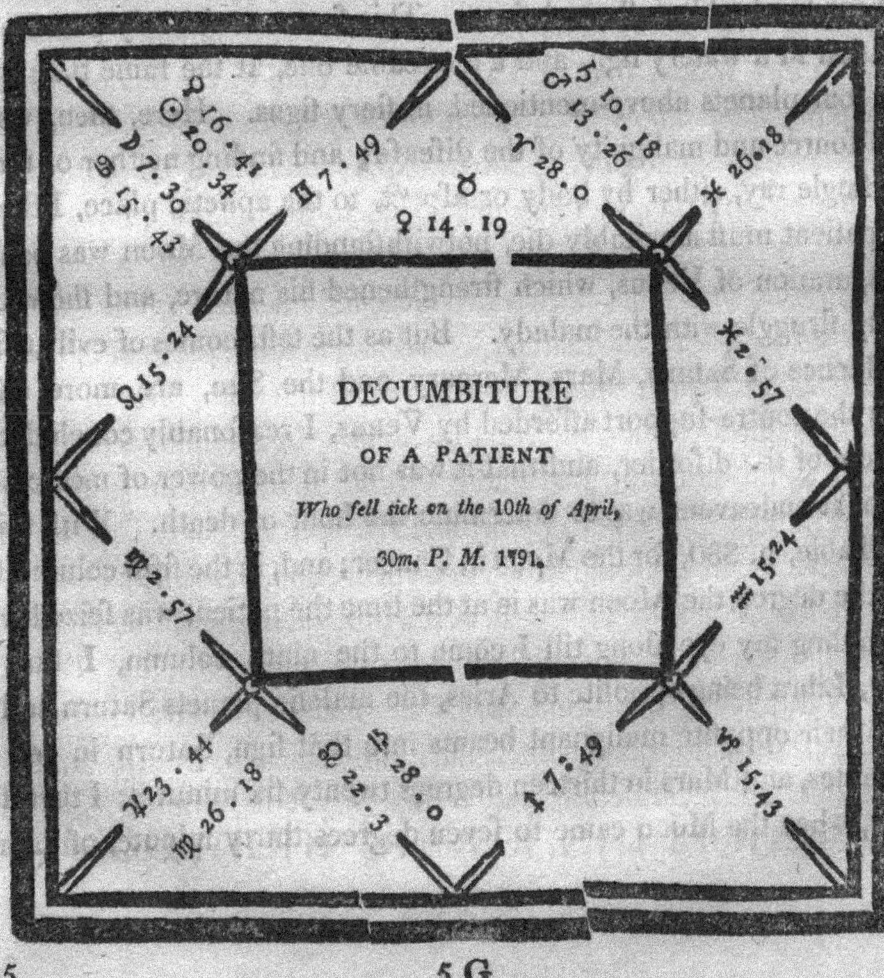

Being ſent for to a perſon who fell ſick on the 10th of April, 1791, and being deſirous to know the event of his diſeaſe, I examined the face of the heavens at the exact time the patient was ſeized, viz. at half paſt twelve o'clock at noon, when the celeſtial intelligencers were poſited as in the above decumbiture, and which are as follow: The Moon, which in all decumbitures repreſents the ſick perſon, is ſituated within the quartile influence of no leſs than four planets, Saturn, Mars, Mercury, and the Sun; and, as they are all within orbs of a conjunction with each other, it follows that the diſeaſe would bear deſignation of their joint pernicious influx, which, ſcientifically conſidered, manifeſts a fever, with putrefaction of the animal juices, as thoſe ſkilled in the aſtral ſcience will quickly ſee. For the Sun's burning influence, in conjunction with Mars, a hot and violent planet, and Mercury being controvertible in his nature, unites in the malefic rays of the Sun and Mars; and, although Saturn is conſtitutionally cold, yet, being alſo dry, his cold quality is over-balanced, inaſmuch as drought participates of the qualities of heat, being fuel for the fire. Yet the cold quality of Saturn ſpecificates the diſeaſe, by ſhowing that it ſprung from a cold cauſe, or deathly chill, extended over the whole circulating ſyſtem, or maſs of blood.

On examining the patient, he informed me he had drunk a quart of cold water, being overcome with heat and thirſt, and in a violent perſpiration, whence his blood muſt have been in a highly-inflamed ſtate. This ſeems eminently prenoted by the Moon's poſition in a watery ſign, and a moveable one, at the ſame time in oppoſition to the four planets above-mentioned, in fiery ſigns. Here, then, we at once perceive the ſource and malignity of the diſeaſe; and finding neither of the benefic ſtars caſt a ſingle ray, either by body or aſpect, to the aphetic place, I thence concluded the patient muſt inevitably die, notwithſtanding the Moon was beheld by a ſextile configuration of Venus, which ſtrengthened his nature, and ſhowed that he would greatly ſtruggle with the malady. But as the teſtimonies of evil ariſing from the joint influence of Saturn, Mars, Mercury, and the Sun, are more and much greater than the contra-ſupport afforded by Venus, I reaſonably concluded the patient would die of the diſorder, and that it was not in the power of medicine to ſave him. My next endeavour was to determine the hour of death. With this view I look to the Table, p. 380, for the Moon in Cancer; and, in the fifth column from the left, I find the degree the Moon was in at the time the patient was ſeized, viz. 7 30; and then, guiding my eye along till I come to the ninth column, I find 7 30 of Libra; now, Libra being oppoſite to Aries, the malefic planets Saturn and Mars of courſe ſend their oppoſite malignant beams into that ſign, Saturn in ten degrees eighteen minutes, and Mars in thirteen degrees twenty-ſix minutes; I therefore concluded, that, when the Moon came to ſeven degrees thirty minutes of Libra in the zodiac,

Dodd del. Pass sculp.

Nativities of the late King & Queen of France.

zodiac, the crisis would take place, as may be seen at the top of the fifth column in the Table; and that when the Moon came to ten degrees eighteen minutes, being the opposition of Saturn, a visible change in the patient would take place for the worse; and that when she arrived at the thirteenth degree twenty-six minutes of the same sign, thereby forming the opposition with Mars, the disease would prove mortal, and terminate in death. If, therefore, the duration of the disease be reckoned by the motion of the Moon, we shall find, without any enquiry from the nurse or doctor, that the patient died about half past four o'clock in the afternoon, on the 17th of April, 1791, at which precise time the critical aspect on which the disease turned was formed. Thus was afforded an incontestible proof of the correctness of my Lunar Table, and of the force and power of the planetary influx on sublunary bodies.

OF THE DIFFERENCE BETWIXT A NATURAL AND VIOLENT DEATH, exemplified by the FATE of the late KING AND QUEEN OF FRANCE.

WITH a view to teach the curious reader how to distinguish the astral testimonies portending a violent death from those which foreshow our natural dissolution, I shall, by way of example, investigate the particular configurations which prenoted the violent death of the late unfortunate Louis XVI. king of France, and his unhappy consort, Marie-Antoinette of Austria. For this purpose I have annexed a copper-plate engraving of their nativities, with figures of the positions of the celestial intelligencers at their coronation, and on the days of their execution. Whoever has perused my ILLUSTRATION OF ASTROLOGY, will have seen, that in my predictions, published in the year 1786, I foretold the revolution in the French empire, and the dethronement and execution of the French king and queen, six years before it happened; with all the dreadful consequences appertaining thereto, exactly as they have since fallen out; and whoever lives to see the upshot of a few years, will also see every other part of my predictions literally fulfilled. I could even now publish to the world the success and termination of many great events, which all men are anxious to know, though few perhaps would believe, were I at liberty to disclose them. The safety of particular individuals, the well-being of the state, the peace of society, the prosperity of empires, hang upon the issue of a few revolving periods; and, though *wise* the age we live in, yet few would brook the admonitions of a friend, though they should lead to a prosperity and riches—to glory and renown. The uncommon positions of the heavenly bodies, which are the second causes in the secret operations of Nature, are at this time well worthy the attentive consideration of the naturalist and speculative philosopher. Their mundane influence is by no means confined to the nations of Europe, but threatens the

the ruder and more widely-extended realms with the effect of a convulsive stroke. God, in his unbounded favour to the British isles, will overshadow and protect them: and it is not impossible but the day may come, when the humble author of these remarks may be at liberty to amplify and develope the subject in some future publication. In the mean time, let Prussia, and all the smaller German states, beware; for the time will soon come, that *Germany shall be no more.**

Louis XVI. was born the 23d of August, 3h. 50m. P. M. 1754. The sign Sagittarius ascended upon the eastern fineter of the horizon, intercepted by Capricorn; wherefore Jupiter and Saturn are the lords of his ascendant, and, with the Moon in the tenth house, represent his person. From the nature and quality of these significators, we may deduce the following inferences: That Saturn being in the ascendant gave the native a wavering and irresolute disposition; the Moon posited in Scorpio gave him a taste for luxury; and Jupiter, being co-significator in the eighth house, in aspect with Mercury, rendered him mild and passive, yet declared that his principal actions shall be attended with disappointment, and produce him much anxiety, vexation, and insult. In the royal hereditary figure of birth, we find four planets occupying the house of death; and as the Moon, giver of life, is disposed of by one of those planets, it is an argument that he would not live to an old age. The precise time of death is only ascertainable by bringing up the several directions of the nativity: and, as the method of doing this is already amply explained in my Illustration of Astrology, we must refer those who choose to work them up, for any assistance they may want, to the rules there laid down. Suffice it here, that we point out those testimonies from the face of the nativity, that are always found to be arguments of a violent death. These are, First, Saturn in the ascendant, posited in a violent sign. Secondly, the Moon, giver of life, configurated with the violent fixed star Chælæ. Thirdly, the lord of the ascendant malevolently conjoined with the lord of the house of death. Fourthly, the two lights of the world depressed in the eighth house. And, fifthly, the Moon elevated in the dignities of an infortune, and Mars, her dispositor, having his fall in a human sign, plainly demonstrate that the native should fall by the hand of man. And this was unhappily verified by his execution, on the 21st day of January, 1793, at twenty-two minutes past ten in the morning; at which time these malefic directions came up, as represented in the figure of his execution, in the preceding plate. At that fatal moment we find the Sun, the light of time, was in his detriment; that Mercury, the significator of

* I presume that the accomplishment of this bold prediction, which the author did not live to see, will be found in the formation of the Confederation of the Rhine, in the suppression of the small states and annihilation of the Germanic circles, and the erection of the kingdoms of Bavaria, Saxony, Wirtemberg, and Westphalia, instead of them; but particularly when Francis II. renounced the title of Emperor of Germany, Aug. 6, 1806. Editor.

the

the French people, occupied the cusp of the tenth house, transiting the place of Saturn, the king's significator, at the time of birth; and that the two malefic planets Saturn and Mars are in reception of each other; and that the Moon, the significator of life, is surrounded with violent fixed stars, in opposition to the benign planet Jupiter, in angles, and in quartile to Mars, her dispositor in the horoscope of birth; all which peculiar configurations are so many strong and irresistible arguments of the resolution of the people to proceed to extremities, and of the irrevocable fate of this unfortunate monarch. For, although Jupiter, his co-significator, is observed to send a friendly ray to the aphetic place, yet having no dignities, and being disposed of by Mars, the significator of the convention, this benefic aspect was depressed, and its influence overcome, by the redundancy of a malefic influx. This admirably points out the struggles of Dumourier, in the hope of being able to preserve the life of the king; but his endeavours were quickly borne down by the violence of the leading faction, and there was not a man to be found who had courage enough to second his heroic intentions. This also, by the rules of the sidereal science, is clearly prenoted by the circumstance of Mars being the dispositor of Jupiter; and that Mars is disposed of by Saturn, the author of pusillanimity and fear. Thus the ascendant of birth shows that want of resolution and intrepidity in the native, which, if exerted in the favourable moment, would have turned the daggers of his enemies towards their own breasts, and have permanently secured himself and his posterity on the throne; and thus the figure of his decumbiture points to the fatal execution of the guillotine, and proves, that although the native sprung from a most illustrious house, having the two superior planets for his significators, and although he was a king, at one time beloved and idolized by his people, yet that he was but a man, subject to the severest reverse of fortune, and doomed to as ignominious an end as the vilest of his subjects!

The elegant and accomplished consort of this unfortunate monarch was born on the 2d of November, 1755, 7h. 23m. P. M. as expressed in the plate. In the figure of her nativity, we find the Moon is lady of the ascendant, rising upon the sign Libra, in the fifth house, the house of pleasure and sexual enjoyment; of which, it is sufficiently evident, she was passionately fond. Mars being posited on the ascendant, in his essential dignities, shows her to have been stately, austere, and proud; yet predicts that she would be unfortunate in her connections, and impatient of controul. This is the more obvious, because Mars, the significator and influencer of her passions, is dignified in her ascendant, though approaching to an opposition of Saturn, lord of the seventh and eighth houses, who is likewise configurated in his essential dignities, and, more extraordinary still, is posited in the

the exaltation of Mars. This gave her an unconquerable ſpirit, and ſtrong natural paſſions, with an inſatiable appetite for intrigue, united to an inconſtant and arbitrary turn of mind. This is ſtill more ſtrongly demonſtrated by the preſence of four planets in the fifth houſe, and two of them aſpected in the ſign Scorpio. That her reputation would be arraigned by the voice of the multitude, is foreſhown by the Dragon's Tail being in the tenth houſe, the houſe of dignity and honour; and the treachery of her confidants is pointed out by Mercury being in the ſixth houſe, in his detriment, and in oppoſition to the Part of Fortune. That her conſort would be involved in misfortunes, and ſuffer greatly on her account, is made manifeſt by Saturn, the ſignificator of the king, being in oppoſition to Mars; and the Moon, lady of her aſcendant, in baneful quartile to both the infortunes. Indeed, there never was a nativity yet made public, wherein the infortunes were ſo miſchievouſly configurated, or wherein the general ſignificators ſo conſpicuouſly denoted individual misfortune and univerſal rage. But I ſhall paſs over, for the preſent, any farther remarks on the unfortunate deſignation of the ſignificators in this nativity, and notice them no farther than as they point out the teſtimonies of a violent and premature death. Theſe are: Firſt, the Sun and Venus, configurated with a violent fixed ſtar, in a violent ſign. Secondly, Mars aſcending to the violent fixed ſtar Hercules; and the ſign aſcending being of a violent nature. Thirdly, the lord of the eighth houſe, the houſe of death, afflicting the Moon in the aphetic place. Fourthly, the lord of the ſixth houſe afflicting the Moon, the lady of the aſcendant, and ſignificator of life, with a malefic quartile ray, the harbinger of violence, and the preſage of death.

Thus we may obſerve, that one of the principal luminaries is afflicted by both the infortunes, and the othe ris poſited in a violent ſign, denoting aviolent death. Again, the lord of the eighth houſe, a malevolent planet, afflicted by the quartile rays of an infortune by nature, is another preſage of untimely death. Alſo the lady of the aſcendant, in a violent ſign, banefully configurated with the infortunes, and the diſpoſitor of the luminaries in a violent ſign, is an irrefragable proof of an approaching untimely death. Now, the Moon, who is lady of the aſcendant, having her fall in a human ſign, portends a violent death by the hand of man; and, if we examine the face of the heavens at the time of her execution, we ſhall find her death proceeded from the violence of an uſurped power, occupying the ſeat of juſtice; for the Moon, elevated in her aſcendant at the time of birth, is moſt remarkably configurated in the fall of the Sun at the time of her execution; and that ſame ſign culminating on the cuſp of the tenth houſe, the houſe of juſtice, and the Sun being poſited there in his fall, in conjunction with Mercury, moſt aptly de-

ſcribes

ſcribes the manner of the native's death. Mercury, who is the natural ſignificator of the French people, being in his eſſential dignities, elevated, and in reception of Venus, lady thereof; and Mars beholding Jupiter with a quartile ray, poſited in the twelfth houſe, and lord of the aſcendant of death; and the Moon, lady of aſcendant of birth, being within orbs of an oppoſition of Mars and Venus, who have their fall in the houſe of dignity and honour; all tend to foreſhow that royalty was for a time to be deſtroyed in France; as is moſt wonderfully prenoted in the horoſcope of the coronation. And what is very remarkable, at the time of the French monarch's death, the Sun, who is king among the planets, was poſited in his own detriment, or in that peculiar point of the heavens which is oppoſed to his own houſe; and at the time of the unfortunate queen's execution, the Sun was in his fall, without a ſingle dignity to ſupport him, as is moſt clearly evinced by the horoſcopes in the preceding plate; ſo that we may ſay, the ſtars in their courſes fought againſt this illuſtrious pair, as they fought againſt Siſera of old; and thus we may perceive, that the moſt valiant and the moſt courageous are not proof againſt the ſhafts of fate; but that the nobleſt, and moſt glorioſly clad, whether in honour, glory, or renown, are but like the offspring of plants, which have their ſpringing up, their flowering, and their fragrant maturity; until, plucked by a rude hand, they wither, fade, and die.

DIRECTIONS FOR PLACING THE CUTS.

INDEX

THE END.

W. Lewis, Printer, St. John's-square, London.

www.ingramcontent.com/pod-product-compliance
Lightning Source LLC
LaVergne TN
LVHW062128200926
848484LV00007B/1781